The Palgrave Handbook of
Critical Thinking in Higher Education

The Palgrave Handbook of Critical Thinking in Higher Education

Edited by

Martin Davies and

Ronald Barnett

palgrave
macmillan

First published in 2015 by
PALGRAVE MACMILLAN®
in the United States—a division of St. Martin's Press LLC,
175 Fifth Avenue, New York, NY 10010.

Where this book is distributed in the UK, Europe and the rest of the world,
this is by Palgrave Macmillan, a division of Macmillan Publishers Limited,
registered in England, company number 785998, of Houndmills,
Basingstoke, Hampshire RG21 6XS.

Palgrave Macmillan is the global academic imprint of the above companies
and has companies and representatives throughout the world.

Palgrave® and Macmillan® are registered trademarks in the United States,
the United Kingdom, Europe and other countries.

ISBN 978-1-349-47812-5   ISBN 978-1-137-37805-7 (eBook)
DOI 10.1007/978-1-137-37805-7

Library of Congress Cataloging-in-Publication Data

The Palgrave handbook of critical thinking in Higher education /
edited by Martin Davies and Ronald Barnett.
   pages cm
Includes index.
   Summary: "Palgrave Handbook of Critical Thinking in Higher
Education explores critical thinking in higher education in all its forms,
from definitions to teaching and incorporating it into the curriculum, its
relationship to culture and the professions, and its social perspectives
and scientific and cognitive manifestations. Davies and Barnett ask
what is critical thinking, especially in the context of higher education?
The handbook explores these questions, with contributors offering their
insight into the contemporary understandings of higher education
worldwide"— Provided by publisher.
   Summary: "What is critical thinking, especially in the context of
higher education? How have research and scholarship on the matter
developed over recent past decades? What is the current state of art
here? How might the potential of critical thinking be enhanced? What
kinds of teaching are necessary in order to realize that potential? And
just why is this topic important now? This handbook explores these
questions, with contributors offering their insight into the contemporary
understandings of higher education worldwide"— Provided by publisher.
   1. Critical thinking—Study and teaching (Higher) I. Davies, Martin,
1960, March 6– editor of compilation. II. Barnett, Ronald, 1947– editor of
compilation. III. Title: Handbook of critical thinking in higher education.
LB2395.35.P35 2015
378.1'7—dc23
                                                      2014040131

A catalogue record of the book is available from the British Library.

Design by Newgen Knowledge Works (P) Ltd., Chennai, India.

First edition: April 2015

10 9 8 7 6 5 4 3 2 1

Contents

Illustrations

Figures

Tables

Introduction

Martin Davies and Ronald Barnett

Questions, context, and challenges

What is critical thinking, especially in the context of higher education? How have research and scholarship on the matter developed over recent past decades? What is the current state of the art here? How might the potential of critical thinking be enhanced? What kinds of teaching are necessary in order to realize that potential? And just why is this topic important now? These are the key questions motivating this volume. We hesitate to use terms such as "comprehensive" or "complete" or "definitive," but we believe that, taken in the round, the chapters in this volume together offer a fair insight into the contemporary understandings of higher education worldwide. We also believe that this volume is much needed, and we shall try to justify that claim in this introduction.

The context here is complex, with strands running in different directions and overlaying each other. Four paths open up. First, there is a sense—especially in the Pacific Rim, notably China but also in other Asian countries—that critical thinking has been given insufficient attention over decades and even centuries. Pedagogies have been influenced by a complex of Confucian and cultural attitudes to the teacher-as-authority and tacit rules of knowledge transmission, buttressed by a sense of education assisting in the building of a national identity. Over recent decades, however, there has developed a sense that modernity requires more questioning stances among students, if higher education is to fulfill its potential in the forming of a changing society. Second, there is a parallel sense—especially in the newer countries of Africa and South America, and also dimly becoming evident in more developed countries of the North—that critical thinking is a necessary part of the formation of critical citizens. Third, there continues to be—especially in the United States—a concern that amid mass higher education students are insufficiently developing their powers of critical thought. Last, especially in the wake of the emergence of "the entrepreneurial university" and the development of market principles in higher education, a concern with critical thinking is apparently being displaced by a determination to raise the levels of skills more obviously suited to the requirements of a global economy. Notably in the United Kingdom and Australia, for example, "critical thinking" has faded from the public debate about higher education, as "employability" has risen.

These currents prompt a number of observations, namely that critical thinking is of worldwide concern, that its presence—for a variety of reasons—may be fragile, and that its interpretation is connected with a range of purposes that are themselves changing. Once, critical thinking was once understood to be the mark of a person who had been in receipt of a higher education. Indeed, there was considerable overlap between the liberal conception of the idea of the university and the idea of critical thinking. The university precisely made available a space in which the mind could be so educated that it was able to form its own authentic judgments. Now, as higher education has both become a mass enterprise and its value to the economy has multiplied, it is an open matter as to whether and in which ways critical thinking might be of economic value. Consequently, it sometimes appears that if the idea of critical thinking is to find broad support across society, then it needs to be reframed so that its social and civic value might become more apparent.

What this book does

This book does more than investigate critical thinking as either a concept or as a set of skills in itself. There are plenty of books that do this already (for a recent example, see Moon 2008). Specialist edited collections have also been created, for example, looking at critical thinking and generalizability (Norris 1992). Rather, this book examines the nature of critical thinking within, and its application and relevance to, higher education. As we shall see, the very idea of critical thinking in higher education has generated profoundly different, and even antagonistic, views among scholars and researchers who have thought deeply about the matter.

The aims of this volume are fourfold:

1. to bring together key papers, or excerpts of key texts, that have already been published in this area;
2. to explicitly focus on the work being done on critical thinking in the particular context of higher education;
3. to provide (in this introduction) an overview of the literature; and
4. to stimulate further interest and debate on the topic.

In selecting contributors, we have been mindful that critical thinking in higher education is a global concern with a potential worldwide audience of millions. All educators across all the disciplines are interested—or should be interested—in critical thinking. It is arguably a central concern of higher education of our time. We have, therefore, been keen in embarking on this volume to solicit contributors from around the world and from all continents, as well as from a range of disciplines and wide perspectives. To this end, this volume includes contributions from five continents, ten countries, and over eighty institutions, making the resulting book a truly global product of the collective efforts of dozens of scholars.

Considerations

"Critical thinking in higher education" is a phrase that means different things to many people. Does it mean a propensity for finding fault? Does it refer to an analytical method? Does it mean an ethical attitude or a disposition? Educating to develop *critical intellectuals* and the Marxist concept of *critical consciousness* are very different from the logician's interest in identifying fallacies in passages of text, or the practice of distinguishing valid from invalid syllogisms. Critical thinking in higher education can encompass debates about critical pedagogy, political critiques of the role and function of education in society, critical feminist approaches to curriculum, the development of critical citizenship, or any other education-related topic that uses the appellation "critical." Equally, it can be concerned to develop general skills in reasoning—skills that all graduates might possess. With all of these multiplying interpretations and perspectives, and after more than four decades of dedicated scholarly work, critical thinking remains more elusive than ever. The concept is, as Raymond Williams has noted, a "most difficult one" (Williams 1976, 76).

Traditional philosophical definitions of the concept of critical thinking do not necessarily inform debates in all of the areas of critical thinking scholarship. Definitions of critical thinking are not central to areas such as critical pedagogy or critical feminism. Learning about such definitions does not help one develop a critical attitude about the society in which one lives. Philosophical definitions of critical thinking do not directly assist—or so many believe—in becoming a critical citizen. It may be that the core attributes of critical thinking will always remain fundamental to what we mean by "critical thinking" since, at a basic level, critical thinking is about having skills of a certain sort (inference making, reasoning, and so on). Yet, critical thinking is also much more than this. Traditional philosophical definitions of critical thinking seem impotent in relation to these wider areas of critical thinking scholarship as they apply to the field of higher education.

There are significant practical matters at stake here. An American book, *Academically Adrift: Limited Learning on College Campuses* (2011), provoked widespread interest and media attention in the United States (Mataconis 2011; NPR Staff 2011; Rimer 2011). The study on which the book was based tracked the educational development of a range of skills of 2322 American college students from 2005 until 2009. It established that 45% of students made *no significant improvement* in their critical thinking or reasoning skills during the first two years of college and 36% made no significant improvement after an *entire* four-year college degree (Arum and Roska 2011). Students tested could not, after completing a university course, sift fact from opinion, nor could they clearly present an objective review of two or more conflicting reports or determine a cause of an imaginary problem without being influenced by persuasive rhetoric and emotional blackmailing. This was a disturbing set of findings, and placed in serious doubt the assumption that critical thinking was being adequately taught, at least on American college campuses.

Further, in a major report by a consortium of US organizations in 2006 (Casner-Lotto and Benner 2006), the employers surveyed articulated the skill set that was needed in the workplace in the new century. The highest ranked skill as rated by employers was "critical thinking," surpassing "innovation" and "application of information technology." The employers were invited to consider recently hired graduates from three types of institutions: high school, two-year colleges, and four-year colleges and made clear their views regarding the skill deficiencies in the new graduates. The proportions of employers evincing such concerns were 57.5, 72.7, and 92.1% respectively (Casner-Lotto and Benner 2006). That is, *92.1% of the employers surveyed regarded graduates from four-year colleges as being "deficient" in critical thinking.* The US business community, it seems, is well appraised of the importance of critical thinking, even if its perceived value may be languishing in the academy.

In the United Kingdom, higher education institutions have now largely "abandoned critical thinking," and turned to "vaguely defined" skills such as "teamwork," "communication" and "leadership." It is these "skills" sets that "lecturers have to tick off as they incorporate them into their lessons…students [become] commodities [which] transforms education into a 'big business' rather than education for education's sake" (*Education News* 2013). There appears something of a paradox in the modern academy. Industry wants more critical thinking, but increasingly—at least in some countries—universities have little interest in providing it, even if they continue to proclaim its value. We may at least conjecture that it is not coincidence that the United States and the United Kingdom are among the countries that have seen the most marked moves in the marketization of higher education, and driving forward its demonstrable economic value, while it is just such countries in which the place of critical thinking appears to be in jeopardy.

All this comes at a time when, globally, universities are more associated than ever before with the business world. In short, universities have never been more aligned with the business sector, and yet (ironically) never less capable of meeting its needs. Critical thinking skills development, among other things, may well have been abandoned as part of the emergence of the entrepreneurial university.

However, whether critical thinking can and should be taught is itself a contested matter; and views here depend in part on what is understood as critical thinking. Many would concur that recognizing and constructing arguments—that is, critical thinking as reasoning skills—is valuable and important. Much less agreement attaches to the idea of educating for radical social and political change ("critical pedagogy"). Others are not happy with the teaching of critical thinking in *any* form. The Texas Republican Party actually tried to ban the teaching of critical thinking in schools (Strauss 2012). But what *exactly* did the Republicans want to ban? This was not obvious. Little progress on the topic of critical thinking in higher education can be made if the concept itself lacks a theoretical and conceptual grounding. Critical thinking surely remains "one of the defining concepts in Western education which enjoys wide endorsement, [and] yet we have no proper account of it" (Barnett 1997, 1)

What should be done?

Our sense is that while the topic of critical thinking in higher education is- and should be—of concern to many, it has, to date at least, typically been addressed in a piecemeal fashion, and within the confines of separate disciplines and fields (philosophy, sociology, psychology, education, pedagogy, management studies, and so on). Few attempts have been made to construct a broad overview of the field, with a focus on how critical thinking should be located, applied, studied, and taught within higher education. This then is a pressing need given the increasing importance of critical thinking in the curriculum, the university, and the world beyond, that of bringing together the key approaches so as to begin to form a unified field for study and practical implementation.

Another outstanding task is that of constructing a *model* of critical thinking as it might apply in higher education. Work has been done for at least forty years on the topic of critical thinking and informal logic, and the skills needed for fine critical thinking. This is a matter of explicitly educating *for* critical thinking (i.e., for developing the skills required by students for critical thinking within higher education). However, there has been very little done on the matter of being critical in the wider world and the ways in which higher education can help here. Providing a model of critical thinking in higher education will go some way toward clarifying its nature and its possibilities.

Initially much of the intellectual running was made by philosophers, especially those working on reason, argument, and the philosophy of the mind. Not surprisingly, they came to associate critical thinking with precisely what it is to reason carefully and soundly. Such legitimate concerns need to be reconciled with another set of very legitimate concerns, namely what it is to be educated in the modern world; and there, in this educational perspective, two camps may be discerned, those who have been interested most in what it is to be educated as an individual, and those who see in education a way of helping to transform society. To our knowledge, a model of critical thinking in higher education that reconciles all these different perspectives has not before been attempted. We attempt to provide such a model in these pages and, in so doing, we shall see points at which the various positions run into each other.

Three rival perspectives

The *philosophical* perspective (on which we have just touched) is principally interested in clear and rigorous thinking. It has a particular interest in logical thinking, including informal and formal logic, in how critical thinking relates to language use in ordinary contexts, how it forms part of metacognitive processing in complex adaptive systems, and so on. The *educational* perspective is interested chiefly in the wider educational development of the individual student and, to that end, is concerned with ways in which critical thinking can benefit the wider society outside the classroom through the development and formation of a critico-social *attitude*. The *socially active* perspective—as we might term it—is itself a complex of positions but is prompted by a concern to see society itself transformed and sees the inculcation of critical attitudes in students as a propaedeutic to that end. It encompasses critical pedagogy (i.e., educating to dissolve habits of thought and to promote political activism) and critical citizenship (i.e., cultivating a critical citizenry).

As we shall see, these three perspectives are by no means entirely separable. Their boundaries are permeable, with commentators, researchers, and scholars taking up all manner of cross-boundary positions. For example, cutting across the latter two perspectives—the educational and the socially active perspectives—is a concern as to how to reconcile tensions that exist in the modern corporate university, with its emphasis on developing technical and work-ready skills in graduates, and the traditional role of the university that aims to prepare thoughtful, well-read critical thinkers who are beneficial to society at large (Daymon and Durkin 2013). There are, though, tensions between the perspectives. For any book attempting to survey this field, the concerns of the educators risk being seen (by philosophers) as tangential and remote while the concerns of the philosophers risk being seen (by educators) as myopic and obscure. Any book on the topic of critical thinking in higher education has to try and address both perspectives without compromising the integrity of each. This book is our attempt to do just that.

Critical thinking movements

Richard Paul (2011) sees developments in understanding critical thinking as occurring in three separate if overlapping waves. These began in the 1970s, with the move to introduce formal and informal logic in the curriculum, a practice dominated largely by philosophers and their concerns. This wave emphasized skills in both (1) the identification of arguments and (2) the evaluation of arguments. It saw identification and evaluation of logical structures, and the awareness and avoidance of fallacies of reasoning, and so on, as largely equivalent to critical thinking. Skills in argumentation, on this view, led to the purportedly laudable aim of producing better critical thinkers. An implication here was that such critical thinking could best be promoted by institutions putting on dedicated courses, being essentially programs designed to develop skills of logic, reasoning, and argument. We still see the influence of this wave today with a number of generalist critical thinking and informal logic courses taught in institutions around the world (but mainly in the United States).

The 1980s saw a second wave, with an introduction of concerns that were much wider than critical thinking as adumbrated by philosophers. This more educational orientation included standpoints of cognitive psychology, critical pedagogy, feminism and other perspectives, as well as discipline-specific approaches to critical thinking (critical thinking in Business Studies, and so on). It had a wider agenda than that of critical thinking as argumentation. It was concerned with the development of the student as a person (rather than as a cognitive machine) and emphasized critical thinking in relation to attitudes, emotions, intuitions, human *being*, creativity, and so on.

The rise of *critical pedagogy* during this period—with its origins in German critical theory, Marxism, phenomenology, and psychoanalysis—resulted in an interpretation of critical thinking far wider than that offered by first-wave theorists, seeing critical thinking as an ideological issue, not one concerned with validity and reliability of arguments. As a point of difference, the first-wave theorists took the adjective "critical" to mean "criticism" (i.e., identifying weaknesses in and correcting some claim or argument). The critical pedagogues, or the second-wave theorists, by contrast, took "critical" to mean "critique" (i.e., identifying dimensions of meaning that might be missing or concealed behind some claim or argument) (Kaplan 1991, 362). This is an important difference, and one that is often the basis of misunderstanding among scholars in this field and that results in scholars talking past one another.

A third wave of the critical thinking movement, Paul (2011) identifies as a "commitment to transcend the predominant weaknesses of the first two waves (rigor without comprehensiveness, on the one hand, and comprehensiveness without rigor, on the other)." Paul sees this third wave as "only beginning to emerge," but he identifies this as one which includes the development of a "theory" of critical thinking, which does justice to the earlier emphasis on structures of argumentation, and yet which does not neglect other important human traits such as the emotions, imagination, and creativity, or, for that matter, the wider educational possibilities within higher education.

All of these waves are on display in many of the chapters in this volume. However, many of the papers here cannot easily be ascribed to any one wave but cut across concerns relevant to more than one wave. Some of the papers, too, are openly reflective of critical thinking itself, and we can see their contributions as a modest step in the direction of third-wave theorizing.

Toward a model of critical thinking

From the overview so far offered, it is evident that any account of the place of critical thinking in higher education needs to make sense, for example, as to how critical thinking is represented in debates about critical pedagogy, the role of education in leading to individual fulfillment and collective sociopolitical activism, the place of critical thinking in educating for citizenship, the role of critical thinking in relation to creativity, and so on. Any such account of critical thinking must also embrace the long-standing focus of critical thinking as a composite of skills and judgments, and as a variety of dispositions as well. A model of critical thinking in higher education is needed that incorporates all these perspectives and approaches.

Critical thinking in higher education has, we contend, at least six distinct, yet integrated and permeable, dimensions: (1) core skills in critical argumentation (reasoning and inference making), (2) critical judgments, (3) critical-thinking dispositions and attitudes, (4) critical being and critical actions, (5) societal and ideology critique, and (6) critical creativity or critical openness. Each of these, we believe, has a particular place in an overarching model of critical thinking. The model we propose here indicates that critical thinking has both an *individual* dimension, as well as a *sociocultural* dimension and incorporates six distinct dimensions of critical thinking, namely skills, judgments, dispositions, actions, critique, and creativity. For reasons of space, we shall not deal with the "creativity" dimension here. For this, and for a more detailed development of the model, see Davies (2015).

The place of critical thinking in higher education

What is the place of critical thinking in higher education? At one level, as noted, critical thinking is all about the development of certain sorts of skills. These include skills in argumentation, and skills in making sound judgments. Employers want evidence of critical thinking skills in their employees, and graduates are assumed to possess these skills. However, skills without the disposition to *use* them are not much use, so critical thinking is about dispositions as well. On this view, critical thinking, as both skills and dispositions, is mainly about the development of the *individual*. We might call this the *individual dimension* of critical thinking. For the most part, it embodies a sense of critical thinking being rather narrowly bounded, working within, say, the frames of thought that characterize different disciplines (and making reasoned judgments within those frameworks).

However, theorists who promulgate what has become known as critical pedagogy think that critical thinking is more about *changing matters,* and here changing society as much as if not more than individual students, Such an approach is fired by concerns about society, its conditions of social oppression (as it advocates perceive them), its ideologies, and its fundamental inequities. They regard truth claims, for example, "not merely as propositions to be assessed for their truth content, but as part of systems of belief and action that have aggregate effects within the power structures of society. It asks first about these systems of belief and action, *who benefits?*" (italics in the original, Burbules and Berk 1999, 47). Their focus is on the social and political *functioning* of arguments and reasoning and their wider frames of thought. Questioning power relationships in society that lie behind forms of thought must, they argue, be considered a central part of critical thinking (Kaplan 1991).

Scholars who write about what has become known as critical democratic citizenship education offer a yet further account of critical thinking. Given that critical thinking has a social and political dimension, it is not unreasonable for it to have a dimension of interpersonal socially appropriate *caring* as well (Noddings 1992). In order to cultivate critical citizens, they argue, "instructional designs are needed that do not capitalize on applying tricks of arguing, nor on the cognitive activity of analyzing power structures, but contribute…in a meaningful and critical way in concrete real social practices and activities" (Ten Dam and Volman 2004, 371). They argue that learning to think critically should—in part at least—be conceptualized as "the acquisition of the competence to participate critically in the communities and social practices of which a person is a member" (Ten Dam and Volman 2004, 375). This kind of educational aim, naturally, has an impact on the development of critical character and *virtue*. A good "citizen," they suggest, should be a socially adept and virtuous person, caring in nature, with the capacity to consider the interests and needs of humanity. On this view, critical thinking has *moral* as well as cultural characteristics. We might call this the *sociocultural dimension* of critical thinking.

Both the individual and the sociocultural dimensions can be given a place, and reconciled, in a single model of critical thinking in higher education. We see here two dimensions as separate and distinguishable axes or vectors that account for very different, equally important, aspects of critical thinking. To

date, much of the scholarly effort has been (rightly) expended on the individual axis, with its emphasis on the cultivation of skills and dispositions.
This is understandable: being an (individual) critical thinker naturally has
many personal and social benefits, not to mention its need in the workplace.
Increasingly, however, over the past twenty years, the sociocultural dimension
has been developed, and it should be accorded an equal place in any model of
critical thinking.

What is critical thinking?

In 1990, the American Philosophical Association convened an authoritative
panel of forty-six noted experts on the subject to produce a definitive account
of the concept. It resulted in the production of the landmark Delphi Report
(Facione 1990). This led to the following definition of critical thinking which
is as long as it is hard to follow:

> We understand critical thinking to be purposeful, self-regulatory judgment
> which results in interpretation, analysis, evaluation and inference as well
> as explanation of the evidential conceptual, methodological, criteriological
> or contextual considerations upon which that judgment was based. Critical
> thinking is essential as a tool of inquiry. Critical thinking is a pervasive and
> self-rectifying, human phenomenon. The ideal critical thinker is habitu-
> ally inquisitive, well-informed, honest in facing personal biases, prudent in
> making judgments, willing to consider, clear about issues, orderly in com-
> plex matters, diligent in seeking relevant information, reasonable in selec-
> tion of criteria, focused in inquiry and persistent in seeking results which
> are as precise as the subject and circumstances of inquiry permit. (Facione
> 1990)

While of undeniable importance as a definition of critical thinking for educational philosophers, this account of critical thinking does not lend itself easily
to educational implementation. How would a dean of a Faculty, for example,
use this definition to further embed the teaching of critical thinking in the
curriculum? How useful is it, in a practical sense, in a higher education context? It is not clear that higher education can benefit from such a definition in
the form it is presented. Nor does it square with the wider concerns about the
nature of criticality. It seems, on the face of it, a definition rooted in *one kind*
of critical thinking, namely, critical thinking as argumentation and judgment
formation.

Among the various threads in the above definition, we can distinguish the
following: critical thinking as skills in inference making and argumentation,
critical thinking as (reflective) judgment formation, and critical thinking as

a variety of dispositions and attitudes. These can be classified into two broad categories: *cognitive elements* (argumentation, inference making, and reflective judgment) and *propensity elements* (dispositions, abilities, and attitudes) (Halonen 1995). Note, however, the phenomenon of *action* is not mentioned in the Delphi definition. It is, in principle possible to meet the stipulated requirements of the definition and not *do* anything.

Strong skills in argumentation are not to be dismissed. They help to provide a sound basis for capable decision making. This is because decision making is based on judgments derived from argumentation. Such decision making involves understanding and interpreting the propositions and arguments of others, and being able to make objections and provide rebuttals to objections. Broadly speaking, then, this sense of the term "critical thinking" is seen as involving skills in *argumentation*. Critical thinking in this sense is a fundamental skill and is one which—on the available evidence—universities have apparently not been teaching as well as they should.

Critical thinking as reflective thinking (the "skills-and-judgments" view)

However, even within the cognitive-philosophical camp, critical thinking is often defined more widely than this, and in practical and instrumental terms, for example, as: "reflective and reasonable thinking that is focused on deciding what to believe or do" (Ennis 1985) or as "thinking aimed at forming a judgment" (Bailin, Case, Coombs and Daniels 1999, 287) or as "skillful, responsible thinking that facilitates good judgment" (Lipman 1988, 39). This definition focuses less on the mechanics of the skill of argumentation and more on the *reflective* basis for decision making and judgment calls. We might call this the "skills-and-judgments" view.

These wider senses of critical thinking are not inconsistent with "critical thinking as argumentation," and are, indeed, in some sense premised on it. Being able to demonstrate "reflective thinking" for the purposes of decision making requires skills in argumentation. However, this account does bring in a different emphasis, focusing less on mechanisms of argumentation qua inference making, and more on judgment formation, which is at a higher cognitive level. (The relationship seems asymmetric: one can engage in idle argumentation without making a judgment toward a decision, but not vice-versa—or at least not *ideally*.)

The definition by Ennis, given above—"reflective and reasonable thinking that is focused on deciding what to believe or do"—is recognized as the leading definition in the "skills-and-judgments" view. However, note that Ennis's definition is somewhat limiting by again not necessitating, for its application, any commitment to *action* on the part of the critical thinker. On this account, a person might exhibit critical thinking, without requiring that a decision so reached actually be implemented.

To sum up the "skills-and-judgments" view, we can think of cognitive critical thinking skills as involving *interpretation, analysis, inference, explanation, evaluation,* and some element of *metacognition* or *self-regulation* (Facione, Sanchez, Facione, and Gainen 1995, 3; Halonen 1995, 92–93). These facets of critical thinking are all in the Delphi list. This is sometimes collectively known as the "skills-based" view of critical thinking.

A taxonomy of critical thinking skills

At this point, categorizing these skills would seem to be useful. We shall use the framework by Wales and Nardi (1984) and borrowed by Halonen (1995). Cognitive critical thinking skills as such can be seen as falling under four main categories: *lower-level thinking skills* (which might be called "foundation" thinking), *thinking skills* (or "higher level" thinking), *complex thinking skills,* and *thinking about thinking* or metacognitive skills. "Identifying an assumption," for example, is clearly less difficult—and requires fewer cognitive resources—than say "analyzing a claim" or "drawing an inference." There might be debate about which skill belongs in which category, but there is little doubt that some cognitive skills are demonstrably more sophisticated than others (see table 0.1):

There is considerable degree of unanimity in the literature on many of the cognitive skills involved in critical thinking, if not the degree of importance accorded to each. In any event, the view that critical thinking involves both (1) rigorous argumentation, assessing propositions, analyzing inferences, identifying flaws in reasoning, and so on and (2) judgment *formation* is pervasive. However, as noted, despite its importance, when applied to the higher education context (as opposed to a philosophical context), there has been a tendency to define critical thinking far too narrowly.

Table 0.1   Critical thinking skills

Lower-level thinking skills ("Foundation")	Higher-level thinking skills	Complex thinking skills	Thinking about thinking
Interpreting	Analyzing claims	Evaluating arguments	Metacognition
Identifying assumptions	Synthesizing claims	Reasoning verbally	Self-regulation
Asking questions for clarification	Predicting	Inference making	
		Problem solving	

Critical thinking as dispositions (the "skills-plus-dispositions" view)

It has long been recognized that the ability to think critically is different from the attitude or *disposition* to do so (Ennis 1985; Facione 1990), and this too needs to be considered in any attempt to define critical thinking. Dispositions have been described as "at least half the battle of good thinking, and arguably more" (Perkins, Jay, and Tishman 1992, 9).

Dispositions are sometimes defined as a "cast or habit of the mind" or "frame of mind" that is necessary for exercising critical thinking. Dispositions are not arguments or judgments, but *affective* states. They include critical thinking *attitudes* and a sense of *psychological readiness* of the human being to be critical. They are equivalent to what Passmore once called a "critical spirit" (1967, 25) and have been defined as a constellation of attitudes, intellectual virtues, and habits of mind (Facione et al. 1995). Correspondingly, we may distinguish between critical thinking in a "weak" sense and in a "strong" sense (Paul 1993). The former consists of the skills and dispositions already discussed; the latter consists of the *examined life* in which skills and dispositions have been incorporated as part of one's deep-seated personality and moral sense—in short, one's *character*.

A taxonomy of critical thinking dispositions

Critical thinking dispositions might be broadly categorized as falling under dispositions arising in relation to the *self*, in relation to *others*, and in relation to the *world*. Again, it might be debated which category a disposition belongs to (and some might belong to more than one), but it is fairly clear that there are at least four dispositional orientations (see table 0.2):

Table 0.2   Critical thinking dispositions

Dispositions arising in relation to self	Dispositions arising in relation to others	Dispositions arising in relation to world	Other
Desire to be well-informed	Respect for alternative viewpoints	Interest	Mindfulness
Willingness to seek or be guided by reason		Inquisitiveness	Critical spiritedness
Tentativeness	Open-mindedness	Seeing both sides of an issue	
Tolerance of ambiguity	Fair-mindedness		
Intellectual humility	Appreciation of individual differences		
Intellectual courage			
Integrity	Skepticism		
Empathy			
Perseverance			
Holding ethical standards			

Critical thinking as a composite of skills and attitudes

Critical thinking has naturally been seen in terms of a composite of skills, knowledge, and attitudes too—including argumentational, reflective, and affective features (Boostrum 1994; Brookfield 1987; Facione 1990; Kurfiss 1988; McPeck 1981; Paul 1981; Siegel 1988; 1991; Watson and Glaser 2008). Most theorists hold a composite account. The composite view includes both the cognitive and propensity elements discussed above. While the ability to argue and make inferences, to reflect and make judgments, and be critically disposed is all important, it is also crucial to recognize that each of these does not occur in isolation. For McPeck, critical thinking involves a disposition and a skill, and "one must develop the disposition to use those skills" (1981, 3), hence, his definition of critical thinking as "a propensity [disposition] *and* skill to engage in an activity with reflective skepticism" (1981, 8).

How the cognitive and propensity elements relate to each other in any definition of critical thinking is subject to much discussion. Facione et. al., for example, postulate an interactionist hypothesis where "the disposition toward critical thinking reinforces critical thinking skills and that success with critical thinking skills reinforces the disposition" (1995, 17).

To conclude here, as it has been traditionally defined—by Ennis, Paul, McPeck, Lipman, and others in the critical thinking movement—critical thinking has been seen largely in terms of cognitive elements, that is, as "reflective and reasonable thinking that is focused on deciding what to believe or do." However, as intimated, this definition is remiss by not including in its scope any sense of actual or potential *action*.

Dimensions of criticality: An axis diagram

Figure 0.1 represents the critical thinking movement as outlined so far. This movement is largely concerned with an individual's *cognitive* qualities, that is, cognitive elements or skills (argumentational skills, skills in thinking) and reasoning and argumentative propensities or character attributes of the *person*. These are inclusive of all the skills and attributes mentioned in figure 0.1 (namely foundation, higher-level, complex, metacognitive skills, as well as critical thinking abilities and dispositions). These skills and dispositions are represented by separate lines radiating out from the bottom of the Y axis. This account of criticality is what might be termed "critical thinking *unadorned*" or critical thinking in its traditional senses. (The X axis will be added in a moment.) For the full development of this diagram see Davies (2015).

Critical thinking as "criticality" (The "skills-plus-dispositions-plus-actions" view)

Following Barnett (1997), the term now most commonly used in relation to critical thinking is that of "*criticality*." Criticality is a term deliberately distinct

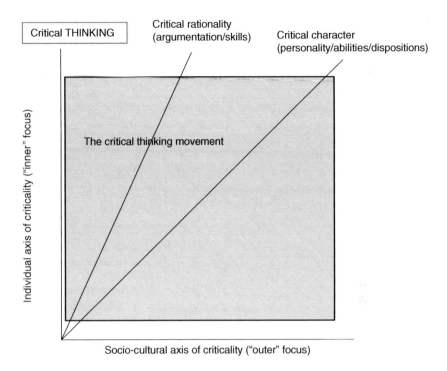

Figure 0.1 Axis diagram: The critical thinking movement.

from the traditional expression "critical thinking," which was felt to be inadequate to convey the educational potential that lies to hand. The term "criticality" attempts to inject a perspective that widens critical thinking to incorporate not only argument and judgment and reflection but also the individual's wider identity and participation in the world. This is a concept of critical thinking involving students reflecting on their knowledge and simultaneously developing powers of critical thinking, critical self-reflection, and critical action—and thereby developing (as a result) critical *being* (Barnett 1997; 2004; Johnston, Ford, Mitchell, and Myles 2011). Now, criticality, not unlike critical thinking, appears, in some quarters, to be gaining its own scholarly industry.

What is "criticality"?

Broadly speaking, criticality comprises—and is a composite of—three things: *thinking, being,* and *acting.* In emphasizing action in addition to thinking (in the form of argumentation and reflective judgment), criticality might be conceived of in relation to established definitions of critical thinking as *trait.* That is, while a critical thinker can be disposed to think critically, criticality points to the way a person is in the world. A critical person exhibits a critical orientation toward the world and has a trait, thereby, to act accordingly. Criticality requires that one be moved to *do* something (Burbules and Berk 1999, 52). While skills and dispositions are crucial for critical thinking, they are not sufficient unless a person is in her- or himself critical and unless she or he is disposed to act in a critical vein. To adapt a famous line from Kant: criticality without critical thinking skills is empty; critical thinking without action is myopic.

An example of "criticality"

The concept of criticality—as a composite of critical thinking, critical reflection, and critical action—has been made concrete by the use of a famous photograph as a frontispiece to Barnett's book *Higher Education: A Critical Business* (1997). The photograph depicts a student in front of a line of tanks in Tiananmen Square in 1989. Most people have seen this photograph; indeed, it is one of the defining photographs of the latter part of the twentieth century. How does the photograph demonstrate critical thinking as "criticality"?

This photograph is intended to imply that higher education should be (if not always in practice) an educational process involving a composite of *thinking, being-in-the-world,* and *action.* Critical thinking, in the established cognitive sense proposed by philosophers such as Ennis, Siegel, Lipman, McPeck, and others, is an important perspective, but by itself inadequate as a way of capturing what higher education can be *at its best.* Higher education can, therefore, potentially do much more than teach students how to demonstrate (for example) critical thinking as analytic skills and judgments. It can also prompt students to understand themselves, to have a critical orientation to the world, and to demonstrate an active sociopolitical stance toward established norms or practices with which they are confronted. This, it is argued, is more than what is offered by the critical thinking movement in relation to skills in critical thinking; it is tantamount to the development of critical *beings.*

This is a sense of "critical thinking" that extends beyond the individual and his or her cognitive states and dispositions to the individual's participation in society as a critically engaged citizen in the world. Note that it also includes a *moral* and *ethical* dimension to critical thinking. After all, critical thinkers do more than reason; they also *act ethically* on the basis of their reasoned judgments.

In this argument for the criticality dimension, *critical reasoning, critical reflection,* and *critical action* could be thought of as three interlocking circles in the form of a Venn diagram (see figure 0.2). It is important, according to Barnett, that they be regarded as interlocking—but not as entirely congruent with each other; otherwise, the space for each of them to work (including critical thinking in the cognitive sense) would be lost.

The respective concerns of educational philosophers and higher education scholars in relation to the topic of critical thinking are then quite different. The work of Ennis, Paul, McPeck, and others aims to identify the philosophical

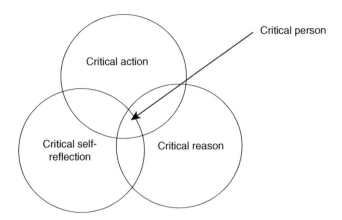

Figure 0.2 The intersection between critical reason, critical self-reflection, and critical action (Barnett 1997, 105).

elements of what a critical thinker *is* or *should be*; the work of those interested in criticality aims to identify what a critical thinker *does* and can *become*. In turn, the implications for higher education on producing critical beings also holds out a promise for what *higher education* can be, which, however, especially given the corporate nature of the university, it seldom is at present (Cowden and Singh 2013)

Criticality, then, is a *wider* concept than critical thinking, as it has been generally defined by educational philosophers. To some extent it subsumes critical thinking. One outcome of this wider concept being taken up, of course, is that it suggests a wider set of responsibilities befalling higher education professionals, that is, teachers and academics, than that of (simply) imparting skills in argumentation, or developing in students a capacity for rational "reflection" or decision making, or even cultivating critical thinking dispositions. Educating for criticality, in contrast, holds out a sense that higher education can become (more) a process of *radical* development than merely a cognitive process. It captures a sense of enabling students to reach a level of "transformatory critique" (i.e., to live and breathe as a critical thinker, to become an *exemplar* of what it means to be a critical being).

The axis diagram revisited

The concerns of the criticality movement arose, as we have seen, in reaction to the narrow emphasis of previous accounts of critical thinking. These previous accounts view critical thinking in terms of individual skills, dispositions, and abilities. While proponents of the criticality dimension certainly do not eschew these important individual facets of critical thinking entirely (indeed, they endorse their importance), the criticality perspective adds something

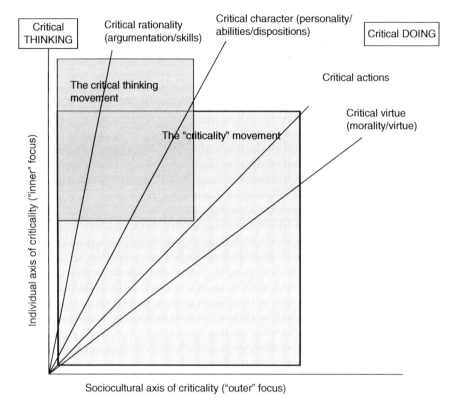

Figure 0.3   Axis diagram: The critical thinking and "criticality" movements (Davies 2015).

new. It adds the dimension of *action* to the mix. This is represented by the addition of the social-cultural axis (the "X" axis) of critical thinking shown in figure 0.3, and here depicted as "critical doing."

However, there is more to it than action. Unlike the views of critical thinking as adumbrated by proponents of the critical thinking movement (CTM), for the criticality theorists the *ethical* dimension is also important to critical thinking. Ethical decisions are, of course, usually (if not always) accompanied by ethical *actions*. This is represented by the *critical virtue* axis below. Note in the diagram that the CTM, with few exceptions, does not include the action and morality dimension.

Critical thinking as critical pedagogy

Critical pedagogy is defined as the use of higher education to overcome and unlearn the social conditions that restrict and limit human freedom. According to one of its major proponents, it is "an educational movement, guided by passion and principle, to help students develop consciousness of freedom, recognize authoritarian tendencies, and connect knowledge to power," and the ability to take "constructive action" in relation to education and society at large (Giroux 2010).

Like the approach taken by Barnett, Johnston, and others in their account of criticality, critical pedagogy takes the view that critical thinking needs to be broadened beyond skills and dispositions. It sees the account of critical thinking as comprising skills-plus-dispositions as very much concerned with the individual. Like the adherents of the criticality approach, the critical pedagogues include the importance of *action*. However, unlike adherents of the criticality approach, they consider *social institutions* (and society more broadly)—not merely individuals' actions—to be a vital factor for critical thinking. This broadens the notion of critical thinking even further than any of the views previously discussed.

This is clearly an extension of the account of the radically transformed student within the criticality perspective; indeed, it extends radical educational transformation to *society at large*. The critical pedagogues see critical thinking to be not about mere argument analysis, or dispositions, or individual actions (although these too are important). They see critical thinking to be principally about "the critique of lived social and political realities to allow greater freedom of thought and action" (Kaplan 1991, 362). Specifically, the critical pedagogues are alert to the presence of ideology in discourse and social institutions and see education as a critical and active engagement with such ideologies.

The key theorists in this area are Freire (1972), McLaren (1989), and Giroux (1994; 2005). In an illuminating article by Burbules and Berk (1999) a number of distinctions are made between the critical thinking movement (incorporating the "skills-based" view of critical thinking and the "skills-plus-dispositions" view) and the critical pedagogy movement.

The critical thinking movement theorists had taken the adjective "critical" to mean "criticism" (becoming aware of weaknesses in some claim or argument). Their aim was putting logic at the service of clear thinking. The critical pedagogues, by contrast, took "critical" to mean "critique" (i.e., identifying dimensions of meaning that might be missing or concealed behind some claim or belief or institution) (Kaplan 1991, 362). Their further understanding is that such concealment serves an ideological function, masking an underlying state of affairs. Their aim puts critical thought at the service of transforming undemocratic societies and inequitable power structures, that is, not simply educating for critical thinking or even enabling individuals to embody a critical spirit, but educating for *radical* transformation in society as well. They see the critical person as resisting the ideological hegemony of capitalism, a hegemony that foists conditions favorable to the maintenance of the capitalist system onto unwitting members of society. Here, higher education becomes a vehicle for combating perniciousness—as they see—inherent in capitalist society. They see advertising, for example, as encouraging and fostering increased material consumption while simultaneously reinforcing the myth that large corporations are there to serve their customers, when they are, in fact, serving their own interests, and maximizing profit, often at the expense of both customers and the social good (Burbules and Berk 1999, 50).

The critical pedagogues accordingly believe that the aim of education should be about turning students against the idea of being trained for the economic needs of large corporations. The followers of the critical pedagogy movement see the role of higher education not as reinforcing but as dispelling these uncritical attitudes and questioning these assumptions. They see the role of higher education as working within higher educational institutions to identify and critique power inequities in society, the myths of opportunity in capitalist economies, and "the way belief systems become internalized to the point where individuals and groups abandon the very aspiration to question or change their lot in life" (Burbules and Berk 1999, 50). Thinking critically, for the critical pedagogists, is a matter of recognizing, critiquing, and combating societal formations (really *de*formations)—including discourses—that maintain the capitalist status quo. This can be achieved by developing students and their teachers not only as critical intellectuals (Giroux 1988) but also as critical activists. This is clearly a very different sense of critical thinking than the other camps identified earlier here.

Like Barnett, the critical pedagogists see action as an intrinsic, not separable, aspect of criticality. However, they take critical action much further. They see action as important not merely for encouraging students' personal individual critical comprehension of, and reaction to, events, but as a justification for wholesale social and political *change*. As Burbules and Berk put it, for them: "challenging thought and practice must occur together…criticality requires *praxis*—both reflection and action, both interpretation and change…Critical pedagogy would never find it sufficient to reform the habits of thought of thinkers, however effectively, without challenging and transforming the institutions, ideologies, and relations that engender distorted, oppressed thinking in the first place—not an additional act beyond the pedagogical one, but an inseparable part of it" (1999, 52). Critical pedagogy, accordingly, becomes a way of alerting students to the indoctrination that is felt here to be endemic in society *and* of combating it—so, deliberately and systematically deploying the potential of higher education as a transforming device in society.

For the critical thinking movement, this is a misguided stance. It amounts to taking for granted and prejudging the conclusions to an issue (that society *is* inequitable, that society *is* ideologically saturated and so on, and that society *is* characterized by undue repression). It is itself equivalent to indoctrination.

However, in the critical pedagogy movement, raising the issue of the social conditions of freedom is *essential* to critical thinking. True critical thinking, for the critical pedagogists, involves liberation from an oppressive system as a condition of freedom of thought. As Burbules and Berk put it: "Critical thinking's claim is, at heart, to teach how to think critically, not how to teach politically; for Critical Pedagogy, this is a false distinction...self-emancipation is contingent upon social emancipation" (1999, 55). In the words of the Critical Pedagogy Collective (echoing Dewey): "Education is not preparation for life— education is life itself" (2013).

The axis diagram revisited again

We can now move to a further refinement of our axis diagram (see figure 0.4); and here we use the term "critical participation" to denote the perspectives that are orientated toward participating critically in society. Note that "critical participation" is oriented in figure 0.4 spatially closer to the category of "critical doing" compared to the category of "critical rationality" (it has a stronger "outer" than an "inner" focus). It is positioned closer to the X axis. However, there is a difference in the degree of commitment here. The "participation" facet of criticality, in turn, has two dimensions: (1) an *awareness* of oppression (known in the literature both as critical consciousness or *conscientization* (Freire 1972; 1973) and (2) a more practical dimension, the *resistance* to oppression (demonstrably, to "resist" something one needs to be aware of what one is resisting). This is known in the critical pedagogy literature as *praxis*. Both these vectors are represented in figure 0.4.

However, this separation of concerns belies deep similarities. As Burbules and Berk note: "each invokes the term 'critical' as a valued educational goal: urging teachers to help students become more skeptical toward commonly accepted truisms. Each says, in its own way, 'Don't let yourself be deceived.' And each has sought to reach and influence particular groups of educators...They share a passion and sense of urgency about the need for more critically oriented classrooms. Yet with very few exceptions these literatures do not discuss one another" (Burbules and Berk 1999, 45). However, there are synergies between the criticality and critical pedagogy movements as indicated by their focus on action.

Conclusion

Attention to critical thinking or criticality, as we prefer it, is in greater need than ever in the contemporary world. There are, though, some challenges in giving it the important place in higher education that we suggest it warrants. Large forces are at work that are tending to diminish a sense of its significance. On the one hand "cognitive capitalism" (Boutang 2011) works—in a digital

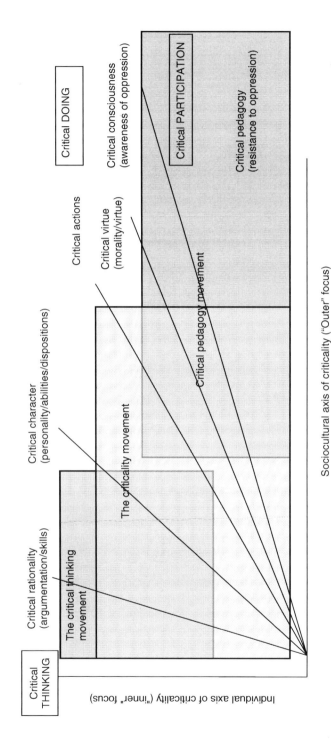

Figure 0.4 The critical pedagogy movement (for an elaboration on this diagram, see Davies 2015.

age—to expand the efficiency with which vast, if not infinite, amounts of data can be assimilated. What counts here is speed of response, measured even in microseconds, with computers programmed to make such responses twenty-four hours per day. This is, as it has been described, an "algorithmic capitalism" (Peters 2014). Critical thinking, on the other hand, betokens a different rhythm, of care, reflection, and repose.

Critical thinking also draws on a particular set of motivations in bringing a critique of forces and institutions that would rather press on, untroubled by critical examination. Ultimately, what is at issue here, in critical thinking, is the concern to enlarge freedom, whether cognitive, discursive, personal, or even societal freedom. But it is at least arguable that educational institutions—including higher education institutions—are being co-opted into the service of the global knowledge economy. So the space for critical thought may be diminishing at precisely a moment when it is especially needed.

But, as we have seen, theorists and educationalists who have given thought to the matter differ profoundly over fundamental aspects of critical thought or criticality. They differ over what is to count as critical thinking, over its purposes and its scope, and the way in which teaching might help to encourage it among students. So any campaign in favor of criticality is—it may seem—bedeviled at the outset by deep schisms within the academic and educational communities.

A first step in the matter must surely be the bringing together of the different points of view, not least to see how they are exemplified in different pedagogical situations—of teaching, learning, curriculum design, and so on. That is what we have attempted to do in this volume. It is a start but no more than that.

References

Arum, R., and Roksa, J. 2011. *Academically Adrift: Limited Learning on College Campuses*. Chicago and London: University of Chicago Press.

Bailin, S., Case, R., Coombs, J. R., and Daniels, L. B. 1999. "Conceptualizing Critical Thinking." *Journal of Curriculum Studies* 31 (3): 285–302.

Barnett, R. 1997. *Higher Education: A Critical Business*. Buckingham: Society for Research into Higher Education and the Open University Press.

Barnett, R. 2004. "Learning for an Unknown Future." *Higher Education Research and Development* 23 (3): 247–260.

Boostrum, R. 1994. *Developing Creative and Critical Thinking: An Integrated Approach*. Lincolnwood, IL: National Textbook Company.

Boutang, Y. M. 2011. *Cognitive Capitalism*. Cambridge: Polity.

Brookfield, S. 1987. *Developing Critical Thinkers : Challenging Adults to Explore Alternative Ways of Thinking and Acting*. 1st ed., Jossey-Bass Higher Education Series. San Francisco, CA: Jossey-Bass.

Burbules, N. C., and Berk, R. 1999. "Critical Thinking and Critical Pedagogy: Relations, Differences, and Limits." In *Critical Theories in Education: Changing Terrains of Knowledge and Politics*, edited by T. S. Popkewitz and L. Fender. New York: Routledge. 45–65.

Casner-Lotto, J., and Benner, M. W. 2006. "Are They Ready to Work? Employers' Perspectives on the Basic Knowledge and Applied Skills of New Entrants to the 21st Century US Workforce." In The Conference Board, Corporate Voices for Working Families, the Partnership for 21st Century Skills, and the Society for Human Resource Management. http://www.p21.org/storage/documents/FINAL_REPORT_PDF09-29-06.pdf.

Cowden, S., and Singh, G. 2013. *Acts of Knowing: Critical Pedagogy in, against and beyond the University.* London: Bloomsbury Academic.

Davies, M. 2015. "A Model of Critical Thinking in Higher Education." *Higher Education: Handbook of Theory and Research* 30: 41–92.

Daymon, C., and Durkin, K. 2013. "The Impact of Marketisation on Postgraduate Career Preparedness in a High Skills Economy." *Studies in Higher Education* 38 (4): 595–612.

Education News. 2013. *Neoliberalism and the Commercialization of Higher Education* (July 29). Available from http://www.educationnews.com/2013/07/29/neoliberalism-and-the-commercialization-of-higher-education/.

Ennis, R. H. 1985. "A Logical Basis for Measuring Critical Thinking Skills." *Educational Leadership* 43 (2): 44–48.

Facione, P. 1990. *The Delphi Report: Critical Thinking: A Statement of Expert Consensus for Purposes of Educational Assessment and Instruction.* Millbrae: California Academic Press.

Facione, P., Sanchez, C. A., Facione, N. C., and Gainen, J. 1995. "The Disposition toward Critical Thinking." *The Journal of General Education* 44 (1): 1–25.

Freire, P. 1972. *Pedagogy of the Oppressed.* Harmondsworth: Penguin.

Freire, P. 1973. *Education for Critical Consciousness.* New York: Seabury Press.

Giroux, H. A. 1988. *Teachers as Intellectuals: Towards a Critical Pedagogy of Learning.* Granby, MA: Bergin and Garvey.

Giroux, H. A. 1994. "Toward a Pedagogy of Critical Thinking." In *Re-Thinking Reason: New Perspectives in Critical Thinking*, edited by K. S. Walters. Albany: SUNY Press. 200–201.

Giroux, H. A. 2005. *Border Crossings: Cultural Workers and the Politics of Education.* New York: Routledge.

Giroux, H. A. 2010. Lessons from Paulo Freire. *The Chronicle of Higher Education* (October 17).

Halonen, J. 1995. "Demystifying Critical Thinking." *Teaching of Psychology* 22 (1): 75–81.

Johnston, B., Ford, P., Mitchell, R., and Myles, F. 2011. *Developing Student Criticality in Higher Education.* Continuum Studies in Educational Research. London: Continuum.

Kaplan, L. D. 1991. "Teaching Intellectual Autonomy: The Failure of the Critical Thinking Movement." *Educational Theory* 41 (4): 361–370.

Kurfiss, J. 1988. *Critical Thinking: Theory, Research, Practice and Possibilities.* Washington: ASHE-Eric Higher Education Report No. 2, Associate for the Study of Higher Education.

Lipman, M. 1988. "Critical Thinking: What Can It Be?" *Educational Leadership* 46 (1): 38–43.

Mataconis, D. 2013. *College Students Lack Critical Thinking Skills, but Who's to Blame* 2011 (May 28). Available from http://www.outsidethebeltway.com/college-students-lack-critical-thinking-skills-but-whos-to-blame/.

McLaren, P., and Hammer, R. 1989. "Critical Pedagogy and the Postmodern Challenge." *Educational Foundations* 3 (3): 29–69.

McPeck, J. 1981. *Critical Thinking and Education.* New York: St. Martin's Press.

Moon, J. 2008. *Critical Thinking. An Exploration of Theory and Practice.* Abingdon: Routledge.

Noddings, N. 1992. *The Challenge to Care in Schools: An Alternative to Education*. New York: Teachers College Press.

Norris, S. P. 1992. *The Generalizability of Critical Thinking: Multiple Perspectives on an Educational Ideal*. New York: Teachers College Press.

NPR Staff. 2011. *A Lack of Rigor Leaves Students "Adrift" in College*. Available from http://www.npr.org/2011/02/09/133310978/in-college-a-lack-of-rigor-leaves-students-adrift.

Passmore, J. 1967. *On Teaching to Be Critical*. Boston: Routledge and Kegan Paul.

Paul, R. 1981. "Teaching Critical Thinking in the 'Strong' Sense: A Focus on Self-Deception Worldviews, and a Dialectical Mode of Analysis." *Informal Logic* 4 (2): 2–7.

Paul, R. 2011. Critical Thinking Movement: 3 Waves. http://www.criticalthinking.org/pages/critical-thinking-movement-3-waves/856.

Paul, R. W. 1993. *Critical Thinking: What Every Person Needs to Survive in a Rapidly Changing World*. Santa Rosa, CA: Foundation for Critical Thinking.

Perkins, D., Jay, E., and Tishman, S. 1992. *Assessing Thinking: A Framework for Measuring Critical Thinking and Problem Solving Skills at the College Level*. Washington, DC: The National Center for Educational Statistics Workshop on the Assessment of Higher Order Thinking and Communication Skills of College Graduates: Preliminary listing of Skills and Levels of Proficiency.

Peters, M. 2014. "The University in the Epoch of Digital Reason: Fast Knowledge in the Circuits of Cybernetic Capitalism." In *The University in the Flux of Time*, edited by P. Gibbs, C. Guzman Valenzuala, O-H. Ylijoki, and R. Barnett. Abingdon: Routledge, Chapter 1.

Rimer, S. 2013. *Study: Many College Students Not Learning to Think Critically* 2011 (May 28) Available from http://www.mcclatchydc.com/2011/01/18/106949/study-many-college-students-not.html-UaVBAr-jPTp.

Siegel, H. 1988. *Educating Reason: Rationality, Critical Thinking, and Education*. New York: Routledge.

Siegel, H. 1991. "The Generalizability of Critical Thinking." *Educational Philosophy and Theory* 23 (1): 18–30.

Strauss, V. 2012. "Texas Gop Rejects 'Critical Thinking' Skills. Really." *The Washington Post*.

Ten Dam, G., and Volman, M. 2004. "Critical Thinking as Citizenship Competence." *Learning and Instruction* 14 (259–379).

The Critical Pedagogy Collective. 2013. Available from http://critped.wordpress.com/.

Wales, C. E., and Nardi, A. H. 1984. *The Paradox of Critical Thinking*. Morgantown, WV: Center for Guided Design.

Watson, G., and Glaser, E. M. 2008. *Watson-Glaser Critical Thinking Appraisal*. Cleveland, OH: Pearson.

Williams, R. 1976. *Keywords*. London: Collins.

Part I

What Is Critical Thinking in Higher Education?

Just what is it that counts as critical thinking, especially in the context of higher education? A belief of ours, and one that has provided much of the motivation for this volume, is that critical thought is a defining condition of higher education. Unless efforts to promote criticality are present in the design of curricula, especially in teaching and in the teacher-student relationship, we cannot say we are espousing the cause of genuine higher education. But then we are faced at the outset with the need to try to give some account of the very idea of critical thought.

It will be noted that, in this opening paragraph, we have used the terms "critical thinking," "critical thought," and "criticality." Are these terms synonyms or are there significant differences between the concepts that underlie them? The welter of different terms is one of the features of the debate here; and to the three terms just used can be added "critique" and "critical pedagogy"; and there are yet others, as the chapters in this opening section display. Such differences—in sheer terminology—across the scholars who have made significant contributions to the debate on critical thinking are not happenstance but, as our Introduction intimated (1–25), spring out of different perspectives and different interests, and those differences in terminology reflect, too, the stages that the debate has traversed.

Some of these different perspectives are apparent in the four chapters in this section. Robert Ennis can be considered to be one of the founders of the field of inquiry into critical thinking and has been working on the topic for some decades. Over time, Ennis has refined his position such that it has evolved into a depiction of critical thinking as residing in certain dispositions and abilities; and Ennis distinguishes twelve dispositions and eighteen abilities. While both the dispositions and the abilities overlap and interact, Ennis holds to their individual distinctiveness and exemplifies them each in turn by recounting the story of a jury faced with a defendant charged with murder and manslaughter. The narrative shows how the jurors, both individually and collectively, displayed to a greater or lesser extent the dispositions and abilities in Ennis's theory and, thereby, demonstrated the significance and practical power of critical thinking.

Richard Andrews takes a different tack. His title, "Critical Thinking and/ or Argumentation in Higher Education?" indicates the direction of travel. For Andrews, critical thinking is a necessary component of argumentation: get the argumentation right—and get students to acquire understanding of and competence in argumentation—and critical thinking will be in evidence pari passu. Here, on this view, the very term "critical thinking" is something of a misnomer, being "tautological," since "to think clearly is to be critical." If students can be helped to think clearly, and to acquire all the self-discipline that accompanies rigorous clear thought, then they will become not only critical thinkers but *self-critical* thinkers.

Notably, both Ennis and Andrews deploy a language of dispositions and abilities (or skills) but they do so through differing perspectives and with differing motives. Partly, going back to our earlier point about perspectives and interests, the difference here can be explained in Ennis coming at the matter primarily through philosophy and Andrews coming at the matter primarily as an educator. For the one, teaching critical thinking is a matter of the application of a philosophical conception of critical thinking whereas for the other, the originating concern is that of enabling higher education students to develop and to realize their potential.

Benjamin Hamby attempts to cut through much of the complexity of this debate by arguing trenchantly for there being one cardinal, "necessary and central," virtue that lies at the heart of critical thinking, namely a "willingness to inquire." Hamby teases out this idea, referring to individuals having "passion," "perseverance," "motivation," and "willing engagement." This virtue "stands behind other critical thinking virtues," such as open-mindedness; after all, "I could be the most open-minded person yet not at all interested in critical inquiry, being open-minded only for the sake of making friends [and other such instrumental reasons]." Equally, the display of critical thinking is not in itself indicative of a critical thinker in our midst. A critical thinker has to want to be critical, has to be stirred up to be critical and be energized so to do, even (so we might add) to do so when no one is watching.

Implicit, therefore, in Hamby's account is a fundamental distinction between dispositions and skills (or abilities), and it is this distinction that Barnett drives forward in introducing the concept of *criticality* into the debate. For Barnett, the full realization of critical thinking resides in three domains, critical thought, critical action, and critical being, which together amount to criticality. The fully critical student not only can think critically but can also exemplify that capacity in action in the world—say in professional life—and also be energized in that way, having the appropriate set of dispositions so to act. These dispositions include, for instance, the virtue of courage since the enactment of critical thinking may run against dominant ideologies and power structures.

Such criticality can be displayed at a number of levels, and Barnett identifies four levels: it might be exemplified in a rather perfunctory way, but at the highest level, it would amount to critique, in which students were able to see their studies in the widest possible way, enact their criticality on the largest and even global stage, and be fully committed to the critical way of life, even at some personal cost. "The critical spirit" (Siegel 1988) is a way of capturing such a large conception of criticality.

Cognition, skills, abilities, dispositions, and ways of being in the world: these are just some of the fault lines that have permeated the critical thinking debate for thirty years. These are apparent in the chapters in this opening section, and these chapters serve, accordingly, to set the scene for the sections to come.

1
Critical Thinking: A Streamlined Conception

Robert H. Ennis

Beginnings

Critical thinking under that name was inspired by pragmatic philosopher John Dewey (1910) and endorsed by analytic philosopher Max Black (1946). Dewey was revered by the progressive educators, who re-labeled his "reflective thinking" as "critical thinking," a name I believe they originated and that persists to the present (see Aiken (1942) for a mixture of both terms). Black (1946), insofar as I can determine, wrote the first college text including the words "critical thinking" in the title.

But critical thinking did not assume extensive prominence until the early 1980s. The California State University system in its "Executive Order 338" (Harmon 1980) required that all students study critical thinking in order to graduate from its units. The Commission on the Humanities ("Rockefeller Commission") asserted, "The Department of Education should define critical thinking as one of the basic skills that provides the foundation for advanced skills of all kinds" (1980, 37). There is not enough space here to report the many other strong expressions of support for critical thinking since then (see Ennis 2011), but I should mention the rapid development of interest in critical thinking worldwide since the 1970s. Also especially noteworthy is the support for critical thinking from both major US political parties in the form of statements by two presidents, George H. W. Bush and Barack Obama. Critical thinking was an explicit expressed goal in *America 2000* (1991, 40), an education policy statement endorsed in its preface by then-president Bush. President Obama, in his State of the Union Address (2014), listed critical thinking as one of the six basic goals of education.

But what is this critical thinking *concept* (using Rawls's [1971] distinction between *concept* and *conception*) that has been receiving increasing support? After careful consideration of its use in the previous century, I proposed this definition of the concept in the mid-1980s and still think it fits what most people were and are talking about:

> *Critical thinking is reasonable reflective thinking focused on deciding what to believe or do.*

Brief listings of dispositions and abilities of the ideal critical thinker

Given this definition, the ideal critical thinker can be characterized in somewhat more detail by the following proposed interdependent and somewhat-overlapping sets of twelve dispositions and eighteen abilities that constitute a streamlined conception. I have modified the organization and wording of critical thinking dispositions and abilities over the years in the direction of theoretical refinement and precision (1980; 1987; 1991; 1996a; 2011; 2013b); and Norris and Ennis (1989), but the basic ideas have not changed.

Critical thinking dispositions

Ideal critical thinkers are *disposed to*

1. seek and offer clear statements of the thesis or question,
2. seek and offer clear reasons,
3. try to be well informed,
4. use credible sources and observations, and usually mention them,
5. take into account the total situation,
6. keep in mind the basic concern in the context,
7. be alert for alternatives,
8. be open-minded
   a. seriously consider other points of view,
   b. withhold judgment when the evidence and reasons are insufficient,
9. take a position and change a position when the evidence and reasons are sufficient,
10. seek as much precision as the situation requires,
11. try to "get it right" to the extent possible or feasible, and
12. employ their critical thinking abilities.

Critical thinking abilities

Ideal critical thinkers have the *ability* to

1. have a focus and pursue it,
2. analyze arguments,

3. ask and answer clarification questions,
4. understand and use graphs and maths,
5. judge the credibility of a source,
6. observe, and judge observation reports,
7. use their background knowledge, knowledge of the situation, and previously established conclusions,
8. deduce, and judge deductions,
9. make, and judge, inductive inferences and arguments (both enumerative induction and best-explanation reasoning),
10. make, and judge, value judgments,
11. define terms, and judge definitions,
12. handle equivocation appropriately,
13. attribute and judge unstated assumptions,
14. think suppositionally, and
15. deal with fallacy labels.

Three nonconstitutive but helpful abilities that ideal critical thinkers possess are to:

16. be aware of, and check the quality of, their own thinking (metacognition),
17. deal with things in an orderly manner, and
18. deal with rhetorical strategies.

More detail in the form of principles and criteria can be found in Ennis (1996a; 2011; 2013b).

Exemplification of critical thinking dispositions and abilities

I shall exemplify these dispositions and abilities, showing the vital role that they can play in dealing with real issues. The main source of these examples is my experience as a juror in a murder trial. This experience was unique, but dealing with the legal system was not. Furthermore, most parts of this experience are similar to many features of our daily lives.

The defendant, Arlene, was charged with murder and voluntary manslaughter in the death of her companion, Al, late at night in her parents' kitchen. Soon after they entered the house through the back door, Arlene stabbed Al through the heart with a kitchen knife. She went to her parents' bedroom and awakened them, whereupon they called an ambulance. The victim was dead when the ambulance arrived. No one except Arlene witnessed the events leading to the killing, or the killing itself.

Although the defendant was charged with voluntary manslaughter as well as murder, I shall simplify by specifying in full only the nature of the charge of murder in the terms that were given in writing to the jurors.

The charge of murder in this trial

This is the charge of murder that we were directed by the judge to address:

> To sustain the charge of Murder, the State must prove the following propositions:
>
> First: That the Defendant performed the acts which caused the death of the Victim; and
>
> Second: That when the Defendant did so she intended to kill or do great bodily harm to the Victim, or she knew that her acts would cause death or great bodily harm to the Victim, or she knew that her acts created a strong probability of death or great bodily harm to the Victim; and
>
> Third: that the Defendant was not justified in using the force which she used.
>
> If you find from your consideration of all the evidence that each of these propositions has been proved beyond a reasonable doubt, then you should find the Defendant guilty.
>
> If, on the other hand, you find from your consideration of all the evidence that any of these propositions has not been proved beyond a reasonable doubt, then you should find the Defendant not guilty.

Critical thinking dispositions exemplified

We used all the specified critical thinking dispositions in dealing with this murder charge:

1. Seek and offer clear statements of the thesis or question. We needed to be clear about what was at issue. If we had not been clear about the murder charge, we might have carelessly assumed that murder in this context *required* intent to kill, but this charge did not so require, we noted, although intent to kill would have been sufficient to establish the second necessary condition for murder. We also had to be clear about the difference between there being a strong probability of great bodily harm and the defendant's knowing that there was such a probability.

This clarity disposition had a more sophisticated application. In our situation, proof was the basic concern. The standard of proof in this situation was proof beyond a reasonable doubt. But one of the jurors acted as if the standard were logical necessity. Assuming the logical-necessity interpretation of proof would have resulted in a different verdict about voluntary manslaughter than the one we produced, though it would not have made a difference in our decision about murder. We needed to be clear about the difference between the two concepts of proof. At one point we sent a note to the judge seeking a definition of "proof beyond a reasonable doubt." This note was part of our effort to secure a clear statement of the question.

Critical Thinking   35

2. Seek and offer clear reasons. When one juror said at the beginning of our deliberations, "She's guilty, let's vote," others asked *why* he thought so. He gave some reasons, and discussion followed. Without clear statements of reasons, it is very difficult to discuss the acceptability of the conclusion. We sought his reasons.

3. Try to be well informed. We listened intently at the trial. When one of us was in doubt during the post-trial deliberations, he or she would ask the others exactly what happened at a particular point in the trial. Amazingly every such event was remembered by a number of jurors. For example one juror was in doubt about the pathologist's demonstration of the force of the knife stroke. Most of us remembered the demonstration and refreshed that juror's memory. The degree of force of the knife stroke was a key feature in judging whether the last subcondition of the second condition for murder had been satisfied.

4. Use credible sources and observations, and usually mention them. We felt that the pathologist was a credible source to vouch for the strength of the knife blow, and that the observations on which she based her conclusion were also credible. These things were important to us in deciding whether Arlene was guilty of murder.

5. Take into account the total situation. This disposition was evidenced by our remembering that the standard for proof was "proof beyond a reasonable doubt," a key feature of the situation in a criminal trial. It was also evidenced when we realized that the defendant had an alternative to stabbing Al. Even if he had threatened her, an escape to her parents' room was an alternative for her. There were other features of the situation we remembered and that were important, such as the configuration of the cabinets and counters in her parents' kitchen (which we visited).

6. Keep in mind the basic concern in the context. It was important and generally easy for us to maintain focus on the question, "Is she guilty of murder?" One crucial subquestion became a main question on which it was less easy to maintain focus. This was whether she knew that her acts created a strong probability of great bodily harm to the victim. Some jurors argued as if they thought the question was whether her acts actually created a strong probability of great bodily harm, rather than whether she knew this. They needed to be reminded of the actual wording.

The disposition to maintain the focus was again evidenced by some of us when we responded to a juror who claimed that the body was probably moved by someone before the photograph of the victim was taken, and claimed this while we were trying to decide whether Arlene knew that her acts created a strong probability of great bodily harm. We asked him how that was relevant to the topic at hand. He said that he thought it might be important, but offered no reason. What he said seemed irrelevant and we acted accordingly. We gracefully put him back on focus.

7. Be alert for alternatives. One alternative that made a big difference in that situation was escaping to her parents' room. It was our realization of this alternative that resulted in our judging her guilty of voluntary manslaughter.

If I am asked to choose one and only one aspect of critical thinking to emphasize in teaching, I pick this one, being alert for alternatives. This is because I have seen so many cases in which this disposition was crucial, and because it does overlap with a number of other dispositions (like open-mindedness).

In this particular case, I learned later, the bearing of the escape alternative was controversial. We assumed without saying so that when there is a nonviolent alternative available, a person threatened with physical violence should pursue it. But a colleague has since objected that women should stop fleeing in the face of violence from men; when forceful resistance is open to them, they should consider and often choose that alternative. In this particular case, she urged, if the victim was threatening the defendant with violence, Arlene would have been justified in stabbing him, and thus should have been judged innocent of both murder and voluntary manslaughter.

8. Be open-minded: (a) Seriously consider alternative points of view. Unfortunately we did not consider the point of view just described. We just did not think of it. We did give serious consideration to the possibility that she was defending herself against violence by Al, but focused on our belief that the alternative of escaping to her parents' room was one she should have taken.

(b) Withhold judgment when the evidence and reasons are insufficient. One of our jurors announced at the beginning of the deliberations that it was obvious that Arlene murdered Al, so we should just vote and get out of there. A juror in the Henry Fonda movie *Twelve Angry Men*, which is a superb movie for a critical thinking class, had the same incautious approach. Fortunately, other jurors rejected the approach, favoring instead continued consideration of the sufficiency of the reasons.

We did consider the alternative possibility that she lured him into the house by taking his keys and putting them in her purse (where they were found). But we did not conclude that she lured him to his death. We felt that the evidence was insufficient.

9. Take a position and change a position when the evidence and reasons are sufficient. The line between Disposition 8(b), which is desirable, and total skepticism is sometimes difficult to draw. One of our jurors was for a time a total skeptic. For him nothing could be proved beyond a reasonable doubt (including the availability of the alternative of fleeing to her parents' room), making it impossible to find anyone guilty of anything when his concept of proof (actually logical necessity) is operative. But we talked him out of that position by using simple everyday examples in which he accepted conclusions as proved beyond a reasonable doubt. Ultimately he was willing to change his position.

10. Seek as much precision as the situation requires. This disposition was exhibited when the pathologist acted out the strength of the knife stroke, as she reconstructed it from the measured depth of the wound and the fact that there were no marks on the bones. She moved her arm saying that the stroke was "moderate, like this." Then she moved her arm much more vigorously, saying "not strong, like this." The pathologist had sought a degree of precision required for the situation. Numbers giving precise amounts of kinetic energy or velocity would have been overprecise and less helpful.

11. Try to "get it right" to the extent possible or feasible. In cases where there is or might be a truth of the matter, it is important to try to determine it. We paid close attention to the pathologist acting out the strength of the knife blow. When we visited the scene of the killing, we noted carefully the position of the cabinets and the counter, which showed the possibility of Arlene's turning around and striking Al with a knife at the angle reported by the pathologist.

12. Employ their critical thinking abilities. It does not help to have critical thinking abilities if we do not use them. This is why some authorities recommend much more emphasis on this critical thinking disposition (e.g., Stephen Norris, personal communication). With very few exceptions, the members of the jury were disposed to use their critical thinking abilities.

In sum, it can be seen that these dispositions are important qualities. It can also be seen that they overlap and are interdependent. Modeling, considering examples, noting the dispositions when needed or used, and engaging students deeply in issues that are real to them are useful approaches for the promotion of these dispositions.

Critical thinking abilities exemplified

The first four listed abilities (1 to 4) involve basic clarification:

1. Have a focus and pursue it. The ability to identify, attend to, and keep track of a focus (the issue, question, or conclusion) is listed first because, unless we know the question on which to focus, we do not know what to do with the rest. We as jurors knew the main focus: to judge whether or not the defendant was guilty of murder and/or voluntary manslaughter. The main focus was easy to identify in that situation, because the judge explicitly told us the basic issues to which we should attend. But identifying the focus is not always so easy. In deciding whether the second proposition (a necessary condition) for murder was satisfied, it was more difficult to keep the current focus in mind. We had to focus on each of the six subconditions in the second proposition in turn, ultimately narrowing the discussion down to a focus on the last subcondition. The focus then became first to determine whether the question was "Did her acts create a strong probability of great bodily harm?" or "Did she *know* that her acts created a strong probability of great bodily harm?" Finally the focus became whether the answer to the latter had been proved beyond a reasonable doubt.

2. Analyze arguments. The judge's written description of the charge of murder made it easier for us methodically to analyze the prosecutor's argument for murder. We had to be able to see that each of the three major propositions was a necessary condition, that any of the subconditions in the second proposition was sufficient to establish the second proposition, but that at least one of them had to be satisfied, and that the prosecutor had to prove this beyond a reasonable doubt. When the defense attorney was arguing that none of the six subconditions for the second necessary condition had been proved beyond a reasonable doubt, we had to be able to pick out this conclusion, and see the bearing on the total charge for murder. And we had to be able to see that he was trying to show that since the blow was only of moderate force, it had not been proved beyond a reasonable doubt that the defendant *knew* that there was a *strong probability* of great bodily harm (the weakest condition). We felt that he had shown this, and voted her not guilty of murder.

Standard advice in argument appraisal is to determine what type argument is being appraised (e.g., deductive, inductive) and then apply the appropriate criteria for the type. The trouble with this approach is that real arguments (like the pathologist's argument) do not come labeled. If there is doubt, I suggest that we first apply whatever set of criteria seems most likely to be satisfied. Then if the argument is found wanting, apply another set of criteria. If the argument satisfies a set, then, to oversimplify a bit, accept the argument and the conclusion (Ennis 2001; Hitchcock 1980). If it satisfies no set of criteria, do not accept it (though it might be acceptable later on if more support turns up). For example, if we apply deductive criteria to the pathologist's argument, we can see that it does not satisfy these criteria. But it probably satisfies best-explanation criteria (to be discussed later), given what I took to be the pathologist's credibility as a source of facts and an expert inferrer in the field.

3. Ask and answer clarification questions. On numerous occasions we had to be able to ask questions of clarification, for example the crucial critical thinking question "Why?" that we asked of the juror who was sure that Arlene was guilty of murder before the deliberations started. Another example: "What does 'proof beyond a reasonable doubt' mean?" As mentioned earlier, we sent a note to the judge requesting the meaning.

4. Understand and use graphs and maths. This ability was not used in this jury case. But ability to understand these things is essential in many contexts, including, for example, reports of research studies on the effects of medications. Two experimental/control group studies of a new medication recommended to me showed that the differences between two pairs of experimental and control groups were statistically significant one month after administration. Grasping the relevance of this information required me to know what statistical significance is and the strengths and weaknesses of such a finding, given the number of subjects in the experiments and the limitation to one specific elapsed time (one month).

The next three abilities (5 to 7) deal with the bases for inference:

5. *Judge the credibility of a source.* We had to judge the credibility of all the witnesses, including the pathologist who judged that the knife blow was only moderate in force, and the defendant herself, who said that she was defending herself against attack. Criteria we used included expertise and lack of conflict of interest. See Ennis (1974a) for a discussion of credibility of sources.

6. *Observe, and judge observation reports.* We made our own observations when we were taken to the defendant's home (a special arrangement for this trial). We used this information to form our own judgments about the defendant's statements, including her claim about the way she swung at the victim. See Norris and King (1984) for a discussion of observation.

We had to judge whether to accept the observation reports on which the pathologist based her judgment: the measurement of the depth of the wound, and the observation that there were no marks on the bones. We also judged the observation reports made by the investigating detective about the position of the body and the location and condition of the knife. All the reports of observations by professionals were based on a written record they made themselves at the time of observation, according to their testimony. The facts that there were records, that they were made at the time, and that they were made by the same person reporting the observation, all added to the credibility of the observation reports.

7. *Use their background knowledge, and their knowledge of the situation and previously established conclusions.* We used our experience of viewing the scene of the crime to infer that the way she described her turning around and stabbing him with an overhand stroke was plausible. We used this experience, together with background knowledge about keys and cars (at that time), to judge that he would need his keys in order to leave. Lack of concern with background knowledge is a common criticism of efforts to teach critical thinking (Ennis 1989; 1992; McPeck 1990).

The next three abilities (8 to 10) involve inference from the above bases:

8. *Deduce and judge deductions.* We used deduction in concluding that she was not justified in using the force that she used. We assumed that if she had an escape alternative, she was not justified in using the force she used; and she did have an escape alternative. The conclusion follows necessarily from these premises.

We also needed and used deductive understanding in interpreting and applying the total charge for murder, containing as it does a variety of intermixed necessary and sufficient conditions.

A mistake made by one juror was in the area of deduction. He asked a reluctant juror, "Have you proved beyond a reasonable doubt that she was justified in using the force that she used?" The answer was negative. The eager juror then said, "So that shows that she was not justified." This mistake might be classified by some as the "either-or" fallacy, and by others as the illicit shifting of a negation. But in any case the eager juror made an error by not seeing that there is a third logical possibility, namely that she actually was justified in using the force she used—without this having been proved beyond a reasonable doubt. See Ennis (1969a; 1969b; 1975; 1976; 1981; 2006; 2007) for discussions of basic deductive logic competence, one basic issue being whether material implication and its cohorts should be included. See Skyrms (1966) and Hitchcock (1980) for interpretation and defense of the inductive-deductive distinction in critical thinking.

9. *Make, and judge, inductive inferences and arguments (both enumerative induction and best-explanation reasoning).* See Ennis (1968) for a defense of the distinction between enumerative induction (generalization) and best-explanation reasoning, and Harman (1965; 1968) for the conflicting view that enumerative induction is a special case of inference to best explanation. General discussions of best-explanation reasoning can be found in Harman (1973), Ennis (1996a), and Lipton (2004).

One enumerative-inductive generalization that we jurors concluded was that the bailiff's behavior was nonresponsive. On this generalization we based our decision to stop asking him for help. The prohibition against hearsay evidence that was prominent in the trial is at least in part based upon another generalization resulting from an enumerative inductive inference: hearsay is often unreliable.

The other type of inductive inference or argument is often called "inference to best explanation" or "argument to best explanation" (Battersby 2006), which I combine under the label "best-explanation reasoning." Part of the pathologist's argument for her conclusion that the knife blow was only moderate in force was that this conclusion *explained* why the depth of the wound was only 2 1/2 inches, given that there were no marks on the victim's bones. This helped to satisfy the first criterion for judging best-explanation reasoning, which is that the conclusion explains key facts. The absence of facts that were inconsistent with the conclusion satisfied the second criterion. Five criteria for best-explanation reasoning are offered in Ennis (2011; 2013b).

Best-explanation reasoning has wide applicability. For examples from English literature (Shakespeare), archeology, French history, psychology, oceanography, and airplane maintenance, see Ennis (1996a, 226–228, 231–237); also see Follesdal (1979), using *Peer Gynt* as the example.

Causation is involved in best-explanation reasoning, but I shall not here go into the concepts of *causality* and *explanation*, assuming for present purposes that our intuitive understanding of causation and explanation will suffice. There are difficult theoretical issues here. See Ennis (1973; 1982b; 2012).

10. Make, and judge, value judgments. Making defensible value judgments is the last of these three basic types of inference. As I indicated earlier, we assumed the value judgment that it is better to flee than respond to violence with violence, but we did not reflect upon this value assumption. Value judging is a particularly difficult area for critical thinking instruction, because of the controversy over how to make defensible value judgments, although it is generally agreed that the alleged facts on which they are based should be true or justified.

The next five abilities (11 to 15) involve advanced clarification:

11. Define terms and judge definitions. For the jury the most troublesome definitional problem concerned the meaning of "proof beyond a reasonable doubt." For voluntary manslaughter the state needed to prove beyond a reasonable doubt that the defendant was not justified in using the force she used. Several jurors felt that without knowing the meaning of "proof beyond a reasonable doubt" we could not decide about this condition. As I noted earlier, we sent a note to the judge for help. The judge sent back a message that there is no definition of that phrase, and to do the best we can! Deliberation was about to collapse.

At this point I proposed a definition that enabled us to proceed: "To prove a proposition beyond a reasonable doubt is to offer enough evidence in its support that it would not make good sense to deny that proposition." The form I used, equivalent-expression definition (sometimes called "contextual"), seemed more appropriate than the more commonly recommended classification form (sometimes called *genus-differentia*). The definitional act that I was performing, reporting a meaning, seemed appropriate for the situation, since standard usage was what the jurors needed. The content of the definition made the jurors comfortable (though it gave them no new information), so we were able to proceed with the discussion. See Ennis (1964; 1969a; 1974b; 1996a; 2013b) for the development of some criteria for, and distinctions among, forms, stances, content, and uses of definitions, sometimes expressed under different labels.

12. Handle equivocation. To equivocate is to take advantage of the ambiguity of a term in order to support a position. This is sometimes done deliberately and knowingly, but sometimes unwittingly. I call the latter "impact equivocation" (1980) because the equivocator does not realize that he or she is taking advantage of a shift in meaning of a word in mid-argument, even though what the equivocator is doing has the *impact* of equivocation.

For example, as I noted earlier, one of the jurors kept insisting that a particular proposition was not proved beyond a reasonable doubt. The proposition was that Arlene could have escaped to her parents' room as an alternative to remaining in the kitchen and defending herself. We challenged him by asking for an example of some factual claim that had in his lifetime been proved beyond a reasonable doubt, and he admitted that he could think of nothing, making proof beyond reasonable doubt an empty concept for him. In other words, he was a total skeptic who in that context would accept nothing empirical as proved beyond a reasonable doubt. His meaning of "proof beyond a reasonable doubt" was, I believe, logical necessity, which is different from the meaning in use in the courts. He was using his meaning of "proof beyond a reasonable doubt" to draw his conclusion that the proposition had not been proved beyond reasonable doubt, and interpreting this conclusion to mean that the proposition had not been proved beyond a reasonable doubt in the standard meaning of the court. His thinking involved impact equivocation because I believe he was not intentionally exploiting the ambiguity. What he was doing had the impact of equivocation, even though he did not deliberately equivocate.

A possible example of equivocation not drawn from the murder trial deals with a topic in this essay, the meaning of "critical thinking." Michael Roth (2010), an intellectual historian, essentially defined "critical thinking" as negative thinking, which he condemned, but then he seemed to be talking about the critical thinking that is getting so much support in academia and the media these days, which sometimes is, and sometimes is not, negative thinking. The success of that condemnation (if he so intended it) seemed to depend on equivocating with the term "critical thinking": condemning critical thinking in one sense and then applying that condemnation to critical thinking in the other sense. A caveat: Cases of equivocation are generally hard to prove. The background is usually complex, and alternative explanations often develop.

13. Attribute and judge unstated assumptions. One assumption that we made was that it is better to flee, if possible, than to respond with violence to a threat of violence, an assumption that, I indicated, has since been challenged by a colleague. Another example: One juror identified the assumption of another when he said, "You're assuming that I have to prove that she was defending herself against attack. Rather, the State has to prove that she was not." A third example: An assumption of the skeptical juror was that proof in this case requires logical necessity of the conclusion. These examples illustrate the importance of the ability to ascribe assumptions in a situation.

See Hitchcock (1985) and Ennis (1982a; 1996a; 2013b) for an extended discussion of the ascription of assumptions, and an outline of types of assumptions and criteria for ascribing assumptions.

14. Think suppositionally. We jurors had to be flexible enough to suppose things, some of which we doubted, to see where they led. We all supposed for the sake of argument that the victim had wanted to harm the defendant, which supposition suggested that she was defending herself from attack. A second supposition was that if he intended to harm her, he would have done damage to her before they entered the house, which did not happen. But we decided that the second supposition was not plausible enough to justify the conclusion that he did not intend to harm her. So it remained an open question whether or not she was defending herself against attack.

15. Deal with fallacy labels. Although using fallacy labels with wisdom is helpful, one must be careful. Popular examples of such labels are *circularity, bandwagon, post hoc, non sequitur, hearsay,* and *appeal to authority.* These labels are often used to communicate the claim that there is a mistake of a certain kind. They have advantages and weaknesses. One major advantage is that they provide efficient ways of describing a complaint.

For example, *post hoc* warns of the mistake of concluding that one thing caused another just because it preceded the other. The label *post hoc* sensitizes people to the fact that showing that one thing came after another does not prove that the first caused the second, a common error. The fact that her killing him occurred after he followed her into the house does not prove that his following her caused her to kill him.

A disadvantage with fallacy labels is that often things that fit a fallacy label are not fallacious. For example, many cases of appeal to authority constitute good thinking. Often appealing to an authority is the appropriate thing to do, as when the defense and prosecution appealed to the authority of the pathologist. Even circularity is not always a fallacy, as in Yogi Berra's "It ain't over 'til it's over." Similarly, deductively valid arguments are circular because the content of the conclusion is also in the premises. So circularity is not fallacious in most deductively valid arguments.

Although courses in, and units of, critical thinking are sometimes organized in accord with a list of fallacies, I have not used this approach partly because of the above disadvantage and because I do not see fallacy labels providing a comprehensive approach.

The last three abilities (16–18) are not constitutive of critical thinking, but are generally helpful for critical thinkers:

16. Be aware of, and check the quality of, their thinking (metacognition). It often helps to be aware of what we actually are thinking and assuming, that is, thinking about our thinking, so we can evaluate and reevaluate what is happening in our thinking, often producing better results. Richard Paul (2012, 7) has advocated this position. For example I thought about my thinking when I produced the definition of "proved beyond a reasonable doubt." However, I often engage in critical thinking without thinking about my thinking. When I realized that the pathologist's conclusion did not follow necessarily, but only beyond a reasonable doubt I did not think about my thinking. I just did the thinking.

17. Deal with things in an orderly manner. This ability was evidenced by the jurors when we went through each of the conditions for murder and voluntary manslaughter, item by item. Following problem-solving steps one by one (regardless of which of the many versions of problem solving one is using) exemplifies this ability.

18. Deal with rhetorical strategies. Rhetorical strategies can be used to deceive, so it helps to have some familiarity with them. But familiarity with them can also help to present a strong critical thinking argument in a way that it is more likely to be accepted.

The defense attorney stipulated in advance that his client did kill the victim, apparently because the evidence was overwhelming, so he avoided looking foolish defending a lost cause—a good strategy. Another example: By challenging in advance the point that the defendant was defending herself against attack, the prosecution reduced the impact of the point, showing that it had taken the possibility into account.

It helps a critical thinker to be familiar with rhetorical strategies, whether to use them in presenting the results of true critical thinking, or challenge them when used to mislead people.

The next step

The conception of critical thinking described in this chapter can serve as a basis for further development of critical thinking teaching, assessment, and curriculum. Critical thinking curriculum development particularly has been for the most part neglected, though we often run into the dispute about whether critical thinking should be taught in a separate course or infused in existing subject-matter courses. This dispute neglects the possibility of the combination of both in coordinated ways that complement each other. This combination, labeled "critical thinking across the curriculum," is rarely seen or even attempted in higher education, though it has the potential to provide a deep and comprehensive grasp of critical thinking by our students.

I have developed a proposal (2013a) for critical thinking across a higher education curriculum that offers a comprehensive plan that addresses teaching, assessment, and especially curriculum organization. Elements of the proposal include coordination; comprehensiveness; avoidance of duplication and neglect; sensitivity to the capabilities of, and differences among, subject matter areas; transfer to the daily civic, personal, and vocational lives of our students; and strong leadership and faculty support.

In this proposal guidance is given not only for a course focused on general critical thinking abilities and dispositions, but also for infusing general critical thinking in subject-specific courses and promoting subject-specific critical thinking dispositions and abilities. An example of a subject-specific critical thinking ability is the ability to handle (calculate, interpret, and use) analysis of covariance in the social sciences.

Although continued work on teaching and assessing critical thinking is still needed, I believe that putting much-increased emphasis on critical thinking across the curriculum should be the next step in incorporating critical thinking into our education system. I invite others who are sympathetic to the conception of critical thinking in this essay to join me in this effort.

Summary

Assuming that "critical thinking," as the term is widely used, means reasonable reflective thinking that is focused on deciding what to believe or do, I have offered a streamlined conception of general critical thinking that consists of twelve dispositions and eighteen abilities. To clarify this general conception and give it vitality, I exemplified these dispositions and abilities, mostly with my experiences as a juror in a murder trial. I call the presented conception "streamlined" because it provides only an outline, and for the most part does not include criteria and principles for making decisions about what to believe or do. The full general conception is in Ennis (1996b, where it is fully exemplified; 2011; and 2013b).

I have also included a brief description of what I believe should be the next step in critical thinking education, that is, development of critical thinking across the curriculum. Given the current state of the field of critical thinking, now is the time to use the streamlined conception as a springboard to this next step.

Acknowledgments

With permission of the Philosophy Documentation Center this essay is an updated version of Ennis (1991). I appreciate the comments of Jennie Berg, John Canfield, Sean Ennis, David Hitchcock, Stephen Norris, and Robert Swartz.

References

Aiken, W. M. 1942. *The Story of the Eight-Year-Study*. New York: Harper & Brothers.
America 2000: An Education Strategy. 1991. Washington DC. http://www.scribd.com /doc/29278340/109-America-2000-An-Education-Strategy.
Battersby, M. 2006. "Applied Epistemology and Argumentation in Epidemiology." *Informal Logic* 26 (1): 41–62.
Black, M. 1946. *Critical Thinking*. New York: Prentice Hall.
Commission on the Humanities. 1980. *The Humanities in American Life*. Berkeley, CA: University of California Press.
Dewey, J. 1910. *How We Think*. Boston: D. C. Heath.
Ennis, R. H. 1964. "Operational Definitions." *American Educational Research Journal* 1: 183–201.
Ennis, R. H. 1968. "Enumerative Induction and Best Explanation." *The Journal of Philosophy* 65: 523–530.
Ennis, R. H. 1969a. *Logic in Teaching*. Englewood Cliffs, NJ: Prentice Hall.
Ennis, R. H. 1969b. *Ordinary Logic*. Englewood Cliffs: Prentice-Hall.

Ennis, R. H. 1973. "The Responsibility of a Cause." In *Philosophy of Education 1973*, edited by Brian Crittendo. Edwardsville, IL: Studies in Philosophy and Education. 86–93.

Ennis, R. H. 1974a. "The Believability of People." *Educational Forum* 38: 347–354.

Ennis, R. H. 1974b. "Definition in Science Teaching." *Instructional Science* 3: 285–298.

Ennis, R. H. 1975. "Children's Ability to Handle Piaget's Propositional Logic: A Conceptual Critique." *Review of Educational Research* 45: 1–41.

Ennis, R. H. 1976. "An Alternative to Piaget's Conceptualization of Logical Competence." *Child Development* 47: 903–919.

Ennis, R. H. 1980. "Presidential Address: A Conception of Rational Thinking." In *Philosophy of Education 1979*, edited by Jerrold Coombs. Bloomington, IL: Philosophy of Education Society. 1–30.

Ennis, R. H. 1981. "A Conception of Deductive Logic Competence." *Teaching Philosophy* 4: 337–385.

Ennis, R. H. 1982a. "Identifying Implicit Assumptions." *Synthese* 51: 61–86.

Ennis, R. H. 1982b. "Mackie's Singular Causality and Linked Overdetermination." In *Philosophy of Science Association 1982*, edited by Peter D. Asquith and Thomas Nickles. East Lansing, MI: Philosophy of Science Association. 55–64.

Ennis, R. H. 1987. "A Taxonomy of Critical Thinking Dispositions and Abilities." In *Teaching Thinking Skills: Theory and Practice*, edited by J. B. Baron and R. J. Sternberg. New York: W. H. Freeman and Company. 9–26.

Ennis, R. H. 1989. "Critical Thinking and Subject Specificity: Clarification and Needed Research." *Educational Researcher* 18 (3): 4–10.

Ennis, R. H. 1991. "Critical Thinking: A Streamlined Conception." *Teaching Philosophy* 14 (1): 5–25.

Ennis, R. H. 1992. "John Mcpeck's Teaching Critical Thinking." *Educational Studies* 23 (4): 462–472.

Ennis, R. H. 1996a. *Critical Thinking*. Upper Saddle River NJ: Prentice Hall.

Ennis, R. H. 1996b. "Critical Thinking Dispositions: Their Nature and Assessability." *Informal Logic* 18 (2 and 3): 165–182.

Ennis, R. H. 2001 "Argument Appraisal Strategy: A Comprehensive Approach." *Informal Logic* 21 (2): 97–140.

Ennis, R. H. 2006. "Probably." In *Arguing on the Toulmin Model*, edited by David Hitchcock and Bart Verheij. Dordrecht, the Netherlands: Springer. 145–164.

Ennis, R. H. 2007. "'Probable' and Its Equivalents." In *Reason Reclaimed: Essays in Honor of J. Anthony Blair and Ralph Johnson*, edited by Hans V. Hansen and Robert C. Pinto. Newport News, VA: Vale Press. 243–256.

Ennis, R. H. 2011. "Critical Thinking: Reflection and Perspective, Part I." *Inquiry: Critical Thinking across the Disciplines* 26 (1): 4–18.

Ennis, R. H. 2012. Analyzing and Defending Sole Singular Causal Claims. Paper presented at Biennial meeting of the Philosophy of Science Association, November 15–17, at San Diego, CA.

Ennis, R. H. 2013a. "Critical Thinking across the Curriculum: The Wisdom CTAC Program." *Inquiry: Critical Thinking across the Disciplines* 28 (2): 25–52.

Ennis, R. H. 2013b. The Nature of Critical Thinking (under "What Is Critical Thinking?"). Available from http://criticalthinking.net.

Follesdal, D. 1979. "Hermeneutics and the Hypothetico-Deductive Method." *Dialectica* 33 (3/4): 319–336

Harman, G. 1965 (January). "The Inference to Best Explanation." *The Philosophical Review* 74 (1): 88–95.

Harman, G. H. 1968. "Enumerative Induction as Inference to Best Explanation." *The Journal of Philosophy* 65 (18): 529–533.

Harman, G. 1973. *Thought*. Princeton: Princeton University Press.

Harmon, H. 1980. Executive Order No. 338. Long Beach, CA: The California State University and Colleges.

Hitchcock, D. 1980. "Deductive and Inductive: Types of Validity, Not Types of Argument." *Informal Logic Newsletter* 2 (3): 9–11.

Hitchcock, D. 1985. "Enthymematic Arguments." *Informal Logic* 7 (2 and 3): 83–97.

Lipton, P. 2004. *Inference to Best Explanation* (second edition). London: Routledge.

McPeck, J. 1990. *Teaching Critical Thinking: Dialogue and Dialectic.* New York: Routledge.

Norris, S., and Ennis, R. H. 1989. *Evaluating Critical Thinking.* Pacific Grove, CA: Midwest Publications.

Norris, S. P., and King, R. 1984. "Observation Ability: Determining and Extending Its Presence." *Informal Logic* 6 (3): 3–9.

Obama, B. *State of the Union 2014 Speech* 2014. Available from http://www.whitehouse.gov/the-press-office/2014/01/28/president-barack-obamas-state-union-address.

Paul, R. 2012. *Critical Thinking: What Every Person Needs to Survive in a Rapidly Changing World.* Anthology ed. Santa Rosa: The Foundation for Critical Thinking.

Rawls, J. 1971. *A Theory of Justice.* Cambridge, MA: Harvard University Press.

Roth, M. 2010. "Beyond Critical Thinking." *The Chronicle of Higher Education,* January 3.

Skyrms, B. 1966. *Choice and Chance.* Belmont, CA: Dickenson.

2
Critical Thinking and/or Argumentation in Higher Education

Richard Andrews

Introduction

Critical thinking and argumentation are closely allied. And yet each field has its own derivation and antecedents, and the differences between these are fundamental not only to debates today about their centrality in higher education, but to the entire history of the relationship (in Europe at least) between thought and language as well. On the one hand, critical thinking is most closely allied to philosophy; on the other, argumentation is allied with rhetoric. The debate about the relationship between philosophy and rhetoric goes back to Plato and Aristotle. It concerns ideas, ideals, concepts, and abstract thought and logic in relation to philosophy and the expression of these categories in verbal and other forms of language. Both critical thinking and argumentation overlap in their territories of engagement, and both have pedagogical implications for learning and teaching in higher education. This chapter explores the relationship, examines some examples at doctoral level (and briefly at undergraduate level), and puts the case for argumentation as the best focus in terms of taking forward practice in higher education. In doing so, it may run counter to the arguments in many of the chapters in this book, but the challenge presented in this chapter may act like the grit in the oyster. In Toulminian terms, the challenge can be rebutted or lead to a more qualified position on the role of critical thinking in higher education.

Critical thinking

The case for critical thinking starts with Lipman, because of his work with primary/elementary schoolteachers and children. The reason for starting here is that the fundamentals of critical and clear thinking are established at that point, and they prepare the ground for consideration of such approaches in higher education in a developmental sense. Lipman's work concentrates on thinking in education in the school, basing its approach on philosophy "when properly constructed and properly taught"(2003, 3). The thinking is to be taught within a community of inquiry and includes critical reflection. Lipman is skeptical of the notion of critical thinking, however, seeing it (as I do) as tautological. To think clearly is to be critical; being "critical" simply adds an epithet that is redundant. He notices a fading away of the critical thinking movement in the first part of the present century. In higher education, there has been a lack of theoretical work on thinking in the wake of mid-century adherence to Piagetian approaches to cognition and cognitive development and the publication of neo-Piagetian and Vygotskian approaches in the 1980s. Lipman argues that without such theoretical or historical examination critical thinking's claim to be a discipline "can hardly be persuasive" (ibid., 4). The present book addresses this perceived gap in theoretical work.

In higher education—and assuming that "thinking" is core to intellectual life—practice can be deemed to be "critical" if one is driven by (a) a spirit of inquiry and skepticism, (b) able to take criticism of one's colleagues and other academics' work as part of the fabric of intellectual exchange, and (c) is self-critical. Half of these concepts I have taken from Lipman (2003, 16–17), while I have rejected others to do with "correction." Lipman's inclusion of categories of correction and self-correction suggests that his model of thinking and argumentation is too closely allied to the field of argument and thinking that concerns itself with fallacies. But the weakness of Lipman's position lies in the foundations of his reflective model of educational practice and its overdependence on Schön's theory. If its key concepts are "inquiry, community, rationality, judgement, creativity, autonomy" (2003, 19), these are at such a level of general acceptance as to be less than useful in distinguishing ideas—and, ironically, in thinking critically in higher education. The strengths in Lipman's work reside in his mission to improve the quality of thinking in schooling.

More substantial than the critical thinking movement to date has been the informal logic movement, emerging in the late 1970s, and embracing theoretical, practical (including how best to structure arguments), and pedagogical issues. In many ways, informal logic provides a bridge between critical thinking and its varied application in higher education institutions on the one hand, and linguistic and discourse analysis on the other. Behind the informal logic movement is a longer tradition of classical and contemporary rhetoric, concerned with the *deployment* of thinking in the real world in the form of argumentation.

Lipman sums up the approaches well:

Informal logicians and rhetoricians attack the same problem from different directions...Both are examining claims to reasonableness (and therefore are concerned with a theory of rationality). But the informal logicians move towards a new conception of reasonableness by broadening and refining the concept of logic, while the rhetoricians do so by examining writing that is not or does not appear to be formally logical, in an effort to determine what justification such prose may claim to being reasonable. Moreover, both are inclined to focus on argumentation, but the one group emphasizes the persuasive force of argument while the other emphasizes its logical force. (2003, 42)

Ennis (1987), suggests Lipman (ibid., 46ff), bases his approach to critical thinking on its power to help us decide what to "believe and do." Lipman is cautious about claims of this kind, preferring to see critical thinking as providing "a tentative scepticism" (47) rather than a justification of a set of beliefs or tools for making judgments and action. But Ennis's distinction between dispositions and abilities is useful in characterizing the field of critical thinking. In terms of dispositions, critical thinkers are concerned that their beliefs are true and their decisions justified; that their positions are reasonable, honest, and clear; and that others' views and feelings are respected. In terms of abilities, critical thinkers have the skills to clarify; justify the basis for decisions; infer, both deductively and inductively; make suppositions; and approach problems with equanimity, due sequence, and propriety with regard to rhetorical strategy.

Paul (1987), in the same volume as the chapter by Ennis, argues that critical thinking involves judgment, and judgment necessitates a consideration of context. Judgment, however, requires a disposition toward dialogue and dialectic.

Critical thinking, as a movement, has had more currency in the United States than the United Kingdom despite the launch of an Advanced Supplementary (AS) level examination for 16–17-year olds in critical thinking in the United Kingdom in 1999. The range of definitions tends to cluster around the notion of robust thinking skills, with the epithet "critical" adding an edge to the activity that suggests both meta-thinking about thinking and the deployment of philosophical analytical processes to the discussion of propositions and evidence. When characterized as a "movement" rather than a discipline or subject, critical thinking can be seen as a loose federation of pedagogical approaches designed not only to improve the quality of thinking generically, but also to raise awareness about the construction of discipline-based or subject-based knowledge. What is of concern with the movement, particularly as far as its application in higher education is concerned, is the lack of sharp focus as to its parameters, the lack of agreement about its internal definitions, and the lack of curricular space in higher education to deploy its insights effectively.

Kuhn (2005) is particularly critical of the broad and overdefined (and thus unclear) nature of the critical thinking movement, preferring to base her own work on empirical evidence (as opposed to most of the work on critical thinking); on cognitive development (often not addressed in higher education); and on life outside the school classroom. She divides her attention between the two elements necessary for more socially based thinking in and outside formal schooling and higher education: skills of inquiry and skills of argument.

A key issue for higher education is therefore: What is the balance between epistemological discipline-based enquiry on the one hand and generic critical thinking education on the other? I will return to this topic toward the end of this chapter in revisiting some of the findings of a study of undergraduate argumentation, undertaken at the University of York in the mid-2000s. In this study (reported in Andrews, Torgerson, Low, McGuinn, and Robinson 2006a; Andrews, Torgerson, Robinson, See, Mitchell, Peake, Prior, and Bilbro 2006b) aspects of the relationship between critical thinking and argument were explored. The focus on argumentational skills enables a return to the main focus of this chapter.

Argumentation

Argumentation—the more technical and process-based term than the more general "argument"—has a number of antecedents—mostly in the form of theories or disciplines. Among these are dialogism, discourse theory, linguistics, logic (especially informal logic), pragma-dialectics, speech act theory, communication theory, and classical and contemporary rhetoric. Such an eclectic pedigree may be due to the fact that argumentation is seen as central to everyday interaction as well as to local, national, and international politics. It is not possible in the confines of this chapter to give a full account of the derivation of argumentational practice in the light of all these theories and disciplines. I will focus particularly on the rhetorical dimensions of argument for the purposes of this chapter, as both classical and contemporary rhetoric have a theoretical as well as a practical application. Furthermore, in terms of the relationship with critical thinking, rhetoric provides a useful counterbalancing set of theories and practices. The present section will focus first on the practical aspects of argumentation in higher education, then come back to theoretical concerns via a look at rhetoric.

Argumentation is indubitably connected with higher education in that the most successful undergraduates tend to be those who can argue well, in speech and/or writing, whatever their discipline. Argumentation is assessable, and it increasingly appears in criteria for success at the highest levels of undergraduation. It is even more integral to masters or doctoral level work, often (now) appearing explicitly as a criterion for success. At the doctoral level, one could say that the thesis *is* the argument, and vice-versa; but the same is true, to different degrees, for masters and undergraduate assignments (and we must not forget that most undergraduate and masters degrees see the dissertation or report as the summation of study, with a significant proportion of the marks attributed to this final piece of work). At all levels, discipline-dependence ("field-dependence" in Toulminian terms) is a significant factor and needs to be balanced with generic argumentational skills ("field-independence"). Such a balance is the basic position taken in *Argumentation in Higher Education* (Andrews 2009a).

Why is argumentation so important to higher education? It is about articulation in both senses of that word: both the *expression* of ideas, thoughts, feelings, and suppositions; the *joining together* of these ideas and notions in logical and quasi-logical sequences, supported (usually and beneficially) by evidence; and also the *positioning* of the student in relation to existing bodies of knowledge.

For the purposes of this section, I will concentrate on the doctoral thesis or dissertation, as work on argument in a range of undergraduate disciplines has been addressed elsewhere (Andrews 2002; 2009a; 2009b) and will be returned to in the final section. Work on masters-level argumentation has been addressed elsewhere (Andrews 2007).

Arguments emerge gradually in doctoral work. A student will first have a hunch that he or she wishes to do research in a particular field. The search then takes place, via the reading of existing literature as well as observation through experience and/or thought, for a problem. From the problem emerges either a research question (or set of related questions) and/or a hypothesis. Once a clear starting point has been established, and taking into account the flexibility needed in adjusting the question or hypothesis, the literature search is mapped and undertaken, the methodology and methods decided upon and piloted, and the work begins of creating a pathway through the field. It is this pathway that in due course becomes the argument.

One of the most critical decisions in composing a thesis or dissertation at this level is the structure of the piece. Is it to be linear (the conventional form) with an introduction followed by literature review chapters, followed by methodological considerations, and in due course by results, discussion, and conclusions? If so, what differences are there between the arts, humanities, and social sciences with regard to conventional linear structures? I have seen theses that follow this classic social science structure. But I have also seen examples of theses that are more narrative and autobiographical in nature; that follow the introduction with a brief methodological discussion; that do not follow the classic structure at all, but take a philosophical or at least reflective excursion through a field; that do not have "results" as such; that are more like the structure of Sterne's *Tristram Shandy* (Sterne 1759–67), a work published over several years with a combination of narratives, poems, illustrations, blank pages, and conventional academic argumentational text. Some of these theses translate readily into book form; others (perhaps the conventional model) do not. And yet linear structure is critical to the argument (*post hoc ergo propter hoc*), and, one could say, it is only proper that the supervisor advise the student on how best to meet the expectations and regulations and criteria of the university in which the thesis is being created. The only thesis I have ever recommended for failure, even at second attempt (there was no second attempt) had no such rationale for a structure of any kind: it was an assortment, a *bricolage*, a series of unconnected notes that could not justify itself, even in a postmodern sense, as having any kind of unity of argument, try as hard as the other examiner and I did to make connections between the various elements.

The elements of argument embedded within the conventional thesis are similar to those of classical rhetoric, the function of which was to ensure the quality of public oratory and debate in democratic society. The expectation is that there will be an introduction; a narration of the existing state of knowledge; a means by which that knowledge can be tested via empirical data gathering and analysis, and/or via reflection; a discussion of the state of existing knowledge in relation to new knowledge created; and a conclusion that looks back at the starting point of the research, and then forward to future research, and to applications of the research in policy and practice. But the argument does not emerge until the researcher is well into the process of research. Once its lineaments become clearer, it affords structure and direction to the thesis. By the time the thesis is submitted, and in the viva voce examination that follows, the argument should be clear, defensible, open to question, and well supported by evidence and/or logical or quasi-logical reflection.

What happens to argument and argumentation when the thesis moves away from linearity toward a more hypertextual, spatial, and generally *nonlinear* format? The creation and submission of alternative forms of thesis, like a website, an art installation, a film—forms which are becoming increasingly common, partly as a result of the digitization of the doctoral research process and product (see Andrews, Borg, Boyd Davis, Domingo, and England 2012)—present interesting challenges to the student, supervisor, and examiner in terms of where the argument sits. If the argument is central to the thesis in the arts, humanities, and social sciences in the European humanistic tradition, where is it in a digitized work like a website? First, it should be acknowledged that such alternative forms of submission for the award of PhD tend to hedge their bets by insisting that the "creative' component" is accompanied by a critical commentary in writing of about half the length of a conventional thesis, and representing about half of the submission itself. Thus an exhibition of paintings or sculptures, a website, or a film would be usually accompanied by a 40–50,000 word catalogue or critical commentary where the argument could be explicitly stated. But we also have to acknowledge that the creative work itself embodies an argument. Let us take a website submission, for example. It might include various sections that are hypertextually related, and all accessible from the front page. These sections might include still and moving images and sound files, hypertextual links to other sites and sources, as well as conventional argumentational and nonconventional written text (remember *Tristram Shandy*). The key point as far as argumentation is concerned is that there is no linear sequence in which one is expected to read the doctoral submission. You can "enter" at any of a number of points; you can read the various elements and sections in any order; and you can construct your own arguments, as a reader, that may or may not reflect the implicit or explicit arguments of the composer. An example of such a work is by Milsom (2008) in which the opening page of the submission is a photograph of the researcher's desk (see also Milsom 2012). The photograph is interactive in that the drawers, folders, books, notebook, computer, Post-It notes, etc. are points via which one can access the various elements that make up the submission.

Panning out from these practicalities of constructing arguments for the doctoral thesis, what are the rhetorical considerations relevant to argumentation in higher education? While rhetoric for Aristotle was defined as the *art of persuasion*, contemporary rhetoric concerns itself more broadly with the *arts of discourse*, and is applicable in a wide array of situations, from the personal to the political, and in a range of modes and media. It basically answers the questions: Who is communicating to whom? What about? What are the available means (modes and media) of communication? Which are best used, in which combinations, to ensure successful communication? How is that communication effected? It embraces many of the considerations listed above in the opening paragraph: dialectics and pragma-dialectics, dialogism, speech act theory, linguistics, and discourse theory. What it does not tend to do, and where the link with critical thinking will be explored later in the chapter, is to ask *why* such communicative exchanges are undertaken.

In the examination of the doctoral thesis, in the United Kingdom at least, the genre (in the sense of genre as social action, not as text type—see Miller 1984) concerns the presentation of a work, either in conventional format or within the regulations of the particular university, and a discussion between the examiners and the candidate (with sometimes the supervisor present as a silent witness) of that work. The examiners are usually asked to submit independent reports, giving a provisional judgment on the submitted work, prior to the *viva*, and which they exchange in order to see to what degree their views are similar or different. They then conduct the discussion accordingly. All the power rests with the examiners—especially the external examiner, who, besides judging the work itself, is judging the standards of the institution itself. The "external," therefore, has the more powerful voice in the final outcome. In rhetorical terms, then, we have a submitted work on the table where the argument is put explicitly, or somewhere on a spectrum from the implicit to the explicit; the researcher is there to defend both his or her decisions in the process of undertaking the work, and the product itself; he or she does so by adding to the written submission with verbal (oral) commentary and responses to the examiners' questions. The whole is framed by the criteria for the award of a doctorate at the institution in question, as well as by the experience of the examiners in seeing other such works recommended for a "pass," both in the present institution, and in others. There are other hidden criteria at play: the elegance of the submission, the appropriateness of the researcher joining the "community" of those with a doctoral degree; and sometimes (unfairly, often) the prejudices of the examiners for particular kinds of work they wish to see exhibited (e.g., preference for a particular methodology or research paradigm; the degree to which they see required corrections as minor or major amendments). The criteria for the award of a doctorate always include some reference to argumentation: a "clear line of argument" or "a coherent and critical argument, well supported by evidence and/or logic," for example. If a candidate can see that argumentation is critical to success, it usually does not matter that the examiners may agree or disagree with his or her argument; the path to success is to argue well, both in the presented text and in the *viva*. Where theses are likely to fail or be referred for further work, the argument is either nonexistent, flawed or half-baked; and the questions of the examiners in this respect are unanswered or answered badly.

The rhetoric of the thesis and of the *viva*, then, is part ritual, part a genuine attempt by the examiners to elicit the argument put forward by the candidate for the award. Understanding the rhetorical context is important for the supervisor and candidate. For example, colleagues and I have noticed a tendency in theses in education and the social sciences over the last ten years or so to have increasingly large sections on methodology, and increasingly small sections on theory, perhaps reflecting tendency to validate generic methodological competence at the end of a period of doctoral study rather than, say, grasp of a particular field. Candidates must be able to defend their thesis by understanding where the examiners are "coming from," that is, what ideologies and values and past experience, expertise, and interests they are bringing to the inevitably intersubjective nature of the examination. Candidates also need to understand that an immediate and unqualified pass at doctoral level is being increasingly replaced by a desire, on the part of examiners, to treat the thesis like a draft book, and to suggest amendments, corrections, and other improvements—within the confines of the regulations and options open to them as examiners. In the United Kingdom, at least, the PhD *viva* is becoming more like a very unbalanced peer-review occasion. Very few candidates ever "fail," but minor corrections and more substantial amendments, including referrals for further work, are increasingly *de rigueur*. The social and political framing of the event is rhetorically informed, and the arguments within it are influenced accordingly.

Argument, within a theory of classical or contemporary rhetoric (Andrews 2014), is a means by which agreement and consensus (or at least, *a way forward*, even if, in the case of the viva described above, the relationship between the parties is unequal) are reached; a form of discourse in which excellence is expected at the summation of degree courses; a process which is undertaken in a range of modes and media; and a process in which difference is explored in order to, in due course, clarify positions and reach consensus for action.

Critical thinking and argumentation

Much of the work on bringing together critical thinking and argumentation has already been done by Walton (2008); indeed, the term "critical argumentation" in the preface to the second edition of *Informal Logic* indicates the nature of the relationship. "Informal" indicates a grounding of arguments "as they occur in natural language in the real marketplace of persuasion on controversial issues in politics, law, science, and all aspects of daily life" (2008, xi)—hence the subtitle of the book: "A Pragmatic Approach." Such an approach sets argumentation within a dialogic frame, assuming that "the concept of question-reply dialogue as a form of interaction between two participants, each representing one side of an argument" (2008, xii) is fundamental to the study of argumentation in society. The notion of *informal* logic has suggested groundedness in specific contexts, rather than informality per se; the aim has continued to be identification of general patterns of logic in everyday argumentational discourse. As Walton notes, "what is happening now could be described as a movement from informal logic to semi-formal logic" (2008, xiii)—in other words, the gradual identification of laws and patterns in argumentation that are emerging in fields as diverse as computing, linguistics, discourse analysis, and dialectics. Informal logic and argumentation are steadily establishing themselves as more reliable and coherent tools for the examination of dialogic exchange in everyday life and in academia.

One particular context for critical argumentation is higher education. In a study undertaken with co-researchers at the University of Illinois at Urbana-Champaign, Queen Mary, University of London, and the University of York (Andrews et al. 2006a; Andrews et al. 2006b) and later written up (Andrews 2009a) as part of a book on argumentation in higher education, investigation was undertaken into first-year argumentation in three disciplines: biology, electrical engineering, history, with subsequent research on educational studies. My own reflections on the project are influenced by reading in rhetoric and argumentation; reading on research in writing development in schools; and experience of teaching within the US and UK higher education systems.

Essentially, the development of argumentational skills in higher education appears to be based on the combination of a number of key elements:

1. a *disposition* on the part of students to be "critical," that is to weigh up different points of view; to be able to separate claims and propositions from

evidence; to question received assumptions; to hold a skeptical attitude toward "facts" and assumptions,

2. a *disposition* on the part of university lecturers and professors to accept and promote such a critical approach,
3. a *knowledge* of some of the theories and models of argumentation that have application in a generic sense,
4. an *awareness* of the way argument is manifested and structures the nature of particular disciplines,
5. the disposition, on the part of lecturers and students, to *"drill down at the points of dispute"* within a discipline where knowledge is contested, and
6. an understanding on the part of lecturers and students that *development* of such argumentational skills is expected within the period of study for a degree, and if not formed at the start, will go through a number of stages.

A disposition to be critical

Criticality in students' work is highly prized, and even if it is not mentioned in criteria for the grading of undergraduate essays, masters assignments and dissertations, and doctoral theses, is always a hidden criterion in the judgment of excellence and—for many students—a key distinguishing feature between work that is mediocre and work that is rated as very good and above. To be critical means drawing on the largely European tradition of *critique*: being driven by "suspicion" and skepticism rather than by obedience or deference to the presented truth; weighing up validity claims against each other; testing the warrants that hold together claims/propositions on the one hand, and data and evidence on the other; and developing a critical stance or position, which often emerges after a great deal of reading and reflection in the field or on the topic. None of these features are easy options for students to take. They require independent thought and hard work in reading and research.

Lecturers' disposition to celebrate the critical

Many lecturers in higher education do not accept that their positions can be criticized; that their take on knowledge is partial; or that their authority in the lecture hall or seminar room can be questioned. And yet teaching a critical approach, and celebrating the critical spirit, requires humility with regard to one's own knowledge and a sense that it could be improved by the application of criticality. Such an open approach to the fostering of criticality requires knowledge to be seen as provisional; expertise as always emergent; and student voices to be heard in the discussion and debate of key topics in the field or discipline.

Knowledge of key theories and models of argumentation

To argue well, it is helpful to know how argument is constructed. There is no single model to fulfill this need, and attempts to apply or adapt Aristotelian and other forms of classical rhetoric to the needs of the modern student only offer partial success. For example, it is helpful to know that arguments are composed of constituent parts; that these parts can be arranged in different sequences (there is no agreement in classical rhetoric as to how many); that proposition must be linked to evidence ("statement" to "proof"); and that weaker parts of the argument should be positioned in the middle of the essay or speech. But the classical emphasis on the character of the speaker, on feeling, and on persuasion in an oral public forum are not always transferable to the seminar presentation or the essay/dissertation. Classical rhetoric needs to be complemented by twentieth- and twenty-first-century thinking on argumentation, like that of Toulmin (1958), Yoshimi (2004), and Kaufer and Geisler (1991), each of whose models offers something to students at different stages in the composition of an argument (Andrews 2005).

Awareness of the way argument manifests itself in different disciplines

There may be generic skills in argumentation that are helpful to teach and learn, but the subject- or discipline-specific elements of argument are likely to be of more immediate interest to the student of a particular discipline. We could say that it is the very nature of disciplines to create parameters within which argumentation of a particular kind is undertaken: where the disputes in the field are played out. For example, what does it take to argue as a student of an undergraduate degree in History, as opposed to one in Literature, Biology, or Engineering? Arguments in History may drill down at the points of dispute in the field; they may move from tertiary evidence to secondary and then primary evidence; in many ways argument is the sine qua non of History. Literature study is different in that arguments are more inductive and lead to appreciation and interpretation. In Biology, at undergraduate level, arguments are less likely to emerge as core to a study of the discipline unless there are sociopolitical dimensions of biology that are addressed; and in Engineering, the argument can be made in at least two forms: one, in the design of a product, and other in the presentation and justification of that product design to an audience.

Drilling down at the point of dispute

It was mentioned in the paragraph above that historians "drill down at the points of dispute" to explore the arguments in the field. But there are such points of dispute in all disciplines in higher education. While knowledge may be presented as noncontroversial in some early stages in the study of a subject, all disciplines provide space for the discussion of key problems. Undergraduate students who do not locate these points of dispute tend to see the field as unproblematic and tend to write middle-ranging essays that are more expository than argumentational. To achieve higher recognition in work at undergraduate, masters, and doctoral levels, the points of dispute—and their precise delineation—must be discovered. Then the student must have the courage and know-how to "drill down" to understand and reveal, if not necessarily solve, the disputes. Such drilling down always opens up ground for discussion and argumentation.

The development of argument skills during a degree

Becoming a critical thinker as an undergraduate is a matter of development. It is not expected that all undergraduates, at the start of their studies, have fully formed argumentational skills in their chosen fields. Those who start with an understanding of and skills in argumentation are at an advantage; they will hone their skills through discussion, feedback from tutors, and through their ability to manage complex aspects of a discipline with agility and depth. But many students will arrive at the beginning of a course of study without such highly developed skills. It is the aim of many undergraduate courses to equip students to be able to argue well. To be a historian, for example, you need to be able to think and argue like a historian. To then undertake a masters in the field, you need to attain "mastery"; and to undertake doctoral work, you will be expected to contribute to the field itself. Each stage of undergraduate and graduate education requires a further improvement in argumentation skill and capability.

Conclusion

We conclude that one way to encourage critical thinking in higher education is through an increased focus on argumentation. Argumentation implies criticality; the one cannot function without the other. But thinking, as those in the field suggest, can take various forms, from critical thinking to creative thinking, from naturally occurring cognitive development to productive thinking (see Moseley, Baumfield, Elliott, Gregson, Higgins, Miller, and Newton 2005). To provide the epithet "critical" to thinking processes and procedures appears tautological, unless one wishes to distinguish such different kinds of thinking. But the categorization of the range of ways of thinking has not been the principal driver behind the critical thinking movement. Rather, the movement has identified generic thinking skills that have been separated from context and taught as an abstract mode of operation.

Recent studies in argumentation, on the other hand, have acknowledged that generic argumentational skills can only provide a framework, and that discipline-specific argumentation in higher education is an essential counterweight to the more general approach. The advantage of taking an argumentational perspective, rather than one on critical thinking, is that argumentation can be concretized; it has a variety of models that can be applied to writing, speaking, and composing; it manifests itself in one or a number of modes; and there are distinct ways in which argued positions can be challenged.

If the aim of a focus on thinking and argumentation is to help students to sharpen the focus of their studies and to improve the quality of their engagement in speech, writing, and other modes, then the provision of short courses in argumentation at the beginning and toward the end of their studies, supplemented by in-depth exploration of how their chosen disciplines work, would seem the best way forward. Ultimately, the provision of such generic and discipline-specific argumentational strands will produce students who, once they graduate, are able to think clearly, argue well, and take their place in a democratic society where difference is tolerated, understood, and, where possible, resolved to allow consensus and action. If they progress to further study at postgraduate level, the aim becomes one in which a deep and synoptic understanding of complexity in a chosen field and topic is attained, coupled with a coherent argument that is able to present and defend such complexity in accessible and contestable terms.

References

Andrews, R. 2002. "Argumentation in Education: Issues Arising from Undergraduate Students' Work." In *Proceedings of the Fifth Conference of the International Society for the Study of Argumentation,* edited by F. H. van Eemeren, J. A. Blair, C. A. Willard, and F. S. Henkemans. Amsterdam: SicSat (International Center for the Study of Argumentation). 17–22.

Andrews, R. 2005. "Models of Argumentation in Educational Discourse." *Text* 25 (1): 107–127.

Andrews, R. 2007. "Argumentation, Critical Thinking and the Postgraduate Dissertation." *Educational Review* 59 (1): 1–18.

Andrews, R. 2009a. *Argumentation in Higher Education: Improving Practice through Theory and Research.* New York: Routledge.

Andrews, R. 2009b. "A Case Study of Argumentation in Undergraduate Level History." *Argumentation* 23 (4): 547–558.

Andrews, R. 2014. *A Theory of Contemporary Rhetoric.* New York: Routledge.

Andrews, R., Borg, E., Boyd Davis, S., Domingo, M., and England, J. 2012. *The Sage Handbook of Digital Dissertations and Theses.* London: Sage.

Andrews, R., Torgerson, C., Low, G., McGuinn, N., and Robinson, A. 2006a. *Improving Argumentative Skills in Undergraduates: A Systematic Review.* New York: Higher Education Academy.

Andrews, R., Torgerson, C., Robinson, A., See, B. H., Mitchell, S., Peake, K., Prior, P., and Bilbro, R. 2006b. *Argumentative Skills in First Year Undergraduates: A Pilot Study.* New York: Higher Education Academy.

Ennis, R. H. 1987. *A Taxonomy of Critical Thinking Dispositions and Abilities.* In *Teaching Thinking Skills: Theory and Practice,* edited by J. B. Baron and R. J. Sternberg. New York: W. H. Freeman and Company. 9–26.

Kaufer, D., and Geisler, C. 1991. " A Scheme for Representing Academic Argument." *The Journal of Advanced Composition* 11 (Winter): 107–122.

Kuhn, D. 2005. *Education for Thinking.* Cambridge, MA: Harvard University Press.

Lipman, M. 2003. *Thinking in Education.* (second edition). Cambridge: Cambridge University Press.

Miller, C. 1984. "Genre as Social Action." *Quarterly Journal of Speech* 70: 151–167.

Milson, A.-M. 2008. *Picturing Voices, Writing Thickness: A Multimodal Approach to Translating the Afro-Cuban Tales of Lydia Cabrera.* unpublished PhD dissertation, Middlesex University, London.

Milsom, A.-M. 2012. *Translating Lydia Cabrera: A Case Study in Digital (Re)Presentation.* In *The Sage Handbook of Digital Dissertations and Theses,* edited by R. Andrews, E. Borg, S. Boyd Davis, M. Domingo, and J. England. London: Sage. 276–297.

Moseley, D., Baumfield, V., Elliott, J., Gregson, M., Higgins, S., Miller, J., and Newton, D. P. 2005. *Frameworks for Thinking: A Handbook for Teaching and Learning.* Cambridge: Cambridge University Press.

Paul, R. W. 1987. *Dialogical Thinking: Critical Thought Essential to the Acquisition of Rational Knowledge and Passions.* In *Teaching Thinking Skills: Theory and Practice,* edited by J. B. Baron and R. J. Sternberg. New York: W. H. Freeman. 127–148.

Sterne, L. 1759–67. *The Life and Opinions of Tristram Shandy, Gentleman.* London: Ann Ward (vol. 1–2), Dodsley (vol. 3–4), Becket & DeHondt (5–9).

Toulmin, S. E. 1958. *The Uses of Argument.* (first edition). Cambridge: Cambridge University Press.

Walton, D. 2008. *Informal Logic: A Pragmatic Approach.* (second edition). Cambridge: Cambridge University Press.

Yoshimi, J. 2004. "Mapping the Structure of Debate." *Informal Logic* 24 (1): 1–21.

3
A Curriculum for Critical Being

Ronald Barnett

Only connect

My argument is that criticality can be distinguished through two axes: first, its levels, ranging from narrow operational skills to transformatory critique, and second, its scope, consisting of the three domains of formal knowledge, the self, and the world. In summary, my schema takes the form shown in table 3.1.

Against this background, a curriculum for critical being presents itself immediately. It has to be one that exposes students to criticality in the three domains and at the highest level in each. Our task in this chapter, therefore, is clear: to examine in broad-brush terms what it would mean in a mass higher education system to construct a curriculum that gives full rein to criticality at all its levels and in its three domains. But, while necessary, that task cannot be sufficient to sustain a higher education in the modern world. For that task would amount to an identification and a bringing together of elements. Still in front of us would be the task of supplying some means of holding the elements together, conceptually and practically.

Unless we are able to supply an account of how these different critical tasks can be held together, the danger looms that we might produce students who are adept at critically evaluating, say, literary texts or other works of humanistic culture in one way, but who adopt quite different powers of critical evaluation in relation to the world. This is the nightmare which Steiner (1984) presents us: a world in which the Nazis might appreciate Schubert or Picasso and then turn to their critique of the Jewish community in the Final Solution.

Such a schizophrenic realization of critical powers has to be avoided by the Western university. Yet just that prospect beckons as we see it: on the one hand, widening its application of critical thinking to embrace domains other than the world of formal knowledge, but, on the other, tending to confine that development to operational demonstrations. A new kind of final solution cannot be ruled out, even if it seems a remote prospect (Bauman 1991).

Table 3.1   Levels, domains, and forms of critical being

	Domains		
Levels of criticality	**Knowledge**	**Self**	**World**
4. Transformatory critique	Knowledge critique	Reconstruction of self	Critique-in-action (collective reconstruction of world)
3. Refashioning of traditions	Critical thought (malleable traditions of thought)	Development of self within traditions	Mutual understanding and development of traditions
2. Reflexivity	Critical thinking (reflection on one's understanding)	Self-reflection (reflection on one's own projects)	Reflective practice ("Metacompetence," "adaptability," "flexibility")
1. Critical skills	Discipline-specific critical thinking skills	Self-monitoring to given standards and norms	Problem-solving (means-end instrumentalism)
Forms of criticality	*Critical reason*	*Critical self-reflection*	*Critical action*

Largely, after all, universities in Germany acquiesced in the activities of the Nazi regime: they were accomplices in the domination of instrumental reason over more humanistic forms of critical reason (Nash 1945; Stryker 1996). At the least, the Western university must strive to avoid producing the fragmented critical consciousness that would again support such a situation. And it can only do so if, through our universities and in our institutions of highest learning, we develop whole persons who integrate all their critical capacities, across all three of the domains and at all their levels.

Students as persons

The emerging determination to see students performing is not totally worthless. A higher education for the new century has to have an eye to the students as actors in the world, not just as thinkers. But the contemporary fragmentation is reducing action to mere performance. We see this in sandwich courses, in which the training element is not properly integrated with the student's core studies; in teacher education courses, in which the new insistence on classroom effectiveness drives even further a wedge between theory and practice (when the school-based regime was intended to unify them); and in the inculcation of so-called transferable skills such as presentational skills, where attention focuses on the overt performance.

These are all signs of the performativity to which Lyotard (1984) was point-ing in his diagnosis of contemporary society. Lyotard's analysis is both fruitful *and* not entirely coherent. Supposedly the seminal text on postmodernity, it actually points up trends that are signs of an excessive modernity alongside those of postmodernity. On the one hand, we are told that postmodernity exhibits an "incredulity towards metanarratives" and, instead, celebrates local discourses. We can see this tendency in mass higher education, which shirks from any attempt to offer a grand overarching account of higher education. Now mass higher education is encouraged to sustain multiple "meanings' and to eschew unifying aims (Goodlad 1995; Halsey 1982; Scott 1995). On the other hand, Lyotard points to a performativity, shorn from a proper interpreta-tion. But this is the technicism to which Critical Theory has long objected, and has done so as an attack on modernity and its narrow interpretation of reason. This separation of reason from morality and understanding led, as per Critical Theory, to Auschwitz and the Gulag.

Understanding, therefore, has to be reunited with performance so as to pro-duce action. Critique in the domain of knowledge has to be brought into a relationship with critique in the domain of the world. Indeed, it is only by being shot through with analytical insight, intentionality, and a wisdom born of the weighing of alternatives that we can talk of action at all. Our students— in a higher education worthy of the name—have not merely to perform com-petently; they have to have an account of what they are doing so articulate that they can offer a rationale for what they are doing and for the discarded alternative actions.

Many lecturers will say that they do this in their courses. However, such reflection will hold students at the lower performative levels of criticality unless those reflections situate the action in the wider world of social arrangements, policies, and public interests, and students are invited to envisage alternative structures, systems, and possibilities for collective action.

Unless reflection rises to these higher levels of reflection, the student's reflection would amount to decisionism and operationalism. Simply being able to identify a range of alternative courses of action and to supply reasons for the chosen course of action does not attain the higher levels of criticality. At the highest levels, these powers of imaginary reflection call up critique in the domain of the self, to accompany critique in the domains of formal knowledge and the world. When we are in the presence of critical being, which connects critical reflection in the three domains of knowledge, the self, and the world (figure 3.1), then we are in the company of critical persons.

It is the concept of the student as person, therefore, that supplies the concep-tual and practical glue in a higher education for critical being. Taking students as persons seriously is a formidable challenge to put to our higher education institutions: not merely the development of critical being in each of the three

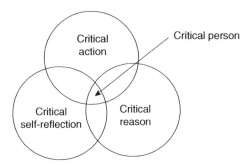

Figure 3.1   Critical being as the integration of the three forms of criticality.

domains but, crucially, their integration, and, within a challenge, they falling short of their responsibilities in the modern world. And yet, if universities do not bring off this challenge, they will be falling short of their responsibilities in the modern world. This would not just be an injustice toward their students, it would be failing society. In modern society, saturated by "manufactured risk" (Giddens 1994), in which our knowledge systems act recursively on society, we shall only be able to purchase some small space for informed and controlled action and for the attainment of a durable self through a critical disposition integrated across all three domains. Otherwise, the world will inevitably run away, since the means to bring it under critical control will have been surrendered.

It will be noted that a conception of higher education of this kind, which seeks to integrate the three domains of critical thought, runs counter to the three dominant contemporary models of higher education. The academic model overconcentrates on critical thinking toward formalized knowledge; the competence model is so focused on effective performance in the world that it does not warrant the title critical action; and reflective practice seeks to unite action and self-reflection but often downplays formalized knowledge if not disparaging it altogether and, as a result, surrenders itself to an overlocalized and operational view of professional action. Students as persons don't get a look in on any of these accounts.

Even before we attempt to fill out this idea of unification, three objections can be envisaged.

Purifying purity

First, it will be said from the traditional wing of the academic community that to widen critical thinking to embrace the domains of the self and the world would be tantamount to weakening and dissipating the force of higher education. The proposals offered here would spell the reduction of academic standards, since they represent a dilution of the purity of the intellect as the overriding value and concern of the university. The university gains its special status and function in society through the critical standards it upholds, precisely in relation to formalized bodies of knowledge. This is what the university is for. Anything else represents "the degradation of the academic dogma" (Nisbet 1971).

The objection holds no water. The so-called purity of the academic calling is a very recent idea, spawned of essentially a post–Second World War period that has seen the academics come to exert monopoly power over the academy and the definitions of education that it entertains. Both the mediaeval universities and the nineteenth-century newcomers were founded to provide the realm with useful skills, Cardinal Newman notwithstanding. The main objection to the purity ideal is not historical but is of the moment.

We cannot understand higher education today unless we situate it in some understanding of modern society—to that extent, Peter Scott (1995) is right. That a mass higher education system, born of a welfare state and now intended to meet the challenges of late modernity, should repudiate legitimate expectations of it to have some sense of the wider society is an incoherent position, to put it charitably. The liberal idea of higher education, insofar as it is being called up to justify a purist definition of critical thought, can be turned against itself; purity does not bring liberty but abandons any such prospect.

Paradoxically, late modernity poses problems not essentially of knowledge, since the world is unknowable both substantively and in terms of the tests of validity by which we come to know the world. Instead, amid discursive challenge and even discursive contradiction, late modernity poses problems of being and of the constitution of the self (Giddens 1991). Accordingly, critical thought in relation both to the self and to the world has to be brought into play. A liberal education resting on critical thought solely in relation to formalized knowledge is no liberal education at all.

The second riposte will be heard from the world of work. It will be said that the main business of business is business. Action, change, profitability, competitiveness, customer satisfaction, just-in-time quality systems, imagination, and creativity, these are the key attributes. Critical reflection, accordingly, is to be valued just so long as it generates these characteristics. This is not to disparage critical thought in relation either to formalized knowledge or to the self; these two domains of critical thought are worthwhile but only insofar as they promote change in the world itself. They are residual domains, brought in only where necessary to shore up the worldly focused forms of critical thought.

This view must also be rejected. It not merely produces a diminution in critical thought in the two domains of knowledge and the self, but it also truncates critical being in the world. Critical action in the world is logically different from critical thinking in the context of formalized knowledge, but it can be enhanced by that form of critical thinking. So, too, with critical thinking in relation to the self. Critical action becomes action rather than mere performance when it is an authentic representation of the self, when it is the self-in-the-world. Graduates will do their professions or their organizations no service if they simply live out, however reflectively, the roles assigned to them.

Critical action demands that persons fully inhabit their actions; that they are brave enough to live out their understandings in the world. Without this bravery, without this living out of one's comprehensions, a person's life is diminished, but so, ultimately, are our institutions. Despite itself, the corporate world ultimately requires real selves, three-dimensional selves inhabiting all three domains of critical being.

A third critical voice on the project before us—the unity of critical thinking and action in the three domains of knowledge, self, and world—can be heard from reflective practice. The objection from this quarter would deny that the world of formal thought is significant in its own right. Such a view amounts, we are informed, to a form of "technical rationality" (Schön 1987). Formal knowledge comes into play, if at all, to assist in the reflective action as part of the armory with which to interrogate action.

Reflective practice is the most sophisticated of the three critical voices since, in its more subtle interpretations at any rate, it finds a place for both understanding and the self. In particular, the self forms an important component of reflective practice, since it has to be a self that both engages in reflection and is, in part, constituted by it. The main problem lies in reflective practice giving prominence to "practice" even if reflective in character and downplaying knowledge as such. The very possibility of an integrated and coherent unity between critical thinking in knowledge, the self, and the world is outlawed by the notion of reflective practice. Reflective practice denies the possibility of unity by denying a serious role for knowledge. The pass is sold at the outset and has to be reclaimed.

All three of these contending voices have their own king of purity, which they would hold on to. But, paradoxically, the upshot of purity in the modern world is fragmentation. A higher education for critical being can only be achieved by accepting the multidimensionality of the task. The three domains of criticality have each to be respected *and* they have to be integrated. Otherwise, we shall see students emerge whose criticality either is stunted in one of more of the three domains or is fragmented or, more probably, is both. We have, simply, to purify our educational ideas of the belief in purity.

For critical persons

What, then, in general terms would a curriculum intended to develop critical persons look like, a curriculum in which critical thinking in the three domains of knowledge, the self, and the world was brought together at all their levels?

At once, the question rules out modes of thinking that overemphasize the different domains. Didactic lectures reflective of a excessive concern with formal knowledge; performance reflective of an exaggerated interest in having effects in the world; self-reflective activities, which stamp in a self-monitoring according to external norms: all of these have to be jettisoned. They have to be jettisoned not just because they represent a lopsided approach to critical thinking, concentrating overly on one of the three domains. They have to be jettisoned not just because they fail even to realize the potential human development in each of the three domains (so that we end up with rote learning, sheer performativity, and introverted reflection, devoid in each case of broad understanding and critical insight). More searchingly, they have to be jettisoned because they fail to offer even the beginnings of a unitary approach to criticality and, thereby, fail to develop critical persons and critical being as such.

Much, therefore, of our contemporary approaches in higher education has simply to be set aside. Quite different pedagogical relationships are required. Students have to be given the space to become themselves, to bring their understandings to bear on situations and, in the process, make them *their* understandings; to understand themselves in relation to situations requiring some insight and learning, *including their own limitations*, and to develop the capacity for critical insight in action. A vocabulary such as student-based learning, problem-based learning, and independent learning does no justice to the magnitude of what is required. On the contrary their development as a strategy for resource economies will produce—unless otherwise checked—a narrowness both in the domains and in the levels through which critical being is realized. If the human learning and development of the kind sketched out here is to come about, we have to dispense with notions not just of teaching but even of learning ordinarily conceived. We need a different vocabulary.

The vocabulary to which we need to have recourse would include items such as self, being, becoming, action, *inter*action, knowing, understanding, risk, exploration, emotion, interpretation, judging, insight, courage, exposure, daring, authenticity, collaboration, and dialogue. A cluster of concepts of this kind is necessary if we are to do justice to an education conceived as the fostering of critical persons We do not have to call up the dramatic picture of the student in front of the line of tanks in Tiananmen Square; we simply have to have in mind the challenge of professional life. If students are to prosper in the modern world, if they are to carry their world forward in worthwhile fashion, they have to become critical persons embodying critique in the three domains of knowing, of self, and of the world at the same time.

The riposte may come, but what is the cash value of all of this? What does a course look like that incorporates such abstract ideas? The riposte is not without substance, but it scuppers itself to some extent. It is right to be asked to indicate the practical insignificance of educational ideas and to map out in general terms what a curriculum might look like if it is to carry off a proposed program. But the demand as stated falls foul of just the kind of thinking that I am challenging. If we are to produce critical persons, a course in physics or in English literature or even shall we say, in timber technology or in landscape architecture will be unlikely to do it.

The problem lies not in any subject or discipline per se, it lies in the preposition "in." A course in physics or even in landscape architecture does not promise to produce critical persons of the kind I am proposing. Landscape architecture could come closer to it than physics, since the former willy-nilly will expose the student to some form of critical refection in each of the three domains of knowing, the self, and the world. Physics, in contrast, does not compel a student to engage with the self or the world except in severely truncated ways. English literature and timber technology both occupy intermediate positions, with English literature inviting critical self-reflection on one's personal values and timber technology requiring engagement with productive processes.

What has to be faced is that a higher education anchored in a discipline, in which critical being is restricted to the domain of formal knowledge *and* is restricted in its scope to technical operations within a single field, cannot supply critical persons for the twenty-first century. Both the domains *and* the levels of critical being have to be incorporated, and those three domains of critical being have to be integrated. Critical persons are precisely those individuals who exert some unity of critical power over their experiences in relation to knowledge, themselves, and the world.

Keep it simple

So what, in general terms, would it mean for higher education to develop critical persons, in whom there was some integration of critical thinking in the three domains of knowing, the self, and the world? The first point is negative in character; there are no generic forms of criticality appropriate to the three domains to be incorporated into curricula. Academics in the disciplines do not have to turn themselves into educationalists who have identified realm of pure critical thinking, independent of the disciplinary base of a site of professional action. What is required is both simpler and much more complex.

Simplicity comes into play because, in the first place, what is called for is for academics to live out their own identities fully and utterly. Rather than imagine afresh a process that we call teaching and learning, the first requirement is that academics reveal themselves to their students as the hard-pressed inquirers that they are. In a genuine process of inquiry, they have to engage in a struggle to formulate their thoughts, to labor to develop their thoughts (whether in the laboratory, the clinical situation, or in the library), to expose their thoughts to others, to encounter critical evaluations of that thinking, to engage in risky undertakings, and to move on in the light of those critical comments. These are not purely cognitive processes. They require of scholars and researchers to give of themselves, to develop themselves as persons, and to engage—albeit in a truncated form—in critical action.

The propositions, theories, findings, proposals, creations, technologies, and methods that they propose embody their beliefs. Belief in turn requires commitment and personal investment: publicly expressed, it is a way of constituting the self (although a limited academic self). But to this existential moment of self-declaration is coupled intersubjective processes of engagement according to the rules within which that academic discourse works. And that process calls both for courage, integrity, and authenticity on the one hand, and for qualities of intersubjective patience, sensitivity, respect, and reciprocity on the other.

The simplicity, therefore, of getting students embarked on the road to a critical consciousness lies in academics avoiding concepts of teaching and learning as such, and setting aside the thought that there are institutionalized roles and relationships captured by conventional terms, such as teacher and student, which do justice to higher education. To put it another way, rather than hypothesizing a conceptual distinction between research and teaching (which then have to be brought together in some way), teaching may be seen as an insertion into the *processes* of research and not into its outcomes. What is required is not that students become masters of bodies of thought, but that they are enabled to begin to experience the space and challenge of open, critical inquiry (in all its personal and interpersonal aspects).

What is being suggested here is the abandonment of teaching as such (Sotto 1994). But this is not to open the door to lecturers to construe their task as one of introducing students to the latest research findings. The educator's task, for a critical consciousness, is to set up an educational framework in which students can make their *own* structured explorations, testing their ideas in the critical company of each other. This is a highly structured process, in which the students are subject not only to the local rules of the particular discipline but also to the general rules of rational discourse as such. Turn-taking, acute listening, respect for the other's point of view, expression of one's ideas in ways that are appropriate to the context, and even the injection of humor: the critical consciousness can be too serious for its own good. More than that, there would have to be elements of genuine openness such that students can feel that their own voice and their own existential claims matter. This means that the lecturer's own position can and will be challenged.

So the simple task of educators adapting their own approaches to scholarship and research to their educational setting turns out to be complex. Pedagogical roles and relationships become uncertain, and necessarily invite risk into the proceedings. If students are to be given the space genuinely to form their own critical evaluations, and to engage in critical acts, the educational process has to become uncertain, A looser framing of the pedagogical relationship between lecturer and students is imperative. An uncertain world requires an uncertain education.

Levels of a critical education

The idea of levels of criticality points to an ever-changing horizon of perception: as the epistemic level rises, the object is viewed against an ever-wider context. Does critical thinking just limit the student to deploying a set of logical moves on the material in front of her? Does it enable the student to evaluate the text or the data in the context of an understanding of the field of study as a whole? Does it invite the student to place the topic in a wider context such as its implications for our understanding of the world? Does it allow the student to come at it from a variety of critical perspectives, such that the field, with its presuppositions, is itself susceptible to critique?

Take, for example, a student taking a degree in tourism studies. Tourism studies is a complex field, potentially incorporating subfields as diverse as finance, accountancy, marketing, business studies, economics, the study of tourism itself as a social phenomenon, geography, history, cultural studies, politics, and ethics. The key word in that sentence is "potentially." Very few degree courses in tourism studies will encompass that range: all will draw on studies of basic business functions such as finance, accountancy, and marketing, but few will draw on the more human and social studies to any significant extent. Accordingly, a higher education for a critical consciousness in tourism studies immediately invites the question: What is the scope of critical thinking that informs its study on this particular course?

The lecturing staff, when interviewed, are likely to say that they warmly endorse the notion of critical thinking as a feature of their course, but what does it mean in practice? Are the students simply expected to acquire elementary skills ("critical thinking skills") of knowing how argument works, of forming valid inferences from the available and, in this case, often incomplete or rudimentary data? What is the range of perspectives through which they are invited to view tourism? Are they encouraged to see tourism purely as an economic matter, including its employment characteristics, or are they invited to explore tourism through a range of perspectives and considerations (such as its effects on indigenous cultures, its ethical components, its impulses toward globalization, and its postmodern character, not to mention its effects on world health with the spread of AIDS). Even more significantly, are the students given the wherewithal to place their own program of studies, to understand its limitations, and its emphases? Are they able to critique that? And yet more fundamentally, are the students offered an educational experience that challenges them to develop their own critical stances in a nonthreatening environment, so that they acquire the dispositions of critical thinking to sustain them beyond their immediate educational framework into their future careers?

There is a double difficulty here for many academics. First, talk of dispositions will be discomforting. How are dispositions to be developed? Academics may feel that they have enough challenges mastering their own field without having to pause to work out how they might foster students' dispositions. Second, critique works—as we have seen—by exposing one framework to the critical interrogation of another, or allowing frameworks to collide (in tourism studies, the cultural with the economic, for example). But that means that academics have to acknowledge the limitations of the very framework in which they have developed their own intellectual identity. They have, if critique is to be fostered, to become other than they are, to pretend that their own framework is not that important after all.

These two difficulties can be addressed at the same time. What is required is an inversion of self-understanding on the part of the academics as educators. The development of dispositions and the capacity to bring into one's understanding a range of frameworks are, ultimately, a responsibility of the student. The role of the educator is to provide the educational space in which those developments are likely to occur. Compiling an agenda of issues that draw upon multiple frameworks, structuring tasks, getting students to collaborate on projects, positing imagination and intellectual range as criteria, and drawing students' attention to a range of relevant literature (not just in one's immediate field): strategies of this kind rather than teaching per se are necessary elements in producing critical persons. But they are not sufficient: for that, students have to take on their own responsibilities for their own continuing explorations. A curriculum for a critical consciousness requires real curriculum space for the students; but it is an existential space in which students can, interactively, form their own critical evaluations from this perspective or that, without any sense of intimidation or of being ruled offside.

The domains of critical being

This sketch, however, only points to necessary conditions of a critical education and, in itself, falls into the trap of embellishing critical thought in relation to formal knowledge. It is the kind of educational approach that many lecturers might feel they can easily accommodate. But much more is required of a critical higher education. A higher education for critical being has to extend across the other two domains of the self and the world. What might this mean in practice?

The most problematic domain is that of the world. What does it mean to take up a critical stance toward the world? How might we bring into our story the image that has formed this book's leitmotif, that of the Chinese student halting a line of tanks? Was that simply an example of an individual expressing himself as a citizen and his being a student was immaterial? In other words, the argument might be that the university, qua university, has no responsibility and even no place in developing a critical stance toward the world among its students. I have argued to the contrary. The university precisely has a responsibility, *qua* university, to develop the capacity within its students to take up critical stances in the world and not just toward the world.

The first question is whether the argument will be general to the university or will be specific to particular fields. It might, prima facie, be more legitimate to develop a justification for a critical stance toward the world built around tourism studies than around mathematics, for example. The argument, however, if it is to work with any force, has to be general in character. It has to be a condition of higher education that all its students are enabled to develop a critical capacity in the domain of the world, if the university is to fulfill its responsibilities toward criticality. But this consideration, that the argument has to have general applicability, only plunges us into further difficulty. What arguments might suffice to work across tourism studies, mathematics, history, and archaeology? What would such an education for this critical thinking look like?

An immediate answer is, perhaps surprisingly, straightforward. Every form of thought, every field of inquiry, every subject has a place in the modern academy because it has some form of social legitimacy. In most fields, there are all kinds of sites of their application in the world, both in school education and in the corporate, professional, and industrial worlds. That, in some fields and disciplines, following the professionalization of knowledge, the links have become buried should not detract from the point. This is not a case of finding spurious "relevance" but of recovering the social interest in each intellectual field (where it is not apparent). An extraordinary feature of modern academic life is that the veil separating knowledge from practice has been drawn down even in relation to professional fields. Intent on winning their intellectual spurs, academics take delight in declaring that they are not in the business of professional preparation.

But even if the dormant social interest in each form of thought could be brought explicitly into the curriculum, we would still be short of a fully critical stance toward the world, much less giving an account of our student in front of the tanks in Tiananmen Square. Did it matter whether he was student of politics or history? Could he equally have been an engineering student?

At this point, we need to draw in the third domain, that of the self. If students seriously begin to reflect on themselves, to understand their own thinking, they might characteristically begin by gaining insight into the frameworks that they typically deploy. Do they enjoy just working things out according to the rules of the game, or are they prepared to be more adventurous by looking at things in different and even in new ways?

In this way, critical thinking in relation to the world of knowledge (CT1) and in relation to the self (CT2) can come into a relationship with each other. But this could still remain a cerebral activity. Skills come into play, but internally, within the rules of the local form of thought. However, once the student is placed in situations where those skills and understandings are exposed to pragmatic situations in the world, then the potential for critical thought to widen to embrace the world arises (CT3). This critical thought in the world, critical action, is not an add-on; it could never be, for example, the demonstration of "enterprise" as such. Rather, it becomes part of an integrated being, in which critical persons in the three domains of knowledge (CT1), or self (CT2), and of action-in-the-world (CT3) are constituted and, through their integration, the possibility of transformations in each domain arises.

We have briefly to backtrack here. It only makes sense to talk of bringing into the curriculum critical thought in relation to the world (CT3) if and when critical thought in the other two domains—knowledge and the self—is present. We get to CT3 through CT1 ad CT2. This seems to run counter to the point made earlier that the three forms of critical thought are independent of each other. But the two positions are entirely compatible. CT in the world (CT3) is sui generis and is not to be reduced to either of the two other forms of critical thought. However, it is enhanced by critical thought both in the domain of knowledge and in the domain of the self. If students are to act authentically in the world, they will need to draw on both their imaginary insights in the domain of knowledge and their own self-understanding. In this integration of the three domains of critical being, they become persons, acting autonomously in the world and taking up a critical stance toward it.

Conclusion

Criticality can be achieved in three domains, those of knowledge, the self, and the world. A curriculum intended to develop critical persons necessarily, therefore, has to find some way of developing critical thinking in the three domains so as to develop critical thought, critical self-reflection, and critical action. However, these three exemplifications of the critical life have also to be brought together, if a unity of the critical outlook is to be achieved and if creative criticality is to be developed. The integration of criticality in the three domains calls for nothing less than taking seriously students as persons, as critical persons in the making. Students come into themselves as persons in command of themselves.

The concept of person, in other words, supplies the conceptual and practical glue required if fragmented critical being is to be avoided. This sounds simple

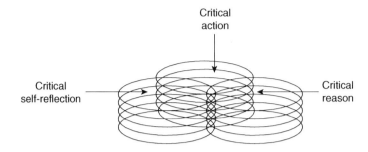

Figure 3.2 Connections between knowledge, self-understanding, and actions at the highest levels of criticality.

enough, but it calls for students to be given the cognitive, personal, and practical space in which they can develop in all three domains, yet be guided in such a way that the three domains are brought into a relationship with each other. Understanding, self-reflection, and action have to be given the space to take off, such that the prospect of independent persons can be present. But the educator's task is not complete unless the student is challenged continually to make connections between her knowledge, self-understanding, and actions at the highest levels of criticality (as in figure 3.2).

Then, and only then, in this integration at the highest levels of creative critique, would we have the prospect of higher education becoming a site where critical being adequate to the wider world might be fostered.

References

Bauman, Z. 1991. *Modernity and the Holocaust.* Cambridge: Polity.
Giddens, A. 1991. *Modernity and Self-Identity: Self and Society in the Late Modern Age.* Cambridge: Polity.
Giddens, A. 1994. *Beyond Left and Right: The Future of Radical Politics.* Cambridge: Polity.
Goodlad, S. 1995. *Conflict and Consensus in Higher Education.* London: Hodder and Stoughton.
Halsey, A. H. 1982. *Decline of Donnish Dominion.* Oxford: Clarendon Press.
Lyotard, J.-F. 1984. *The Postmodern Condition: A Report on Knowledge.* Manchester: University of Manchester.
Nash, A. S. 1945. *The University and the Modern World.* London: SCM Press.
Nisbet, R. 1971. *The Degradation of the Academic Dogma.* London: Heinemann.
Schön, D. A. 1987. *Educating the Reflective Practitioner.* San Francisco, CA: Jossey-Bass.
Scott, P. 1995. *The Meanings of Mass Higher Education.* Buckingham: Open University Press/SRHE.
Sotto, E. 1994. *When Teaching Becomes Learning: A Theory and Practice of Teaching.* Trowbridge: Redwood books.
Steiner, G. 1984. "To Civilise Our Gentlemen." In *A Reader.* London: Penguin.
Stryker, L. 1996. "The Holocaust and Liberal Education." In *The University in a Liberal State,* edited by B. Brecher, O. Fleischmann, and J. Halliday. Aldershot: Avebury.

4

Willingness to Inquire: The Cardinal Critical Thinking Virtue

Benjamin Hamby

Introduction

Every craft has its set of tools, which the expert craftsperson uses adeptly in her creative efforts. Critical thinking is a craft no different than others in this respect: through guided practice and refinement, the expert craftsperson of *reasoned judgment* has developed a set of cognitive tools she uses in her reflective endeavors. But how are the skills of a critical thinker connected to the character of such a craftsperson? I argue that the critical thinker's skills are suggestive of the person she must be to properly employ them. These critical thinking virtues are the motivations, dispositions, and values that animate her skilled thinking, and this *willingness* to think critically is what drives any appropriately applied skill. Rather than interpretive charity, open-mindedness, valuing nonfallacious reasoning, or any of the other virtues that connect to skills, I conclude that willingness to think critically is the most fundamental critical thinking virtue.

Most mainstream conceptions of critical thinking tend to focus on teaching transferable thinking skills that can be applied across a variety of curricular and real-life contexts. While my treatment is congruent with such common approaches, I follow Bailin and Battersby's (2010) conceptualization. They equate critical thinking with "critical inquiry," the process of "carefully examining an issue in order to reach a reasoned judgment" (4). Since I have argued elsewhere that this conceptualization is defensible (Hamby 2013, 45–46), I argue here that critical thinkers must have what I call a "willingness to inquire": the firm internal motivation to employ one's skills in the process of critical inquiry, seeking reasoned judgment through careful examination of an issue.

I make my case by first presenting what I take critical thinking skills to be. Then I explain how associated critical thinking virtues are suggested by those skills, when we imagine the kind of person who would consistently and competently use them to the ends of critical inquiry. The person who would tend to fail to use those skills appropriately, or who would not use them at all, is not someone whom we should call a critical thinker. I then argue for why willingness to inquire is the cardinal critical thinking virtue, after which I respond to one detractor, whose view I find ironically corroborates my own. I conclude that critical thinking virtues are important if a thinker is to be the kind of person who appropriately applies her skills in critical inquiry. We should seek to foster such virtues in our students, and we should only call a person a critical thinker if she possesses those virtues, especially the central motivating virtue, that is, a willingness to inquire.

Critical thinking skills

Critical thinking skills are those cultivated abilities that a person must have to engage in critical inquiry. By conceiving of critical thinking skills in this way I follow the consensus found in Facione (1990), thinking that critical thinking skills are like any other kind of skill: special kinds of purposeful abilities "to engage in an activity, process, or procedure" (14). What makes critical thinking skills special? They are those particular abilities that contribute to the process of critical inquiry, and to achieving its end, which is reasoned judgment.

For instance, one important critical thinking skill that often contributes to the process of critical inquiry is the ability to interpret the core of an argument: its main conclusion, and its supporting reason or reasons (cf. Blair 1995; Johnson 2000). Arguments are relevant considerations requiring interpretation and evaluation in reasoned judgment making, so the ability to interpret an argument is properly considered a critical thinking skill when it is employed toward that end. This is true even while the skill of argument interpretation could be put to other uses, such as in a debate, where one party attempts to dogmatically or otherwise single-mindedly defend a view from alternative perspectives. But then, in being employed toward this end, a person would not be thinking critically, though she would still be performing a skilled thinking activity in her argumentative defense.

Argument interpretation is only one specific skill that is usually brought to bear in the process of critical inquiry, however. There are many other skills as well, including argument evaluation, clarifying meanings of terms and statements, evaluating authorities and sources, and examining plausible alternatives, among others. These are all activities that contribute to critical inquiry, because they are all especially relevant to the process of carefully examining an issue in an effort to reach a reasoned judgment. So, if a thinker has not succeeded in cultivating abilities to engage in these intellectual activities, she is not a critical thinker.

Critical thinking virtues

For any skilled activity I might engage in, however, sheer technical ability is never enough for me to actually engage in that activity on a consistent basis. The reason is that I might have an ability to do something, but without a corresponding inclination to do it, my ability will not characteristically be employed toward achieving the ends proper to it. This is true for skilled thinking activities that contribute to a process of critical inquiry aiming toward reasoned judgment, just as it is for other skilled activities that achieve other ends.

For example, if I am a musician and want to be an excellent and convincing performer, I must be technically adept at playing my instrument, but I also must actively want to perform, having a certain *passion* for the stage. If I am an athlete and want to compete at an elite level I must have the skills and conditioning required, but without the *perseverance* to play through the pain of extreme physical exertion, my skills will never be pushed to the limit when I need them most. And if I am to be a well-trained and combat-ready Marine, I must be skilled in the use of a rifle, able to reliably hit my mark; but without the *relentless motivation to kill* my enemy and the *selflessness* and *obedience* to follow orders even at the cost of my own well-being, I will not be a well-trained Marine that will fulfill my duties in combat.

In these and all other skilled activities, more is required than the sheer ability to perform some technical task to some end. What is required in addition is a certain commitment to and willing engagement in that activity: those personal qualities that drive a person to employ her skills, and to employ them appropriately, aiming toward the ends that are proper to them. Without those qualities, a person with ability will not be the kind of person who consistently employs that ability. A musician without passion will not be a convincing performer; an athlete without perseverance will not be a worthy competitor; and a Marine without bellicosity will not be a valorous warrior. The basis of consistently produced skilled activity, such that we would identify a person by her habitual and competent performance of it, is therefore a matter of character: those firm personal qualities that consistently contribute to a person producing that activity's ends.

So, even though to be a critical thinker one must be a skilled thinker, being a skilled thinker is not enough. One must also be a virtuous thinker, possessing and manifesting the critical thinking virtues: those cultivated excellences of a person's character that consistently guide her in the skillful process of critical inquiry. They are the motivations, values, dispositions, goals, and other habits of mind that, being connected to purposeful abilities, lead a thinker to be the kind of person who engages appropriately in the interpretive and evaluative process of critical inquiry.

There are many reasons why critical thinking virtues are required for someone to be a critical thinker. For instance, I could be a highly skilled thinker yet not aim toward reasoned judgments when I employ my skills. Or I could be a skilled thinker aiming toward reasoned judgments yet never be disposed to employ my skills. Or I could be disposed to use my skills aiming toward reasoned judgments, but be compelled, coerced, or otherwise improperly disposed to employ my skills. Imagine the following scenarios.

In the case of someone who is highly skilled in debate, such a person will most likely have a mastery of various skills that are brought to bear in that process. However, even if debating involves marshalling evidence and making the best case for some position, it is still more about defending a view from alternatives than it is about reaching a reasoned judgment offered with "full justification" in light of other alternatives (cf. Goodwin 2013). In this sense, debate winning is a kind of (improperly) motivated thinking that tends toward error as compared to reasoned judgment. Since the ends a debater has in mind when she employs her skills are *not* to reach a reasoned judgment, the critical thinking virtues will not come into play, and her skills coupled with her desire to win will be enough for her to fulfill her ends. Such a person is not thinking critically, and if such thinking is characteristic for her, she is not a critical thinker.

Now imagine the case of a person who is a skilled thinker and who *does* aim toward reasoned judgment: but we can think of her not employing her skills to that end because of her intellectual laziness or carelessness. In that case she might haphazardly and prematurely latch onto a claim that is not the main claim being supported. This would constitute an example of a skilled thinker who has the right ends in mind, but who nevertheless employs her skills in an inappropriate enough way such that they are not put to proper use considering her ostensible aims. This person also is not thinking critically. If such intellectual behavior is characteristic of her, she is not a critical thinker.

Finally, imagine a person who is a critical thinking student in a classroom where employing her skills aiming toward reasoned judgment is required of her: we can imagine her employing her skills aiming toward reasoned judgment *only* in that context. Such a student might be thinking critically in the classroom, but she is doing so because she knows she must in order to earn a passing mark in the course. Outside of the classroom she might not value critical inquiry such that she is disposed to engage in it when the opportunity presents itself. Such a person is not a critical thinker, even if she thinks critically when she is compelled to do so.

I have so far presented examples of thinkers who are skilled, but who lack the proper virtues to put those skills to appropriate use in efforts at critical inquiry. But the same cannot be said for someone who is an excellent critical thinker: such a person not only possesses relevant skills, but also relevant personal characteristics that make her the kind of person who employs those skills appropriately in critical inquiry. For instance, given the skill of argument interpretation, a critical thinking virtue animating that skill in a process of critical inquiry is charity. A charitable paraphrase of a text is one such that it *fairly* represents the argument or arguments it contains, accurately and most plausibly articulating the conclusion, subconclusions, and supporting premises that comprise it. Charity is important when employing the skill of argument identification in critical inquiry, because an unfairly or implausibly interpreted discourse will stand in the way of the ends of critical inquiry, tending to confound a proper evaluation from being carried out and a reasoned judgment from being reached. Therefore, to be able to perform the skill of argument interpretation in the service of critical inquiry in a consistent and appropriate way, one needs more than sheer ability to render the arguments of others. In addition, one needs to interpret arguments in an appropriate way: *attentively* considering the context, *carefully* and *accurately* extricating the argument at hand, and *fairly* representing what is concluded, and what ostensibly provides the support for that conclusion.

One can be skilled at argument interpretation without being charitable, paraphrasing a text with ulterior motives, or more overtly in an effort other than to reach a reasoned judgment. But since it is not reasoned judgment one is aiming toward when one uses that skill uncharitably, it cannot be said one is thinking critically. For instance, a politician who casts her opponent's reasoning in an subtly unfair way might be very skilled at argument interpretation, but because her purpose is to sway public opinion away from her opponent, her skill need not be employed charitably, even though it still may be used in an effective way to achieve her ends.

The same can be said for other critical thinking skills: associated critical thinking virtues are indicated when these skills are put to use in the service of critical inquiry. For instance, to evaluate an argument as part of a process of critical inquiry one must be open-minded: ready to entertain an argument on its merits without deciding beforehand whether it is cogent or not; ready to revise one's view should the argument prove to be stronger than, yet contrary to, a prior view one holds (cf. Hare 1979); and ready to admit one's own fallibility (Riggs 2010). One can evaluate arguments skillfully without being open-minded, but then one will usually have other aims when using that skill. A defense attorney, for instance, might evaluate the prosecution's arguments with great skill, but since the end in view is to acquit her client, an open-minded approach to the opposing arguments is not required for her to be effective in her aims.

As a final example, take the critical thinking skill of making good inferences and correcting for bad ones, as well as being able to recognize good and bad inferences that are made by others in various contexts. This is a key critical thinking skill a person will tend to need in a process of critical inquiry. However, inferences can be made fallaciously yet be perfectly efficacious in helping one to achieve some end other than reasoned judgment. For instance, an appeal to popularity is an effective way for an advertiser to sell a product. Appealing to popularity, however, tends to be an ineffective way of going through a process of inquiry. A critical thinker must therefore make inferences in a way that avoids fallacies such as an unwarranted appeal to popularity, whereas an advertiser need not so long as it helps her to sell her product. Valuing thinking that is nonfallacious is therefore an important critical thinking virtue that enables a person who is skilled at inference-making and identification to put those skills to appropriate use in the process of reaching reasoned judgment.

Willingness to inquire

There is one critical thinking virtue, however, that moves every critical thinking skill. Without this virtue, my skills will tend not to be employed consistently and appropriately in critical inquiry, and others should resist calling me a critical thinker. For, regardless of the critical thinking skill, without a motivation to use it and a tendency to actually employ it in a process of inquiry, I will either not employ it at all, or not employ it appropriately. In the case of the single-minded debater, the politician, the defense attorney, and the advertiser, whatever skills they employ in their intellectual efforts, they are surely not employing them in an effort to reach reasoned judgment, but for other ends. This disqualifies them from being considered critical thinkers if such thinking is a characteristic part of their intellectual orientation.

As a critical thinker, the end of my skilled thinking is reasoned judgment reached through a process of critical inquiry. Therefore, the necessary and central virtue I should possess and manifest before all others is willingness to inquire: an internal motivation to employ my skills appropriately, aiming toward reasoned judgment. Such a person recognizes the value of going through that process aiming toward that end, and characteristically engages in that process in her intellectual life. She appreciates the power of critical inquiry enough so that she is willing to seek reasoned judgment thorough a careful examination of an issue, whenever there is room for reflection. Returning to the case of the student who is required to think critically: her skills are wasted outside of the classroom because she does not appreciate inquiry or think it important enough for her life outside of class. Without willingness to inquire, this student will never be a critical thinker.

This is therefore the critical thinking virtue without which a person will not characteristically put her critical thinking *skills* to appropriate use aiming toward reasoned judgment. But it is also the basic orienting virtue that stands behind other critical thinking *virtues*, such as charity, open-mindedness, and valuing nonfallacious reasoning. These other critical thinking virtues are important, and *inter alia* contribute to the fundamental attitude that moves a person to be a critical thinker; but in the ideal critical thinker the general drive to engage in critical inquiry aiming toward reasoned judgment stands behind the other virtues in their particular manifestation in that process.

Of course, one can be open-minded in particular, context-specific situations. In addition, one can exhibit in specific instances any of the other virtues without having willingness to inquire. I could be the most open-minded person yet not at all interested in critical inquiry, being open-minded only for the sake of making friends, changing my opinion to curry favor, admitting fallibility with an unreflective congeniality. But then, as this kind of an open-minded person I could hardly be said to be a critical thinker, even though I might evaluate arguments in the most open-minded of ways. If one is to be open-minded *in the process of critical inquiry*, then one must of necessity also have willingness to engage in that process. So, following Siegel (2009), while open-mindedness is required, it is not enough for critical thinking. I would also add, however, that willingness to inquire is also required if one is to be open-minded in critical inquiry: without willingness to engage in that process, one cannot be open-minded in that process (Hare 1979). The same can be said for the other critical thinking virtues: willingness to engage in critical inquiry ranges over the *virtuous* application of skills that are used in that process, by people who are rightly deemed critical thinkers.

A response to one detractor

My formulation of the critical thinking virtues is consonant with most theorists' take on what are commonly called "critical thinking dispositions." Some, such as Siegel (1988), following Passmore (1980), call this constellation of virtues "the critical spirit." Some, such as Bailin and Battersby (2010), follow Glaser (1941/1972), calling it "the spirit of inquiry." Whatever the label, this commonality of sentiment gives my formulation some plausibility. For while all theorists are not agreed on exactly how to formulate the educational ideal of critical thinking, nor precisely how to conceive of how critical thinking virtues connect to critical thinking skills, what most implicitly, and sometimes explicitly (e.g., Hare 1979), agree upon is that in order to think critically one must be both *willing* and *able* to do so. My formulation connecting the skills of critical thinking to associated virtues specifies just how this willingness and ability are linked: given the special aims of critical thinking, and the special skills that are required to pursue those aims, associated virtues are indicated. Without such firm aspects of a person's character, the exercise of some skills, when they occur at all, will tend to confound critical thinking. In the pragmatic endeavor of teaching students how to properly employ their skills, we should therefore foster in them the associated virtues. Only then will they have a better chance to become true critical thinkers: people who are committed to using their skills appropriately in the practice of critical inquiry.

What remains in this paper is for me to briefly show how my formulation can accommodate the view of one significant detractor. Ironically, I find that even though Missimer (1990; 1995) is the most outspoken critic of critical thinking virtues, with her ostensible rejection of critical thinking "character," her articulation of that critique nevertheless concedes my basic point regarding willingness to inquire. I argue that Missimer's view is consistent with the mainstream view: critical thinking skills require the internal motivation to use them in the person who is truly a critical thinker.

Missimer attacks the "character view" of critical thinking, and defends the "skill view." The character view is the view that "the critical thinker has certain character traits, dispositions, or virtues" (Missimer 1990, 145), in line with the consensus view of the ideal critical thinker, and the skill view, by contrast, is the view "that critical thinking is a skill or set of skills" (145). Missimer complains that different versions of the character view are "advanced without much analysis"; "are inconsistent"; are such that "historical evidence can be brought against" them; are not as "exciting" as the skill view; and that they "smuggle in moral prescriptions" (145). Rather than rehearse these arguments against the critical thinking virtues and reply to each of them in turn, I will stress how Missimer's overall attack rings of incoherence, as she concedes that the critical thinker must possess the habit of thinking critically, implying what she herself calls a "minimalist" character-view position (Missimer 1990, 149). While Siegel (1997) has adequately responded to the other charges Missimer levels against the character view, he only briefly mentions this incoherence as a reason to reject Missimer's position, but does not elaborate (65). In line with Siegel's critique, I wonder how Missimer's is a skills-alone view when for her doing critical thinking is always "a result of having wanted to" (Missimer 1990, 149).

The incoherence of Missimer's position comes when she says that "to be (thought) a mathematician, historian, [or] sailor, you must do a lot of mathematics, history, [or] sailing as a result of acting on your disposition to do so" (149). Likewise, so Missimer admits, to be thought a critical thinker, one must do a lot of critical thinking as a result of acting on a disposition to think critically. While Missimer thinks this sounds tautological and trivial (149), rather than also concede that her concept of the critical thinker involves character in an important way, she goes on to say that "one could think of these dispositions to do critical thinking (or mathematics, sailing, etc.) as character traits; one could as easily think of these activities as habits born of skill and enthusiasm to keep up the habit" (149).

However, this denial that the disposition to think critically is an important character trait is unconvincing, because Missimer describes the habitual critical thinker as one who has a certain "enthusiasm" to think critically. And enthusiasm, especially when it is manifested habitually, is surely an important character trait of a person. I have in mind Hare (1993), who explicitly stresses that enthusiasm is a central intellectual virtue that good teachers should possess and seek to foster in their students (24). Being characteristically enthusiastic about sailing or thinking critically is certainly not a skill, even if through the skillful practice of thinking critically or sailing an enthusiasm for those activities is nurtured. So we should not take that enthusiasm as a result of practice for granted, as Missimer does when she says that "if you get [students] to practice the skills in a myriad of areas but do not tell them that they should have the disposition to do critical thinking, you are teaching the disposition in the sense that you are instilling the habit of critical thinking" (24). This claim seems highly doubtful in light of the discussion above. Even in the habitual exercise of some skill it is not a foregone conclusion that the disposition or motivation to use that skill (or an "enthusiasm" to use it, or as I would call it, a "willingness") will thereby be developed. A person might sail every day because her rich father insists on it, and might as a result become quite a skillful sailor. Still, she might at the same time never wish to sail on her own, or develop the enthusiasm for sailing that would lead her to value and appreciate her sailing skills, and to willingly use them away from her father's insistence that she sail. The same can be said for any skilled activity, including critical thinking: just because we drill the skills of critical thinking does not mean we instill the will to think critically. Therefore, nurturing in students a respect for inquiry and a willingness to go through that process with others is not an exercise in triviality based on a mere tautology, but is vitally important for their development as critical thinkers. I submit that even in her attempt to deny that her view involves intellectual character, Missimer reveals that she thinks it plays an important role in the skillful habit of thinking critically.

In addition, the presence of this internal motivation to think critically and its connection to skilled thinking is corroborated by empirical evidence of the sort that Missimer (1995) demands should be present if we are to accept that the critical thinker is someone who is skilled, and who has a certain type of virtuous character. That evidence is found in a recent study by Nieto and Valenzuela (2012), which investigates the internal structure of critical thinking dispositions. They hypothesize a "motivational genesis of the dispositions of critical thinking" (36), finding that the "motivation to think critically continues to be an important factor in the deployment of critical thinking skills, even though certain mental habits or attitudes associated with performing them have become consolidated" (36). They furthermore argue "that mental habits or attitudes come from the exercise of *motivated* skills" (37, emphasis added). For Nieto and Valenzuela this has the pedagogical implication that we should work toward "increasing the value [students] assign to critical thinking" (37).

Their study parallels my conceptual claim that the internal motivation to think critically is the primary virtue of the critical thinker, and that the habit of skillful thinking is in part born of this virtue.

In sum, Missimer's attack against the character view fails. Apart from the ways that her attack fails on its merits, which Siegel (1997) has done well to enumerate, her view is also incoherent, as she admits a minimalist position regarding the character view: that the critical thinker is a habitually skilled and enthusiastic critical thinker who wants to think critically. This conceptual connection between skills and virtues is not merely tautological and trivial, but has pragmatic significance when one considers the classroom context: student motivation to engage in critical inquiry should not be taken for granted, but stimulated if it is such an essential ingredient in habitually thinking critically. The same is true for the other virtues, none of which always come so naturally to thinkers. In addition, the evidence Missimer asks for regarding the character dimension has been supplied in the intervening years since she demanded it, from Nieto and Valenzuela (2012). In short, Missimer's skills-alone view remains an implausible conceptualization of critical thinking, the character view remains the most plausible, and is ironically suggested by her own view. This is important because it solidifies the consensus among critical thinking theorists and pedagogues that the critical thinking virtues are a constituent aspect of a person being a critical thinker. It helps my case because by singling out the importance of an enthusiasm and a desire to think critically, Missimer echoes the idea that willingness to inquire is the cardinal critical thinking virtue.

Conclusion

Critical thinking skills indicate critical thinking virtues that a person must have if she is to be a critical thinker: someone who characteristically uses her skills appropriately in efforts aiming toward the end of critical thinking, which is reasoned judgment. Critical thinking virtues such as open-mindedness, charity, and valuing fallacious-free reasoning are important, but willingness to inquire stands behind their manifestation in the ideal critical thinker. It is therefore willingness to inquire that is the fundamental critical thinking virtue. Pedagogues should not neglect teaching and modeling the other virtues in their classrooms. However, our pragmatic attempts at teaching students how to be better critical thinkers should focus on fostering willingness to inquire, without which our students will not be the kind of people who are properly motivated to use their skills in critical inquiry. People who lack this basic drive and who lack other important virtues are not the kinds of thinkers we aim to educate, and should not be called critical thinkers, whatever their intellectual skills.

Acknowledgments

For their helpful feedback on earlier drafts of this essay I would like to express my thanks to Katharina von Razdiewsky, Sharon Bailin, Andrew Pineau, Fàbio Shecaira, William Hamby, Allen Pearson, Mark Vorobej, David Hitchcock, Frank Fair, and Martin Davies.

References

Bailin, S., and Battersby, M. 2010. *Reason in the Balance: An Inquiry Approach to Critical Thinking*. Whitby, ON: McGraw-Hill Ryerson.

Blair, J. A. 1995. "Premise Adequacy." In *Proceedings of the Third ISSA Conference,* edited by F. H. v. Eemeren, R. Grootendorst, J. A. Blair, and C. A. Willard (Vol. 2). Amsterdam: Sic Sat. 191–202.

Facione, P. 1990. *The Delphi Report: Critical Thinking: A Statement of Expert Consensus for Purposes of Educational Assessment and Instruction*. Millbrae: California Academic Press.

Glaser, E. 1941/1972. *An Experiment in the Development of Critical Thinking*. New York: AMS Press.

Goodwin, J. 2013, August, 2013. *Norms of Advocacy.* Paper presented at the Virtues of Argumentation Conference: Ontario Society for the Study of Argumentation (CD-ROM), Windsor, Ontario.

Hamby, B. 2013. "Libri ad Nauseam: The Critical Thinking Textbook Glut." *Paideusis* 21 (1): 39–45.

Hare, W. 1979. *Open-Mindedness and Education*. Kingston: McGill-Queen's University Press.

Hare, W. 1993. *What Makes a Good Teacher: Reflections on Some Characteristics Central to the Educational Enterprise*. London, ON: The Althouse Press.

Johnson, R. H. 2000. *Manifest Rationality: A Pragmatic Theory of Argument*. Mahwah, NJ: Erlbaum.

Missimer, C. 1990. "Perhaps by Skill Alone." *Informal Logic* 12: 145–153.

Missimer, C. 1995. "Where's the Evidence?" *Inquiry: Critical Thinking across the Disciplines* 14: 1–18.

Nieto, A., and Valenzuela, J. 2012. "A Study of the Internal Structure of Critical Thinking Dispositions." *Inquiry: Critical Thinking across the Disciplines* 27 (1): 31–38.

Passmore, J. 1980. "On Teaching to Be Critical." In *The Philosophy of Teaching,* edited by J. Passmore. London: Duckworth.

Riggs, W. 2010. "Open-Mindedness." *Metaphilosophy* 41(1 and 2): 172–188.

Siegel, H. 1988. *Educating Reason: Rationality, Critical Thinking, and Education*. New York: Routledge.

Siegel, H. 1997. "Not by Skill Alone: The Centrality of Character to Critical Thinking." In *Rationality Redeemed: Further Dialogues on an Educational Ideal,* edited by H. Siegel New York: Routledge. 55–72.

Siegel, H. 2009. "Open-Mindedness, Critical Thinking, and Indoctrination: Homage to William Hare." *Paideusis* 18 (1): 26–34.

Part II

Teaching Critical Thinking

Assuming that one considers that critical thinking lies at, or close to, the heart of higher education, how might it be developed among students? Between them, the seven papers in this section offer answers to this question and, in so doing, reflect several of the fault lines in the debates over critical thinking.

Perhaps the dominant fault line has been that of what might be termed the "general/contextual" controversy. Are we to understand critical thinking as a set of skills or cognitive processes or states of human being that have a kind of integrity in themselves and so might be found exemplified across—and so be transferable across—all manner of settings? *Or*, on the contrary, does critical thinking only reveal itself in particular contexts and take on its character partly as a result of the influence of that particular context? This is a crucial matter insofar as higher education is concerned, giving rise to both views that critical thinking can be taught and acquired by students as a set of tasks in its own right—in specialist courses of critical thinking—*and* that critical thinking has no such general character and so cannot be taught abstractly but rather has to be acquired within the separate challenges posed by each discipline.

Several of the papers here take up this issue. In each case, a judicious mix of understanding critical thinking as having both cross-disciplinary aspects as well as disciplinary specific coloration is proposed. However, the balance of that relationship is seen in different ways.

Anna Jones reflects on substantial empirical work, including interviews with disciplinary specialists, and carefully dissects the nature of critical thinking across physics, economics, history, law, and medicine. Jones develops the view that general attributes of critical thinking can be said to exist but that they take on their character within the different disciplines. Indeed, "even within one discipline critical thinking takes many forms." Accordingly, "generalisable critical thinking is a useful foundation for disciplinary critical thinking but will not substitute for it." In contrast, on the view of Bailin and Battersby, "reasoning and argumentation are generally not a focus of disciplinary pedagogy."

They therefore look to a teaching approach that promotes "aspects of argumen-tation which transcend disciplinary boundaries." And having noted that stu-dents' critical thinking skills "are much weaker in cases of far transfer, where the context of application is very different from the context of learning," Paul Green explicitly "assume[s]...that there are general critical thinking skills." This move then allows Green to develop an argument around critical thinking as a form of and as a means of lifelong learning.

Another feature of the critical thinking debate is that of *levels* at which it might be exemplified. Jones notes that Barnett (1997) pointed out this feature of critical thinking, with the ultimate level residing in "metacritique," namely, the capacity to stand outside a framework (such as a discipline) and critically evaluate *that*. Jones further observes that "the question of whether disciplinary thinking can equip students to think beyond a disciplinary frame is a difficult one." Consequently, in order to "enable metacritique," Jones goes on to suggest "a curriculum model that allows for multi- or interdisciplinary study." Thomas and Lok also deploy the idea of levels, picking up on Ramsden's (1992) idea of qualitatively different levels of learning, wherein thinking critically is situated at the most abstract level. For Thomas and Lok, critical thinking has three dimensions or "thematic groups," namely cognitive skills, dispositions, and knowledge. This matrix—of levels and thematic groups—provides the basis for Thomas and Lok to go on to suggest that critical thinking resides in "rising levels of intellectual and ethical competence, and self-awareness," involving not merely a "doing component" but a "being component" as well.

Wendland, Robinson, and Williams also draw on the metaphor of levels (or "stages"), in pointing to critical thinking as an ability to handle evermore com-plex understandings of the world. They begin from a concern that the use of debates as a pedagogical technique may emphasize dualistic thinking and look to what they term "thick critical thinking." In such a form of thinking, the student will progress above and beyond a simplistic form of critical thinking to gain an appreciation of the complexities of situations and their associated debates. Students will form a sense not merely of differing views as such but will gain a nuanced sense of their historical development, their context, and their "dialectic," as such views contend with each other. Here, students come to see into argumentative positions, "examining what is at risk for each side." The ultimate pedagogical move is to bring students to position themselves in a debate and "to articulate their own positions." The metaphor of levels is per-sistent here for "the entire process is cumulative, each step building on the one before it."

Present in all of the papers in this section are insights into ways in which critical thinking offers educational value. Just some of the suggested educa-tional benefits that we glimpse here are the following: a sense of students being able to form reasoned judgments (van Gelder; Wendland, Robinson, and Williams); the development of capacities of critical self-reflection and metacognition (Green; Thomas and Lok); a disciplining of one's thinking (Llano); and the ability to live with contingency, complexity, and "multiplicity" (Llano; Wendland, Robinson, and Williams; Thomas and Lok). Perhaps it is this last that, more than any other virtue, stands out as a feature of critical thinking in higher education. For, at its best, critical thinking might just enable students to grow into a way of engaging with the world that Sinclair Goodlad once termed "authoritative uncertainty." Critical thinking can offer students resources for recognizing "multiplicities" in the world and yet not to be paralyzed into inaction. Rather, students can emerge from such a higher education being able to think deeply in the presence of such intractable complexities and form their own beliefs and actions, which they can buttress with their own reasoned arguments.

5
Teaching Critical Thinking: An Operational Framework

Keith Thomas and Beatrice Lok

Introduction

Critical thinking has attracted considerable focus in recent years in both high school and tertiary curricula (Ku 2009). While the concept has a long tradition and both philosophers and educators agree on its importance, there is a reported lack of agreement on what the concept involves (Cheung, Rudowicz, Kwan, and Yue 2002; Green, Hammer, and Star 2009; McMillan 1987). There is even less agreement on how to teach it (Noddings 1995). Described as a most difficult term in education (Moore and Parker 2011), it is not surprising that critical thinking has also taken on a narrower focus than when earlier conceived (Davies 2011). This chapter presents an operational framework for teaching critical thinking and illustrates its application in an educational setting.

The need for conceptual clarity

Concerned as critical thinking is with discerning and recognizing faulty arguments, generalizations, assertions, and the like, teachers and academics can have a practical understanding of the many skills associated with this cognitive capability. However, given the lack of agreement on what critical thinking involves, their conceptual understanding can be less clear. This lack of clarity is a problem, as the way we view critical thinking has a major bearing on curriculum design and on the educational approaches adopted (Barnett 1997; Barrie and Prosser 2004). Consequently, as others have suggested, there is a need for conceptual clarity (Green, Hammer, and Star 2009). The following section outlines the process used to identify the attributes and processes discussed in the literature.

Process adopted

While described as a purposeful and goal-directed activity that involves reasoned and reflective thought processes (Halpern 2003), what are the important
attributes associated with this complex capability? Three broad approaches to
understanding critical thinking are evident in literature: philosophy, psychology, and education (Lai 2011). Based on a review of this literature, we initially
identified and sorted the common attributes associated with critical thinking
into thematic groups or components. Tables 5.1, 5.2, and 5.3 present a consolidated summary of processes and attributes associated with critical thinking
skills, disposition, and knowledge, respectively (where a √ indicates the process
or attribute mentioned).

Fleshing out this material sourced from literature into a conceptual framework was an iterative process. The first stage involved identifying and synthesizing the many descriptive elements in literature. Terms and attributes
highlighted in each description were coded and analyzed progressively, with
some terms merged when their described meanings were regarded to be very
similar. It was clear from the outset that much of the literature described critical thinking from the perspective of skills or as a combination of disposition
and skills. A third large thematic group was initially called "other items." It was
subsequently labeled knowledge. This third group encompassed intellectual
development (Kurfiss 1988)—an important determining factor to thinking
critically—as well as general and context-specific knowledge, and experience
that itself spans items such as experiential learning, life experience, the age of
thinker, personal exposure, and maturity.

The follow-on stage in the process toward devising a conceptual framework was to recode all descriptive attributes and items identified based on
the three thematic groups identified, with subset labels used to categorize
items that looked interesting or that appeared repeatedly. This process served
to confirm the thematic labels and articulate the primary subset items illustrated within each thematic group. Importantly, the process also revealed
the seemingly limited attention given to critical thinking performance. The
next section outlines and discusses the operational framework. It is followed
by a review of some challenges to performance, that is, to being a consistent
and effective critical thinker in actual practice, in either the classroom or the
workplace.

Conceptualizing an operational framework

Figure 5.1 is a conceptualization of critical thinking attributes. Conceived as
three interconnected sets of attributes, critical thinking can be described as

Table 5.1 Summary of functional attributes comprising critical thinking skills

Author(s)	Interpretation	Explanation	Analysis	Inference	Evaluation	Self-regulation
J. Dewey (1909)		√		√		√
E. Glaser (1941)	√	√	√	√		√
J. Sternberg (1986)	√		√	√	√	√
S. Brookfield (1987)				√	√	
B. Beyer (1988)			√	√	√	√
R. Ennis (1989)			√	√	√	
Delphi report (1990)	√	√	√	√	√	√
J. Chaffee (1992)	√		√	√	√	
R. Paul (1993)			√	√	√	√
A. Freeley (1993)			√	√	√	√
Fischer and Scriven (1997)	√			√	√	√
Pithers and Soden (2000)	√	√		√		
A. Fisher (2001)		√		√	√	
Watson and Glaser (2002)	√			√	√	
Simpson and Courtney (2002)			√	√	√	√
P. Facione (2009)	√	√	√	√	√	√

Table 5.2   Summary of dispositions associated with critical thinking

Selected scholar	Brookfield (1987)	Costa (1991)	Paul (1993)	Ennis (1994)	Perkins, Jay, et al. (1994)	Facione et al. (1995)	Banning (2006)
Disposition items: Is clear about the intended meaning				√			√
Is systematic		√	√	√	√	√	√
Takes the total situation into account				√			√
Is analytical	√		√	√	√	√	√
Is inquisitive	√		√	√	√	√	√
Looks for alternatives	√			√			
Seeks precision as the situation requires	√			√			√
Is conscious (aware)		√	√	√			
Is open-minded		√	√	√		√	
Is truth-seeking	√		√	√		√	√
Uses one's critical thinking abilities				√			
Is intellectually careful				√	√		
Is metacognitive	√	√	√	√	√		
Seeks efficacy		√					
Is self-confident	√		√			√	√
Shows maturity			√			√	

a composite of certain skills, dispositions, and knowledge. Illustrative subset items associated with each theme are also identified. This is not to suggest that these items define the respective groups exclusively. Rather, teachers and practitioners should see these items as needing to be affirmed and/

Table 5.3 Summary of elements comprising knowledge

Selected scholars / Knowledge elements	Kurfiss (1988)	Thoma (1993)	Noddings (1995)	Kimmel (1995)	Barnett (1997)	Pithers and Soden (2000)	Moore et al. (2003)	Halpern (2003)	Ruggiero (2004)	Ramsden (2007)	Paul & Elder (2012)
Background knowledge			✓								
Communicate effectively									✓		
Critical reflection			✓		✓					✓	
Cultural knowledge								✓	✓		
Disciplinary knowledge							✓	✓			✓
Ethical competence		✓	✓								
General knowledge	✓		✓		✓	✓		✓	✓	✓	✓
Intellectual development	✓	✓		✓	✓	✓				✓	
Life experience					✓					✓	✓
Political knowledge						✓	✓				
Reflective awareness					✓			✓	✓	✓	

or supplemented as appropriate to their specific learning and disciplinary context.

Before discussing each thematic group, we review this framework in the broader context of student learning in higher education. Described as a series of qualitatively different levels (Ramsden 1992), at its most abstract, learning is said to involve general abilities and personal qualities such as being able "to think critically" and "communicate effectively." A second level involves content-based understandings related to specific disciplines or professions. A third and final level involves categorical proficiencies—of factual information, technical skills, and problem-solving techniques. Assuming an educational objective "to analyze ideas or issues critically," as Ramsden (1992) noted, students need to acquire knowledge at each level and be able to connect knowledge at each level to the other levels. This integrative aspect to learning is important in higher education that has as its central purpose the imaginative acquisition of knowledge (Ramsden 1992, citing Whitehead). Moreover, to enable Whitehead's suggested need for creative dissent over uncritical acceptance of orthodoxy (or the modern variations of the same sentiment—the ability to look at problems from different perspectives and respond flexibly to changing circumstances), there is an evident interdependence in the component attributes and processes identified. For critical dissent, as an example, there is a clear need for some relevant knowledge, as well as self-regulation (shown in figure 5.1 as a skill) and the willingness in attitude to offer critique (a dispositional element).

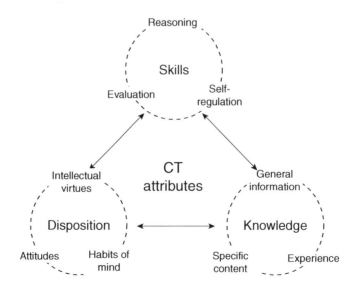

Figure 5.1 An operational framework.

A brief elaboration of each thematic group and its relationship with the other component groups follows.

Skills and critical thinking

Critical thinking involves a number of cognitive skills. Although there are varying descriptions, we suggest there is a general consensus around three composite subset skills. The first is reasoning, which comprises the ability to identify and explore evidence using particular generic methods such as reading and discussion, as well inference and explanation. A second skill set involves evaluation, which comprises the skills of interpretation and analysis, including technical analysis using tools appropriate to the discipline (Ennis 1992). A third skill set involves the capacity for reflection or self-regulation that supports a "disciplined" ability to seek out knowledge and evidence (Ikuenobe 2001), as well as minimize assumptions and biases and see weaknesses even in one's own thinking (Ruggiero 2004).

Disposition and critical thinking

Disposition implies a willingness to do something given certain conditions (Ennis 1992). Personal disposition is an implicit component that can either enhance or hinder critical thinking and while literature tends toward positive dispositions as a key determinant of performance, some disagreements between the normative and laudatory roles of disposition in defining critical thinking are also noted (Lai 2011). Three main subset dispositions are identified: attitudes such as being open-minded and fair-minded; intellectual virtues such as truth seeking and curiosity; and habits of mind that include cultural- or trait-induced bias and the tendency toward black-and-white (dichotomous) thinking. Linking skills and disposition, studies suggest that being positively disposed toward critical thinking does not ensure that a person will be skilled. Equally, being skilled does not ensure that a person is disposed to use critical thinking (Facione, Facione, and Giancarlo 2000). For these reasons, as Facione et al. say, school curricula and educational outcome assessments need to focus on both developing skills and positive dispositions.

Knowledge and critical thinking

The implicit connection between knowledge and critical thinking is embedded in most literature. For example, as various authors have noted, knowledge and thinking are closely related (Bereiter 2002; Pithers and Soden 2000). Similarly, as Halpern noted, it is clear that knowledge is acquired through thinking, while the process of thinking requires knowledge (Halpern 2003). In effect, critical thinking is not possible without knowledge, and intellectual maturity is a prerequisite for critical thinking (Kurfiss 1988). These remarks in turn identify the strong enabling role of education in developing critical thinking (Kimmel 1995), while experience linked to knowledge is reported to determine the overall quality of reasoning (Paul and Elder 2012). Three categories of knowledge are identified: general information and basic facts to enable valid evaluation; specific content-based knowledge related to discipline-specific and contextual information; and experience, which includes intellectual development and knowledge gained from life and work experiences.

Applying theory to practice

Critical thinking ability is demonstrated by a rising level of intellectual and ethical competence, and by a shift from black-and-white views to multiplicity (Thoma 1993). There is, however, somewhat less discussion of performance in terms of the disciplined ability for evaluative reasoning and self-awareness in practice. The point is skills (and knowledge) acquisition is necessary, but not sufficient. Rather, consistent application of skills and knowledge when forming judgments and making decisions, to the problems of everyday life (Noddings 1995), is also required.

Assuming intellectual maturity, there are a number of evident challenges to applying critical thinking in practice. One challenge is the reality that while people are generally well able to criticize propositions they oppose, what is difficult is to think critically about one's own favored ideas. This ability, described as being a strong-sense thinker (Paul and Elder 2012), can be acquired by getting people to simply "do it" according to Bereiter (2002). The reality though is that dispositional and practical issues, such as cultural- and trait-induced bias, can get in the way. Another challenge is in the successful transfer of learning, because behavior can be bound to situations in which it was learned (Bereiter 2002). For teachers, on the one hand, the task is to enable learning at an abstract enough level for behavior to be transferred. On the other hand, learning must also be assimilated into the way individuals respond to new situations. Thus, for the learner, it is a matter of developing self-awareness—of knowing themselves and their learning preferences, and of knowing their biases and shortcomings. It will require engaged partnerships between the student and the institution (Ramsden 2008), in order to foster these metacognitive skills that are also necessary for lifelong learning.

Human (behavioral) factors and, particularly, the application of knowledge in collective activity are a somewhat more complex challenge. We know, for example, that decision making cannot be divorced from context, and situational factors such as time pressure, fatigue, and stress can moderate effective reasoning (Kahneman 2002). Similarly, other studies show that cultural beliefs and inborn traits can support or stifle performance (Lunney 2013). In sum, the capacity for reason and self-regulation invites a corresponding task for teachers and students—to develop the ability to ignore subjective anomalies (bias) associated with uncertainty or incomplete information, while looking also to moderate the effects of other situational influences.

Given this brief exploration of the challenges to applying critical thinking in practice, we turn our attention to supporting critical thinking performance. The issue is of some importance, given the evidence that some 45% of students (in US-based colleges) showed no significant gains in critical thinking, analytical reasoning, and problem solving after their first two years (Arum and Roska 2011). There is a clear need for greater academic rigor in teaching critical thinking.

Supporting critical thinking performance

While the ability to communicate and persuade by structure of argument and quality of reasoning is important (Ruggiero 2004), what of "disciplined" performance and what is staged development of this capability? Performance is disciplined because critical thinking requires effort and self-discipline. It is staged (in development) because critical thinking ability involves rising levels of intellectual and ethical competence and self-awareness. Effective performance is founded on disposition, practice, and time. Being a purposeful and goal-directed activity, it involves a *Doing* component. Incorporating an internalized ability for reflective awareness, it also involves a less tangible internal journey of "becoming" a critical thinker—a *Being* component.

In the context of addressing everyday problems, performance will vary across a situational continuum from routine to novel. Typically, familiar or routine actions will be performed with minimal attention, the most familiar actions perhaps automatically. Novel or complex actions, however, will require a more reasoned approach wherein critical thinking comes to the fore. This behavioral tendency has been modeled using three qualitatively different performance levels—skills-based, rule-based and knowledge-based levels (Reason 2000). These levels are somewhat of an industry (safety and risk) standard in aviation, nuclear power generation, and oil exploration and production. Assuming the purpose of critical thinking is to support the quality of reasoning and subsequent judgments, and not to determine the most effective action, these performance levels can be incorporated to support performance.

The simplest form of behavior is skills based. Skilled performances (SP) are highly integrated, automatic routines that deal with the familiar and non-problematic. They develop through experience. The next level of complexity, rule-based performance (RP) is based on subroutines that follow stored rules: if X (situation) do Y (action rule). Acquired through training or experience, rule-based performances apply in less routine situations, when automatic behavior must be modified. Finally, knowledge-based behavior is what you turn to when SP and RP are not adequate. Knowledge performances (KP) deal with unfamiliar or novel conditions such as complex fault diagnosis. In KP mode, tasks will be performed in a conscious manner, although the process is likely to also be slow and error prone.

Table 5.4 is an illustrative checklist based on items in the operational framework and applied across the three performance levels and disposition. This checklist is constructed for use by students participating in a group-based project on a graduate program. In this activity, students are invited to self-assess on a scale of 1–4 across four progressive levels *"naïve, novice, apprentice and master"* (emphasis added, Wiske 1998, 172). The open columns allow for up to five students to participate in completion of a project. The aim is to build self-awareness and self-monitoring capacity. The checklist can be tailored into an evaluative rubric using an appropriate scoring scale to suit disciplinary specific considerations in reasoning and problem solving. In an academic context, the Blooms or SOLO (Biggs's structure of observed learning outcomes) taxonomies can be substituted, as preferred.

Applying these performance levels can be best understood by reference to a familiar activity such as driving a car. All three performance levels are engaged at the same time, as required. Steering is an automated skill-based (SP) activity developed through experience; managing space (distance) between moving vehicles and negotiating traffic lights are largely rule-based (RP) matters. However, driving in an unfamiliar environment will involve skill and rules, as well as some necessary improvisation and reference to a body of knowledge (KP), perhaps gained from experience. As lessons from industry suggest, knowledge or training cannot immunize a person or system against performance error. Hence, the focus is on making errors less probable and their consequences less severe. For critical thinking, it suggests inquiry-based practices and other systemic solutions to support practice and desired performances, while looking also to minimize the number and gravity of violations.

Assuming a satisfactory learning environment, the operational framework (figure 5.1) can be used to support development of critical thinking ability and the application of this ability in disciplined performances over a course or across the duration of a study program. High-leverage thinking processes in a classroom context include inviting students to: observe and describe what's there, build explanations, reason with evidence, consider different viewpoints and perspectives, capture the core concept, ask questions, and generally go below the surface to uncover complexity. This short list of processes is a starting point for mapping activities and learning outcomes for the respective thematic groups identified in the operational framework. Once learning outcomes are identified, teaching staff can select learning activities suited to developing functional attributes across the performance levels discussed earlier. For example, learning journals can encourage reflection (Costa 2001), in-class discussion can support reasoning skills, and a problem- or inquiry-based learning

Table 5.4 Checklist for the three performance levels and disposition (requires a scoring scale ideally attached as an appendix)

Add reflective comment: (e.g., outline group study process and any insights)					
Critical thinking performance items	**Insert name**	**Insert name**	**Insert name**	**Insert name**	**Insert name**
Disposition:					
1. Displays consistent motivation					
2. Is inquisitive					
3. Is open-minded					
4. Is truth-seeking					
Knowledge-based performance:					
1. Applies discipline content					
2. Avoids assumptions (or ambiguity)					
3. Considers rival causes					
4. Is able to improvise					
5. Ensures validity of conclusion					
6. Thinks systemically (not event based)					
Rules-based performance:					
1. Quality of evidence (source)					
2. Applies appropriate techniques					
3. Avoids generalizations					
4. Able to problem solve known issues					
Skills-based performance:					
Thinks reflectively					
1. Questions					
2. Self-questions					
3. Link ideas to previous experiences					
4. Accepts and learns from feedback					
5. Manages biases					
Problem solving					
1. Generates ideas					
2. Generates solutions					
3. Cooperates effectively					
4. Avoids reasoning fallacies					
Thinks critically/metacognitively					
1. Formulates questions for inquiry					
2. Can sort fact from opinion					
3. Can justify opinions					
4. Shows originality in thinking					
5. Can self-assess (aware of own biases)					
6. Thinks about personal learning/ thinking					

approach can help foster all three critical thinking attributes (Tiwari, Lau, So, and Yuen 2006).

Conclusion

The operational framework conceptualizes critical thinking as three interconnected composite attributes. The challenge of applying this capability is discussed and the chapter incorporates three performance levels from industry to support effective performance over time. The framework is suited for use in general education and discipline-specific subjects, as well as in workplace-based professional development activity. Academic staff can use it to identify learning outcomes and take a proactive role in curriculum development. The sample checklist is intended to assist students to self-assess. It can also be used by teachers' to focus on critical thinking development across the three performance levels (SP, RP, and KP) and disposition. From a research perspective, this framework is a useful basis to explore extant conceptions of critical thinking and perhaps also to examine university practices in teaching and learning. The limitation of the framework is that it is conceptual and generic. It needs examination in specific subjects and disciplines to confirm its utility. However, the checklist is a useful starting point for teachers and academics seeking to connect theory with practice and so support the development of critical thinking that is useful in addressing the problems of everyday life.

References

Arum, R., and Roksa, J. 2011. *Academically Adrift: Limited Learning on College Campuses.* 3 vols. Chicago: University of Chicago.
Barnett, R. 1997. *Higher Education: A Critical Business.* Buckingham: SHRE & Open University.
Barrie, S. C., and Prosser, M. 2004. "Generic Graduate Attributes: Citizens for an Uncertain Future." *Higher Education Research & Development* 23 (3): 243–246.
Bereiter, C. 2002. *Education and Mind in the Knowledge Age.* Mahwah, NJ: Lawrence Erlbaum Associates.
Cheung, C., Rudowicz, E., Kwan, A., and Yue, X. 2002. "Assessing University Students' General and Specific Critical Thinking." *College Student Journal* 36.
Costa, A. 2001. *Developing Minds: A Resource Book for Teaching Thinking* (third edition). Alexandria, VA: Association for Supervision and Curriculum Development.
Davies, M. 2011. "Mind Mapping, Concept Mapping, Argument Mapping: What Are the Differences and Do They Matter?" *Higher Education* 62 (3): 279–301.
Ennis, R. 1992. "Critical Thinking and Subject Specificity: Clarification and Needed Research." *Educational Researcher* 18 (3): 4–10.
Facione, P. A., Facione, N. C., and Giancarlo, C. A. 2000. "The Disposition Towards Critical Thinking: Its Character, Measurement and Relationship to Critical Thinking Skill." *Informal Logic* 20 (1): 61–84.
Green, W., Hammer, S., and Star, C. 2009. "Facing up to the Challenge: Why Is It So Hard to Develop Graduate Attributes?" *Higher Education Research and Development* 28 (1): 1–29.

Halpern, D. F. 2003. *Thought and Knowledge: An Introduction to Critical Thinking.* Edited by Lawrence Erlbaum Associates. New Jersey.

Ikuenobe, P. 2001. "Teaching and Assessing Critical Thinking Abilities as Outcomes in an Informal Logic Course." *Teaching in Higher Education* 6 (1): 19–35.

Kahneman, D. 2002. "Maps of Bounded Rationality: A Perspective on Intuitive Judgement and Choice."January 26, 2014. New Jersey. http://wisopsy.uni-koeln.de/uploads/media/kahnemann_Nobelpreisrede_20.pdf.

Kimmel, P. 1995. "A Framework for Incorporating Critical Thinking into Accounting Education." *Journal of Accounting Education* 13 (3): 299–318.

Ku, K. 2009. "Assessing Students' Critical Thinking Performance: Urging for Measures Using Multi-Response Format." *Thinking Skills and Creativity* 4: 70–76.

Kurfiss, J. 1988. "Critical Thinking: Theory, Research, Practice, and Possibilities." *ASHE-ERIC Higher Education Report No.2.* Washington, DC.

Lai, E. 2011. *Critical Thinking: A Literature Review Research Report.* Hong Kong.

Lunney, M. 2013. *Critical Thinking to Achieve Positive Health Outcomes.* Edited by Nursing Case Studies and Analyses. Ames, IA: John Wiley & Sons.

McMillan, J. H. 1987. "Enhancing College Students' Critical Thinking: A Review of Studies." *Research in Higher Education* 26 (1): 3–29.

Moore, B., and Parker, R. 2011. *Critical Thinking* (eleventh edition). New York: McGraw-Hill.

Noddings, N. 1995. *Philosophy of Education.* Boulder, CO: Westview Press.

Paul, R. W., and Elder, L. 2012. *Critical Thinking: Tools for Taking Charge of Your Learning & Your Life.* (third edition). Boston: Pearson Education.

Pithers, R., and Soden, R. 2000. "Critical Thinking in Education: A Review." *Educational Researcher* 42 (3): 237–249.

Ramsden, P. 1992. *Learning to Teach in Higher Education.* London: Routledge.

Ramsden, P. 2010. *The Future of Higher Education: Teaching and the Student Experience* Gla. ac.uK 2008 [March 7, 2010].

Reason, J. 2000. "Human Error: Models and Management." *British Management Journal* 320 (7237): 768–770.

Ruggiero, V. R. 2004. *The Art of Thinking: A Guide to Critical and Creative Thought.* New York: Pearson.

Thoma, G. A. 1993. "The Perry Framework and Tactics for Teaching Critical Thinking in Economics." *The Journal of Economic Education* 24 (Spring): 128–136.

Tiwari, A., Lau, P., So, M., and Yuen, K. H. 2006. "A Comparison of the Effects of Problem-Based Learning and Lecturing on the Development of Students Critical Thinking." *Medical Education* 40: 547–554.

Wiske, M. S. 1998. *Teaching for Understanding.* San Francisco, CA: Jossey-Bass.

6
Teaching Critical Thinking for Lifelong Learning
Paul Green

Introduction

The task of improving critical thinking skills in students at American universities has typically been addressed by offering a stand-alone course in critical thinking, often taught by philosophy or psychology departments. In this paper I argue that critical thinking courses of this type need to be substantially redesigned before they can meet the appropriate critical thinking learning goals for their students. I begin my argument by noting that students in American higher education are currently not improving their critical thinking skills. The solution, I argue, will not lie in better pedagogy; even in the best-taught classes students will not achieve the level of critical thinking that is necessary. Instead, we must redesign our courses to help students become lifelong learners who can then continue the task of improving their critical thinking skills long after the course ends. I discuss in some detail some of the more significant ways courses will need to be redesigned to achieve these student learning outcomes.

The problem identified

For over three decades, the vast majority of faculty across American higher education have identified developing critical thinking skills in their students as one of their top pedagogical priorities (Paul 2004). For example, "a 1972 study of 40,000 faculty members by the American Council on Education found that 97 percent of the respondents indicated the most important goal of undergraduate education is to foster students' ability to think critically" (Paul 2004). Thirty-seven years later, a nationwide study by the Higher Education Research Institute reported that 99% of college faculty "say that developing students' ability to think critically is a 'very important' or 'essential' goal of undergraduate education" (Arum and Roksa 2011, 25).

There is, of course, some disagreement about what critical thinking is. My own preferred account connects it with the capacity to make good decisions based on relevant evidence. But I think my argument in this essay will also apply to those who accept any of the following definitions (all cited in Hatcher 2000):

1. "reasonable, reflective thinking about what to do and believe"(Ennis 1962),
2. "being appropriately moved by reasons" (Siegel 1988),
3. "skillful, responsible thinking that facilitates good judgment because it 1) relies on criteria, 2) is self-correcting and 3) is sensitive to context" (Lipman 1988), and
4. "thinking that attempts to arrive at a judgment only after honestly evaluating alternatives with respect to available evidence and arguments" (Hatcher and Spencer 2000).

In addition to subscribing to critical thinking as a goal of undergraduate education, most faculty members report they aspire to teach critical thinking skills in their own courses (Paul 2004). However, rather than encourage widespread instruction in critical thinking across the curriculum (similar to the "writing across the curriculum" movement), at the institutional level colleges and universities seem to have focused on requiring undergraduates to pass a stand-alone course in critical thinking, usually taught by faculty from the philosophy or (less often) psychology departments. (See Lipman [2003] for a survey of the history of critical thinking courses in higher education.) Though there does not seem to have been any systematic study on the extent to which stand-alone courses in critical thinking are required, my own unscientific survey of forty private liberal arts colleges and small universities shows that 80% of them have adopted this requirement.

The content varies across different critical thinking courses offered by philosophy departments though my sense is that they usually emphasize argumentation and informal logic—but it is fair to say that they all have the same primary learning objective. They—really, *we,* since I am one of these instructors—want their students to become better critical thinkers. This does not mean that instructors do not differ about how to accomplish this objective, but it does mean that the success of our course is determined by whether our students become better critical thinkers.

If that is the case, then there is powerful evidence that we are not succeeding. There are several disturbing findings that support this conclusion:

1. The most prominent critique in recent years is the widely discussed *Academically Adrift* (Arum and Roksa 2011). One of their most startling conclusions was that the students in the study had an average gain in critical

thinking skills of only .18 of a standard deviation from the beginning of their freshman year to the end of their sophomore year. Another (more intuitively dramatic) way of putting it: 45% of students have no statistically significant improvement in their critical thinking skills after two years of college.

This is especially worrisome because Arum and Roksa (2011) probably *understate* the problem in two ways. First, their conclusions are based upon a study of about 2,300 students at 24 four-year institutions who took the Collegiate Learning Assessment (CLA) as first-year students in the fall of 2005 and again in the spring of 2007 at the end of their second year. Thus, students who left the university sometime during their first two years are not represented in the study. Yet presumably a good portion of these students left because of academic difficulty, and thus were even less likely to have improved the critical thinking ability over the course of two years than their peers who were retained by the institution.

The second reason Arum and Roksa's findings may understate the lack of improvement in critical thinking has to do with the way the CLA is scored. The CLA administered for the study is a 90-minute test "in which students must write an argumentative proposal for solving an issue presented in the prompt, using various resources they have access to onscreen" (Possin 2008). Though this instrument design makes for an "impressive" test, which "most closely matches the open-ended test of students' CT skills in solving real-life's fuzzy problems which many are looking for," there is some evidence that, when a student's test is scored, the scorers give credit for any rationale for action in the student's answers, rather than a rationale that actually justifies the student's choice of action (Possin 2008). If critical thinking requires the production of *good* reasons—rather than just the production of *any* reasons—the CLA would give to at least some students scores higher than they deserve.

2. Despite the reams of data collected for the study in *Academically Adrift*, nowhere are the students asked if they ever enrolled in a critical thinking course. So it is possible that the 45% of students who have failed to improve their critical thinking are students who have not taken a class in critical thinking. And indeed there is some evidence that, contrary to the general failure to thrive documented in *Academically Adrift*, students who take stand-alone critical thinking courses do improve their critical thinking skills (Possin 2008). What evidence we have indicates that dedicated critical thinking classes with computer-assisted instruction are most effective at increasing student learning, with reported effect sizes of about .50 and up to .79 (Possin 2008).

What does this mean? Well, any educational intervention is considered *effective* if it has an effect size of .50 and *extremely effective* if the effect size is above .80 (Marzano 2001). These effect sizes translate into percentile gains of 19 and

29 points respectively (Marzano nd). So the most effective reported percentile gain for stand-alone critical thinking courses is about 29 points. This means that a student who at the beginning of the course ranked in the 50th percentile of her peers in critical thinking ability would, at the end of the course, stand at most in the 79th percentile. In other words, in very rough terms, a student performing acceptably (i.e., one who would earn a letter grade of "C" on the traditional American A-B-C-D-F scale, assuming a normal distribution) would now be a good student (i.e., now earning a B+).

Admittedly, this is remarkable improvement within a semester-length class, and any pedagogy that is responsible for this level of improvement merits its description as "extremely effective." But is it enough? That is, if our students who were performing at an acceptable (i.e., C) level have achieved a very good level of expertise (i.e., B+) in critical thinking at the end of the course, are we satisfied that they have achieved all they need as critical thinkers?

I would argue that we should not be satisfied. When I think about my students who are very good—even my excellent students—I do not think they have the critical thinking skills that they need to sustain them for the rest of their life. I think they have made a good start—that they have mastered the basics of an effective approach to use in trying to decide whether they should believe a claim. But they are nowhere near as proficient as I would wish them to be.

Put more formally, students from my Critical Thinking class become fairly proficient at near transfer, that is, at applying their skills learned in one context (i.e., class) to another, similar context (i.e., exams). But their skills are much weaker in cases of far transfer, where the context of application is very different from the context of learning. For example, my students do well on sections of exams asking them to identify whether a text is an argument, averaging about 80%. But, in a far transfer assignment, where I ask them to find arguments in the mass media, they have less than a 50% success rate.

(The problems of transfer are well discussed in Ambrose, Bridges, DiPietro, Lovett and Norman [2010, 108–112]. For an important argument that, in effect, far transfer is impossible because there are no general critical thinking skills to apply across different contexts, see McPeck [1981]. In this paper I will assume *contra* McPeck that there are general critical thinking skills.)

Given that we want them to use these newly acquired critical thinking skills for the rest of their lives, we faculty are most interested in far transfer. So the probable failure of students in Critical Thinking classes to achieve this level of expertise severely undermines the ultimate effectiveness of these courses; even though the students are making great progress in the class, a one-semester course is just not enough time for them to develop far transfer of their critical thinking skills. As we see in the next section, this is because improving critical thinking skills to the level they need to be is very hard.

Why is it difficult to improve critical thinking skills?

Why is it so hard to improve critical thinking skills? One potentially helpful answer is identified in the research summarized by Kahneman (2011). (See also Gladwell [2005].) What we call "thinking"—that is, processing information and making decisions—can be conceptualized as two very different activities using two very different mechanisms. Kahneman (2011) calls these two mechanisms "System 1" and "System 2." (For Kahneman, System 1 and System 2 are "fictitious characters," [408] articulated as shorthand indicators of two different types of processing.) System 1 is an intuitive "system that operates automatically and quickly, with little or no effort," while System 2 is devoted to "effortful mental activities…often associated with the subjective experience of agency, choice, and concentration" (20–21). System 2 is the "conscious, reasoning self that has beliefs, makes choices, and decides what to think about and what to do" (21).

Tasks characteristic of System 1 include the following (2011, 21):

1. detecting that one object is more distant than another,
2. making a "disgust face" when shown a horrible picture,
3. noticing the anger in a person's voice,
4. answering a simple math problem, such as "What is 2+2?", or
5. understanding simple sentences.

Also included in System 1 are tasks that are automatic for experts (22)—for example, when a grandmaster sees a good move in a chess game, when you drive your usual route home, or, in an example discussed at length in Gladwell (2005), when art experts detected that an ancient Greek statue was in reality a forgery because it just looked wrong.

Examples of System 2 activities are (22):

1. focusing attention on the words of your companion in a crowded bar,
2. monitoring your behavior when speaking with the president of your college,
3. searching your memory to remember the name of a former student,
4. counting the occurrences of the letter *a* in a page of text,
5. filling out a college application, or
6. performing a more complicated math problem, such as "What is 25 × 18?"

In general, both systems are active whenever we are conscious. But System 1 runs continuously and actively, while System 2 is typically in a "low-effort mode" receiving input from System 1—"impressions, intuitions, intentions, and feelings" (Kahneman 2011, 24). Most of the time System 2 accepts the input and generates beliefs and actions based upon them.

It is only when System 1 is aware of difficulty that it calls on System 2 for more active and detailed processing to solve the problem. Thus, for example, if System 1 fails to generate an answer as when you are asked to multiply 25×18), System 2 kicks in with its multiplication algorithm. Or, if the data do not fit the model of the world that System 1 expects—such as when we are surprised— then System 2 is also invoked (Kahneman 2011, 24).

There are clear benefits to this division of labor, most obviously in terms of efficiency of performance. Kahneman (2011) emphasizes that usually this dual processing mode works well because System 1 is (perhaps surprisingly) quite reliable in ordinary, familiar circumstances.

Critical thinking, with its emphasis on reflection, deliberation, and careful thought, is clearly a System 2 activity. So when we talk about teaching critical thinking, we are talking about getting students to engage in System 2 activity. But getting anyone—including students—to engage in System 2 activity is a challenge for three reasons.

First, as we have seen, System 1 does most of our thinking, or, in Kahneman's memorable phrase, "System 2 is lazy" (2011, 21). As Willingham (2009) puts it, "The brain is not designed for thinking. It's designed to save you from having to think, because the brain is actually not very good at thinking. Thinking is slow and unreliable" (3). So our brain's default setting is to avoid thinking criti- cally; the need for critical thinking is only invoked when System 1 is at a loss. Thus, when we try to teach students to be critical thinkers even under normal circumstances—that is, when we try to enlarge the sphere of critical thinking instead of shrinking it—we are working against the way human brains nor- mally operate.

(I want to be clear here: Kahneman's claim that System 2 is lazy is a metaphor for the circumstances under which System 2 actively engages. It is not a descrip- tion of the students whom we teach in our critical thinking classes; in my expe- rience, they are, on the whole, engaged, curious, and hopeful about learning. Thus, Kahneman's point might be less tendentiously expressed as "System 2 normally only engages when invoked by perceived failures of System 1.")

Second, System 1 provides much of the data that System 2 uses in its reflec- tive decision making. Thus, if System 1 is inaccurate, System 2 is unlikely to yield an optimal result. ("Garbage in, garbage out.") Furthermore, System 1 has systematic biases, which it is blind to—biases that can seriously undermine its effective operation and that are fundamentally ineliminable if System 1 is to maintain its efficiency.

For example, System 1 suffers badly from the "halo effect" where first impres- sions substantially influence later impressions. To illustrate: In a classic experi- ment by Solomon Asch—reported in Kahneman (2011, 82)—participants were asked to evaluate Alan (who was described as "intelligent, industrious, impul- sive, critical, stubborn, and envious") and Ben (who was described as "envious, stubborn, critical, impulsive, industrious, and intelligent"—i.e., as having the same characteristics as Alan, but listed in reverse order). Most viewed Alan much more positively than Ben, even though they have the same personal qualities. Misleading though the halo effect may sometimes be, Kahneman posits that it is obviously efficient; once the first impression establishes a baseline evaluation, we only need to use later impressions to modify the baseline, not to establish a new evaluation from scratch each time we encounter new information about a subject.

Third, System 2 is itself not immune to error, even when used carefully and purposefully. System 2 relies on heuristics that are generally accurate, but very poor in certain important areas. Most prominently, System 2 thinks narratively and causally, so it cannot think statistically. Thus, stochastic reasoning is very difficult to develop (Kahneman 2011).

For all these reasons, then, it is unlikely that meeting for three hours a week over a fifteen-week semester would allow students to become proficient at all the high-level skills required for critical thinking (described in Ennis [1991]). There just would not be enough time to practice the skills to attain sufficient mastery.

Three potential solutions

So what is to be done? As far as I can see, there are three ways to solve this problem in the American system of higher education.

The first way is the most straightforward. If the problem is that requiring *one* critical thinking class is not enough to help students sufficiently develop their critical thinking skills, then we should require *more* critical thinking courses— say, one for each semester of college. Thus, students would typically have eight semesters to grow as critical thinkers, accumulating 24 credits in critical thinking courses. In effect, everyone would be getting a minor in critical thinking. Thus, they should be able to develop their critical thinking skills to a sufficiently high level of expertise.

While this solution has the advantage of being straightforward, it has the decided disadvantage of being completely unachievable given the current structure of general education in the liberal arts in many institutions of higher education, where students take courses in a variety of disciplines to satisfy distribution requirements across the arts and sciences. No college is going to add another 21 units to their general education requirements without also reducing other course requirements, and no department is going to accept a reduction in its share of the general education coursework without a fight. Thus, the prospects for implementing this solution are dim.

The second way to enhance the critical thinking training given students would be to design a program in "critical thinking across the curriculum," modeled after programs like "writing across the curriculum," in which writing is consciously taught in a wide range of courses in many different disciplines instead of just being the responsibility of the composition program. A "critical thinking across the curriculum" program would enlist faculty from a variety of disciplines to intentionally teach critical thinking skills along with the particular content of their courses.

There is much to be said for this approach. It certainly has widespread faculty support; as we saw above, the vast majority of faculty rate teaching critical thinking as a top priority. Unfortunately, however, there is strong evidence that in general we faculty do not do a very good job of *actually* teaching critical thinking. For example, a survey by Richard Paul showed that fewer than 10% of faculty "clearly teach for critical thinking on a typical day in class," and about 77% had "little, limited or no conception of how to reconcile content coverage with the fostering of critical thinking" (Paul 2004). If Paul's survey is representative of the state of American colleges and universities, it appears that any meaningful effort to actually teach critical thinking across the curriculum would require massive retraining of the vast majority of faculty in the fundamentals of critical thinking pedagogy. Though I believe this is actually a good idea, it is also impractical given the limited financial resources of higher education (not to mention the general lack of any real incentive at most American colleges and universities for faculty to change their pedagogy to include more effective practices).

Thus, if any significant change is to occur in our students' critical thinking skills, we will have to manage it without additional coursework or resources. And that is the centerpiece of the third possible solution to the problem.

To see how the solution might work, it might help to first notice that critical thinking is not the only area in which our students are exposed to high-level skills that they cannot possibly hope to master in one semester. A similar observation can be made of any introductory philosophy class. Doing philosophy is a very high level skill (or set of skills), and there just is not enough time to develop these skills in a fifteen-week semester class that meets for three hours a week.

The difference, of course, is that we do not expect (or, at least, *should* not expect) students to develop mastery of even basic philosophical skills in an introductory course. The course is meant to accomplish just what its name suggests—to *introduce* students to the practice of philosophy. Some of them will be enchanted by that practice and desire to spend much more time developing those skills. Those are the people who become philosophy majors.

Most often, other students will find other subjects to enchant them, but it is enough—at least it is enough for me—if they exit the class with a good understanding of what the practice of philosophy involves and an appreciation for the value of philosophy in a well-lived life. But under no circumstances is my introductory class designed to produce a skilled philosopher in one semester.

The difference between an introductory philosophy class and an introductory critical thinking class is mainly in the importance of the subject. Philosophy is important, of course, but critical thinking is vital. Ultimately I want all of my students to be economically productive, engaged in their communities, enmeshed in a supportive network of family or friends, and working to make the world a better place. (This, I believe, is the best recipe for a meaningful life.) Philosophy is necessary for none of these (though I believe it is helpful for all of them—but that's another argument). Critical thinking, I would argue, is necessary for them all. Thus, while I am content to use an introductory philosophy class to entice those who are willing to heed the call of philosophy and allow the rest to move on with their lives, I do not have the same laissez faire attitude toward my critical thinking students. A critical thinking class should lead everyone to continue developing their critical thinking skills long after the course has ended.

In other words, a critical thinking course should be designed to make students lifelong learners in critical thinking, that is, people who continue to develop their ability to think critically even when their formal schooling has ended. This is the only way we can get students to develop their critical thinking skills to the necessary level, given that the academy is unwilling or unable to devote sufficient additional resources to the problem.

So what does it take to be a lifelong learner? The key characteristic of successful lifelong learners as identified in the literature seems to be that the learners take responsibility for their learning, that is, the learner initiates and sustains the learning process (London 2011). To be a successful at this requires at least the following two characteristics:

1. Lifelong learners are self-motivated. They want to learn.
2. Lifelong learners are metacognitive. They know how to monitor how much they are learning and have a variety of learning strategies they can use to learn.

Thus, if we want to produce lifelong learners in critical thinking, we need to be using our critical thinking classes to produce students who have these two characteristics. In other words, these need to be two important learning outcomes for critical thinking courses. Fortunately, there is much useful research in both motivation and metacognition that we can draw upon to achieve these learning outcomes.

Lifelong learning and motivation

What does it take to produce self-motivated learners? Motivation is affected by both extrinsic and intrinsic factors (Nilson 2003, 73). In intrinsic motivation, the motivating factors are tied intimately to the activity; one cannot usually attain the goal without engaging in the activity. Solving a math problem because one likes the challenge is an example of intrinsic motivation. By contrast, extrinsic motivation is purely instrumental; the activity is valued only for the rewards that are contingently associated with the activity, such as when one solves a math problem because the teacher promised a candy bar to the first student with the correct answer. It is only because of the highly contingent connection between action and reward (under most other circumstances, solving the math problem would not earn a candy bar) that the person is motivated to act.

Applying this distinction to our problem, in order to make our courses support lifelong learning in critical thinking, there are at least two minimal conditions that must be met:

1. We must increase the intrinsic motivation of our students to become better critical thinkers.
2. We must reduce (or at least be more judicious and selective in our deployment of) the extrinsic motivation we provide to students in our critical thinking classes.

Increasing intrinsic motivation is important because intrinsic motivations are the ones that endure over time through changes in circumstance. The incentive to engage in extrinsically motivated actions ends as soon as the connection between activity and reward is severed. As teachers, our main extrinsic motivator is the grade, and the power of that to motivate our students ends precisely when they turn in the final exam. When students are finished with our class, we have precious few—if any—carrots and sticks to use to extrinsically motivate them.

What are the great intrinsic motivators? Pink (2009) summarizes them as (1) autonomy, (2) mastery and (3) purpose. That is, people find an activity intrinsically rewarding when (1) they have control over doing the activity, (2) they are good at it, and (3) they see the value in the activity. How might we incorporate each of these in a critical thinking class?

Fortunately, motivating students through autonomy and mastery are prominent topics in learner-centered pedagogy, so, to the extent that we are using this pedagogy, we already are encouraging lifelong learning. Thus, for example, if we are using strategies to give students more choices about and more control over their learning (Doyle 2008, 7), their engagement in their learning might be sustained after the end of the semester.

Similarly for mastery. There is a great deal of research on how to help students develop mastery (Ambrose et al. 2010, 91–120) and other research documenting the pleasure we get from mastery—the sense of "flow" first identified by Mihaly Csikszentmihalyi where one is lost in one's activity (Pink 2009, 111–113).

However, the experiences of autonomy and mastery that we can give students in a critical thinking course do not seem sufficient to motivate lifelong learning. They are effective in making students enjoy the activity *currently*, that is, they keep students intrinsically motivated while taking the course. (As we shall see below, relying on intrinsic motivators like autonomy and mastery rather than the extrinsic motivator of a good grade is a significant accomplishment.) But if we wish to intrinsically motivate students after our courses end, it is the third component—purpose—that I think we shall need to rely on.

This means that we have to consciously aim to get students to value critical thinking. Purpose, as an intrinsic motivator, involves connecting the activity with some greater and highly valued objective (Pink 2009, 131). Thus, to intrinsically motivate students to be lifelong learners, we must convince them of the *long-term* value of critical thinking.

Of course, to me the long-term value of critical thinking is obvious; it is intimately connected with a meaningful life, as I argued above. And I suspect that it is equally obvious to other critical thinking teachers. But the value is not obvious to our students, and so it is important to get them to see the connection between critical thinking and the things that give meaning and purpose to life. In particular, because critical thinking requires a lot of System 2 work that our brains do not want to do, we need to make sure they see that it is worth the effort it will take to be proficient at it.

One way to do this is by showing the bad decisions that we make when we fail to think critically. What might work (and we'd learn by trial and error here, so here is a fruitful area for further research) is to present them with decisions or occasions that expose the ways in which our ordinary thinking practices conspicuously fail us. There is a lot of helpful research by psychologists that we can draw upon.

For example, we might show our students Simons and Chabris's (1999) "invisible gorilla" video to demonstrate the phenomenon of selective attention. The video instructs students to watch a group of basketball players (half of whom are in white shirts and half in black shirts) and to count the number of times the white-shirted players pass the ball. (The black-shirted players are also passing a basketball to one another, but the subjects simply ignore this.) In the middle of the video, an actor clad in a gorilla costume walks right through the crowd of basketball players, pausing to thump its chest. Fifty percent of subjects do not see the gorilla.

After we watch the video in class, we might talk about why many of the students failed to see the gorilla. Students might then be asked to write about (and share if they like) a time in their life when selective attention caused them to miss the equivalent of the gorilla in the room.

It is important to realize that matters are more delicate than merely giving students an experience that shows their cognitive shortcomings. One necessary component for improvement is possessing the confidence that one can improve through effort (Ambrose et al. 2010, 200–202). So we will need to present these experiences in the context of teaching them that their ability to think critically is not fixed, and that they can, through effort, become better critical thinkers.

To produce students motivated to be lifelong learners, it will not, however, be enough to merely incorporate intrinsic motivation into our critical thinking classes. To motivate our students to be lifelong learners, we must also be very careful about how we use extrinsic motivators. This is because extrinsic motivation can destroy the desire to engage in activities that previously were intrinsically motivated.

The classic study showing this counterintuitive claim is a 1973 study by Lepper, Greene, and Nisbett (discussed in Pink 2009, 35). Lepper, Greene, and Nisbett studied preschool children whom they had identified as ones who enjoyed drawing. The preschoolers were divided into three groups. Children in the first group—the "Expected Reward" group—were offered a "Good Player" certificate if they drew a picture. Children in the second and third groups were not offered a reward; however, children in the second group— the "Unexpected Reward" group—were offered the chance to draw, and those who engaged in drawing were handed a "Good Player" certificate at the end of the session. Children in the third ("No Award") group were asked if they wanted to draw (just as the children in the other two groups were), but were neither offered (as in the first group) nor given (as in the second group) a certificate.

Two weeks later, teachers set out drawing supplies in the children's free play period (when they could choose to draw or engage in other activities) while the researchers observed them clandestinely. Children in the second and third groups—the "Unexpected Reward" and the "No Award" groups—drew just as much (and with just as much enjoyment) as previously. But the children in the first group—the ones who had engaged in drawing after being offered the "Good Player" certificate—spent significantly less time drawing, and were much less interested in it. The conclusion drawn by Lepper, Greene, and Nisbett is that the addition of external motivation, rather than overdetermining the students' engagement in an intrinsically rewarding activity, actually reduced their intrinsic motivation (Lepper, Greene, and Nisbett 1973).

The phenomenon of extrinsic motivation destroying intrinsic motivation is a very robust finding, confirmed by three decades of research and over 100 experiments (Pink 2009, 37). It is so robust, in fact, that it has a name: the Tom Sawyer effect. Tom Sawyer, you might recall in Mark Twain's 1876 novel of the same name, was punished for his bad behavior by being forced to whitewash a fence. But Tom was able to convince his friends to whitewash the fence for him—indeed, to pay him for the privilege – by presenting it as an intrinsically rewarding activity. In short, by changing the extrinsic motivation to intrinsic motivation (by shifting from external to internal rewards), *work* became *play*.

(Strictly speaking, Lepper's study supports the *reverse* Tom Sawyer effect, because, by adding extrinsic motivation, *play* became *work*. But the fundamental lesson is the same; sustained engagement in an activity can only be maintained if intrinsically motivated. Extrinsic motivation is the enemy of intrinsic motivation.)

Lifelong learning and metacognition

Besides being motivated to be lifelong learners, as we have seen, students must also become metacognitively proficient. That is, they must be acutely aware of how they learn. Fortunately, as with motivation, there is a great deal of useful research into metacognition that we can draw upon. Metacognition requires that a person *monitor* and *control* their own learning (Ambrose et al. 2010, 192ff). This involves a complicated cycle of basic processes that the student must master.

According to Ambrose et al. (2010), to be metacognitively proficient, students must be able to

1. assess a learning task,
2. evaluate the knowledge and skills they bring (or lack) to succeed in the learning task,
3. plan how to approach the learning task based on their previous assessment of the task and their knowledge and skills,
4. apply appropriate strategies to carry out the plan, monitoring their success along the way, and
5. adjust and restart the cycle as necessary, depending on their degree of success.

The fundamental point I want to make here is that skills like these must be explicitly taught; they cannot be absorbed automatically by the students as they are focusing on course content. Thus, a critical thinking class that focuses on lifelong learning will look different from one that focuses on teaching basic critical thinking skills. But it is important to note that metacognitive skills are not necessary only for lifelong learning. They also are an important way for novice learners in any area to become experts (Doyle 2008, 135). What distinguishes novices from experts is not so much their content knowledge (though that is important) but their inability to see whether they are scratching the surface of a problem or examining it in depth (Doyle 2008, 135). Experts in a field know when they need to keep digging—when they need to check for errors and why these errors are occurring.

Conclusion

To summarize my argument in this paper: In the face of (1) ample evidence that our students are not improving their critical thinking skills sufficiently and (2) a convincing explanation of why their skills do not improve sufficiently, we should no longer think in terms of a one-semester course in critical thinking in which we introduce basic critical thinking skills and trust that the skills take hold and grow over students' lives. What we should focus on is producing students who can develop as critical thinkers (just as an introduction to philosophy class might aim at producing students who can develop as philosophers). Thus, one important learning objective for a critical thinking course is that students become lifelong learners who are intent on growing as critical thinkers.

Achieving this objective requires incorporating two important outcomes into our critical thinking classes (perhaps replacing other worthy outcomes):

1. that students will be motivated to be lifelong learners in critical thinking, and
2. that students will have the metacognitive skills to grow as critical thinkers.

There is much work that we can draw on to help us achieve these outcomes, but also many questions still to be answered. (What, if any, of the traditional critical thinking course content tends to be intrinsically motivating? Does a fifteen-week course give students enough time to see meaningful and motivating progress as critical thinkers? How do we effectively teach metacognitive skills in the context of a critical thinking course?) There is much to do, so we had better get started. Our students are depending on us.

References

Ambrose, S. A., Bridges, M. W., DiPietro, M., Lovett, M. C., and Norman, M. K. 2010. *How Learning Works: Seven Research-Based Principle for Smart Teaching.*San Francisco: Jossey-Bass.

Arum, R., and Roksa, J. 2011. *Academically Adrift: Limited Learning on College Campuses.* Chicago and London: University of Chicago Press.

Doyle, T. 2008. *Helping Students Learn in a Learner-Centered Environment: A Guide to Facilitating Learning in Higher Education.* Sterling, VA: Stylus Press.

Ennis, R. H. 1962. "A Concept of Critical Thinking." *Harvard Educational Review* 32 (1): 81–111.

Ennis, R. H. 1991. "Critical Thinking: A Streamlined Conception." *Teaching Philosophy* 14 (1): 5–25.

Gladwell, M. 2005. *Blink: The Power of Thinking without Thinking.* New York: Back Bay Books/Little, Brown and Company.

Hatcher, D. 2013. *Critical Thinking: A New Definition and Defense* 2000 [Web Page 2013]. Available from http://www.bakeru.edu/crit/literature/dlh_ct_defense.htm.

Hatcher, D., and Spencer, L. A. 2000. *Reasoning and Writing: From Critical Thinking to Composition*. Boston: American Press.

Kahneman, D. 2011. *Thinking, Fast and Slow*. London: Allen Lane.

Lepper, M. P., Greene, D., and Nisbett, R. E. 1973. "Undermining Children's Intrinsic Interest with Extrinsic Reward: A Test of the 'Overjustification' Hypothesis." *Journal of Personality and Social Psychology* 28 (1): 129–137.

Lipman, M. 1988. "Critical Thinking: What Can It Be?" *Educational Leadership* 46 (1): 38–43.

Lipman, M. 2003. *Thinking in Education*. (second edition) Cambridge:: Cambridge University Press.

London, M. 2011. "Introduction." In *The Oxford Handbook of Lifelong Learning*, edited by M. London. New York: Oxford University Press.

Marzano, R. 2001. *Classroom Instruction That Works: Research-Based Strategies for Increasing Student Achievement*. Alexandria, VA: ASCD.

Marzano, R. *What Is an Effect Size?* nd. Available from http://www.marzanoresearch. com/media/documents/pdf/AppendixB_DTLGO.pdf.

McPeck, J. 1981. *Critical Thinking and Education*. New York: St. Martin's Press.

Nilson, L. B. 2003. *Teaching at Its Best: A Research-Based Resource for College Instructors*. Vol. 2.Bolton, MA: Anker Publishing.

Paul, R. 2004. *The State of Critical Thinking Today: The Need for a Substantive Concept of Critical Thinking*. Available from http://www.criticalthinking.org/pages/the-state-of-critical-thinking-today/523.

Pink, D. H. 2009. *Drive: The Surprising Truth about What Motivates Us*. New York: Riverhead Books.

Possin, K. 2008. "A Field Guide to Critical Thinking Assessment." *Teaching Philosophy* 31 (3): 201–228.

Siegel, H. 1988. *Educating Reason: Rationality, Critical Thinking, and Education*. New York: Routledge.

Simons, D. J., and Chabris, C. F. 1999. "Gorillas in Our Midst: Sustained Inattentional Blindness for Dynamic Events." *Perception* 28: 1059–1074.

Simons, D. J. "Selective Attention Test." Available from https://www.youtube.com /watch?v=vJG698U2Mvo.

Willingham, D. T. 2009. *Why Don't Students Like School?*San Francisco: Jossey-Bass.

7
Teaching Critical Thinking as Inquiry

Sharon Bailin and Mark Battersby

Introduction

This paper will describe and argue for an approach to foster critical thinking in higher education based on inquiry. This approach encompasses both critical thinking in everyday contexts and critical thinking within the disciplines.

On the one hand, a common approach to teaching critical thinking in higher education in North America is through separate courses. The focus tends to be on the evaluation of individual arguments typically found in everyday contexts (e.g., newspaper editorials). It is assumed that such a focus will result in students being able to think critically in real contexts. It is often also assumed that acquiring the skills of argument evaluation in these contexts will transfer, where relevant, to critical thinking in particular disciplinary areas. On the other hand, the assumption of traditional teaching in the disciplines has generally been that the modes of argumentation and reasoning of the discipline will be acquired automatically by students through learning the discipline.

We argue in this paper that these assumptions are unfounded. Focusing on the evaluation of individual arguments is problematic, based as it is on a faulty model of critical thinking, which neglects the dialectical and contextual dimensions of reasoning. Knowledge of the arguments on various sides of an issue as well as of the historical, intellectual, and social contexts is essential to making a reasoned judgment on everyday issues as well as in disciplinary contexts.

The assumption that critical thinking will be acquired automatically through disciplinary pedagogy is also unfounded. Reasoning and argumentation are seldom a focus of disciplinary pedagogy. Moreover, this approach neglects the common aspects of argumentation that transcend disciplinary boundaries.

What we propose as an alternative is an *inquiry approach* to critical thinking pedagogy that focuses on the comparative evaluation of competing arguments with the goal of making reasoned judgments (Bailin and Battersby 2010). This approach emphasizes both the aspects common to inquiry across a range of areas and the modes of argumentation that are specific to the area. This approach can be the focus of a separate course and can also be integrated into disciplinary instruction.

Critique of conventional critical thinking courses

It is often the case that the only concerted, overt attempt to teach critical thinking and argumentation at the postsecondary level takes place through separate courses on critical thinking. Such courses are commonly offered in philosophy departments (at least in North America) and generally limit their focus to logic, formal or informal, and the evaluation of individual arguments. The arguments used are usually taken from the media, political speeches, and other sources of "everyday" arguments and are often presented out of context. Although some popular critical thinking texts with many editions (e.g., Moore and Parker 2012; Vaughn and MacDonald 2012; Waller 2011) have started to make some move away from a sole focus on analyzing decontextualized arguments, these efforts are episodic (e.g., a section on analyzing longer arguments). None of these represents a unified focus on developing the abilities and habits of critical inquiry (Hamby 2013; Hitchcock 2013).

We therefore believe that this approach is inadequate(Bailin and Battersby 2009; 2010). In our view the goal of critical thinking instruction is to provide students with the understanding and skills necessary for thinking critically in real contexts. And the kind of critical thinking that actually takes place in real contexts, both in the disciplines and in everyday life, centrally involves making reasoned judgments on complex issues. The focus on reasoned judgments marks an approach to critical thinking that can be seen as epistemological (Lipman 1991; Paul 1990; Siegel 1988; 1997). An epistemological conception views critical thinking in terms of the quality of and criteria for good reasoning, and focuses less on arguments per se than does a more logically oriented conception.

Indeed, it is our view that arriving at reasoned judgments on complex issues involves more than the evaluation of individual arguments. It involves a process that is dialectical (Blair and Johnson 1987, 45–46). To say that the process is dialectical means that it takes place in the context of some controversy or debate. This implies that it is initiated by some question, doubt, challenge, and that there is a diversity of views on the issue, arguments both for and against (if the controversy is genuine, then it is likely that there will be at least some plausible arguments on both sides (Johnson 2003, 42). The dialectical aspect also means that there is an interaction between the arguers and between the arguments involving criticism, objections, responses, and, frequently, revisions to initial positions (Bailin and Battersby 2009; Johnson 2000).

An implication of this view is that it is seldom the case that reasons and arguments can be evaluated individually in any comprehensive or significant manner. It is possible to evaluate individual arguments in a preliminary, prima facie manner, discovering fallacies or errors in reasoning and evaluating the reasons or evidence in support of the conclusion (Bailin and Battersby 2010). In order to reach a reasoned judgment on the issue in question, however, we must go beyond this prima facie evaluation and evaluate the arguments in the context of this dialectic, of this historical and ongoing process of debate and critique. In order to reach a reasoned judgment, arguments need to be evaluated comparatively, in light of alternatives and competing arguments and views (Bailin and Battersby 2009, 4; 2010; Johnson 2007, 4; Kuhn 1991, 201f).

A major weakness of traditional critical thinking courses is that they do not focus on the kind of comparative evaluation that we make in actual contexts of disagreement and debate. It is this dialectical and contextual dimension that is largely missing from traditional critical thinking instruction.

Critique of conventional disciplinary teaching

A different sort of problem arises in the context of attempting to develop critical thinking in the disciplines. The assumption of traditional teaching in the disciplines has generally been that the modes of argumentation and reasoning of the discipline will be acquired automatically by students through learning the discipline. Yet this assumption appears to be unfounded. Much research has indicated that even postsecondary students studying a discipline do not necessarily reason well in that discipline (Ferraro and Taylor 2005; Hestenes, Wells, and Swackhamer 1992; Jungwirth 1987). This should not be particularly surprising given the fact that reasoning and argumentation are generally not a focus of disciplinary pedagogy. While many instructors admit the need to emphasize critical thinking, this concern is often overridden by the need to cover disciplinary content.

Another problem with leaving the acquisition of reasoning to the vagaries of disciplinary teaching is that this approach neglects the aspects of argumentation that transcend disciplinary boundaries. To the extent that the reasoning in the discipline is a focus of study, it is likely to be limited to the type of reasoning and argumentation characteristic of the particular discipline, for example "scientific method" in the sciences. The aspects of argumentation common to various disciplines and to nondisciplinary contexts such as the procedures for conducting an inquiry, the logical analysis of arguments, fallacies, and common errors in reasoning, the evaluation of sources, and those criteria for evaluation that are common across domains are not likely to be included. Thus the connection between inquiry in the particular discipline and the larger enterprise of inquiry is not likely to be made.

An additional problem with much traditional disciplinary teaching is that it tends to neglect the dialectical dimension of argumentation. But, as was pointed out above, reasoning and argumentation need to be evaluated in the context of the dialectic in which they arise and are embedded. This is equally the case for making a reasoned judgment in a discipline as it is for making judgments in everyday contexts. Making such judgments involves weighing and balancing competing arguments and so requires an understanding of the dialectic and a grounding in the debates within the discipline.

However, simply introducing students to a variety of competing theories is insufficient. They also require the resources for comparatively evaluating these theories and judging among them. One of the requirements for comparatively evaluating competing theories and views is an understanding of discipline-specific modes of argument and criteria, for example causal reasoning in science, statistical reasoning in the social sciences, or historical reasoning in history, which may not be addressed in separate critical thinking courses. Without a grounding in the debates within the discipline and without an explicit focus on the modes of argumentation and the evaluation criteria that are specific to the area, the modes of argumentation and reasoning in particular disciplines are not likely to be learned.

Teaching critical thinking as inquiry

What we propose as an alternative is an inquiry approach to critical thinking pedagogy. We use the term inquiry to refer to the careful, critical examination of an issue in order to come to a reasoned judgment. While the term inquiry is not common in the critical thinking literature, Hitchcock's notion of argumentative discussion has considerable overlap with our notion of inquiry: "An argumentative discussion is a sociocultural activity of constructing, presenting, interpreting, criticizing, and revising arguments for the purpose of reaching a shared rationally supported position on some issue" (Hitchcock 2002, 291).

There are several aspects of inquiry that are significant in this approach. The first is that inquiry requires focus on an issue. An inquiry is initiated by some challenge, controversy, or difference of view that is in need of resolution. The second aspect of significance is that inquiry involves a critical examination of evidence, arguments, and points of view. It is not just an information-gathering enterprise but involves, centrally, a critical evaluation according to relevant criteria. The third significant aspect is that inquiry aims toward a reasoned judgment. By a reasoned judgment we mean not simply a judgment for which one has reasons, but a judgment for which one has good reasons, reasons that meet relevant standards. Making a reasoned judgment is not simply a matter of evaluating individual arguments, however. Rather, it requires the comparative evaluation of competing arguments and views (Bailin and Battersby 2010).

An inquiry approach emphasizes both the aspects common to inquiry across a range of areas and the aspects and modes of argumentation that are specific to an area. Conducting inquiries on relevant topics can be used as a focus for and way of structuring free-standing critical thinking courses, and it can also be integrated into subject-area instruction. Thus critical thinking pedagogy is structured around complex, authentic tasks. The various aspects that go into the process of inquiry are learned not as decontextualized "skills" but rather in the context of coming to reasoned judgments on complex issues.

Teaching inquiry in separate courses

How might one teach critical thinking as inquiry in a separate course? Our critical thinking text, *Reason in the Balance: An Inquiry Approach to Critical Thinking* (Bailin and Battersby 2010), provides one example of an inquiry approach to teaching critical thinking. The text uses dialogues among an ongoing cast of characters involved in realistic situations as a context for discussing the various aspects that go into the practice of inquiry, including identifying issues, identifying the relevant contexts, understanding the competing cases, and making a comparative judgment among them. These aspects are instantiated in inquiries on topics such as vegetarianism, the minimum wage, the legalization of marijuana, the regulation of dangerous dogs, the evaluation of a film, the bombing of Hiroshima, and the right of hate groups to speak. These various aspects are also applied to inquiry in specific contexts, including science, social science, philosophy, and the arts. There is also considerable emphasis placed throughout on the habits of mind that are essential for inquiry, including (among others) open-mindedness, fair-mindedness, the desire to act on the basis of reasons, the acceptance of uncertainty, and respect for others in dialogue—habits of the mind that we characterize as the spirit of inquiry.

The following set of guiding questions is used to structure inquiry throughout the text:

1. What is the issue?
2. What kinds of claims or judgments are at issue?
3. What are the relevant reasons and arguments on various sides of the issue?
4. What is the context of the issue?
5. How do we comparatively evaluate the various reasons and arguments to reach a reasoned judgment?

The text devotes chapters to each of these questions, with the students developing an understanding of each, applying them in practice contexts, and then using each one in turn to progressively develop an inquiry on a topic of their choosing. Through this process, the various aspects of inquiry are integrated and students gain proficiency in conducting inquiries.

We have reproduced here an excerpt from one of a series of dialogues between two students, Phil and Sophia, on capital punishment. We shall use this example (the present excerpt and the dialogues that follow it) to illustrate each of the aspects of inquiry.

Capital punishment

Phil has been reading an opinion piece in a newspaper in which the chief of police of his town is arguing for capital punishment for murder:

> **Phil**: Hey, Sophia—let me read you something interesting:
> *"Society has an obligation, first and foremost, to protect its citizens from harm. And the most serious form of harm is murder. Protecting citizens from murder involves ensuring that murderers don't repeat the offence. It also involves dissuading others from committing murder. Now I and other law enforcement officers know from a vast amount of first-hand experience with criminals that the only form of punishment that can effectively achieve both goals is the death penalty. Capital punishment involves taking the life of a person who has committed murder in order to save the lives of innocent people, and so is the best option under the circumstances."*(italics in the original).
> **Phil:** What he says makes a lot of sense. After all, society needs to do whatever it can to protect innocent people. And murderers have really given up their right to be protected because they've taken someone else's life. So killing them to save innocent people seems OK.
> **Sophia:** Hold on a minute, Phil! Not so fast. You're leaping to conclusions again. You haven't even thought the issue through.
> **Phil:** But what this guy says seems right.
> **Sophia:** So are you just going to believe what he says without checking it out? What else would you expect a police chief to say?
> **Phil:** Well, he does have a lot of experience with crime.
> **Sophia:** But you haven't considered the other side. Your police chief certainly hasn't given us any of the arguments against capital punishment.
> **Phil:** But what about his argument?
> **Sophia:** I think that there's a lot more that we need to know before we can decide whether his argument is any good. We need more information. We need to know some facts about capital punishment. We need to look at all the arguments on both sides... We need to... I know. What we need to do is...
> **Sophia and Phil:**... conduct an inquiry!
> **Sophia:** Now the first step, if I remember right, is to be clear about what the issue is.
> **Phil:** That's pretty easy. The issue is whether we should have capital punishment.

Sophia: For what crimes? We need to be specific. In some countries, there's the death penalty for adultery.

Phil: No, no...I wasn't suggesting that. I'm only thinking about cases of premeditated murder.

Sophia: I'm glad you're clear about that.

Phil: OK...next question—what kind of judgment does this involve?

Sophia: Well, since we're talking about what we "should" or "should not" do, then I guess it's an evaluative judgment. But I can see already that we'll also need to look at some factual claims on the way—like whether capital punishment really does help prevent murders. (Bailin and Battersby 2010, 138–139)

What is the issue?

In order to even begin to inquire, it is of vital importance to be clear about the issue, which is to be impetus for the inquiry. Among the characteristics of an appropriate issue are that it be sufficiently focused to allow for productive inquiry; precisely and neutrally framed, avoiding vague, ambiguous, or biased formulations; and controversial, evoking genuine disagreement.

In the dialogue excerpt, Sophia notes that Phil's original formulation of the issue, whether we should have capital punishment, is too vague as it does not specify for which crimes.

What kinds of claims or judgments are at issue?

It is important to understand what types of judgments are called for by the inquiry that we are undertaking because different types of judgments are supported by different types of reasons and arguments and are evaluated by different criteria. For example, while a judgment in science will appeal to the criterion of fit with observations, a moral judgment will appeal to reasoning according to moral principles. Although there is a range of types of judgments, they can be categorized broadly into three types: factual, evaluative, and interpretive.

In the dialogue, Phil and Sophia recognize that their inquiry calls for an evaluative judgment about whether capital punishment *should* or *should not* be practiced, but that it will also involve factual judgments, for example with respect to whether capital punishment really does act as a deterrent. As the inquiry proceeds, they also recognize that their inquiry will require moral judgments, for example, with respect to the state executing innocent individuals.

As another example, if students wished to address the issue of climate change, they would need to be able to distinguish among the kinds of judgments required by different questions about climate change, for example: "Is the climate changing significantly?" (factual descriptive); "Is climate change humanly caused?" (factual causal); "What, if anything, should we do about climate change?" (evaluative).

What are the relevant reasons and arguments on various sides of the issue?

A key aspect of inquiry involves laying out the arguments on various sides of an issue. This will include the various positions on the issue in question that have been offered; the evidence that has been brought forward and the arguments that have been made in defense of the various positions; the objections that have been leveled against the positions and the responses that have been made to these; and the alternatives that have been put forth.

In the dialogue, Phil is initially inclined to accept the one argument in favor of capital punishment, which he reads, but Sophia recognizes the need to look at the whole debate and to evaluate the arguments on both sides of the issue before making a judgment. In a subsequent dialogue, after doing some research, they discover a number of arguments that are generally offered both in favor of capital punishment (e.g., arguments from deterrence, incapacitation, retribution, and cost) and against (e.g., arguments focused on the immorality of taking a life, the immorality of executing innocent individuals, rehabilitation, and the social causes of crime). They also find the various objections and responses that have been offered to these arguments. Making a judgment on the issue of capital punishment will ultimately require them to be aware of this entire dialectic.

What is the context of the issue?

Finding out about the contexts in which issues are situated can provide valuable information when conducting an inquiry. There are three aspects of context that we believe need to be considered: the state of practice, the history of the debate, and the intellectual, social, political, and historical contexts.

The state of practice refers to how things currently stand with respect to the issue. An understanding of the state of practice can provide information necessary for making a reasoned judgment. For example, in order for students to make a reasoned judgment regarding the raising of the minimum wage, they would need to know information such as the wage in other jurisdictions, when the minimum wage was last raised, the effect of inflation on wages, costs of living, and so on.

The history of the debate refers to the history of argumentation and deliberation that has led to current practice or thinking about the issue. Knowledge of the history of the debate can be helpful and is in some cases essential to understanding what is significant or contentious about an issue and in understanding the various positions that are contesting for acceptance. Knowing the history of a debate is also important in determining where the burden of proof lies.

Understanding the intellectual, political, historical, and social contexts sur-rounding an issue is also important in that it can aid us in understanding and interpreting arguments and can reveal assumptions underlying arguments and positions that may be important for their evaluation. For example, in making judgments about the legalization of marijuana in North America, it would be important to understand aspects of the history and social context of marijuana prohibition, including the fact that there is an enormous governmental and police investment in drug prohibition.

In a dialogue subsequent to the one reproduced above, Sophia and Phil inves-tigate each of these aspects with respect to the capital punishment debate. They discover the current state of practice in their location (that there is no capital punishment) as well as the situation worldwide—a general trend toward aboli-tion, and recognize the argumentative implications of these facts in terms of which views carry the burden of proof (those which go against current prac-tices). Looking at the history of the debate, they discover that some of the arguments (e.g., retribution, deterrence, and incapacitation) have very ancient roots and also that the primary argument offered in favor of capital punish-ment has changed recently from deterrence to retribution in light of the lack of evidence of a deterrent effect. With respect to the intellectual, social, political, and historical contexts, they recognize that the pro and con positions on capi-tal punishment tend to be associated with different worldviews with respect to issues such as tradition versus change in society, individual versus societal responsibility, and social order.

How do we comparatively evaluate the various reasons and arguments to reach a reasoned judgment?

The following are the methods by which a reasoned judgment can be arrived at:

1. *Evaluating individual arguments.* The core of an inquiry is the evaluation of the various views and arguments in order to reach a reasoned judgment. A crucial aspect involves the evaluation of the individual arguments made. It is here that the usual criteria for evaluating arguments come in. Undertaking a prima facie or preliminary evaluation of the arguments for fallacies or errors of reasoning is an important first step. In addition, the various claims need to be evaluated according to the relevant criteria—factual claims by looking at evi-dence in support of claims and the credibility of sources and evaluative claims by assessing the argumentation.

In conducting their prima facie evaluation in a subsequent dialogue, Phil and Sophia do encounter fallacies of anecdotal evidence and improper appeal to authority, as well as possible bias in the police chief's argument. They real-ize, however, that the fact that there are fallacies in the arguments does not invalidate the views which he is defending. What it does mean is that they must go on to evaluate the various claims.

With respect to the factual claims, after extensive investigation, they succeed in determining that there is a consensus in the research that capital punishment does not act as a deterrent to murder. The also discover that the claim that capital punishment is less costly than life imprisonment is false. With respect to the moral arguments, they decide that there is a morally appropriate desire for justice behind the retribution argument for capital punishment, but that the concern about the state executing innocent people constitutes a very strong moral argument against capital punishment.

2. *Comparative evaluation*. The evaluation of individual arguments is necessary, but it generally cannot on its own lead to the making of a reasoned judgment. In order to come to a reasoned judgment, we need to perform a comparative evaluation of the arguments in order to determine their weight in terms of the overall case, and then combine the various evaluations in order to make a final judgment. This process involves balancing the various considerations that have come to light.

In their final dialogue on capital punishment, Sophia and Phil summarize their evaluation of the various arguments and weigh their comparative strength. In terms of the pro arguments, they conclude that there is no support for the deterrence or cost arguments, that incapacitation can be achieved by less drastic means than putting the perpetrators to death, and that there is some moral legitimacy to the retribution argument in terms of the desire for justice, but that it can be achieved through life imprisonment. In terms of the con arguments, they conclude that the risk of the state killing innocent citizens is a very strong argument that overrides the retribution argument, especially as there are less morally problematic alternatives to capital punishment that can achieve retribution. Their anticapital punishment judgment is strengthened by the worldwide trend toward abolition, which places a burden of proof on the pro side.

Inquiries in specific areas

It is our belief that if our goal is to foster students' critical thinking in the range of contexts that they will encounter, then it is important in a critical thinking course to include inquiries that focus on disciplinary knowledge and criteria in areas such as science, social science, philosophy, and the arts. Thus, in addition to focusing on topics such as capital punishment, our text also focuses on topics that require a knowledge of discipline-specific procedures and criteria, for example, polygamy (philosophy), the effects of violent video games (the social sciences), interpreting a challenging work of art (the arts), and some historical examples of inquiries in geology, epidemiology, and evolutionary theory (the natural sciences). These inquiries exemplify both how the guiding questions, procedures, and criteria apply in various areas and also the criteria that are specific to the discipline.

Integrating inquiry into subject area instruction

The inquiry approach can also provide a method for instilling critical thinking into discipline-focused courses while still providing adequate coverage of course material. Organizing teaching around inquiries can serve to illuminate the common structure and aspects of inquiry as well as illustrate how this structure and these aspects are manifested in the particular area. This approach also highlights the specific concepts, forms of reasoning, argumentation, and criteria that are particular to and dominant in the particular discipline. Nosich's recommendation to focus student thinking on a deep understanding of the central concepts of a discipline is very much in consonance with an inquiry approach (Nosich 2012).

For an inquiry approach to be successful, the instructors need to be clear about the long-term learning goals of the course. This is especially important for introductory or general education courses where students are unlikely to go on further in the discipline. Presumably, the goals should include not only engaging the student in the subject and the disciplinary approach to subject matter, but also empowering students to use the methods and information produced by the discipline to make thoughtful and reasonable decisions as individuals, citizens, and workers. As long as the primary goal of a course, especially an introductory course, is to lay down a basic vocabulary or get students to retain abundant factual information, it will be difficult to devote enough time or student energy to learning how to inquire and to understanding argumentation in the discipline. But if the primary outcomes include an understanding of the issues and claims in the discipline and the ability to make reasoned judgments using disciplinary criteria, then the inquiry approach can be used both to reinforce the learning of subject material and to develop those abilities and habits of mind that lead to reasoned judgment. For example, students in an ecology course could be asked to assess local laws governing logging. Through engaging in this inquiry, students would learn not only the requisite ecological concepts of forest development and sustainability, but also what is involved in coming to a reasoned judgment on the issue.

To illustrate how one might integrate the inquiry approach into disciplinary teaching, we will show how each of the guiding questions could be used to address the questions of logging and forest management.

What is the issue?

In order to pursue this inquiry, students would need, first, to be clear about the issue or question. They need to know whether the question is about what regulations would provide for sustainable logging? Or is the question how to protect ecosystems for animal conservation?

What kinds of claims or judgments are at issue?

It would be important, for this inquiry, to distinguish between normative claims and judgments about the value of forests, and scientific claims about the consequences of logging on fishing or ecosystem health. The idea of "ecosystem health" is a good example of a concept that students would need to grapple with in trying to sort out value and factual questions. "Health" is a complex concept including both norms and facts and getting clear about what is at issue is an important intellectual challenge.

What are the relevant reasons and arguments on various sides of the issue?

Ecological issues are often characterized by bias, and getting a full range of views with their attendant arguments is obviously important for making a reasoned judgments. Students must have adequate conceptual knowledge and be able to apply an understanding of the scientific approach to these issues to evaluate the debate. In addition students would need to understand the economic pressures that are part of this debate as well as the normative questions that are involved.

What is the context of the issue?

There are a number of ways in which understanding the history of ecological debates is important for coming to reasoned judgments. For example, one well-known debate surrounding logging of old growth forests in the United States is the spotted owl debate. Because the spotted owl's habitat is old growth forest and because the United States has strong endangered species legislation, preservation of the spotted owl has involved protecting large areas of old growth forest from logging. If one does not know this background, the intensity of the current debate over strategies to preserve the owl (including the idea of culling competing species) would be incomprehensible. It would appear to be about owls, but it is actually about logging old growth forests.

Understanding the history of a debate is also important for determining the burden of proof on an issue. At any historical moment in most disciplines there are accepted theories or factual claims that are supported by a wide consensus, and these constitute the default views. Anyone wishing to refute these views bears the burden of proof. Determining where the burden of proof lies with respect to the issue of logging regulation would form an important aspect of this inquiry, although the fact that ecology is a relatively young discipline makes this determination particularly challenging. For many introductory students the default view is whatever they have learned from their upbringing or even perhaps from their own experience. It is interesting to invite students to reflect on the question of who bears the burden of proof and to consider whether their position can be appropriately treated as the default view.

How do we comparatively evaluate the various reasons and arguments to reach a reasoned judgment?

While the issue of logging regulation involves numerous ecological questions, it also involves economic and ethical ones. How do we weigh short-term economic benefits against long-term ecological sustainability? There are no easy answers, but explicitly addressing these issues and attempting to balance competing values and interests are crucial to making a well-informed and reasoned judgment.

The preceding is but one example of how an inquiry approach can be used in disciplinary teaching, in this case with respect to an interdisciplinary issue having a strong scientific component. We would like to stress, however, that this approach can be used in virtually any subject area, for example, in the social science (e.g., Should we allow our children to watch violent video games?), in the arts (e.g., Is Duchamps's urinal really art?), or in philosophy (Should polygamy be legal?) (Bailin and Battersby 2010).

Fostering inquiry across the disciplines

Some more general strategies can also be employed in all areas in order to foster inquiry across disciplinary areas. The goal is to promote an understanding of the process of inquiry practiced in the particular area as just one example of the enterprise of inquiry more broadly, involving a similar aim, namely to reach a reasoned judgment, common guiding questions, some common or overlapping concepts and criteria, and the same habits of mind (e.g., open-mindedness, fair-mindedness, a commitment to reason, an inquiring attitude) (Bailin and Battersby 2010).

One particularly important habit of mind that is central for inquiry in all areas is the propensity to always consider alternative views and theories. In order to develop this habit of mind, students can be required to defend competing theories with which they disagree and attempt to come to reasonable conclusions despite conflicting evidence and theories. It is often an illuminating experience for students to understand their resistance to evidence and argument for a theory with which they have a prior disagreement.

Many key concepts are used widely in many areas (e.g., concepts common in the sciences such as the distinction between correlation and causation, the problem of getting reliable data, the question of experimental validity, the problem of confirmation bias). All these widely shared concepts can be reinforced in almost any subject. Even such subjects as literary or artistic analysis can be shown frequently to involve reasoning to the best explanation while considering alternative points of view.

The ideal situation for teaching inquiry across the disciplines would be one in which instructors were aware of how faculty in other disciplines presented the key concepts of inquiry and critical thinking so that these concepts could be reinforced in all courses. This is a lot to hope for, but the notions of seeking alternative explanations, weighing competing arguments, and coming to a reasoned conclusion are sufficiently applicable across a range of subject areas that parallels can usefully be drawn. It is useful to ask students whether they recognize that argumentative and evaluative approaches in one course have analogies with those approaches used in other courses.

Because many of the problems of the real world involve interdisciplinary or multidisciplinary inquiries, there is a wealth of topics and issues that may be of genuine interest to students and that could be used to illustrate how some of the relevant evaluative criteria can be applied across disciplines.

Conclusion

We believe that an inquiry approach to teaching argumentation and reasoning is to be recommended for several reasons. First, in broadening the focus from the evaluation of individual arguments to the making of reasoned judgments, it aims to foster the kind of critical thinking that takes place in real contexts of disagreement and debate. This changed emphasis brings to the fore the dialectical and contextual dimensions of argumentation, which are central to the making of reasoned judgments. An inquiry approach also makes room for the inclusion of disciplinary criteria and modes of argumentation when dealing with everyday issues, the knowledge of which is often essential for making judgments with respect to complex, real-world issues.

There are also dispositional benefits to an inquiry-based approach. The requirement to actively seek information and arguments in order to resolve an issue or puzzlement may foster habits of mind such as intellectual curiosity, truth-seeking, self-awareness, and intellectual perseverance. In addition, an open-minded, fair-minded, and flexible attitude is much more likely to be encouraged by an approach that focuses on inquiring through the evaluation of competing cases rather than on one focused exclusively on the evaluation of individual arguments (Bailin and Battersby 2009).

With respect to teaching within the disciplines, an inquiry approach has the advantage of putting an explicit focus on disciplinary reasoning and argumentation, making reasoning a central part of what it means to learn a discipline.

By highlighting the aspects of argumentation that are distinctive to particular disciplines, it gives students the tools to reason well within those disciplines and with respect to issues that call on disciplinary understanding. But it also has the additional merit of highlighting those aspects of argumentation that are common to inquiry across disciplines. In so doing, it makes explicit the connection between disciplinary inquiry and inquiry more broadly, enabling students to view reasoning and argumentation in any discipline not as an isolated activity but rather as connected with other critical practices of investigation, discovery, and creation.

To date, our main basis for evaluating an inquiry approach is personal experience. We have been teaching using this approach for several years, both in undergraduate critical thinking courses and in an MEd program for practicing educators, and our results have been extremely promising in terms of students' ability to conduct reasoned inquiries. In addition, Hitchcock (2013) has collected data on more than 400 students over the three occasions in which he used *Reason in the Balance*. What he found was that although students did worse than previous students on some types of multiple-choice exam questions and on tests of micro-skills of argument analysis and evaluation, they did noticeably better on items testing their ability to identify a counterexample to a generalization, judge the trustworthiness of a source of information, and analyze and evaluate causal arguments. Their performance was comparable on items involving supplying missing premises, evaluating conditional arguments, judging deductive validity, and identifying fallacies. These multiple-choices exams did not, however, test the ability of students to conduct inquiries leading to reasoned judgments. More systematic evaluation of the approach, especially in terms of the extent to which it enhances the making of reasoned judgments, would be an important subject for further research.

References

Bailin, S., and Battersby, M. 2009. "Inquiry: A Dialectical Approach to Teaching Critical Thinking." In *Argument Cultures*, edited by Hans V. Hansen, Christopher W. Tindale, J. Anthony Blair, and Ralph H. Johnson. Windsor, ON: OSSA.

Bailin, S., and Battersby, M. 2010. *Reason in the Balance: An Inquiry Approach to Critical Thinking*. Whitby, ON: McGraw-Hill Ryerson.

Blair, J. A., and Johnson, R. H. 1987. "Argumentation as Dialectical." *Argumentation* 1 (1): 41–56.

Ferraro, P. J., and Taylor, L. O. 2005. "Do Economists Recognize an Opportunity Cost When They See One? A Dismal Performance from the Dismal Science." *Contributions to Economic Analysis and Policy* 4 (1): Article 7.

Hamby, B. 2013. "Libri Ad Nauseam: The Critical Thinking Textbook Glut." *Paideusis* 21 (1): 39–45.

Hestenes, D., Wells, M., and Swackhamer, G. 1992. "Force Concept Inventory." *The Physics Teacher* 30: 141–158.

Hitchcock, D. 2002. "The Practice of Argumentative Discussion." *Argumentation* 6 (3): 287–298.

Hitchcock, D. 2013. "Review of Reason in the Balance: An Inquiry Approach to Critical Thinking." *Cogency* 5 (2): 193–202.

Johnson, R. H. 2000. *Manifest Rationality: A Pragmatic Theory of Argument.* Mahwah, NJ: Erlbaum.

Johnson, R. H. 2003. "The Dialectical Tier Revisited." In *Anyone Who Has a View: Theoretical Contributions to the Study of Argumentation,* edited by Frans H. van Eemeren. Dordrecht: Kluwer. 41–53.

Johnson, R. H. 2007. "Anticipating Objections as a Way of Coping with Dissensus." In *Conference: Dissensus and the Search for Common Ground,* edited by Hans V. Hansen, Christopher W. Tindale, J. Anthony Blair, and Ralph H. Johnson. Windsor, ON: OSSA. 1–16.

Jungwirth, E. 1987. "Avoidance of Logical Fallacies: A Neglected Aspect of Science Education and Science Teacher Education." *Research in Science and Technology Education* 5 (1): 43–58.

Kuhn, D. 1991. *The Skills of Argument.* Cambridge: Cambridge University Press.

Lipman, M. 1991. *Thinking in Education.* Cambridge: Cambridge University Press.

Moore, B., and Parker, R. 2012. *Critical Thinking.* (tenth edition). New York: McGraw-Hill.

Nosich, G. 2012. *Learning to Think Things Through: A Guide to Critical Thinking across the Curriculum.* Tomales, CA: Foundations for Critical Thinking.

Paul, R. 1990. *Critical Thinking: What Every Person Needs to Survive in a Rapidly Changing World.* Edited by A. J. A. Binker. Rohnert Park, CA: Center for Critical Thinking and Moral Critique.

Siegel, H. 1988. *Educating Reason: Rationality, Critical Thinking, and Education.* New York: Routledge.

Siegel, H. 1997. *Rationality Redeemed? Further Dialogues on an Educational Ideal.* New York: Routledge.

Vaughn, L., and MacDonald, C. 2012. *The Power of Critical Thinking* (second Canadian edition). Don Mills, ON: Oxford University Press.

Waller, B. 2011. *Critical Thinking: Consider the Verdict.* (sixth edition). Upper Saddle River, NJ: Pearson.

8
Debate's Relationship to Critical Thinking

Stephen M. Llano

How should we imagine critical thinking? Briefly searching the Internet reveals a large number of activities and classroom concepts that treat critical thinking as a skill. Many of the best activities and approaches start with the classroom and the material of the course, developing the critical thinking skill set that should transfer to the activity of daily life.

Marshall Gregory writes about the difficulties in teaching British poetry to undergraduate students. The subject—much like critical thinking—is estoteric, remote, and difficult for students to see as valuable outside of something they must do in a classroom just to satisfy a teacher. Gregory's teaching of a difficult subject improved when he reframed the issue around the terms of means and ends: "Students do not get educated because they study our beloved content. They get educated because they learn *how* to study our beloved content, and they carry the how of that learning with them into the world as cognitive and intellectual skills that stick long after the content is forgotten. In short, the curriculum is not an end it itself" (emphasis mine)(Gregory 2005, 97). It is easy for us as teachers of many subjects, including critical thinking, to mistake means for ends. It comes from a deep passion for critical thinking, and oftentimes we forget that students need to be deeply connected to the material in order to apply it in their lives past our classrooms.

I am concerned that often when we use debate as a way of teaching critical thinking, we fall into the trouble pointed out by Gregory. We might use it as a test, some way to see if critical thinking happened in the past, instead of using it as a lesson in how to generate critical thinking as a practice. I suggest in this chapter that we should consider debating as the teaching of critical thinking itself instead of just a tool of evaluation. I believe that considering debate in this way can help students see critical thinking as a set of cultural practices rather than as a particular skill useful only in certain situations.

Why do I suggest debate as a method? Debate requires careful and thorough thinking in a specific context to reach the minds of others—the audience.

Debate also requires thought to be articulated, oftentimes quite spontaneously, as the product of a critical response to someone else's utterance. Debate is a good practice arena for the timely production of thoughtful, persuasive messages meant for an audience—and the audience is the key factor. Debate forces students to make hard choices with little certain facts in a short amount of time, then present them to interested parties for assent.

What happens when we develop debate as a method of teaching critical thinking? I hope to show that debate encourages students to conceive of critical thinking as a culture rather than a skill set. As a culture critical thinking becomes a normative system of actions and beliefs that one performs in order to participate in valuable interactions within society—interactions that are key to healthy coexistence. As a cultural practice, it is similar to studying language or music—something that can appear to be a technical skill, but manifests itself more as a part of one's life, finding connections between the practice and the more mundane aspects of living. The conception of having students debate instead of other traditional means of coursework is the exercise that can help students bridge the gap between the way they conceive and are taught to conceive of critical thinking and daily, lived life full of evaluative possibilities. Focusing on the practice of critical thinking within the general context of university culture can help students understand better how critical thinking is vital for social living.

Critical thinking as culture

There are many definitions of critical thinking available in numerous books, essays, and other resources. With so many definitions vying to get all of the specifics of what critical thinking could be into a couple of sentences, we might lose sight of what matters most about the skill of critical thinking. Pedagogical scholar bell hooks offers a simple definition—"finding the answers to those eternal questions of the inquisitive child, and then utilizing that knowledge in a manner that enables you to determine what matters most" (hooks 2010, 9). In order to get there, hooks offers a number of practices—everything from asking the well-known "five Ws" students learn in the United States for descriptive writing, to the college classroom where the professor must ensure his or her mind to be open to not knowing everything about the subject. In short, hook's definition is valuable due to its openness, and the way it calls our attention to critical thinking as a living process.

Approaching critical thinking this way avoids some of the murky pitfalls of conceptualizing critical thinking as a set of things to know, processes to deploy, or boxes to tick off when reading someone's argument. Nobody strives to teach critical thinking in this manner. What I am suggesting is that our passion for the subject might blind us from connecting with students on a very basic level as to what critical thinking is and what it can do for society.

I believe that critical thinking only has meaning if it is recognizable to students as an approach one takes within the contours of a situation. Critical thinking—if it is as fluid as hooks is suggesting—might have more in common with the art of cooking or the practice of yoga. Approaches to critical thinking that treat it as engineering, mathematics, or science might fall flat in student retention because of this. Critical thinking must be practiced regularly instead of rigorously. It is something that does not have a correct answer in itself. It is something that requires daily "work outs" in order to maintain its benefits. Critical thinking reminds me of learning a language, or studying a productive art like public speaking. This is why the art of debating—the very same contest debating that clubs on your campus practice annually—might be the best critical thinking activity we could marshal for the benefit of all of our students. Debating is contextual, temporal, critical, and artistic—and traditionally it has been associated with the ability to think clearly. Whether one participates in a debate as a speaker or as an audience member, debate has a way of calling to our critical sides in ways that are unique.

Understanding debating

Debating might be understood as a competitive platform or perhaps a test of critical thinking ability—some dazzling words offered in a short amount of time on a controversial issue, and then set before judges to vote for a winner. Competition debates are very much like this, but this is not all there is to debating.

Debating in a classroom setting can also be problematic. Consider this description from Deborah Tannen who observes a classroom where "the students are engaged in a heated debate," and "the very noise level reassures the teacher that the students are participating, taking responsibility for their own learning." But all is not rosy for Tannen: "The majority of the class is sitting silently, maybe attentive but perhaps either indifferent or actively turned off. And the students who are arguing are not addressing the subtleties, nuances, or complexities of the points they are making or disputing. They do not have that luxury because they want to win the argument—so they must go for the most gross and dramatic statements they can muster" (Tannen 1998, 256). The risk involved here is of using the debate without critical engagement of the topic. Instead of using debate as an activity like this, I suggest using debate as an orientation to critical thinking—as an act with students that generates an experience, or a "text," which then can be evaluated by all who participated, either as an audience member or as a speaker.

Understanding this definition of debating requires understanding rhetoric. Rhetoric, for most of human history, has been understood as more than a derogatory term for unfounded speech that causes erroneous beliefs. Before this modern definition took hold, rhetoric was taught and understood as the art of making things—facts, people, events—make sense within contexts for particular audiences. Reading and writing, both thought of traditionally as subsets of the grander art of rhetoric, are today where we concentrate the most on the idea of teaching a practiced art, that is, critical thinking.

It was Aristotle who claimed that rhetoric was the art of seeing all available means of persuasion in "a given instance" (Aristotle 1984). In the modern era, scholars of rhetoric have pushed with critical acumen the idea of rhetoric to cover what we would consider to be the realm of critical thinking: "Rhetoric happens in unfinished historical episodes, wherein urgent circumstances require that we act, even though we lack complete, reliable grounds for determining what the best action might be" (Farrell 1995, 277). This could also define those situations where critical thinking is paramount. Scholars have argued for many years now that rhetoric is also a perspective, a way of crafting and making meaning. Scholars have written that rhetoric can be used as a curriculum toward the development of other skills and knowledge (Leff 2006). It has also been suggested that rhetoric is an epistemic frame, a way of knowing the world (Scott 1967). Rhetoric serves as a perspective in online searches, where the classification, organization, and algorithmic management of results persuades the researcher that the information offered is comple (Johnson 2012). Rhetoric has been treated as an aesthetic perspective as well (Whitson and Poulakos 1993). Rhetoric has been defined in modern scholarship to be the study of identification between humans and the subsequent division that arises from each identifying move (Burke 1968, 23).

Debate, from the rhetorical perspective, gives students a chance to assess not just the work of others, but of each other as well. They become responsible for the crafting of persuasive messages about uncertain issues, and they become responsible for evaluating those messages when they hear them from each other. We can think of critical thinking with rhetoric as a softening force on both the unknown—the issue facing us—and the "known"—the world, the facts, or whatever people face in their decision-making process. Softening means that alternatives—or room to maneuver—are provided to the decision makers, advocates, or stakeholders.

Many people approach an issue with fixed ideas about it—certain that they are right, they believe that charging into the fray, arguments blazing, will help the truth prevail. Countering this approach on these terms begins with a point of clear agreement: It is certain that the issue before us is uncertain, otherwise we would not be having this discussion. It is on that point that rhetoric comes into play. As Chaim Perelman and Lucie Olbrects-Tyteca point out, "For argumentation to exist, an effective community of minds must be realized at a given moment" (Perelman and Olbrechts-Tyteca 1969, 14). The students using debate must begin thinking critically right away: Who are these minds who are concerned about this issue? The focus changes from the correctness of the facts—verified elsewhere, perhaps in the lecture or another class—and the material concerns of the people to whom the debate might be addressed. Highlighting uncertainty about the audience requires students to return to the question with new eyes and analyze the information with the view they imagine the audience might have on the controversy.

This softening influence is apparent when advocates reexplain either the facts or the problem in different terms, helping the decision makers see the issue(s) from all available positions. It is this softening effect of rhetoric as culture that debate allows individuals to practice. This practice is both a coming into familiarity with critical thinking as well as becoming better at it. In short, critical thinking helps people see the world as more contingent, and more changeable, because they are attending to the connections making the position "hard." Rhetoric and critical thinking serve as a highlighter to these hard points, forcing new explanations of their truth, and encouraging rearticulation.

This idea is similar to a critical thinking exercise suggested by Peter Elbow as a way to teach the art of writing: the "doubting game," which we all learn very well to play when we encounter arguments from others, and the "believing game," which is not well taught—how to find ways to support or believe the arguments we hear from others. He writes, "the function of a group in the believing game is for people to help each other believe more things, experience more things, and thereby move away from the lowest-common-denominator tendency in a majority conclusion" (Elbow 1998). This sort of brainstorming is a necessary part of the generation of new knowledge and is one of the better ways to discover if an idea warrants rejection. Elbow continues: "Doubting is my shorthand for criticizing, debating, arguing, and trying to extricate oneself from any personal involvement with ideas through using logic. 'Believing' is my shorthand for listening, affirming, entering in, trying to experience more fully, and restating—understanding ideas from the inside" (Elbow 1998). Elbow believes, as I do, that building both sides is key to the complete critical thinking practice. We need to encourage students to build up reasons as well as take them apart. Debate does both these things masterfully.

Debate is one of the best activities in the classroom for firming up both the believing and doubting games. Students participate in the debate whether they speak or listen. Since each idea in a debate is presented softly—meaning with as many connections of the shell exposed—the audience can begin to investigate what is holding that shell together. But how do we get engaged, valuable debate in our classrooms and on our campuses?

What debating looks like

On your campus, wherever you might be reading this, there is a vibrant and energetic debate club. This club is run by student volunteers most likely, or in rarer cases has a faculty member who is responsible for the club's activities. It is my strong suggestion that you reach out to this resource in order to help you plan to use debate for teaching critical thinking in your particular course. These students love debating, and they are quite good at putting on an engaging show for your class as to what a debate can look like and sound like. The only negative is that they might be so good they might provide a bit of an uneasy chilling effect for your class to start debating—they may believe that they could never be that good.

Debate comes in many different formats and many different varieties, but for my purposes, that is to teach critical thinking in a rhetorical context and soften positions for examination, some basics must exist. It does not matter what particular debating format you choose for your class as long as it has a few important features. There should be a topic selected—not a question, but a statement that one can agree or disagree with. Second, there should be time limits to the speeches, something reasonable, and there should be a defined order of speakers. Questioning of one speaker by another should also be a part of the debate, but whether it takes the form of an American courtroom cross-examination or a parliamentary session's points of information does not matter. Asking the students what they might like to do also helps in creating an engaged atmosphere.

Assigning a debate on a relevant public issue as an activity in a course allows students to practice both the believing and doubting game. First, a motion must be selected. It should be a statement that can clearly be proven or disproven, given good arguments for that side. Students first must read and research, preparing the case for either side. This exercise forces them into groups to try to encourage one another what proof and evidence will be most persuasive. Most importantly, students should not be allowed to select what side of the issue they will represent. This should be done randomly. The reason behind this is to keep their attention on the construction and deconstruction of positions rather than on the passion students have for their own beliefs. Generally, this is not much of an issue for students as they might not have passionate feelings about the issues you are discussing in your class. After having a debate, this might change—be warned.

Let me take you through a detailed example to demonstrate how this approach works. I was asked to consult a scientific reasoning course instructor who wanted to have a debate on the question of invasive species. First, I suggested framing the issue as a clear statement—this way everyone can see what is at stake. We chose the following motion: "Invasive species are more harmful than they are beneficial." Immediately, the students began to come up with arguments on both sides from the material they had read in the course. I asked them who they thought might be concerned about this topic. After that question was discussed, the arguments became less about the unassailable nature of the evidence, and more about how and why that evidence might be compelling to different groups. Instead of a heated focus on who had the "best evidence," students had to critically consider the information and the person receiving the information. They thought about reasons both for and against the supporting material, and why certain audiences would accept or reject different sources.

Here we see the debate's unique ability of combining both the believing and doubting game into two sides of one practice. In debating, one must doubt in order to believe, and vice-versa. The believing game and doubting game become parts of a complete process. Likewise, students understand as they prepare to speak and reflect on their speech the softness of the connections that make up belief and doubt.

Students who are not presenting on a particular day serve as the audience. Creating a panel of judges from them or putting them in groups of decision makers to confer and offer critique of the debate are obvious tasks that help further critical thinking instruction among the speakers as well as the audience. Less obvious would be to put students in teams to compare and contrast what was said with what they thought should have been said. This puts the whole classroom into direct engagement with the debate topic at hand. This engaged form of classroom teaching is what hooks would call essential to the practice of critical thinking. Instead of judging the truth of the question, students judge the persuasive effort, and evaluate whether or not the speakers did well in approaching the topic for the various individuals who may be concerned about it.

Critical thinking, seen as a daily commitment rather than a hammer to be used when a nail appears, allows for critical thought and analysis on more occasions than our pedagogy currently allows. Of course, there are still objections. Some might claim that the time limits and the focus on winning the debate will detract from the ability of debating to teach critical thinking. I would agree, if debating is not situated in a rhetorical context—that is, a context where persuasion is viewed to be many things, and responsible for why we may call some phenomena information and other phenomena nonsense. The rhetorical perspective is concerned with the making and unmaking of meaning. Debate, practiced within this context, encourages students to critically explore why and how facts garner meaning for audiences.

Toward critical thinking as culture

The musician and the polyglot must know the culture in order to make their art meaningful. They must know situation, timing, variances in performance (accents come to mind as well as interpretations of classical works), as well as audience tastes. Debating, I argue here, is what makes critical thinking a meaningful art. Debating provides timing, situational contours, and variance in performance. All of these things make the assignment challenging and interesting, and all of them together help students create an important emotional and intellectual connection to the art.

Why should we use debating to teach critical thinking? I believe it is because we should think of critical thinking as a culture, and not just a set of skills. I am not making the argument that debate is the best way to teach critical thinking, nor am I saying that it is the only valid method. It does have one distinct advantage in the fact that debate lays bare not just the issue and the facts behind the issues, but focuses attention on the connections humans make between these elements. Also, debate disconnects us from our normal processes of evaluating what is right and makes us focus on the processes themselves. Robert Branham writes:

> Willingness to subject one's opinions to disputation is a necessary but not sufficient condition for the achievement of true debate. For an opinion to be truly tested, it must first be given the strongest possible expression. The best available arguments for the opinion must be advanced, and supported by the most powerful evidence and reasoning that can be mustered. It is not enough to hold an opinion that turns out to be true; one must have come to that opinion for the best reasons. (1991, 3).

Debate becomes the exercise of separating being right from proving right. This exercise is one of exposure and articulation. It is an art of drawing out connections and questioning them. The audience and the debaters evaluate the speaker's persuasiveness not just on the luck of having stated or stumbled onto a correct position, but on how that position is expressed and supported. It is like the age old adage that mathematics teachers use when they tell their students, "show your work" or receive no credit. Debate forces students to show their work, which in turn exposes the rhetorical connections between culture, issue, and proof. This is the operation of softening the controversy so it can be examined. Instead of focusing on finding the facts that prove the other side completely wrong, debaters must identify, describe, and ascertain the connections made between evidence, subject matter, and claims for different sides of any issue.

Oftentimes in our current political climate, controversies are far too polarized and past the point of critical thinking doing any work. The cry for "critical thinking" becomes merely a weapon in the arsenal of one side or another to try to coerce agreement. Critical thinking becomes a trope of one side of the controversy, opposed to a process by which polarization can be undone. Critical thinking, taught through debate, confuses traditional "right answers" with the vexing conclusion—"What if there is no right answer?" Critical thinking, as uncomfortable as it may make us feel, is about these very moments. What do we do if we don't know what to do?

Debate helps with this question. Debate softens the outer shells of controversial issues, allowing the participant to see into the interior workings of a position on an issue. The softening effect of debate helps students explore multiple convictions of things taught in the classroom and gives them free rein to try to amplify or discredit different views. This way they can explore the connection to their own view on the issues being debated.

The unique provision of debate—providing both opportunities to play the doubting and believing game—can also be a source of danger to the student. Ronald Walter Greene and Darrin Hicks criticize the practice of switch side, or assigning sides in a debate, claiming it produces a liberal democratic exceptionalism. This method becomes the value-added form of debating, preparing the students for an imagined role as the purveyors of proper liberal values:

> Debating both sides as a technique of moral development works alongside specific aesthetic modes of class subjectivity increasingly associated with the efforts of the knowledge class to legitimize the process of judgment. Debating both sides reveals how the globalization of liberalism is less about a set of universal norms and more about the circulation and uptake of cultural technologies. (Greene and Hicks 2005)

The result of this is the creation of liberal subjects who can use the technology of seeing both sides to make debate seem as an end in itself—as part and parcel of being a good democratic citizen and therefore can subvert issues such as class and race to the idea that giving up one's personal conviction to explore ideas is the ideal form of citizenship. "To set aside one's convictions and present the argument for the other side demonstrates that the citizen has forsaken her or his personal interests and particular vision of the good for the benefit of the commonweal. That is, the citizen recognizes the moral priority of democratic debate when she or he agrees to be bound by its results regardless of personal conviction" (Greene and Hicks 2005, 126). This recognition is only possible when using certain class privileges, which are obfuscated by the moral act of taking the opposite side for the benefit of others. This verbal alchemy is attractive in manipulating standards of judgment in the name of democracy.

Their critique of debating is a cautionary tale for those of us interested in teaching critical thinking. They warn that the most seemingly "neutral" forms of education and learning can easily serve larger power systems that deserve critique as well. Debate is not immune to forces that would make it a servant of other systems of power and authority under the guise of freedom. However, as far as debate is concerned, Greene and Hicks localize their critique to the particulars of tournament debating:

To be sure, debate was, and continues to be, an effective way to channel and form competitive instincts to improve critical thinking skills by helping students learn to analyze a case. It may also be true that debating both sides is an important part of the value of debate to help students analyze a case. The ethical status of the technique, however, is not guaranteed by this success. Instead, the ethical status of debating both sides increasingly depended on how tournament debating was envisioned: namely, as a pedagogical technique or as public speaking (Greene and Hicks 2005, 103).

The exercise that I am suggesting is not tournament competition, but could be seen as channeling competitive drives in order to get students to play both the believing game and the doubting game as part of critical thinking. As a culture, critical thinking should be characterized by norms of behavior by the members of that culture. Without a tournament, debating can then set its goal as audience education or exposure to the various claims students have constructed as they play Elbow's games with one another during the assignment. For the audience, these norms will come across as missing puzzle pieces to a picture that they may believe has been completed. As one audience member remarked after a public debate I hosted, "I came here to see some blood, to see a fight, but I'm leaving here unsure as to my own opinion on these matters." This is the result that is possible when critical thinking is taught through debating as if it were a set of norms—or a culture.

Debating as the pedagogy of critical thinking culture

Replacing one or two assignments in a course with a public debate can go a long way toward the goals we seek with critical thinking—confidence in questioning the premises of a claim within the situation that claim is raised. Our current pedagogical norms are not up to the challenge of teaching critical thinking. As Mitchell observes, "Many of the received approaches to pedagogy are not up to the task of energizing students to play positive roles as public deliberators" (2000, 124). A supporter of a more "dialogic" approach in the Frierian sense, Mitchell argues that role playing positions in controversial, debatable topics open up alternative classroom roles for students and teachers, thus making the students the producers and editors of a knowledge that typically would flow "downhill" from the instructor providing readings or a lecture on the subject. Mitchell is arguing for activities where students take on the perspective of a particular stakeholder in a controversy and play out what that group might say. Students imagine, through brainstorming and reading, what things each person or group might argue in a "real" case out in the world.

Although not a critique of the power of role play, Mitchell does seem to leave out that many students engaging in traditional debate are "role playing" themselves, quite convinced on an issue. For public debate purposes, taking on a role, like an actor playing a part, is not necessary. Students generate arguments for the audience to evaluate, so playing themselves as a much more convincing version of the self allows them the same ability to see how soft positions can become hard when clashing with opposing ideas. Mitchell's argument would work just as well if students were coached during the term about how they are learning the way a scientist, anthropologist, or literary critic thinks and speaks. This responsibility could convey to the student the importance of being in a position of importance—playing themselves as future experts in a field—and have them restrict their utterances accordingly. This also has the unique advantage of encouraging undergraduate or graduate students to imagine themselves in careers or societal positions that they never considered before as valid options in their lives. Teaching critical thinking through debate mentors students into fields they may have been unaware of otherwise.

Opening up debate events to the general student body may be one way of enhancing the practice of critical thinking as culture on a campus. A public debate is one that is put on for a larger community, about a controversy that contains polarizing positions. The idea is that the debate, conducted by well-prepared advocates who fairly represent their side, will loosen the audience's grip on extreme positions and allow discourse to emerge again where once silence fostered by rigorous commitment reigned. The discussion after the debate, even among audience members, is the goal of such an exercise.

I conducted one such debate with argumentation students several years ago. It attracted a number of undergraduate students, as well as top levels of the university administration (the provost was our invited respondent). (A video of the public debate can be found here: http://vimeo.com/36258385.) This debate represented a real challenge to students not just in synthesizing and understanding the course material, but also in developing the acumen of critical thinking—for the information must not just be presented, but re-presented to an audience of people who are not directly involved in the course. Such an exercise is not only good pedagogy, but an event that will live in the mind of the student for a long time after university is over, and if utilized well, can be a moment they understand as deeply entwined with the art of critical thinking.

Public debate centers on the audience. Students must present well-formed arguments in order to allow the audience to benefit from disagreement, or what in competitive debating slang is called "clash." Without clash, a debate becomes a public speaking contest without either side engaging the arguments of the others involved in the debate. Without engagement, audiences are left no opportunity to test their own beliefs against the strength of the presented arguments and responses.

There is no set model for how a debate should be structured. Generally, each side of the issue (however many your students determine there are) should be given equal time to speak in the course of the debate, and there should be some time given for questioning of one side by the other. The only thing that all debates should contain is a controversy with no clear solution, and each side should establish a clear thesis for their case. This will create a clash—or a clear disagreement—between the teams, which the audience can then use to evaluate their own belief systems and how they were constructed.

Conclusion

Understanding critical thinking as a lifelong practice—following bell hook's definition—is enhanced by offering debate as a way of teaching critical thinking practices. Instead of focusing on the evaluation of truth, debating allows students to focus on the evaluation of evaluation, softening their own positions as they work to try to see through the eyes and mind of interested audiences. Adapting their speeches to these audiences can then be evaluated by the class to see how well they did in the consideration of many critical perspectives.

Adding rhetorical debating to the curriculum could be a way to bring critical thinking culture to the university as a whole. Admiration for someone providing good reasons and dismissing irrelevant evidence would get more excitement from the audience than the happenstance that the speaker supports the side that any audience member already supported. Deciding to turn examinations into campus-invited events opens up the whole university as a place where the cultural literacy of critical thinking can begin, be reinforced, and be celebrated by the students, who then go on to long for this culture and perhaps reinforce it in their daily lives once they have graduated.

Replacing one traditional assignment per term with a debate where the public is invited to attend would be an excellent first step helping students both in the audience and at the podium explore sides of an issue related to any course including, but not limited to, what it is to speak on behalf of a discipline. The power of debate as critiqued by Greene and Hicks can be channeled to discipline student thinking and speaking to be representative of what it is to be a member of a field. Imagining themselves as scientists, historians, literary critics, or engineers is a valuable form of role play where critical discourse and thought are shaped by the norms of the discipline and the connections that can and should be made between facts taught in the classroom and advocacy in the convention chamber. Ultimately, we could see debating on most campuses helping keep the habit and practice of critical thinking alive not just in select classrooms but as part of what makes the campus experience as intellectually challenging as it is special.

References

Aristotle. 1984. *On Rhetoric*. Vol. 2, *The Complete Works of Aristotle*. Princeton, NJ: Princeton University Press.

Branham, R. J. 1991. *Debate and Critical Analysis: The Harmony of Conflict*. Hillsdale, NJ: Lawrence Erlbaum Associates.

Burke, K. 1968. *A Rhetoric of Motives*. Berkeley and Los Angeles: University of California Press.

Elbow, P. 1998. *Writing without Teachers*. (second edition). New York: Oxford University Press.

Farrell, T. 1995. *Norms of Rhetorical Culture*. New Haven, CT: Yale University Press.

Greene, R. W., and Hicks, D. 2005. "Lost Convictions." *Cultural Studies* 19 (1): 100–126.

Gregory, M. 2005. "Turning Water into Wine." *College Teaching* 53 (3): 95–98.

hooks, b. 2010. *Teaching Critical Thinking: Practical Wisdom*. New York: Routledge.

Johnson, N. R. 2012. "Information Infrastructure as Rhetoric: Tools for Analysis." *Poroi: An Interdisciplinary Journal of Rhetorical Analysis & Invention* 8 (1): 1–3.

Leff, M. 2006. "Up from Theory: Or I Fought the Topoi and the Topoi Won." *RSQ: Rhetoric Society Quarterly* 36 (2): 203–211.

Mitchell, G. R. 2000. "Simulated Public Argument as a Pedagogical Play on Worlds." *Argumentation & Advocacy* 36 (3): 134.

Perelman, C., and Olbrechts-Tyteca, L. 1969. *The New Rhetoric: A Treatise on Argumentation*. Notre Dame: University of Notre Dame Press.

Scott, R. L. 1967. "On Viewing Rhetoric as Epistemic." *Central States Speech Journal* 18 (February): 9–17.

Tannen, D. 1998. *The Argument Culture: Stopping America's War of Words*. New York: Ballantine Books.

Whitson, S., and Poulakos, J. 1993. "Nietzsche and the Aesthetics of Rhetoric." *Quarterly Journal of Speech* 79 (2): 131.

9

Thick Critical Thinking: Toward a New Classroom Pedagogy

Milton W. Wendland, Chris Robinson, and Peter A. Williams

Introduction

It is no novel technique to use current events as the source of classroom debates designed to develop the critical thinking skills of students. Such exercises typically divide the class into "pro" and "con," having students prepare arguments for both sides and then pick a side to debate in front of the class. As we will argue in this essay, while this exercise might provide insight into an issue of private and public concern and serve as a convenient classroom model, it may not adequately engage the complexity of the issue at hand and may be at risk of not fully engaging the possibilities of classroom exercises in the development of critical thinking skills.

Pro/con debating as a method of teaching critical thinking—what we refer to in this essay as the "traditional debate model"—can be useful, but it has three possible main limitations. First, this traditional exercise can conflate binary thinking with critical thinking, substituting an either/or framework for the holistic approach required by true critical thinking. Second, it can lead to labeling or "othering" those who hold the opposite viewpoint. Third, this model can possibly fail to recognize how power relations shape not just the issue at hand but also the mechanics of the activity itself. As such, the traditional debate model can tend to reproduce the facile stances that characterize these debates as seen on television news talk shows, on blogs, and in governmental proceedings; these public debates fall into a pro/con binary and often result only in heated and unproductive quarreling. The logic undergirding the traditional debate model reproduces the tendency in popular discourse to see others (or oneself) through a single identity category in order to avoid or end discussion and foreclose productive conversations. For example, simply asserting that those who oppose same-sex marriage are essentially hateful or ignorant people or that homosexuals and their allies are essentially immoral or perverse eventually leads to a rhetorical impasse where both essentialized sides of an argument cannot agree on anything, let alone engage one another on the grounds of their disagreement. However, foundational scholarship on race, gender, and sexuality (Butler 1990; Foucault 1980; Omi and Winant 1994; Said 1978)—not to mention our own lived experiences—has taught us that identity is mutable, multifaceted, and socially constructed, as are, therefore, our positions on any given topic.

To counteract these limitations and to move toward a pedagogical method that more effectively teases out the complexity of any issue, we propose a method called "thick critical thinking." Our method builds on the pro/con debate model and pushes students to see past widely circulated binaries to the many intersecting positions of any social issue, allowing them eventually to position themselves more effectively, convincingly, and compassionately. We borrow from anthropologist Clifford Geertz's concept of "thick description," which assumes that any culture consists of "a multiplicity of complex conceptual structures, many of them superimposed upon or knotted into one another" (1975, 10). The task of the ethnographer is to "contrive somehow first to grasp and then to render" the complexities (1975, 10). Informed by pedagogical and feminist theories, we translate Geertz's concept to the pedagogical realm, suggesting that thick critical thinking recognizes the complexity of human identity and avoids a reductionist either/or approach. After examining traditional critical thinking activities and explaining this new pedagogical form in more detail, we use recent debates over same-sex marriage in the United States as an example of how to deploy thick critical thinking in the college classroom. The authors propose that thick critical thinking can help to produce better students, writers, teachers, thinkers, citizens, and more informed and compassionate human beings.

The traditional modes of teaching critical thinking

Trying to define critical thinking is a bit like trying to teach it—a Sisyphean task that seems never to be fully complete. Inquiries into the precise nature of critical thinking have already been conducted in a range of fields, including cognitive psychology, pedagogy, philosophy, and child development (see, inter alia, Bailin 2002; Kuhn and Dean 2004; Lutz and Keil 2002; Paul and Elder 2006). While the purpose of this article is not to redefine the concept itself, we do operate under a general notion that critical thinking is "that mode of thinking—about any subject, content, or problem—in which the thinker improves the quality of his or her thinking by skillfully taking charge of the structures inherent in thinking and imposing intellectual standards upon them" (Paul and Elder 2008). Despite the rather expansive nature of this definition, we find that in many humanities and social science classrooms, critical thinking is still too often equated with a traditional debate model, and classroom activities designed to develop critical thinking skills rest almost entirely on the idea that "critical thinking consists of seeing both sides of an issue" (Willingham 2007, 8). Our critique is about more than whether or not a classroom exercise meets the specific time-bound requirement of a particular lesson unit. We are instead concerned that overreliance on the traditional debate model of enacting critical thinking in the higher education classroom has implications beyond the immediate classroom experience and its outcome and disadvantages both instructors and students.

Foundational research into human cognition and ways of knowing has generally held that individuals develop thinking skills through three stages. The first is "dualism/received knowledge" in which a student largely views the world through an either/or lens and relies heavily upon external authorities for guidance in choosing a side. The second stage is "multiplism/subjective knowledge" in which a student can recognize that multiple points of view exist but may have difficulty discriminating among them. The final stage is "relativism/procedural knowledge" in which a student recognizes not only that multiple points of view exist but also that multiple pieces of evidence of differing qualities may support different points of view (Belenky, McVicker Clinchy, Rule Goldberger, and Mattuck Tarule 1997; Grasha 1996, 217–220; Perry 1970; 1981). This three-stage model, while revised and extended by other theorists, remains the foundational way of understanding and modeling critical thinking development, although Perry himself included a fourth stage—commitment—in which students are able to commit to a position while still remaining open to revising that position as new information arises and are open to engagement with other thinkers in ways that produce both individual and collaborative changes to the commitment (Kitchener and King 1981; Perry 1970; 1981). Further ahead we will discuss how our thick critical thinking model aligns with the four stages outlined here.

It is no new observation to note that educators of all sorts understand the goals of critical thinking and recognize the limitations and difficulties of developing critical thinking skills in their classrooms. They observe that "forced dichotomous choices" are unsatisfactory and recognize the importance of students realizing "how difficult it is to try to resolve hotly contested, emotional issues or to remain neutral during revolutionary times" (Frederick 2002, 63). Yet class sizes, time limitations, and other pressures often cause instructors to revert to the standard argument that "debates in class are a good way to challenge students to move beyond their dualism." Instructors then design classroom exercises that wholly rest on dualistic thinking: "Burke or Paine on the French Revolution? Karl Marx or Adam Smith on the industrial revolution? Evolution or creationism?" (Frederick 2002, 62). Or, as in our driving example: for or against same-sex marriage?

Indeed, this debate model structure is so endemic to pedagogical thinking that although this sort of exercise is sometimes disguised under different

names—for example, town hall discussion, active debate, or critical debate—
the structure is still largely framed around a binary approach. For example, in
a "four corner debate" each corner of the classroom takes on a degree desig-
nation—strongly agree, agree, disagree, and strongly disagree—and students
congregate physically to work out the strongest arguments for their position
and present them to the entire class (Hopkins 2003; Kennedy 2007). Role play
debates, fishbowl debates, and meeting house debates similarly attempt to confront the dualism of the traditional debate setup, and while each certainly
has value in the classroom, each fails to address the basic conflict this article
identifies, simply multiplying points of view or opinions rather than doing the
thicker work of investigating the context and milieu in and from which such
opinions are produced (Brookfield and Preskill 1999, 114–115; Lowman 1995,
169; Silberman 1996, 84–85).

Even when scholars enliven the pedagogy by analogy to other fields, as when
Timpson and Burgoyne liken teaching in the classroom to performing on a
stage, the underlying assumptions of the exercise remain the same. For example, Timpson and Burgoyne argue that the conflict in a dramatic enactment of
opposition in the classroom produces engagement and excitement. However,
the examples Timpson and Burgoyne offer are again limited largely to binary
conflicts: "opposition . . . over courses requiring animal dissection, oppositions
pitting new theories against the traditional canons, conflicts between envi-
ronmentalists and developers" (2002, 93). Although the preferred exercises are
all worthwhile and indeed quite useful for early development of critical thinking skills, the use of dramatic opposition to heighten student excitement in
the material still fails to address fully how to move students beyond binaristic
approaches.

As Nancy Tumposky has aptly observed (2004, 53–54), the traditional debate
model constructs an epistemological context that "reinforces a Western bias
toward dualism," reducing complex issues into pro/con positions. It also creates a classroom dynamic that rewards winning and favors students who excel
in structured conflict with little or no real recognition of how such a model
replicates binary notions of the self and other. Scholars of critical thinking
such as John McPeck have called for a reexamination of the traditional debate
approach to critical thinking, recognizing that the "real difficulty in assessing these questions has little to do with assessing fallacies and validity, and
almost everything to do with understanding complex information" (McPeck
1984, 35). Classroom exercises organized around a binary approach to thinking about an issue can become problematic when they "validate a point of view
that most reasonable people would find lacking legitimacy" by placing it on
equal ground with other points of view and failing to interrogate that positioning, creating the false impression that all points of view are necessarily equally
valid or logical (Tumposky 2004, 53–54).

Additionally, the traditional debate model can suppress insight and creativity by restricting students to a pro/con approach to critical thinking, thereby narrowing the fields of possibility for achieving the very ends it claims to seek. For example, grouping a multitude of reasons that might influence or inform an opinion on the issue at hand into only two categories (pro/con, yes/no, for/against), rather than productively addressing multiplicity as constitutive of critical thinking around any issue, is ultimately an "attempt to control thinking and action because it asks [students] to conform to the known world" (Hughes 2003, 59). The binary approach also has the effect of creating artificial boundaries (often spatially enacted in the classroom) between two sides, which implicitly leads each side to "other" the "opposite" side. In this process, the "other" ("the other side," "the opposing view," "the other person") is not fully recognized on its own terms with its own sound reasoning but rather as a projection based on how it does or does not match up with students' own senses of the issue; in other words, the "other side" is seen as opponent, adversary, or rival to be fought against and conquered and not necessarily seen as a separate although equally viable position vis-à-vis the issue at hand.

As we noticed in our own classrooms and conversations, a traditional debate approach to same-sex marriage tends to perpetuate the idea that those opposed to it are hateful or ignorant and that those in favor are progressive and equality oriented. While in part these characterizations may be true, they fail to delve into the far more complex reasons that anyone may hold a position on the issue. In the example of the amendment to the North Carolina state constitution to define marriage as a union between heterosexuals, someone who opposes same-sex marriage may still oppose the amendment, arguing that such private matters should be managed by state legislatures rather than constitutional mechanisms. Conversely, someone in favor of same-sex unions may argue in favor of the amendment because of a belief that traditional marriage itself is a social institution that needs to be reconsidered legally and culturally. The classic pro/con debate may not delve into these deeper, thicker issues.

Finally, the traditional debate model runs the risk of veering remarkably close to re-enacting the very privilege that it is often aiming to expose and critique. The traditional debate model leaves unquestioned the liberal humanist notion that human beings exist as individuals wielding both self-control and vaguely equal avenues of opportunity. This ideology goes unquestioned because the traditional debate model does not afford students the opportunity to recognize that choosing between two sides is a privilege possessed only by a few. The privilege to choose a side in the first place—as if all issues can be examined and debated in a vacuum separate from lived experiences, the vagaries of necessity, or the immediate requirements of survival independent of any system of ethics or morality—relegates contextualized human experience to the margins. Furthermore, the choice of a side may be arbitrary or artificial (students assigned by name, randomly, or just by on the spot choice) rather than because a student has even a passing interest in the issue. These arbitrary choices may often result in lack of authentic student engagement, distancing them from the very real implications of such choices.

Our critique of the traditional debate model joins the ongoing pedagogical search for ever-improved ways of teaching deeper critical thinking skills and is informed by work in higher education, feminism, and critical pedagogy. We now turn to some of these theoretical and practical underpinnings.

Foundations of thick critical thinking

In order to address the potential pitfalls and limitations of using the traditional debate method of teaching critical thinking, we propose the thick critical thinking method as an alternative. We suggest that this method will not only enrich the immediate classroom learning experience but also allow students to develop skills over time. Thick critical thinking is rooted in multiple peda-gogical and theoretical foundations: general philosophical and pedagogical debates about the best methods to teach critical thinking, feminist standpoint theory and intersectionality, critical pedagogy, and the work of anthropologist Clifford Geertz.

The thick critical thinking method is in part informed by the "infusion" approach to critical thinking as described by Davies (2006). The infusion approach stems from debates concerning whether a generalist or specifist method is the most effective way to teach critical thinking in higher education (Davies 2006; Ennis 1992; McPeck 1984; 1992; Moore 2004; 2011; Robinson 2011). The generalist approach proposes a broadly applicable, structured critical thinking framework that can be applied to any issue in any field or discipline. The specifist approach, contra, holds that critical thinking skills are specific to each discipline, tending toward over-reinforcement of disciplinary bound-aries, and, thus, inhibit interdisciplinary conversation. The infusion approach operates productively among the two, favoring general, transferable skills in recognizing and constructing sound reasoning but which can be utilized by and among multiple disciplines. As we will demonstrate here, our method involves implementing a general framework to foster critical thinking that can be adapted and used in other disciplines. Thick critical thinking is not meant to break the "impasse" that Moore (2011, 264) notes in the "generalist/specifist" debate; rather, our method extends the "infusion" approach. It does so insofar as it requires students to engage in the transferable process of thick critical think-ing within the context of a specific topic, issue, or discipline, and attend to the contextual elements of varying points of view and how they are produced.

Another foundation of the thick critical thinking method arises from our own interdisciplinary commitments in the fields of American studies, women's studies, and cultural studies, all of which recognize the multivalent nature of identity as well as the many different ways in which subjectivities shape and are shaped by public debates. Two of the preeminent forms of understanding this process come from feminist academic work. Standpoint theory addresses the ways in which one's point of view depends upon one's cultural positioning and hence how knowledge is socially constructed. Intersectionality recognizes that for any single person, multiple axes of identity (e.g., race, class, gender, and sexuality) may intersect in different ways to produce different vantage points (Crenshaw 1991; Haraway 2004; Harding 1993; Hartsock 1998). In our example of the same-sex marriage debates, both standpoint theory and intersectionality suggest that simply identifying a person as a conservative heterosexual would not necessarily guarantee or predict that person's stance on the legality of same-sex marriage. Standpoint theory and intersectionality inform thick critical thinking in that they emphasize the need to recognize the complexity and interconnectedness of any given issue. Whereas the traditional debate model, if done poorly, can lead to simple reductions and categorizations, thick critical thinking can lend itself better to answering the calls that standpoint theory and intersectionality make.

Feminist standpoint theory and intersectionality also align thick critical thinking with the goals of critical pedagogy, which is highly influenced by the work of Paulo Freire (2000). Stephen Sweet defines critical pedagogy as "an outlook that questions the legitimacy of existing systems of hierarchy as related to issues of race, gender, class, disability, sexual orientation, or other socially constructed divisions between people." Critiquing hierarchical systems also means critiquing existing forms of knowledge; therefore, Sweet notes, critical pedagogy is also interested in social change (1998, 101). Critical pedagogy encourages the valuing of alternative perspectives, which, Collin Hughes (2003) insists, must be part of critical thinking. We envision our model of thick critical thinking as continuing critical pedagogy's project of challenging assumptions, questioning existing binaries, and encouraging students to understand multiple and alternative viewpoints.

We borrow the "thick" in thick critical thinking from anthropologist Clifford Geertz. In his seminal essay "Thick Description: Toward an Interpretive Theory of Culture," Geertz (1975) responds to the idea, widely held among anthropologists of the time, that a cultural analyst or ethnographer can adequately map out and explain the behaviors, motivations, and interactions of a particular group of people. Geertz criticizes the "hermetical approach," which requires the ethnographer to lift and isolate elements from their context, deduce relationships extra-contextually, and then to retroactively "characteriz[e] the whole system in a general way" (17). Instead, Geertz argues that ethnographers should attempt "thick description" as a method of understanding the "multiplicity of complex conceptual structures, many of them superimposed upon or knotted into one another, which are at once strange, irregular, and inexplicit" (1975, 10). The ethnographer's task is thus "first *to grasp and then to render*" that multiplicity by being able to write or speak about it, represent it, or re-present it (1975, 10). Such is the analogy with how critical thinking is handled in the higher education classroom today, in which the traditional debate model can tend to bracket off and isolate issues from the complex lived experiences of those participating in the debate. In response, thick critical thinking requires students to more fully comprehend any given phenomenon and to articulate its complex nature, going beyond taking a position on two sides of a debate.

The goal of thick critical thinking lines up with the goal of the analysis in thick description. Geertz argues that cultural analysis is "guessing at meanings, assessing the guesses, and drawing explanatory conclusions from the better guesses" (20). Students who engage in thick critical thinking learn to be comfortable with complexity, openness, and a paucity of tidy explanations for any viewpoint, but as with Geertz, they learn to be detailed and specific in their analysis and their discomfort. Thus thick critical thinking is characterized by five primary qualities:

1. Context: All debates are situated in social and historical contexts; they never arise spontaneously.
2. Complexity: Thick critical thinking aims at complex description, not simplistic categorizing or generalization.
3. Difference: Valuing complexity also means valuing difference.
4. Reflexivity: While understanding multiple points of view would seem to imply a certain "objective" stance, it actually admits that the perspective of the scholar is also tied up in power relations and identity.
5. Power: Thick critical thinking is also interested in power—the ways that current debates over any issue, whether moral panics, hot topics, or crises, represent struggles over meaning and representation in society in a larger sense.

With these points in mind, we now pose an example of the thick critical thinking model at work, using same-sex marriage in the United States as a test case.

Thick critical thinking in the classroom

Thick critical thinking as a pedagogical approach for the classroom consists of four steps, as the authors see it:

1. identifying a debate,
2. listing the standard pro/con arguments for both sides,
3. complicating the binaries by fleshing out the stakes on both sides, and
4. attempting to render the complexity of the debate in some assessable form.

The first two resemble the traditional debate model, and the second two build on and complicate that model. Our model is not intended to map exactly on to the framework of stages of human cognition mentioned above, but it does relate to that framework generally. Steps one and two correlate roughly to the first stage of human cognition mentioned above, dualism/received knowledge. Step three matches roughly with the second cognitive stage, multiplism/received knowledge, and step four aligns with the final cognitive stage, relativism/procedural knowledge. The fourth cognitive stage, commitment, is also captured in our fourth step of thick critical thinking. After elaborating on each of our steps, we will provide a brief example.

Step one is common to both the traditional debate model and to thick critical thinking—identify a matter of contention or an issue of concern. While the instructor might present a problem of her own choosing, students should work toward being able to identify problems themselves. These could be "hot topic" current debates occurring in public forums or long-standing ethical, philosophical, or historical issues in particular fields or disciplines. Students should work toward describing the debate as specifically and objectively as possible. In step two, instructors and/or students identify the binaries that exist in the current debate and create a list of arguments on both sides. Here, the binary arguments often found in these debates are shown to be simplistic and reductionary, yet ubiquitous; students should have little difficulty identifying the ostensible two sides because binaristic approaches are commonplace in both the classroom and the larger world. The instructor might brainstorm with students these opposed arguments; instructors or students could also present examples from newspaper editorials, blogs, cable news discussions, or other public forums.

Step three moves beyond the binaries to identify and flesh out a more complex view of the problem or debate. This involves examining what is at risk for each side. In this step, students are asked to consider who has a vested interest in the debate and why they would be interested. This requires that both the students and instructors develop a sensitivity to the difference between generalizations and stereotypes. The former can be helpful in understanding social location of actors in a debate, but the latter—a product of binary thinking—tend to shut down useful discussion. Specificity in the form of a "thick description" may help overcome this tendency. The instructor could ask students to identify what groups or subgroups might take a particular position on the debate and what identities would be associated with those groups. If students use greater specificity here, they will have more difficulty relying on stereotypes.

Step four asks students to "begin to render"—that is, to translate or depict in some form—the complexity of the issue and to articulate their own positions. Students demonstrate their grasp of the previous three steps by using them to analyze a given argument taken from any public source, like a periodical, blog, or television news debate. In addition, they can demonstrate reflexivity by positioning themselves in this debate. This might be through an exploratory essay, a journal entry, a series of essay questions, or short answer questions on an examination, or through an imaginative writing piece.

The entire process is cumulative, each step building on the one before it. Instructors should evaluate students' understanding at each step. The steps could be implemented in a week-long intensive study, a course unit of several weeks, or stretched out across an entire course or even an entire undergraduate curriculum. A single class period of any undergraduate level would probably be inadequate to the entire process, since going through it from start to finish requires considerable mental effort, especially for students being introduced to such complex thinking for the first time. No matter how the process is implemented, the instructor can begin by modeling the process for the class, gradually enabling students to work through the entire process themselves. The process can be carried out in a variety of ways, including instructor-led class discussion, small groups, lectures, individual writing assignments, or short-answer tests.

Following is an abbreviated illustration of how this four-step process might work, using as an example the recent debates in the United States over same-sex marriage:

Step 1: Identify the debate. Students should move beyond "whether same-sex marriage is right" to issue identification more akin to "whether marriage between two men or two women ought to be granted the same legal status as marriage between a man and a woman."

Step 2: List the standard pro/con arguments for both sides. These sides are arbitrary and subjective, but may appear to be absolute, given the pervasiveness of binary thinking.

For same-sex marriage:

(a) a civil rights issue, analogous to the civil rights movement for racial equality in the United States;
(b) homosexual and heterosexual orientations are innate; thus, banning gay marriage constitutes a denial of human rights; and
(c) same-sex marriage opponents are hateful and ignorant religious zealots.

Against same-sex marriage:

• traditional and historical marriage is heterosexual, intended for reproduction and for the stability of society and the state;
• homosexuality is a choice and therefore gays and lesbians should not receive special consideration; and

- homosexuals are immoral people who choose an "alternative lifestyle" while allies are un-patriotic (and are closeted homosexuals themselves).

Step 3: Complicate the binary by fleshing out stakes on both sides. For same-sex marriage:

(a) homosexuals who want their long-term monogamous love relationships to be granted the same rights as those of heterosexual couples;
(b) heterosexuals who see gay marriage as progress toward equality of all people;
(c) atheists, secular humanists, or others who eschew established religions; and
(d) religious, social, and/or political liberals.

Against same-sex marriage:

- people who are disempowered in other ways (class, gender, race, ability, age) may rely on religion or notions of "tradition" as social anchors;
- people who are uncomfortable with sexuality as a range of practices and ideologies or who have experienced sexuality as a form of domination;
- religious, political, and/or social conservatives, heterosexual males and females, state and federal governmental representatives, pundits and talk show hosts, editors and other writers for the sake of individual, corporate or political gain; and
- anyone who questions the validity of the institution of marriage in general, the discourse of "rights" granted by the state, or the idea that either the state or religion should have anything to do with sexuality/love relationships.

Step 4: Render the complexity of the debate into an assessable form. Designed as an assignment that could be assessed according to course grading rubrics, students might be provided a newspaper editorial or other position piece, and charged with evaluating the issue using their thick critical thinking skills. One possible example might be the following scenario: "The owner of a chain of fast food restaurants espouses conservative religious beliefs and donates corporate money to conservative causes that oppose same-sex marriage. A gay rights organization pickets the company and the media picks up on this debate."

A less successful response to this exercise, or one that comes in step one or step two, would go no further than identifying common binaries (e.g., pros and cons of same-sex marriage)—the first or second stage of human cognition we mentioned earlier. A student successfully engaging in thick critical thinking— one who is further along in the four steps—might demonstrate an understanding of other related issues that exist alongside the same-sex marriage debate, such as religion, national belonging, class, race, and freedom of speech, and the interplay of personal beliefs and occupational responsibilities. Thick critical thinking involves realizing and articulating that any single debate rarely concerns only two opposing sides but almost always involves several complex, overlapping issues. The student would also reflect on how an individual might find herself situated among, and be required to negotiate, these many issues. Similarly designed classroom activities offer students the chance to question assumptions that have not been questioned before entering our classrooms. It also demonstrates—and offers the chance to practice—ways to converse with a range of stakeholders in any issue and to work through a process of compromise and mutual benefit. The process of thick critical thinking asks students to move beyond arbitrarily choosing and arguing a position to interrogate how the very act of choosing between two ostensibly opposed sides shapes the debate itself. This interrogation helps students further see how resulting characterizations of the other side as, for instance, "hateful" or "opposed to progress" rely heavily on notions of normalcy, deviance, and the "other."

The process of thick critical thinking outlined here can apply to a wide variety of past and present debates, including climate change, gun control, the rights of indigenous peoples, economic inequality, or, as in our example, gay marriage. It can be used across disciplines for issues like animal subjects testing in the life sciences or teaching "intelligent design" in education, and also in ethics courses in professional instruction from journalism to business. All academic disciplines hold debates of some sort over issues important to those disciplines and to the world, and our process assumes that those debates are complex and involve a range of opinions and subject positions.

One assumption underlying this method is that the personal is political, that individuals and groups have personal stakes in public debates. These stakes may make discussion awkward and incendiary, but they are the bedrock upon which legitimate, reasoned discussion rests. The thick critical thinking model attempts to develop the ability of students to recognize that these issues are deeply personal and political for everyone. In order for students to situate themselves in a diverse society and world, they must figure out how to discuss differences productively, with an eye toward full understanding. Thick critical thinking helps students understand that even apathy—the decision to "not care" about an issue—is itself a political stance, often in the defense of privilege or status quo. Ignoring the personal stakes for ourselves and others only takes us away from civil, compassionate, and productive discussion. But considering stakes and attempting to understand why issues are important to others can help raise empathy among students as well as hone their critical thinking.

One of the most challenging areas in which to apply critical thinking is that of power relations—the complex notion that different social positions offer historical advantages in different ways. It also means understanding the complexity of human identity, and therefore the ways that a single person may benefit from privilege in one area and suffer oppression in another. Examining our own arguments, tendencies, and reactions for evidence of binaristic thinking can help us recognize how we might benefit from certain power relations or how we reinforce binary thinking by accepting the notion of the "other" forced upon us by dominant groups. These are some of the hardest positions to take cognitively and culturally, and they represent a rather high order of thinking. Not all undergraduates will be prepared to examine their own assumptions in such a radical way, but we can hope as teachers that introducing them to thick critical thinking may begin a lifelong process.

Conclusion

We have proposed a pedagogical method for using critical thinking in the classroom, which they call "thick critical thinking," that goes beyond the traditional debate model so often used to teach critical thinking. Our method borrows from anthropologist Clifford Geertz's concept of "thick description," which assumes that any culture consists of "a multiplicity of complex conceptual structures, many of them superimposed upon or knotted into one another" (1975, 10). The task of the ethnographer is to "contrive somehow first to grasp and then to render" the complexities (1975, 10). Thick critical thinking has students address an issue as such as an engaged ethnographer. Informed, too, by feminist standpoint theory, intersectionality, and critical pedagogy, thick critical thinking tries to get students to the radical position of trying to understand the viewpoint of the "other" as just as complex, personal, and nuanced as one's own.

While thick critical thinking activities, as in the traditional debate-style classroom activities, help students identify and understand two sides to any argument, its seeks to push students beyond such binary thinking and to begin to see the levels of complexity that undergird any issue. Students not only become practiced at recognizing and understanding "both sides" of an issue, but also are able to demonstrate how such binaristic viewpoints are reductive. The intention for students is not to arrive at a supposedly objective point of view on an issue or to attempt to distance themselves from a debate; on the contrary, students practicing thick critical thinking learn to formulate well-reasoned, informed, and nuanced opinions on a given issue. They eventually become more engaged and better able to argue their position after understanding its complexity.

The complications of adhering to a thick critical thinking method in the higher education classroom are, of course, many. Least of these is the time commitment required to prepare the activity and the instructor's own facility in balancing what could easily be dozens of different threads in the course of a brief course meeting. While we the authors have employed this method successfully in their own humanities-based classrooms, further research and refinement is required to determine how such an approach fits into different disciplines and specific unit lesson plans. Additionally, anecdotal and controlled research will produce adjustments to this approach to account for class sizes and class compositions.

It is no new observation that critical thinking remains central to the pedagogical endeavor at all levels and certainly in the higher education classroom; yet, critical thinking remains difficult to define, assess, and teach. Our hope in developing the method of thick critical thinking is not to resolve these issues but rather to offer a new way of approaching this vital skill and how we can help our students develop it.

As is clear from our essay, thick critical thinking is in its theoretical stage. Further research into its efficacy as a pedagogical model that helps students improve their critical thinking skills is needed. This research can be done in several ways. Both thick critical thinking and the traditional debate models can be implemented in a course in which there are two sections. All sections would pick the same issue, as in step one of thick critical thinking, but then the sections would split into their respective pedagogical models. The instructor and/or researcher(s) would then compare the critical thinking results between both sections. This test can be implemented for the duration of a single unit, multiple units, or throughout an entire semester or quarter. More work is also needed at developing proper assessment methods with which to properly gauge the effectiveness of thick critical thinking.

References

Bailin, S. 2002. "Critical Thinking and Science Education." *Science & Education* 11 (4): 361–375.
Belenky, M. F., McVicker Clinchy, B., Rule Goldberger, N., and Mattuck Tarule, J. 1997. *Women's Ways of Knowing: The Development of Self, Voice, and Mind*. New York: Basic Books.
Brookfield, S., and Preskill, S. 1999. *Discussion as a Way of Teaching: Tools and Techniques for Democratic Classrooms*. San Francisco: Jossey-Bass.
Butler, J. 1990. *Gender Trouble: Feminism and the Subversion of Identity, Thinking Gender*. New York: Routledge.
Crenshaw, K. W. 1991. "Mapping the Margins: Intersectionality, Identity Politics, and Violence against Women of Color." *Stanford Law Review* 43 (6): 1241–1299.
Davies, M. 2006. "An 'Infusion' Approach to Critical Thinking: Moore on the Critical Thinking Debate." *Higher Education Research & Development* 25 (2): 179–193.
Ennis, R. 1992. "The Degree to Which Critical Thinking Is Subject Specific: Clarification and Needed Research." In *The Generalizability of Critical Thinking: Multiple Perspectives on an Educational Ideal*, edited by Stephen Norris. New York: Teachers College Press. 21–37.

Foucault, M. 1980. *The History of Sexuality*. (first edition, Vintage Books) New York: Vintage Books.

Frederick, P. J. 2002. "Engaging Students Actively in Large Lecture Settings." In *Engaging Large Classes: Strategies and Techniques for College Faculty*, edited by Christine A. Stanley and M. Erin Portered. Bolton: Onker Publishing.

Freire, P. 2000. *Pedagogy of the Oppressed*. (30th anniversary edition). New York: Continuum.

Geertz, C. 1975. *The Interpretation of Cultures: Selected Essays*. London: Hutchinson.

Grasha, A. F. 1996. *Teaching with Style: A Practical Guide to Enhancing Learning by Understanding Teaching and Learning Styles*. Pittsburgh: Alliance Publishers.

Haraway, D. 2004. "Situated Knowledges." In *The Feminist Standpoint Theory Reader: Intellectual and Political Controversies*, edited by Sandra Hardinged. New York: Routledge. 81–102.

Harding, S. 1993. "Rethinking Standpoint Epistemology: What Is Strong Objectivity?" In *Feminist Epistemologies*, edited by Linda Alcoff and Elizabeth Potter. New York: Routledge. 49–82.

Hartsock, N. C. M. 1998. *The Feminist Standpoint Revisited & Other Essays*. Boulder, CO: Westview Press.

Hopkins, G. 2012. *Four Corner Debate*. Education World, Inc 2003 [December 22, 2012].

Hughes, C. 2003. "Some Thoughts on Critical Thinking." *Rocky Mountain Review of Language and Literature* 57 (2): 57–61.

Kennedy, R. 2007. "In-Class Debates: Fertile Ground for Active Learning and the Cultivation of Critical Thinking and Oral Communication Skills." *International Journal on Teaching and Learning in Higher Education* 19 (2): 183–190.

Kitchener, K. S., and King, P. M. 1981. "Reflective Judgment: Concepts of Justification and Their Relationship to Age and Education." *Journal of Applied Developmental Psychology* 2: 89–116.

Kuhn, D., and Dean, D. 2004. "A Bridge between Cognitive Psychology and Educational Practice." *Theory into Practice* 43 (4): 268–273.

Lowman, J. 1995. *Mastering the Techniques of Teaching*. (second edition). *The Jossey-Bass Higher and Adult Education Series*. San Francisco: Jossey-Bass.

Lutz, D. J., and Keil, F. C. 2002. "Early Understanding of the Division of Cognitive Labor." *Child Development* 73 (4): 1073–1084.

McPeck, J. E. 1984. "Stalking Beasts, but Swatting Flies: The Teaching of Critical Thinking." *Canadian Journal of Education / Revue canadienne de l'education* 9 (1): 28–44.

McPeck, J. E. 1992. "Thoughts on Subject Specificity." In *The Generalizability of Critical Thinking: Multiple Perspectives on an Educational Ideal*, edited by Stephen Norris. New York: Teachers College Press. 198–205.

Moore, T. J. 2004. "The Critical Thinking Debate: How General Are General Thinking Skills." *Higher Education Research & Development* (1): 3–18.

Moore, T. J. 2011. "Critical Thinking and Disciplinary Thinking: A Continuing Debate." *Higher Education Research & Development* 30 (3): 261–274.

Omi, M., and Winant, H. 1994. *Racial Formation in the United States: From the 1960s to the 1990s*. New York: Routledge.

Paul, R., and Elder, L. 2006. "Critical Thinking: The Nature of Critical and Creative Thought." *Journal of Developmental Education* 30 (2): 34–35.

Paul, R., and Elder, L. 2008. *The Miniature Guide to Critical Thinking Concepts and Tools*. Tomales, CA: Foundation for Critical Thinking Press.

Perry, W. G. 1970. *Forms of Intellectual and Ethical Development in the College Years: A Scheme*. New York: Holt.

Perry, W. G. 1981. "Cognitive and Ethical Growth: The Making of Meaning." In *The Modern American College*, edited by Arthur W. Chickering. San Francisco: Jossey-Bass.

Robinson, S. R. 2011. "Teaching Logic and Teaching Critical Thinking: Revisiting McPeck." *Higher Education Research & Development* 30 (3): 275–287.

Said, E. W. 1978. *Orientalism*. New York: Pantheon Books.

Silberman, M. 1996. *Active Learning 101 Strategies to Teach Any Subject*. Boston: Allyn & Bacon.

Sweet, S. 1998. "Practicing Radical Pedagogy: Balancing Ideals with Institutional Constraints." *Teaching Sociology* 26 (2): 100–111.

Timpson, W. M., and Burgoyne, S. 2002. *Teaching and Performing: Ideas for Energizing Your Classes*. (second edition). Madison, WI: Atwood Pub.

Tumposky, N. R. 2004. "The Debate Debate." *The Clearing House* 78 (2): 52–55.

Willingham, D. 2007. "Critical Thinking: Why Is It So Hard to Teach?" *American Educator* 31 (2): 8–19.

10
A Disciplined Approach to Critical Thinking

Anna Jones

Critical thinking appears in university curricula, course outlines, and statements of graduate attributes, and yet there is both uncertainty about what it entails and passionate debate about its essential nature. This chapter argues that while there are common elements of critical thinking, it is, in its practice and teaching, a disciplined act. Critical thinking occurs within the conventions, methodologies, and knowledge bases of particular disciplines and fields and within the structures that they provide. Thus it is disciplined in both its subject specificity and its orderliness. This is not to suggest that critical thinking cannot interrogate the subject area in which it resides, or that it cannot transcend disciplinary boundaries. Rather, it is to suggest that one needs to learn to think critically in an organized manner, that this can be done by following a particular intellectual tradition or discipline, and that critical thinking needs to have some content. As Smith (1992) points out, knowledge is central to critical thinking and one cannot think critically unless one has knowledge of the topic. For McPeck (1981), critical thinking is shaped by the "particular problem under consideration" (7). In other words, we think critically about *something.*

There is an influential, long-running and continuing debate in the literature regarding the generalizability of critical thinking (Ennis 1992; McPeck 1981; Norris 1992), with some research pointing to the importance of the interaction between discipline knowledge and critical thinking (Alexander and Judy 1988; Baron and Sternberg 1987; Nickerson, Perkins, and Smith 1985). Elsewhere I have explored the ways in which critical thinking takes on the particularities of the discipline in which it resides (Jones 2004; 2007; 2009) and related studies have also been conducted by Moore (2011). My position is not an absolute one—there are commonalities in critical thinking across disciplines. Yet there are important differences in emphasis, analytical tools, organizing principles, and thinking patterns used.

This chapter discusses ways in which critical thinking is defined in five disciplines—medicine, physics, law, economics, and history—exploring the contextual nature of critical thinking. It suggests that a disciplined approach to critical thinking is not constraining but rather that a rigorous understanding of a particular body of knowledge can enable thinking. This chapter argues that by understanding the contextual nature of critical thinking we can have a better understanding of the ways in which it can be taught.

There are a number of definitions of critical thinking (Ennis 1987; Facione 1996; Halpern 1996; Kalman 2002; Kurfiss 1988; McPeck 1981; Norris and Ennis 1990; Paul 1989) that describe it as an organized or disciplined form of thinking encompassing some or all of the following: analysis of arguments; considering truth, validity, soundness, and fallacies; and making judgments, evaluating claims, inferences, assumptions, explanations, and relevance. In addition, critical thinking requires dispositions or willingness to think in particular ways (Ennis 1996; Verlinden 2005). The Delphi Report (Facione 1990) identifies six key elements of critical thinking: interpretation, analysis, evaluation, inference, explanation, and self-regulation. It requires both cognitive skills and the disposition to utilize those skills.

The aspects of critical thinking outlined here are present in each of the disciplines discussed in this chapter. However, they take different forms, have different emphases, and require different technical skills. As a consequence, applied critical thinking at anything other than a general level also requires an understanding of the knowledge, language, technical tools, and conventions of the particular subject matter in question. This is not to say that the critical thinker must accept these conventions and traditions—quite the reverse, thinking in a critical way requires knowledge and understanding and demands that one thinks both within *and* beyond disciplinary frameworks.

The debate between generalists and specifists has tended to focus too sharply on an either/or position between general and specific in critical thinking. While there are very important differences between the ways in which critical thinking is understood and practiced in different disciplines, there are cross-cutting elements. Critical thinking can be both general and specific, depending on what it is required to do. Generalizable critical thinking is exactly that—critical thinking that can be generalized. It is thinking that can be used in a range of contexts and is useful for precisely that reason. However, it is not the only form of critical thinking. Robinson (2011) suggests that general critical thinking courses such as those based on informal logic provide a valuable step in the mastery of critical thinking and that training in logical thinking must also be scaffolded by disciplinary studies. The critical thinking provided by generalist courses is focused thinking that can be applied to a range of contexts that do not require specialist knowledge. Specific critical thinking requires some of the same skills as generalizable critical thinking, but is highly contextual and the emphasis and application or the "shape" of critical thinking is different.

Critical thinking in its disciplinary context requires particular knowledge, understanding, techniques, language, and disciplinary grammar. Disciplines have rules for the seeking and testing of knowledge, and Donald (2002; 2009) has outlined in detail the patterns of thinking in a range of disciplinary contexts, pointing out that an expert thinker in each area must understand the organizing patterns in order to navigate a way through them. So critical thinking is learned in particular contexts, according to particular rules (Kreber 2009). It may be possible that skills associated with one subject can inform another (Hounsell and Anderson 2009), but this must be treated with caution rather than assumed. The question of transfer is contentious (Perkins and Salomon 1994), and we should not assume that skills taught in one context can easily transfer to another without careful scaffolding, that is, overt unpacking of ideas, clarity of expectations, transparent examples, and opportunity for rehearsal.

Arguing for critical thinking in a disciplinary context makes the assumption that it is actually valued and taught in every context, and this is of course not always the case. There will always be teaching situations that are less than ideal, cases where students are required simply to acquire information rather than learn to think critically. However, this does not negate the argument that critical thinking, within a disciplinary context, when taught well is the essence of "higher" education because it requires a fusion of knowledge, understanding, the rigor to master a body of knowledge, and the willingness and skill to challenge it in a meaningful and sophisticated manner. This requires both humility and confidence in both the teacher and the taught. It also requires assured understanding and a preparedness to consider other possibilities.

Developing a mastery of disciplinary thinking can and should be a lengthy and challenging process. Physicists in this study, for example, suggest that it is not until students near the end of postgraduate study that they have reached anything resembling disciplinary maturity. In any field it is the intellectual struggle with sophisticated ideas, the mental organization required, an understanding of an integrated system of thinking and body of ideas, flexibility of thought, confidence to make judgments, and the openness to consider these reflectively that enable critical thinking.

The following sections explore critical thinking in five disciplines in order to examine the ways in which it is a highly contextualized practice. The findings were gathered from interviews with 37 academic staff in five disciplinary areas (history, economics, law, medicine, and physics) in two large, research-intensive Australian universities. Interviews were semistructured and audio recorded and transcribed in full (for details see Jones 2006). Analysis was emergent and coding involved re-reading and validation through cross-checking across all transcripts. From this coding, themes or patterns were identified and refined. All the material presented in the next section is from the study. Direct quotes have been used judiciously; however, because discussions were lengthy and there is repetition between participants, much of the material has been synthesized as part of the thematic analysis process (Miles and Huberman 1994; Silverman 2005).

Physics

Physics admits both a high level of certainty and yet fundamental uncertainties within the epistemology of the discipline, which raises some very significant questions regarding the nature of critical thinking, particularly for undergraduates. Much of the physics taught in the early years of an undergraduate degree is both very well established yet counterintuitive:

> Physics is so bizarre. This is not to suggest that we do not welcome questions but in order to study physics you have to put aside common-sense views of the world.

In physics, critical thinking has five central dimensions. Each is interlinked although students may not become fluent in all—at least as undergraduates. The first is knowledge of key principles, an understanding of how they work, and an ability to examine a phenomenon in the light of these principles. This also encompasses a consideration of the possibility of quantification, predictive power, whether the principle will work, and whether the question is actually more complex than it appears. Learning physics requires mastery of a vast and technically difficult body of knowledge and mathematical skills, and so a considerable amount of time and effort on the part of both students and teachers is occupied with coming to grips with this material in ways that constrain some forms of critical thinking in the undergraduate years. Participants were clear that while it was necessary to teach critical thinking that considered mathematical logic and scientific experimentation, more far-reaching critical thinking that challenged established thinking was difficult until a really sound knowledge base had been established; yet it was still something they aimed for.

The second and central dimension is problem solving. This requires knowledge since as one participant pointed out, "If they [students] do not have the ability to do the physics then problem solving and critical thinking are meaningless." Problem solving involves first understanding the problem, second, devising a plan, third, carrying out the plan, and finally, checking the solution. Participants suggested that it was not possible to solve problems without thinking about them critically or the reverse, and this required mastery of a body of knowledge.

A third and very closely connected dimension is a sound but questioning understanding of mathematics, including the ability to identify hidden assumptions in a physical model or elements that will cause a model to be mathematically illogical, inconsistent, or incapable of describing the physical world. It also includes understanding which model is appropriate to a particular situation, whether it is valid, sound, and logical, and then determining its accuracy. This includes understanding whether the uncertainty is too great for a result to be accurate.

The fourth dimension could be characterized as "scientific thinking" including an understanding of experimental design, research protocol, and deductive and inductive reasoning. Because physics involves a high degree of precision, one physicist characterized critical thinking as avoiding a sloppy "that kind of fits, we will call it good enough" way of thinking. Thus it is necessary to have the skill and persistence to really push an idea and to do so in a rigorous, transparent way. Some describe their teaching of critical thinking as encouraging students not to take a formula as it stands but to interrogate it. This is (or should be) part of what it means to be a scientist, either experimental or theoretical.

Finally, critical thinking is an understanding of the uncertainty of knowledge, and participants cited quantum theory of gravity as one example. They pointed out the importance of an understanding of the boundaries of a theory. Moreover, they suggested that it was necessary to be aware that there may be something fundamental that could turn all of physics upside down. The physicists in this study spoke of the importance of teaching undergraduates not only how thinking has advanced knowledge in the discipline but also how the paradigm has changed so that knowledge is not seen as static. The physicists in this study had an awareness of the dynamic, constructed, and tentative nature of knowledge, tempered by the obvious certainties that physics provides.

History

Critical thinking is viewed by historians as a complex and multilayered entity. It is conceptualized as having a number of dimensions utilizing a consideration of logic, evidence, difference, ambiguities, power, gaps, and the nature of history itself. These dimensions will be examined separately although they are not necessarily separate activities but merely different angles on the notion of critical thinking.

First, critical thinking is an ability to examine the logic of an argument and the closely related ability to examine evidence. This means understanding and discussing evidence in its context. It also means the ability to take a text apart and explore its language, relevance, author, audience, purpose, the claims made on knowledge and truth, and then to examine the significance of this. Further, critical thinking involves examining the biases of the text in question.

Next, critical thinking introduces an element of "otherness." This means first seeking other evidence, other voices, and other perspectives. It is also a bigger project as it requires openness to other ways of seeing the world and so is both directed at the evidence or task at hand and at the historians' worldview. History is always about otherness as it is about people who usually cannot explain their position to us and who lived in times that were different from our own.

Critical thinking also involves exploring contradictions, ambiguities, and ambivalence. This requires not only finding a way through conflicting sources, opposing viewpoints, or inconsistent actions, views, or statements but also being conscious that there may not be a definitive answer. Historians argue that students need to learn to appreciate contradictions and uncertainty rather than aim to reduce them. They also see critical thinking as an ability to challenge one's own presuppositions and examine one's own biases.

Critical thinking contains a political dimension, comprising an understanding of the nature and structures of power and notions of ideology, for instance, the ways in which ideas become taken for granted and are assumed to be "natural." Critical thinking involves an awareness of gaps and silences, the people who were not speaking, things that were not said, things assumed not to be important, and evidence that is difficult to find. Historians are aware of the unspoken and what that can tell us about what is important and valued both by past societies and contemporary historians. This allows historians to consider what has become established and why certain perspectives are valued while others are marginalized and to explore ways of "telling the story."

The final form that critical thinking takes is related to the sense that historians are self-conscious about their craft. Their awareness of the notion that historians "make history" means that they aim to be honest about the limits of their own theorizing. Their eclectic practices and interdisciplinarity mean that there is a degree of examination of the nature of history, its power, and its constraints.

Economics

In economics there are two central dimensions to critical thinking although these were not identified or supported by all participants in the study. The first is the logical and analytical use of economic tools to solve problems. This is also expressed as examining accuracy, applicability, and adherence to economic principles. This involved the use of economic tools to solve practical and theoretical problems and an understanding of economic models.

However, there was some disquiet among a minority of the economists in the study, who suggested that this view of critical thinking is limiting since critical thinking in its narrowest sense is, as one economist expressed it, "paring things down to their bare minimum." As she went on to argue, "if you are in mathematics, that is great but the world we live in is more complex than that." She voiced concern that in economics, critical thinking is very limited and sometimes did not encompass a scrutiny of assumptions upon which much of the thinking was based:

> What they [economists] mean by critical thinking is logical analysis, so what you do is check the internal consistency of a particular model rather than whether the assumptions it is based on are right, so there is a fundamental difference in what is meant by critical thinking. We teach the assumptions such as profit maximizing behavior and initially we wave our hands and say of course these assumptions don't hold but we never take students back to critically analyse this. Which is a fundamental problem. Other disciplines would say let's not even worry about building this fantastic edifice; let's look at the foundations first. I think that is the real problem because after three years no one is talking about them [the assumptions] and students take it for granted that they must be right and then go out into the world and start making decisions.

A second dimension of critical thinking is a broader notion encompassing skepticism, lateral thinking and creativity. This comprises an ability to take a contrary view or as participants referred to it, "thinking outside the frame," or "thinking sideways." Included in this understanding of critical thinking is an examination and evaluation of policy. This entails an examination of instances where, for example, a model did not work or where the assumptions are indefensible. One economist argued that the factors that drive human decision making are contextual, and so it is important to examine the societal context in which these decisions are being made. However, this was not a view of critical thinking in economics that was subscribed to by all participants in this study.

Law

In law there are five dimensions of critical thinking: argument evaluation, challenging assumptions, consideration of the social context, examination of law as a profession, and flexibility of thought. These are not separated hierarchically but for ease of analysis as these notions are interconnected.

The first dimension, argument evaluation, is the application of information, developing an argument, examining an argument, or solving a problem. This involves "going beyond the what to the why" as one person expressed it. This means examining the following: the source of information; consistency;

logic; the basis of the argument; evidence; assumptions; consistency; implications. Taking this one step further, critical thinking also requires not accepting something for what it claims to be but examining the underpinnings, be they normative, ideological, or philosophical and examining what has been left unsaid. One participant argued that law had its own "grammar of problem solving" based on rules and facts. Without a good understanding of the organizing principles of law and the technical rules, he suggests, students cannot engage in critical thinking.

Second, critical thinking involves challenging assumptions, thinking outside traditional boundaries, and questioning received wisdom or social structure. Law academics acknowledge that while law tends to be a conservative profession, it is the responsibility of universities to introduce students to thinking that challenges accepted knowledge and beliefs.

The next dimension of critical thinking is law in its policy and social context. This means exploring injustice (a complex issue explored by jurisprudence and dealing with either particular or systemic failures to address problems or cases where harm is sanctioned by the legal system), inefficiency, inconvenience, or expense. This is important since the basis of law is its social function.

Critical thinking also requires carefully examining the legal profession, thinking about one's individual and the profession's responsibility to clients and to society. This involves consideration of ethical questions and reflection on one's own actions. One participant commented:

> I talk to students about honesty, about not stealing from the library, not hiding books in the stairwell, or razoring out important sections. I talk about plagiarism. All this relates to their responsibilities as future professionals.

Finally critical thinking in law involves openness and an understanding that there often is no single clear answer since "in law there is often a good deal of murkiness" and "students should be able to accept the complexity and uncertainty of law." Law is "complicated, ambiguous and unclear, there are uncertainties and room to manoeuvre." Precision in language is highly valued and participants suggest that law students need to understand that language is open to multiple interpretations and can be used both to illuminate and to hide.

Knowledge of the law requires an ability to read a case or statute, and this usually begins by considering the legal principle. It requires understanding how the legal system is constantly evolving and the history behind the law, since what now constitutes the law depends upon what has happened previously. Students of law need to navigate their way through the language of statutes and when considering a matter; they need to identify the material facts, the legal issue, judicial reasoning, and the underlying principle, and then form an opinion.

Medicine

Central to critical thinking in medicine are the questions one asks, as one participant explains, "What are the questions, how would you explore them, where would you go with the evidence, how would you confirm something?" In medicine, critical thinking has five dimensions. The first is clinical reasoning, the second evidence-based medicine, the third consideration of ethical questions, the fourth reflection, and the fifth questioning of the status quo or received wisdom. These are overlapping rather than discrete entities.

Critical thinking in a clinical context or clinical reasoning is probably the most fundamental and is also referred to as medical problem solving:

> It is something that we have to do on a daily, hourly basis. When a patient comes in with a problem you have to be able to hypothesise about the probable cause, do a problem-oriented history and examination, draw that information together, synthesise it and decide what the next step will be. Then you need reflective skills to examine your own diagnosis.

Thus, critical thinking is the construction of a clinical argument and the weighing of evidence and planning next actions. It can be used in diagnostic or therapeutic thinking. It is highly complex as there are so many variables, including the psychosocial context and subjectivity of symptoms. In clinical reasoning, critical thinking is an integration of communication, analysis, and problem solving. As one clinician pointed out, it not only requires the ability to make connections, to see patterns, and implications but also involves "knowing what you don't know." It requires skills of observation, communication, and expert medical knowledge.

Although evidence-based medicine is now central to the discipline, participants are aware of the ambiguities of much of the evidence. "Medicine has to be based in science, on evidence and if something works it has to be demonstrated that it works. But some of the evidence is pretty suspect." However, all considered that it is important for students to learn to examine evidence and to be aware that medicine has its basis in research. Thus students should be able not only to understand research practice but also to examine cases where standard practice relies on little supporting evidence or cases in which evidence actually refutes the efficacy of standard practice.

Ethics is a very important aspect of medical education. Students are explicitly given the opportunity to consider difficult open-ended problems. Ethical questions are raised in relation to the cases being examined, tricky issues are deliberately written into cases students consider from first year on. The questions compel them into a situation where they have to take action on the best possible evidence (although often imperfect).

Reflection is another dimension of critical thinking and is perhaps a more personal approach requiring analytical examination of one's own diagnosis or treatment of a patient. It is self-analytical rather than descriptive. Students are encouraged to reflect on their own health care, for example, the difficulties of changing behaviors (such as eating habits or exercise) and to consider the implications this has for patients. Reflection is an important critical skill because it is part of the autonomy and self-regulation of the profession. It is taught through reflective writing exercises, for example, following community placements or field trips, or as part of communication skills training.

The final way in which critical thinking is conceptualized is questioning received wisdom or established knowledge. Although many of the medical academics remark that medicine is slow to change and is a conservative profession, all are very definite in their belief that it is crucial to teach students to question at a fundamental level. Critical thinking requires one to examine assumptions (one's own as well as those of others), to dispense with false assumptions, and to discriminate between ideas and practices. It requires one to "ask the hard questions, especially when the answer seems obvious." The participants in this study acknowledge that questioning received wisdom can be confronting both to students and to staff and is difficult in a hierarchical hospital environment but is an essential part of the profession.

Discussion

Critical thinking is a disciplined act (or set of acts), first, because it requires an orderliness of thinking and, second, because this order is contextual. There are some aspects of each of the disciplinary descriptions of critical thinking that are relevant across disciplinary contexts, such as use of logic and evidence, evaluation of claims and explanations, analyzing arguments for clarity and precision, and making reasoned judgments. However, the ways in which they operate in each discipline and the knowledge required in order to think critically mean that a course of general critical thinking skills may not equip students to think critically in a disciplinary context. For example, in physics some critical thinking is done using mathematical reasoning—proof is a deductive argument that can be traced back to a generally accepted axiom. A proof must demonstrate a statement to be always true and is carried out using a sequence of formulae, each one of which is the logical consequence of the preceding. This is done using mathematical notation rather than English or another spoken language. Experimental physics requires laboratory technique and conventions, again shaping critical thinking in particular ways as the physicist needs to consider ideas such as controlling variables or fine-grained measurement down to a particular number of decimal places. So even within one discipline, critical thinking takes many forms. For medics, critical thinking must use medical knowledge, patterns of clinical reasoning, an understanding of physiology, diagnostic procedures, and therapeutic possibilities. Furthermore, much clinical thinking must be done under conditions of imperfect knowledge, time constraints, and great uncertainty and must take account of the complex psychosocial dimension of patient care. In history and to some extent law, critical thinking has a political dimension not overtly present in physics.

These are just three examples from the disciplinary outlines of critical thinking presented, but they suggest that general critical thinking skills may be useful for generalizable contexts but only where highly specialized knowledge is not required. Generalizable critical thinking is a useful foundation for disciplinary critical thinking but will not substitute for it. This is not to suggest that disciplinary critical thinking is the only valid form of critical thinking since both disciplinary and generalized thinking have uses and limitations. Generalized critical thinking is the skeleton of critical thinking, but it is "bald" because it is decontextualized and so does not have the subtlety of a historian's understanding of the difficulty of presenting an argument from incomplete evidence, or an economist's defense of a set of assumptions made in order to produce a model to explain an economic phenomenon. This process of learning to think like a physicist, a medic, or a historian provides learners with a rigor, framework, and knowledge base to think in a thorough, orderly, and informed way. This orderly thinking (when done well) contains within it the elements of critical thinking identified by Facione (1990), albeit in differing ways. There is strength and value in having different forms of critical thinking because they have different purposes and do different work; therefore they offer us a wider scope in ways of thinking.

Elsewhere, I have examined the ways in which critical thinking is taught (rather than simply defined) in various disciplines (Jones 2005; 2007), and there is a need for further research in this area but a detailed examination of this is beyond the scope of this chapter. However, what is important for teaching critical thinking is a careful examination of what exactly is meant by critical thinking in each particular context, how this is enacted in the curriculum, how it is articulated to the student and taught, how it develops throughout a degree program, and how it is assessed. As Barnett (1997) points out, critical thinking can exist at a number of levels and can occur as an examination of internal logic, as an ability to explore the debates and controversies within a disciplinary context, and as the capacity to critique the assumptions within a discipline (what Barnett refers to as metacritique).

The question of whether disciplinary thinking can equip students to think beyond a disciplinary frame is a difficult one and one only partially answered by my study. Elsewhere (Jones 2004; 2005) I have used these three views of critical thinking to consider teaching in economics. I argue that at least in the particular context of that study, while critical thinking was taught at the level of problem solving and there was some consideration of the debates within the discipline, there was less evidence of a critical examination that extended beyond the bounds of economics. My later work across a range of disciplines suggests that critical thinking as an examination of internal logic (using very particular knowledge or technical tools) is taught in each area (although not necessarily in every subject or by every teacher). A consideration of the debates, controversies, and uncertainties is taught in each area although perhaps to differing degrees. Metacritique, however, is a much more difficult prospect. This requires the ability not only to have a specialist's understanding of the disciplinary context but also to critically examine its assumptions and look beyond them. However, many (but not all) academics in all disciplines I considered expressed the conviction that as researchers and teachers they must critically interrogate their disciplines and welcome external critique. I suggest that one of the central roles of universities is (or should be) to scrutinize—including the disciplines and the social contexts within which they reside.

A curriculum model that allows for multi- or interdisciplinary study may enable metacritique. For example the "general education" program used in in many US universities allows students to develop specialist disciplinary knowledge and also to consider other points of view in general or interdisciplinary subjects. This approach is also used (albeit differently) at some Australian universities. In the case of the participants in this study, those academics who engaged in inter- or cross-disciplinary research spoke of the ways in which it enabled them to critically consider the parameters of their own fields. For example, developmental economics utilizes social and political research (unlike much mainstream economics) and engages in metacritique. However, these economists did also describe themselves as "heretics," and suggested that their work was rarely published in the prestigious economics journals. Some of the physicists spoke of the challenges provided by working with engineers or musicologists. One physicist also spoke of the ways in which talented physicists were trying to push the frontiers of their discipline. She commented that she dreamed of the "whole [knowledge] edifice tumbling down" because of the exciting prospects this might bring. Many of the medics in this study took the idea of metacritique very seriously in curriculum planning. Their curriculum included teaching staff from disciplines such as the social sciences in order to introduce a breadth of perspectives that would challenge medical assumptions. The medical academics reflected upon the nature of medicine and its place as both an art and a science.

The idea of metacritique does not presuppose generic critical thinking but the ability both to skillfully engage in specialist thinking and to consider other possibilities. As Brookfield (1987) suggests, critical thinking does not just require questioning everything and taking nothing for granted, it also requires a more constructive ability to explore and to imagine alternatives. Rowland (2006) refers to the concept of inquiry as seeking, suggesting "perhaps the most important task of the teacher is to develop an atmosphere or an attitude in which students seek" (109). Seeking, I suggest, goes beyond the mere acquisition of "facts" and relates to a questioning attitude that always asks "why," and "how," and "what does this mean." This requires teachers and institutions that are open and tolerant and have time and space for questions, and this may not always be the case in the contemporary university.

Teaching critical thinking may require teaching staff to (critically) examine what is meant by critical thinking in their areas and how it can best be taught. This would not be a bland box-ticking exercise but one that would examine what was valued, and what the essence of being an historian or economist really was. It may also require what Rowland (2006) refers to as "critical interdisciplinarity." He describes this as challenging dialogue between academics from different disciplines that forces each to defend and hence challenge their own assumptions. This both sharpens disciplinary thinking as well as enables the possibility of metacritique.

References

Alexander, P. A., and Judy, J. E. 1988. "The Interaction of Domain-Specific and Strategic Knowledge in Academic Performance." *Review of Educational Research* 58: 375–404.

Barnett, R. 1997. *Higher Education: A Critical Business*. Buckingham: Society for Research into Higher Education and the Open University Press.

Baron, J., and Sternberg, R. 1987. *Teaching Thinking Skills: Theory and Practice*. New York: W. H. Freeman and Company.

Brookfield, S. 1987. *Developing Critical Thinkers: Challenging Adults to Exlore Alternative Ways of Thinking and Acting*. Milton Keynes: Open University Press.

Donald, J. G. 2002. *Learning to Think: Disciplinary Perspectives*. San Francisco: Jossey-Bass.

Donald, J. G. 2009. "The Commons: Disciplinary and Interdisciplinary Encounters." In *The University and Its Disciplines: Teaching and Learning within and beyond Disciplinary Boundaries*, edited by Carolin Kreber. New York and London: Routledge. 35–49.

Ennis, R. H. 1987. "A Taxonomy of Critical Thinking Dispositions and Abilities." In *Teaching Thinking Skills: Theory and Practice*, edited by J. Baron and R. Sternberg. New York: Freeman.

Ennis, R. H. 1992. "The Degree to Which Critical Thinking Is Subject Specific: Clarification and Needed Research." In *The Generalizability of Critical Thinking*, edited by S. Norris. New York: Teachers College Press.

Ennis, R. H. 1996. *Critical Thinking*. Upper Saddle River, NJ: Prentice Hall.

Facione, P. A. 2000. *Critical Thinking:What It Is and Why It Counts*. California Academic Press, 1997 1996 [16/11/00 2000]. Available from http://www.calpress.com/critical.html.

Facione, P. A. 1990. *Critical Thinking: A Statement of Expert Consensus for Purposes of Educational Assessment and Instruction "the Delphi Report."*Millbrae, CA: American Philosophical Association, The California Academic Press.

Halpern, D. 1996. *Thought and Knowledge:An Introduction to Critical Thinking*. (third edition) Mahwah, NJ: Lawrence Erlbaum Associates.

Hounsell, D., and Anderson, C. 2009. "Ways of Thinking and Practicing in Biology and History: Disciplinary Aspects of Teaching and Learning Environments." In *The University and Its Disciplines: Teaching and Learning within and beyond Disciplinary Boundaries*, edited by Carolin Kreber. New York and London: Routledge. 71–83.

Jones, A. 2004. "Teaching Critical Thinking: An Investigation of a Task in Introductory Macroeconomics." *Higher Education Research and Development* 23 (2): 167–181.

Jones, A. 2005. "Culture and Context: Critical Thinking and Student Learning in Introductory Macroeconomics." *Studies in Higher Education* 30 (3): 339–354.

Jones, A. 2006. *Re-Disciplining Generic Skills: An Examination of the Relationship between the Disciplinary Context and Generic Skills in Higher Education*, Centre for the Study of Higher Education, University of Melbourne, unpublished PhD thesis.

Jones, A. 2007. "Multiplicities or Manna from Heaven? Critical Thinking and the Disciplinary Context." *Australian Journal of Education* 5 (1): 84–103.

Jones, A. 2009. "Re-Disciplining Generic Attributes: The Disciplinary Context in Focus." *Studies in Higher Education* 34 (1): 85–100.

Kalman, C. S. 2002. "Developing Critical Thinking in Undergraduate Courses: A Philosophical Approach." *Science and Education* 11: 83–94.

Kreber, C. 2009. "Supporting Student Learning in the Context of Diversity, Complexity and Uncertainty." In *The University and Its Disciplines: Teaching and Learning within and beyond Disciplinary Boundaries*, edited by Carolin Kreber. New York and London: Routledge. 3–18.

Kurfiss, J. 1988. *Critical Thinking: Theory, Research, Practice and Possibilities*. Washington, DC: ASHE-Eric Higher Education Report No. 2, Associate for the Study of Higher Education.

McPeck, J. 1981. *Critical Thinking and Education*. New York: St. Martin's Press.

Miles, M. B., and Huberman, A. M. 1994. "Data Management and Analysis Methods." In *Handbook of Qualitative Research*, edited by Norman K Denzin and Yvonna S. Lincoln. Thousand Oaks, CA: Sage.

Nickerson, R. S., Perkins, D. N., and Smith, E. E. 1985. *The Teaching of Thinking*. Hillsdale, NJ: Lawrence Erlbaum Associates.

Norris, S., and Ennis, R. H. 1990. *Evaluating Critical Thinking*. Edited by R. J. Swartz and D. N. Perkins. Cheltenham, VIC: Hawker Brownlow Education.

Norris, S. P. 1992. *The Generalizability of Critical Thinking: Multiple Perspectives on an Educational Ideal*. New York: Teachers College Press.

Paul, R. 1989. "Critical Thinking in North America: A New Theory of Knowledge, Learning and Literacy." *Argumentation* 3: 197–235.

Perkins, D. N., and Salomon, G. 1994 "Transfer of Learning." In *The International Encyclopedia of Education*, edited by T. Husen and T. N. Postlethwaite (second edition). Oxford: Pergamon, 11.

Robinson, S. R. 2011. "Teaching Logic and Teaching Critical Thinking: Revisiting McPeck." *Higher Education Research and Development* 30 (3): 275–287.

Rowland, S. 2006. *The Enquiring University: Compliance and Contestation in Higher Education*. Maidenhead, Berkshire: Open University Press McGraw-Hill Education.

Silverman, D. 2005. *Doing Qualitative Research*. (second edition). London: Sage.

Smith, F. 1992. *To Think: In Language, Learning and Education*. London: Routledge.

Verlinden, J. 2005. *Critical Thinking and Everyday Argument*. Belmont, CA: Wadsworth.

11

Using Argument Mapping to Improve Critical Thinking Skills

Tim van Gelder

Introduction

The centrality of critical thinking (CT) as a goal of higher education is uncontroversial. In a recent high-profile book, *Academically Adrift*, Arum and Roksa report that "99 percent of college faculty say that developing students' ability to think critically is a 'very important' or 'essential' goal of undergraduate education" (2011, 35), citing (HERI 2009).

However a major message of their work is that college education generally makes little progress toward this goal: "Many students are only minimally improving their skills in critical thinking, complex reasoning, and writing during their journeys through higher education" (35). Indeed for many students college education appears to be failing completely in this regard: "With a large sample of more than 2,300 students, we observe no statistically significant gains in critical thinking, complex reasoning, and writing skills for at least 45 percent of the students in our study" (36).

Their message is barely more positive than H. L. Mencken's acerbic comment, over a century ago: "Certainly everyday observation shows that the average college course produces no visible augmentation in the intellectual equipment and capacity of the student. Not long ago, in fact, an actual demonstration in Pennsylvania demonstrated that students often regress so much during their four years that the average senior is less intelligent, by all known tests, than the average freshman" (Mencken 1997, 98).

Yet we also know that college education *can* positively impact CT; simply put, CT can be taught. In a meta-analysis of 117 studies of college-level efforts to teach critical thinking, Abrami et al. found "a generally positive effect of instruction on students' CT skills" (2008, 1119).

However the amount of gain found in these studies varied widely, and Abrami et al. concluded that it makes quite a difference *how* CT is taught. They say "both the type of CT intervention and the pedagogical grounding of the

CT intervention contributed significantly and substantially to explaining variability in CT outcomes" (1120).

Therefore an important challenge in improving critical thinking is clearly identifying the types of CT instruction that have the most impact on AM skills. One type of instruction that seems to be showing significant promise in this regard is argument mapping (AM). This chapter briefly reviews AM-based instruction and the evidence that such instruction is an effective way to improve CT skills.

Argument mapping

Argument mapping, also known as argument diagramming or argument visualization, is visually depicting the structure of reasoning or argumentation (Davies 2011; Macagno, Reed, and Walton 2007; van Gelder 2013). Typically an argument map is a graph-type or "box and arrow" diagram, with nodes corresponding to propositions and links to inferential relationships. For an example see figure 11.1.

AM's roots reach back into the nineteenth century, but it has only become popular in the last decade or two, primarily as a tool to help students build

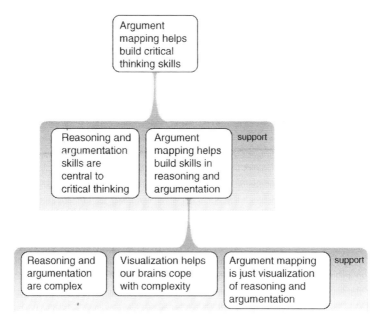

Figure 11.1 A map of an argument for the proposition that argument mapping helps build critical thinking skills. Map produced using the Rationale software (van Gelder 2007).

reasoning and CT skills. Indeed its immediate precursor was the kind of argument diagramming found in many introductory textbooks (see e.g., Fisher 1988; Govier 1988).

A key factor in the recent growth in popularity of AM has been the development of software tools designed specifically to support it. Previously, argument diagrams would have to be hand-crafted (whether on paper, or on a computer using generic drawing software), which made producing maps of any complexity both tedious and time-consuming. New software packages eliminate much of the "futzing around" with boxes and arrows, and provide varying amounts of guidance, scaffolding, and inbuilt exercises.

Using AM in CT instruction

As mentioned, argument diagramming of some sort is frequently found in introductory critical thinking textbooks, though it is generally treated as just one useful technique among many. AM-based instruction goes further in making argument mapping the primary or central method used to develop CT skills. Typically, this involves setting lots of AM exercises using dedicated AM software (though see Harrell 2008). A range of concepts and principles have been developed to help students map arguments properly (e.g., ter Berg, van Gelder, Patterson, and Teppema 2013).

The most common type of exercise involves providing a short text and requiring the student to identify and map out the argument it contains, that is, to produce an argument diagram faithfully representing the reasoning in the text. This can be surprisingly difficult, even for philosophers and others with prior training in argument analysis.

Another common type of exercise is requiring students to develop an AM representing an argument of their own creation, which may be preparatory to drafting an argumentative text. A third type of exercise is taking an argument map and rendering it into fluid argumentative prose.

Good AM-based instruction, like good instruction generally, presents a carefully graduated sequence of exercises of increasing difficulty. Also, as with instruction generally, good AM-based instruction requires students to receive good-quality feedback on their work. This requires human instructors with strong AM skills. Such people are in short supply, so this is a key obstacle to wider uptake of AM-based instruction.

Does it work?

It is prima facie plausible that learning and practicing AM would help students build their critical thinking skills. Reasoning and argumentation are not the entirety of critical thinking, but they are central to it; and AM should help build skills in reasoning and argumentation. There is a simple and compelling, almost syllogistic, argument for the latter point, with two uncontroversial premises: reasoning and argumentation are complex, and visualization, in general, helps our brains cope with complexity.

Note that in figure 11.1 the overall argument has two component arguments (reasons), each of which is made up of multiple premises. This argument has prima facie plausibility but requires buttressing with empirical evidence.

For many AM instructors this kind of general argument draws apparent support from what they see in the classroom. Students' attempts to think critically are frequently confounded by an inability to disentangle the threads of inference running through disputation on any given topic. To AM instructors it seems obvious that the diagrams help students grasp what is going on, and over time, help build logical acuity and facility.

However, as good critical thinkers we also know informal observations are suspect on matters of any subtlety or complexity. After all, medical practitioners for centuries thought they could see the benefits of bleeding patients. What we'd like is more empirically rigorous substantiation or validation of the "prima facie" case.

We also want to better understand just how well AM works. The claim that AM "helps build" CT skills is disturbingly vague. We would much prefer to have rigorous quantitative insight into the extent to which AM builds critical thinking skills, and how it compares with other instructional approaches. But how can this be obtained?

Empirical research on impact of AM instruction

Most readers would be familiar with the idea that the "gold standard" in social scientific research is the large, randomized controlled trial. Applied to evaluating the effectiveness of AM as an instructional method for CT, this would mean taking a large number of students and randomly assigning them to two CT subjects. One subject would make substantial use of AM, and the other would be similar in all significant respects except that it does not use AM, using instead some more traditional form of instruction. At the end of the instruction period, students in both subjects would be tested for their CT skills, using the same good-quality test for both groups. AM would then be deemed effective just to the extent that the students in the AM-based subject score more highly.

Unfortunately, no such study has ever been conducted. There are a number of reasons. Numerous practical challenges stand in the way, such as the bureaucratic difficulties involved in setting up two versions of a subject and making a genuinely random assignment of students to one or the other. There is also the difficulty of ensuring that the two subjects are sufficiently close to identical in all significant respects other than the use of AM. For example, do the two subjects cover essentially the same content, despite the difference in method? Are the students equally motivated to perform?

For these reasons, most efforts to rigorously evaluate the impact of AM on CT have taken a different approach, seeking to understand the impact of AM-based instruction by testing students at the start (pre-testing) and at the end (post-testing) of the instruction, and comparing the results.

Although pre- and post-testing is far more feasible than conducting a full-scale RCT, it is not without practical challenges of its own. For example, there is the problem of ensuring that students put proper effort into the tests, and that they are equally motivated to perform on both tests. Degree of motivation can make a huge difference to performance (Liu, Bridgeman, and Adler 2012), and if students slack off on the post-test, the real gain might be seriously underestimated (or vice versa).

Examples of the pre- and post-testing approach are the studies reported by van Gelder and colleagues (van Gelder, Bissett, and Cumming 2004). Starting in the late 1990s, the "Reason Project" at the University of Melbourne developed a radical alternative to traditional CT instruction, based on extensive deliberate practice (Ericsson, Krampe, and Tesche-Römer 1993) using AM. A dedicated AM software package, *Reason!Able* (van Heuveln 2004), was developed to support the approach. (Note that *Reason!Able* was the precursor to *Rationale* [van Gelder 2007].) Aided by a grant from the Australian Research Council, in two subjects students exposed to the approach were pre- and post-tested using the California Critical Thinking Skills Test (Facione 1991). To handle the motivation issue, students were assigned 5% of their overall score for the subject for their best performance on the two tests.

The data indicated that students had improved their CT skills by around 20%. Since CT is a generic cognitive skill, and since it is notoriously difficult to raise performance in such skills, this sounds like a substantial and worthwhile improvement, suggesting that the new approach works quite well. However this conclusion would be a bit hasty. For proper interpretation, the results need to be put in context.

First, we have to consider how much the students' CT skills would have improved anyway, due to factors such as maturation and just being at university. Second, and similarly, we have to consider how much the students would have improved due to the fact that they were taking a CT subject. Perhaps all or most of the 20% gain was due to experiencing CT instruction of some kind, rather than AM-based instruction specifically.

Given the simple pre- and post-test study design without a control group, both these issues need to be addressed by looking at the results of other studies. We need a good general estimate of how much we would normally expect students to gain in CT over one semester at university, and a good general estimate of how much we would expect students to gain over one semester in a CT subject.

Now there are plenty of studies that address these questions, particularly the former (i.e., typical gain over a university semester). But here we face more problems. First, there are quite a few tests of critical thinking, and some are more difficult than others. A 20% gain on one test may not be equivalent to a 20% gain on another. Second, the available studies are heterogeneous, differing in many key aspects such as the size of the study (number of participants), the size of the gains (or losses) they found, the quality of the instruction in the subject being assessed, and level of care and rigor involved in the assessment. In the face of all this, how does one know, or estimate, what the "true" gains are?

Taking a meta-analytic approach

The best approach to handling these issues is to use a procedure called meta-analysis. In essence, meta-analysis is a way of pooling studies together to identify common trends or effects. Meta-analysis is a complex topic, but fortunately there are excellent introductions available—for example Cumming (2012). Here I'll describe the bare minimum required to understand the empirical results we have obtained for studies of AM-based instruction.

For current purposes, meta-analysis has three main steps:

1. *Select studies.* The first step is to determine which studies should be pooled together. This involves searching far and wide for potentially relevant studies, including unpublished studies, then using a set of criteria to determine which of these "make the cut," that is, are included in the data analysis.
2. *Convert results to effect sizes.* As indicated above, studies use a variety of different tests of critical thinking. They also report their results in a variety of ways. To enable pooling, these results need to be made commensurable. A common way to do this is to express the gain (or loss) found in a particular study as a proportion of the extent of the variability in performance on the test (technically, "Cohen's *d*" [Cohen 1969]; see also [Cumming 2012], Chapter 11) a figure often referred to as the "effect size." For example, in the van Gelder et al. studies mentionedearlier, a gain of around 20% converted to an effect size of around 0.87.
3. *Calculate pooled effect sizes.* Finally, results are thrown into the pool. This is not just a matter of finding the average effect size. Rather, effect sizes of individual studies are weighted by the size of the study (i.e., the number of participants), then the average is calculated. This gives the results of larger studies more weight, on the grounds that they are less susceptible to statistical noise and so more likely to indicate the true gain.

What do we find if we apply meta-analysis to studies of AM-based approaches to CT instruction? For some years, we have been gathering relevant studies and pooling their results. At time of writing, we obtained results twenty-six pre- and post studies of AM-based instruction in a one-semester CT subject, from institutions in Australia, Europe, and the United States. Many of these are unpublished, but published studies include those found in Butchart (2009), Dwyer, Hogan, and Stewart (2011; 2012), Harrell (2011), Twardy (2004), and van Gelder, Bissett, and Cumming (2004).

This is work in progress, but it currently appears that the weighted effect size for AM-based CT instruction is around 0.7. This effect size is based on all studies that meet the basic criteria for inclusion in the meta-analysis. However within that set there are clear differences in what we call the "intensity" of AM-based learning activities. Dividing the studies into high, medium, and low-intensity groups, we find a clear relation between intensity and amount of gain. In a high-intensity study, students took a subject in which AM was the primary or central activity, with lots of homework activities, and with instructors with high proficiency in AM. Fifteen of the twenty-six studies were high-intensity, and the weighted average effect size for these studies is 0.85.

However, as compared with a 20% gain, talking of an effect size of 0.85 means little to most people and may sound negligible. How do we gauge its significance? One approach is to use the rule of thumb recommended by one of the pioneers of meta-analysis, according to which 0.2 is a small effect, 0.5 is medium, and 0.8 is large (Cohen 1969).

This makes it seem like AM-based instruction has a "large" effect, but we haven't yet taken into account how much students would have gained anyway, even without AM. To estimate these, we turn to other meta-analyses. For example Alvarez conducted a meta-analysis of studies of gains in CT over one semester at college or university (Alvarez 2007), and found an effect of around 0.11 over one semester just due to maturation and being at college generally. This is a little larger than the 0.18 gain over two years of undergraduate education found identified in Arum and Roksa's large study (2011). The conclusion we can safely draw from these numbers is that the "value add" of AM-based CT instruction, relative to just being at college, is around 0.6 (or 0.7 for high-intensity AM), which is somewhere between a medium and a large effect size. Or, put another way, AM-based CT instruction yields many times the gain in CT skills over one semester than is normally achieved by just being at college.

Using a similar approach we can estimate the benefit of AM over other forms of CT instruction. In their meta-analysis (mentioned earlier) Abrami et al. found an effect size of 0.34 for college-level CT instruction generally (Abrami 2008), so AM-based instruction appears substantially more effective than other forms of CT instruction generally. (This doesn't rule out the possibility of some other particular form of instruction being at least as effective as AM-based instruction.)

Future directions

The upshot of the previous section is that increasing amounts of empirical research have been lending convergent support to the intuitively plausible idea that AM can substantially enhance critical thinking skills. Indeed, at this stage it seems fair to say that high-intensity AM-based instruction is one of the most effective techniques we know for accelerating CT skill gains in higher education.

Why does it work?

Insofar as AM does accelerate CT skill gains, why is this? What are the causal mechanisms? Little research has been done on this. The question was partially addressed by van Gelder, who asked why a specific AM software package might facilitate better thinking *performance* (van Gelder 2007). He canvassed three potential causal mechanisms:

1. that such software is more "usable" than the standard technologies we use for representing and manipulating reasoning;
2. that such software complements the strengths and weaknesses of our inbuilt cognitive machinery; and
3. that AM represents a semiformal "sweet spot" between natural language and formal logic.

It is not hard to imagine how each of these mechanisms may also play a role in facilitating not just performance on a given task, but also learning of CT skills. Another potential causal mechanism is that working with argument maps builds, in the learners' minds, mental templates or schemas for argument structures, making it easier for them to critically evaluate argumentation.

What dimensions of CT are being enhanced?

CT is multidimensional. For example, the Halpern Critical Thinking Assessment has five "subscales" for different dimensions of CT: verbal reasoning, argument analysis, thinking as hypothesis testing, likelihood and uncertainty, and decision making and problem solving (Halpern 2010). It is plausible that AM-based instruction will be more effective in enhancing some dimensions—say, verbal reasoning and argument analysis—than others. Closer analysis of data from existing and future studies may shed some light on this.

How much CT gain can be generated?

The meta-analysis suggests a strong relationship between intensity of AM and CT gain. Could even greater gains be achieved by even more intense training?

Even the most intense AM regimes in the studies included in this meta-analysis were not particularly demanding, being only somewhat more challenging than typical undergraduate subjects, and certainly much less intensive than, say, college athletics training. Thus it is plausible that substantially higher gains could be achieved, though of course there must also be practical limits. Given that high-intensity AM-based instruction is already showing gains of around 0.85 standard deviations, it is a reasonable conjecture that this practical limit would be somewhere between one and two standard deviations—which does not of course rule out even larger gains from exceptionally intense instruction.

What would it take to achieve gains of this order?

1. *Combining AM with other general approaches.* AM techniques should be used in conjunction with other techniques known to enhance learning, such as mastery learning (Kulik, Kulik, and Bangert-Drowns 1990) and peer instruction (Crouch and Mazur 2001), as suggested by Neil Thomason.
2. *Developing and deploying automated feedback.* One of the enabling conditions for rapid skill acquisition, in general, is timely, good-quality feedback. Having human instructors provide sufficient feedback of adequate quality is a very substantial challenge for AM-based CT instruction under normal resource constraints. Thus we must develop and use rich automated feedback systems of various kinds (Butchart 2009).
3. *Improved mapping tools.* The AM software in use today, while better than nothing, is much less sophisticated than it could be. In particular, improved educational mapping tools will need to integrate automated feedback.

To the extent that conditions such as these can be satisfied, the prospects for very substantial gains in CT being reliably achievable via semester-length instruction using AM are very good.

References

Abrami, P. C., Bernard, R. M., Borokhovski, E., Wade, A., Surkes, M. A., Tamim, R., and Zhang, D. 2008. "Instructional Interventions Affecting Critical Thinking Skills and Dispositions: A Stage 1 Meta-Analysis." *Review of Educational Research* 78 (4): 1102–1134.

Alvarez, C. 2007. *Does Philosophy Improve Reasoning Skills?* Masters Thesis, University of Melbourne, Melbourne, Australia.

Arum, R., and Roksa, J. 2011. *Academically Adrift : Limited Learning on College Campuses.* Chicago: University of Chicago Press.

Butchart, S., Forster, D., Gold, I., Bigelow, J., Korb, K., Oppy, G., Serrenti, A. 2009. "Improving Critical Thinking Using Web Based Argument Mapping Exercises with Automated Feedback." *Australasian Journal of Educational Technology* 25 (2): 268–291.

Cohen, J. 1969. *Statistical Power Analysis for the Behavioral Sciences.* Hillsdale, NJ: Lawrence Erlbaum Associates.

Crouch, C., and Mazur, R. 2001. "Peer Instruction: Ten Years of Experience and Results." *American Journal of Physics* 69: 970–977.

Cumming, G. 2012. *Understanding the New Statistics: Effect Sizes, Confidence Intervals, and Meta-Analysis*. New York: Routledge.

Davies, M. 2011. "Concept Mapping, Mind Mapping, Argument Mapping: What Are the Differences and Do They Matter?" *Higher Education* 62 (3): 279–301.

Dwyer, C. P., Hogan, M. J., and Stewart, I. 2011. "The Promotion of Critical Thinking Skills through Argument Mapping." In *Critical Thinking*, edited by C. P. Horvath and J. M. Forte. New York: Nova Science Publishers. 97–122.

Dwyer, C. P., Hogan, M. J., and Stewart, I. 2012. "An Evaluation of Argument Mapping as a Method of Enhancing Critical Thinking Performance in E-Learning Environments." *Metacognition & Learning* 7: 219–244.

Ericsson, K. A., Krampe, R. T., and Tesche-Römer, C. 1993. "The Role of Deliberate Practice in the Acquisition of Expert Performance." *Psychological Review* 100: 363–406.

Facione, P. A. 1991. "Using the California Critical Thinking Skills Test in Research, Evaluation, and Assessment." ED337498.

Fisher, A. 1988. *The Logic of Real Arguments*. Cambridge, England; New York: Cambridge University Press.

Govier, T. 1988. *A Practical Study of Argument*. (second edition). Belmont, CA: Wadsworth Pub. Co.

Halpern, D. F. 2010. *Halpern Critical Thinking Assessment*. Vienna: Schuhfried.

Harrell, M. 2008. "No Computer Program Required." *Teaching Philosophy* 31 (4): 351–374.

Harrell, M. 2011. "Argument Diagramming and Critical Thinking in Introductory Philosophy." *Higher Education Research and Development* 30 (3): 371–385.

HERI. 2009. *The American College Teacher: National Norms for 2007–2008*. Los Angeles: Higher Education Research Institute, University of California.

Kulik, C., Kulik, J., and Bangert-Drowns, R. 1990. "Effectiveness of Mastery Learning Programs: A Meta-Analysis." *Review of Educational Research* 60 (2): 265–306.

Liu, L., Bridgeman, B., and Adler, R. 2012. "Measuring Learning Outcomes in Higher Education: Motivation Matters." *Educational Researcher* 41 (9): 352–362.

Macagno, F., Reed, C., and Walton, D. 2007. "Argument Diagramming in Logic, Artificial Intelligence, and Law." *Knowledge Engineering Review* 22 (1): 87–109.

Mencken, H. L. 1997. *Minority Report : H. L. Mencken's Notebooks*. (Johns Hopkins paperbacks edition). *Maryland Paperback Bookshelf*. Baltimore, MD: Johns Hopkins University Press.

ter Berg, T., van Gelder, T. J., Patterson, F., and Teppema, S. 2013. *Critical Thinking: Reasoning and Communicating with Rationale*. Seattle, WA: CreateSpace.

Twardy, C. 2004. "Argument Maps Improve Critical Thinking." *Teaching Philosophy* 27 (2): 95–116

van Gelder, T. J. 2007. "The Rationale for Rationale™." *Law, Probability and Risk* 6:23–42.

van Gelder, T. J. 2013. "Argument Mapping." In *Encyclopedia of the Mind*, edited by H. Pashler. Thousand Oaks, CA: Sage.

van Gelder, T. J., Bissett, M., and Cumming, G. 2004. "Cultivating Expertise in Informal Reasoning." *Canadian Journal of Experimental Psychology* 58: 142–152.

Part III

Incorporating Critical Thinking in the Curriculum

If we assume that critical thinking is or should be an important part of higher education, how might it be incorporated into the curriculum? This is the matter under consideration in the chapters of this section.

Like van Gelder (section 2), Harrell and Wetzel place much weight on argument mapping since, for them, argument mapping—or argument analysis—skills lie at the heart of critical thinking. However, they acknowledge that "teaching students to read for an argument" is "notoriously difficult," and they propose a specific method. According to them, computers can help considerably since they offer students the capacity to quickly visualize the structure of an argument. Accordingly, at their university (in the United States), they have developed a first-year writing program along these lines. The theoretical backing is derived from a critical analysis of Stephen Toulmin's influential text on *The Uses of Argument*. While finding great merits in Toulmin's analysis of arguments, they come to the conclusion that it is overly complex for their educational situation and they accordingly offer a simpler structure, around which their program is developed. This simplicity allows for application "across a variety of texts from the wild (i.e.,...from a variety of contexts)." They also contend that their favored method improves students' capacities to improve their writing, including offering "metacommentaries" on their own analyses.

While beginning from quite a different starting point, a parallel approach is provided by Hammer and Griffiths. They observe that essays have been central to undergraduate education, at least in many subjects. They contend, however, that essays fall short of their potential and, in particular, often fail to promote critical thinking. Accordingly, they seek to revamp essays by having students—in their essays—focus on the structure and character of arguments of (other) authors; in short, essays are reshaped as a form of argument analysis. Such a reconfiguration of the essay can be "effective in developing students" critical evaluation skills because it situate(s) their essay writing within a debate with multiple positions on a topic and no right/wrong answers.

While not disagreeing that argument analysis is a key part of critical thinking, three of the papers here seek to widen discussion about the curriculum. Indeed, they at least implicitly contend that our thinking—about curricula for critical thinking—has been unduly limited. Eva Brodin goes even further, explicitly contending that doctoral students are dogged by perniciously narrow assumptions about critical thinking, and that these act to constrain and even deny students' capacities to express their creativity and their own voice. "The academic community can be a powerful force in choking the individual voice of a doctoral student." In the process, doctoral study can often be reduced to the "drudgery" of mere labor, as students learn defensive scholarly techniques "necessary for survival." Drawing on the writings of Hannah Arendt, a thesis is developed that looks to doctoral work incorporating but going well beyond "labor" so as to contain significant elements of creative and self-authored "work" and "action," where students find the courage to put themselves forward in relation to others, even at some risk to themselves.

A plank in Brodin's argument is that "creativity is seldom educationally encouraged in students at deeper levels." As a consequence, the student is often unable to identify with his or her efforts: even "the doctoral student (can run) the risk of becoming alienated from his or her own...activities." Much, therefore, depends on the pedagogical environment that provides the context for the student experience. This matter is taken up directly both by Iris Vardi, and by Justine Kingsbury and Tracy Bowell.

Vardi's particular concern is that higher education should enable students to come fully into the criticality that disciplines open, namely those experiences in which students identify with their studies, and feel able to take up their own critical stances. This entails, for Vardi, elements of "self-regulation" and of "personal epistemology." In short, criticality at its fullest involves critical self-reflection and a wish to go on improving one's own thinking by and for oneself. This though requires a pedagogical environment that "provides the opportunity to self regulate in the first place. Some contexts do not allow for self-regulated critical thinking processes." And what then might be the conditions that encourage such self-regulation? In short, it is a matter both of "control and support," the student being set "complex tasks with multiple goals" that not only "allow for personal choice and self-direction" but also for being provided with sensitive and helpful "feedback." The student-tutor relationship then—the pedagogical relationship—is crucial.

Another way of putting some of these reflections is that the development of criticality among students amounts to the formation of what Brady and Pritchard (2003) termed "epistemic virtues." It is this angle that Kingsbury and Bowell seek to open. Virtues are "dispositions to act in certain ways" and 'include a normative element," which is to say there is an onus on both educators and students that they be acquired. Kingsbury and Bowell distinguish three kinds of virtue—reliabilist (skills that can be deployed whenever required), regulatory (capacities to determine if, when, and to what extent such skills should be deployed), and motivational (having the wherewithal to deploy those skills even under taxing conditions).

We may note that there is a fundamental difference of view in these chapters. Kingsbury and Bowell contend that "traditional critical thinking teaching tends to focus on enabling students to gain the skills associated with identifying and appraising weak arguments rather than on constructing strong arguments." Even more, they claim that the "visual laying out of the structure of an argument" in particular and the "standard critical thinking course" more generally "does not live up to its billing" (thereby implicitly combating the positions both of Harrell and Wetzel and of Hammer and Griffiths). For Kingsbury and Bowell, there are three problems here. First, students are too often positioned as an audience for the arguments with which they are presented (a contention with which Brodin would surely agree); second, this positioning cannot readily help the development of students' own regulatory and motivational virtues, the virtues that are part of the student becoming her own person (a contention with which Vardi would surely agree); and third, without the development of the full range of epistemic virtues, students are unlikely to acquire either the ability or the propensity to carry over their criticality into their personal lives and the broader society. They go on, accordingly, to offer some suggestions as to curricula interventions that may help in fostering such epistemic virtues.

Whichever curriculum is adopted (with the intention of encouraging forms of criticality), how might its effectiveness be ascertained? This is the topic of the chapter by David Hitchcock, from whom we urgently need "well-designed studies of the effectiveness of undergraduate instruction in critical thinking." This is however far from easy, due to practical, methodological, and indeed ethical constraints in setting up controlled tests. Hitchcock argues that the most effective way forward lies in a combination of "computational assistance with built-in tutorial help" plus subject-matter instruction, backed by "special training of the instructor for teaching critical thinking."

We should observe, however, that there is much agreement across these chapters that the development of criticality among students is both a demanding and a lengthy processs. Iris Vardi notes that students' "beliefs about thinking are notoriously difficult to change"; Kingsbury and Bowell comment that "becoming an excellent critical thinker takes hard work" and "long hours of practice"; Hammer and Griffiths contend that "essay-writing (of the kind that sponsors criticality) is a complex writing genre that continues to evolve over time"; and Harrell and Wetzel observe that "teaching students to read for an argument is notoriously difficult." Indeed, following Brookfield's chapter elsewhere in this volume, we can surely say that the development of criticality is a lifelong process. This is hardly surprising, given the scope of criticality and its various levels, as revealed in the chapters here. Perhaps, then, curricula and pedagogies might be highly ambitious but yet be understood as a crucial part of a lifelong endeavor in the shaping of students' and graduates' criticality. A key task, therefore, is that of encouraging in students the wish and the will to develop and express their criticality evermore fully through their efforts in all aspects of their lives.

12

The Relationship between Self-Regulation, Personal Epistemology, and Becoming a "Critical Thinker": Implications for Pedagogy

Iris Vardi

Introduction

Developing critical thinkers is an important remit for universities, ensuring graduates who can improve our lives and our understanding of the world. So it is no wonder that universities spend much time and effort ensuring that critical thinking skills are embedded and assessed in the disciplinary curriculum. But is requiring students to use a set of skills in their studies sufficient for developing the critical thinkers society needs? This chapter goes beyond incorporating skills in the curriculum to examining the development of "critical thinkers": people who of their own volition approach problems, issues, learning, and the process of critical thinking itself through critical eyes. It does so by examining the development of critical thinkers through the lens of self-regulation and personal epistemology. Viewing development through this lens provides important insights for pedagogy. It highlights the importance of creating teaching environments that support and develop the goals, beliefs, attitudes, language, behaviors, and ways of doing things of a critical thinker. Further, it shows how such environments can be created to develop the competent, self-determined, critical thinkers we need for today and the future.

When students come to university, they usually have in mind what they want to become: a "historian," an "economist," a "physicist," or a "psychologist"—to name but a few. This ambition to move from what they are now to what they can become is very powerful. It results in a commitment to several years of study and can involve financial hardship and future financial burden. Yet students see a value in it. And they do much more than "go through the steps." They put in effort, persevere when the going gets tough, and make many changes to who they are through the process. They start to write like a "historian," view world events through the eyes of an "economist," evaluate physical phenomena like a "physicist," or observe others' reactions and behaviors like a "psychologist." They start this process of change by mimicking the language, behaviors, ways of "doing things,"and the world perspective of the members of their chosen discipline or profession (Bartholomae 1985). Their transformation is such that by the time they graduate, they *are* "economists" or "physicists." This change is not lost on graduates who often note how the university has changed their lives (Barnett 2009). However, for many this is not where it ends. After graduation, they often make the effort to become more skilled, knowledgeable, and influential in their field. The impact of the university on "becoming" a professional or disciplinary expert is evident in their student life and their later life.

The contrast between those who have determined what they want to become and those who haven't can be stark. Those who have not yet made the determination often change courses and programs, move in and out of university, and their effort can be lackluster. Without the desire to "become," there is no underlying reason to put in the effort and to persevere either now or in the future, and the university's impact is diminished.

How can we harness the power of "becoming" and apply that to becoming a "critical thinker," not only in students' chosen disciplines and professions, but also in their broader approach to life? Choosing to become a critical thinker involves much more than a set of skills. Much like becoming a "business person" or a "psychologist," it involves adopting the language, behaviors, ways of doing things, and perspectives of a "critical thinker." This chapter explores what it means to be a critical thinker, the factors impacting on students becoming critical thinkers, and the implications this has for developing and impacting on students' lives both in the university and beyond. In doing so, it draws on (1) the critical thinking literature, (2) the self-regulated learning literature on how students determine, control, and monitor their development, and (3) the personal epistemology literature on students' beliefs about how one "knows" something, their role in coming to "know" this, and how these beliefs relate to becoming a critical thinker.

In drawing these perspectives together, this paper aims to provide another lens through which to view critical thinking and its development within the disciplines and provide universities and teaching staff with contemporary directions for pedagogy.

What it means to be a critical thinker

The critical thinking literature points to a particular approach, set of behaviors, and attitudes that characterize "critical thinkers" and influence how they seek, interact with, and talk about knowledge and understanding. In broad terms, critical thinkers take the view that knowledge is worth pursuing, and that sound understanding is achieved through reasoning. This process of rea-soning can reveal itself in various ways in the disciplines such as judgment, skepticism, originality, rationality, and activist engagement with knowledge (Moore 2011). In whichever way critical thinking manifests or is conceived, being a critical thinker involves being proactive in both behavior and thoughts where needed. This proactivity is revealed in a number of ways. First, it is revealed in the effort to be well-informed (Ennis 1987) where knowledge, data, and information are sought and questioned. Critical thinkers recognize the need for a sound knowledge base with which to reason so that they can come to sound understandings, conclusions, and decisions.

Second, it is revealed in the behaviors, thoughts, and language associated with thinking and reasoning: a process of interpreting, analyzing, and evaluating one's own thinking and others' thinking in relation to specific questions, prob-lems, and issues (Facione 1990). This process can include examination of the matter at hand and its purpose; frame of reference; information, data, and meth-ods; claims and reasons; conclusions; implications; and consequences (Paul and Elder 2001). Being a critical thinker involves taking the necessary steps in the process, without being prompted or assessed, to sound independent thought.

However, the critical thinking literature points to more than just undertak-ing certain steps. It points to enacting these behaviors and thoughts in a cer-tain way and within a certain frame of mind. First is being self-disciplined and self-managed (Paul 1993), a set of behaviors referred to in the psychology lit-erature as being self-regulated. This is revealed through a methodical, diligent, and persistent approach (Facione 1990) to the entire process of critical think-ing. Second is being aware of and overcoming the many biases that can corrupt one's own thinking and the conclusions, understandings, and judgments that one draws. These include biases arising from beliefs, mental short-cuts, social influences, and personal motivations (Hilbert 2011). This involves having an open, fair, and reasonable mind, a preparedness to identify and face one's own biases, and a preparedness to reconsider one's own views where warranted (Facione 1990). Third is being committed to ongoing self-improvement (Paul and Elder 2001). This involves keeping abreast of knowledge and knowledge debates, ongoing evaluation of one's own ideas and thoughts, and improving one's own clarity of thought.

The role of goals and beliefs in becoming a critical thinker

Deciding to become a critical thinker clearly takes time, effort, and a persis-tence that goes well beyond learning a set of skills. What underlies this choice? To answer this question, it is useful to consider critical thinking within the framework of self-identity, goals, and beliefs arising out of the self-regulatory and personal epistemology literature.

Self-identity, who you believe you are, is acknowledged across theoretical traditions in self-regulation as enormously important to determining how you control or regulate your behaviors (Zimmerman and Schunk 2001). Self-identity determines students' motivation and the actions they take both in the short and long term. Seeing oneself as being "strategic," "artistic," "fun loving," or "academic," for instance, impacts on what one does and how one does it. However, this is more than simply situated in the present. What you *intend* to become and what you desire also affects what you do and how you feel about events and conditions around you. It is these intentions that provide commitment to actions in the future (Bandura 2001), hence the power of students deciding to become, for instance, "historians" or "economists."

Carver and Scheier (2000) argue that a person's concept of their "ideal self," the person that they aspire to be, influences a hierarchy of goals from higher-level goals, articulated as abstract principles, to lower-level goals, articulated as actions. Applying this concept to becoming a critical thinker reveals critical thinking principles at the higher level and associated critical thinking actions at the lower level. Examples of this are depicted in figure 12.1.

The higher-level goals or principles give one's life sense and provide direction for action, which is revealed in the lower-level practical goals (Carver and Scheier 2000). Such higher-level future goals have been found to enhance motivation, persistence, and performance, impacting on what people choose to do in the present and how they do this in order to achieve their aspirations (Simons, Dewitte, and Lens 2004).

Literature from the critical thinking movement and personal epistemology research suggests that becoming a critical thinker and adopting such high-level goals and principles is underpinned by a number of interrelated beliefs. The first, from the critical thinking literature, is that the quality of one's *own* thinking and understanding can affect the quality of one's life, others' lives, and society more broadly (Paul 1993). This belief in the power of thought and

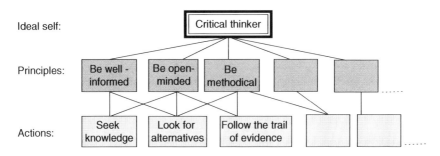

Figure 12.1   Illustrating the concept of a hierarchy of goals in becoming a critical thinker.

Source: Based on Carver and Sheier's hierarchy of goals (2000).

understanding provides a strong sense of purpose. This is augmented by the belief that with practice one *can* improve the quality of one's own thinking (Paul 1993), thereby developing the intellectual means to improve the quality of life. In other words, critical thinkers believe that they can learn to think better and, furthermore, that this has a benefit. Personal epistemology research, which examines students' beliefs about "knowing" and their role in coming to "know," shows that believing that one can learn and improve is critical to taking the necessary steps (Schommer-Aikins 2004). Where students believe that intellect and the ability to learn is fixed, they are less likely to persist and less likely to value education (Schommer-Aikins 2004).

Alongside these beliefs is the belief that understanding comes *through* critical thinking. This belief is based on a number of further epistemological beliefs individuals can hold about the nature of knowledge and understanding. According to Kuhn (1999), where people hold immature beliefs about knowledge they will not see the need to critically think. She argues that where one believes that knowledge is certain and conveyed by an authority, critical thinking skills will be seen as unnecessary, or at best, a means for comparison with the "correct' answer. Even when people believe that thought generates understanding and knowledge is uncertain, Kuhn highlights that they may perceive critical thinking to be irrelevant, coming to the view that knowledge claims and assertions are opinions and that everyone is entitled to their own opinion. It is only when people perceive assertions to be judgments that can be subject to evaluation will they see the need to create their own knowledge and understanding through critical thinking. According to Kuhn (1999), "Epistemological understanding may be the most fundamental underpinning of critical thinking" (23). Believing that one is the owner and creator of one's own understandings and changing knowledge base allows for the intellectual independence that critical thinking affords. These beliefs "give permission" to change one's mind, question authority and received wisdom, and take a position that is different to commonly held beliefs where reasoning and the available evidence indicates that this is reasonable or necessary.

These beliefs about knowledge, learning, thinking, ability, and quality of life give purpose and reason to pursuing future goals to become a critical thinker. It is these types of beliefs that lay the basis for being inquisitive and well-informed and aiming for what Ennis (1996, 171) terms "wanting to 'get it right.'" Future aspirations to become a critical thinker, coupled with these beliefs, provide direction, reason, motivation, and persistence.

Self-regulation in the context of becoming a critical thinker

Aspiring to and reaching the goal of becoming a critical thinker, however, requires successful self-regulation. While very important, self-regulation is much more than motivation and persistence in the pursuit of such a future goal. It is a process of control and monitoring that is interactive and adaptive (Zimmerman and Schunk 2001). For each person, these interactions and adaptations are different. Achieving the goal of becoming a critical thinker can depend, for instance, upon the available opportunities, how a person determines to take advantage (or not) of these opportunities, the social context, competing demands, their beliefs, their knowledge application and adaptation of strategies, and their emotions and feelings at the time. These factors, arising from the environment, personal characteristics, and self-regulatory processes, are not static. They influence each other as shown in figure 12.2.

Successful self-regulators interact with and adapt to the environment to generate thoughts, feelings, and behaviors that will result in achievement (Zimmerman 2000). Future goals on their own, however, make self-regulation difficult to sustain: progress appears slow, goals are not specific enough, the path to take can be unclear, and feedback related to goal progression can be a long time in coming (Zimmerman and Schunk 2001). It is therefore important to turn future goals of becoming a critical thinker into achievable short-term goals that allow for success. These focus attention, lay the basis for step-wise progression toward the ultimate goal, and allow for flexibility and adaptability as one progresses toward one's "ideal self."

Two types of short term goals are of key importance to successfully self-regulating in order to become a critical thinker. First are short-term goals for developing a sound knowledge base. As discussed earlier, being well-informed in a domain is essential to be able to soundly reason in that domain. Further, domain knowledge has been found to be a key condition for effective self-regulation (Greene and Azevedo 2007). Research into expertise suggests that a deep, well-organized knowledge base reduces mental load, allowing for automatic knowledge activation and ease of knowledge growth (Pintrich and Zusho 2002).

Second are short-term goals for developing the ability to reason in an open and fair minded way. Just as automaticity is needed for knowledge activation so automaticity through learning and practicing the process of thinking critically

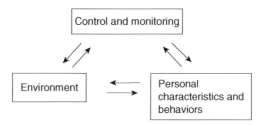

Figure 12.2 Interactions between self-regulatory processes, the environment, and personal characteristics and behaviors.

is needed. To this end, Paul and Elder (2001) exhort learners to develop as critical thinkers by first recognizing the need to improve and then making plans of their own for practice to attain various levels of automaticity including "advanced thinker" and "master thinker" (22), in essence developing short-to-medium-term goals that drive plans for reaching their future "ideal self."

Together, the future goals for becoming a critical thinker and the short-term goals for developing a sound knowledge base and sound reasoning ability provide direction for personal control or "agency." Personal agency creates success by taking advantage of opportunities; "making things happen"; planning; managing and monitoring motivation, affect, and behavior; and examining one's functioning (Bandura 2001). Successful self-regulators take action and influence their environment in ways that support achievement of their goals: seeking out advice and information, teaching themselves, initiating activities for practice and improvement, using effective strategies to reach their goals, and reviewing how they are going (Zimmerman 1990)—all important for becoming an independent critical thinker.

Reviewing or self-evaluating is particularly important and is emphasized in both the self-regulatory (Zimmerman and Schunk 2001) and the critical thinking literature (Facione 1990; Paul and Elder 2001). The ability to evaluate one's own efforts and improve is dependent upon having standards, benchmarks, or specific goals against which to judge how one is going. When successful self-regulators see a discrepancy between what they do and the standard or goal they are trying to achieve, they persevere until this discrepancy is resolved (Zimmerman and Schunk 2001). For some learners, meeting a particular short-term goal or standard can result in them setting a new more challenging goal or benchmark for themselves to meet (Zimmerman and Schunk 2001). This is the type of approach to becoming a critical thinker that is emphasized by Paul and Elder (2001).

Given the personal effort, direction, and focus entailed in self-regulation, it is not surprising that many have found that when students self-regulate, academic achievement, self-satisfaction, and motivation increase (Vrieling, Bastiaens, and Stijnen 2012; Zimmerman 2002). As this happens, the cycle is reinforced. With achievement through self-regulation, successful self-regulators believe they will be successful, that they are responsible for and capable of their own development, and that efforts to self- regulate will work (Greene and Azevedo 2007; Zimmerman 1990). This self-reinforcing cycle is fundamental to becoming an accomplished self-regulated critical thinker.

However, not all learners hold these types of beliefs; understand this interactive process and the personal control or agency that they have; or have the self-regulatory attitudes, skills, strategies, and behaviors required. This affects perseverance and success and can result in inhibiting or self-defeating behaviors and even abandonment of goals such as becoming a critical thinker.

Environmental impacts on becoming a
self-regulated critical thinker

Becoming a self-regulated critical thinker requires individuals to take personal control over such aspects as goals, plans, actions, the pace in which these actions are undertaken, environmental conditions, and with whom one collaborates. However, environments interact with personal characteristics and behaviors, such as self-control, in various ways that can enhance or inhibit self-regulation.

Fundamental to assuming responsibility for self-control is whether the environment provides the opportunity to self-regulate in the first place. Some contexts do not allow for self-regulated critical thinking processes. How and when things are done are controlled or directed by others. Such environmental control does not encourage initiative, the development of plans, or the testing and developing of processes, skills, and strategies for self-regulation of critical thinking. It has been found that when students are given increased opportunities to self-regulate, their use of self-regulatory processes and strategies increases (Vrieling, Bastiaens, and Stijnen 2012). Similarly, where students perceive that self-regulation is not required, is not preferable, or that they will not gain from self-regulating, they will not use the process (Zimmerman 2001).

The opportunity to self-regulate as a critical thinker is enabled in a number of key ways. First, it is enabled by complex tasks with multiple goals (Greene and Azevedo 2007). Such tasks allow for multiple decisions to be made and for control over these to be exerted. For critical thinking, these are typified by appropriately challenging open-ended questions, problems, or issues, also known as "ill-structured," "messy," or "authentic,"which do not have "one right answer" (Facione 1990) such as feasibility studies, design briefs, research projects, systems analysis, argumentative essays, and the like (Vardi 2013a). Second, it is enabled by an environment that allows for personal choice and self-direction, or autonomy, in the questions, problems, or issues to be addressed and the manner in which they are tackled (Ryan and Deci 2000).

However, simply providing the opportunity to take control and self-regulate when confronted with a complex critical thinking task may not result in personal success, satisfaction, or persistence toward the future goal of becoming a critical thinker. According to Self-Determination Theory, humans possess a fundamental need for autonomy *coupled with* competence, and when both of these needs are supported and met by the environment, motivation, initiative, curiosity, and effort increase, particularly when learning is conceptually and creatively challenging (Ryan and Deci 2000). This desire for competence in challenging situations where students are required to self-regulate was observed by Vrieling et al. (2012). They found that when students were given increased opportunities to self-regulate, they wanted and needed instruction in strategies and knowledge of the domain to be successful. In situations requiring critical thinking, this includes the need for competence and instruction in the relevant critical thinking strategies, skills, and dispositions.

Within an educational setting, the balance between personal control, or autonomy, and environmental control and support is complex. While some students are willing and able to assume responsibility and have the necessary personal resources (e.g., skills, strategies, knowledge, behaviors), others do not have these personal resources, and therefore need more environmental control, support, and intervention in the process (Garrison 1992). The need for balance between personal and environmental control has led Vrieling et al. (2012) to suggest that as competence in using the necessary skills, strategies, and dispositions is progressively developed, control is progressively handed over. The question is the nature of that control. Eshel and Kohavi (2003) have shown that control, when it manifests as a well-organized and well-structured supportive environment, can coexist with high levels of student autonomy, and further that this coexistence leads to high levels of student satisfaction and achievement. Perception plays an important role. Where students perceive that their level of control is, on-balance, higher than the control exerted by the environment, there is a positive impact on self-regulation and achievement (Eshel and Kohavi 2003).

Control and support are different. The literature points to two key environmental factors that are of particular importance in supporting students in becoming self-regulated critical thinkers. First is the provision of externally provided feedback. Externally provided feedback is particularly important for self-regulation as it provides grounds for re-interpreting and calibrating performance against standards and goals (Butler and Winne 1995). The second involves sharing a good quality relationships with mentors, such as their teachers, whereby they tend to internalize their mentor's beliefs (Martin and Dowson 2009), including those about knowledge, one's role in acquiring knowledge, and one's ability to do so.

These findings point to the need for an environment that supports and develops both self-regulation and critical thinking for success. Such environments are well-organized and well-structured for developing competence in self-regulation and critical thinking, as well as for providing appropriate levels of personal control or autonomy. These types of environments increase motivation, self-efficacy, and ability, and support personal agency in reaching one's goals of becoming a critical thinker.

Implications for critical thinking pedagogy in higher education

Applying the lens of self-regulation and personal epistemology to developing critical thinkers reveals the need for students to have: (1) the aspirational goals

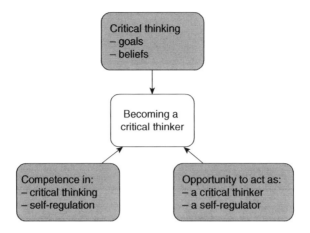

Figure 12.3   What students need to become a "critical thinker."

and supporting beliefs of a critical thinker; (2) the opportunity to act as a self-regulated critical thinker; and (iii) the competence in using the skills, dispositions, strategies, and processes of a self-regulating critical thinker. This is depicted in figure 12. 3.

Providing an educational environment that supports and develops these needs is important for developing critical thinkers. This has significant implications for practice at universities.

Goals and beliefs

Often students come to university knowing what they want to become in a disciplinary sense, such as a "medical researcher" or a "biologist." They have some idea about what this might entail and, through their experiences with the university, these aspirations may remain, change, or become more focused so that now their goal may be to become, for instance, a "geneticist." In other words, universities have a strong influence on aspirations. This influence on aspirations can go well beyond a professional identity to include aspirations for the type of person one wants to be, including a "critical thinker": a person who adopts a critical thinker's language, behaviors, and "ways of doing things."

Influence can occur by highlighting to students why becoming a critical thinker matters and what it entails: in essence acting to make the broader societal goals of a university education the personal goals of the student. Giving meaning and a broader purpose to life provides for powerful personal goals (Carver and Scheier 2000) that can influence students' behaviors and actions in the future. Further, showing students that the university will help them achieve this aspiration by breaking it down into achievable steps will help students identify and determine how they take advantage of the opportunities provided by the university.

Influencing students' aspirations and future actions may for some students, however, require engagement with fundamental beliefs that they have about their personal abilities and the potential they have to make a meaningful impact. Further, as shown through the personal epistemology literature, it may require engagement with fundamental beliefs about the nature of knowledge, the role one plays in gaining understanding, and the personal control one needs to invoke. As beliefs about thinking are notoriously difficult to change (Schraw, McCrudden, Lehman, and Hoffman 2011), influencing these beliefs and aspirations may well take several years and be affected by students' experiences in the university, their developing competence, and the expectations set by the teaching staff. Research shows that that while beliefs about the nature of knowledge do change as students progressively seek higher qualifications, this progress is gradual (King and Kitchener 2002), with the nature of the disciplinary learning environment impacting significantly on development (Palmer and Marra 2004). This is where providing opportunity and developing competence in the disciplinary context come into play.

Opportunity

What a student might aspire to and what he or she is expected and allowed to do are two different things. Opportunities for students to act as critical thinkers need to be frequent throughout the educational experience: in the class, in activities and tasks, in assessments (Vardi 2013a), and in the total campus experience (Tsui 2000). Embedding such opportunity provides immersion in a culture of critical thinking. Opportunity lies in the way teaching staff engage in discussion with students and make clear their expectations for students to engage with and interrogate knowledge on their own and with their peers (Hofer 2004; Maclellan and Soden 2012). It also occurs through providing students with challenging, meaningful, complex tasks that require sound reasoning based on investigation of the literature, data collection, interpretation, analysis, synthesis, and evaluation (Facione 1990; Paul and Elder 2001; Vardi 2013a) both within and external to the curriculum. These are the types of opportunities that help develop those personal epistemological understandings and beliefs that are so important to critical thinking development (King and Kitchener 2002).

As discussed earlier, however, opportunity is more than immersing students in critical thinking. It is also about allowing students to self-regulate (Zimmerman 2002) by providing them with autonomy, particularly with tasks that require persistence over time and that utilize the entire process of critical thinking. Several ways have been suggested for increasing autonomy. These include providing choice, for instance, in tasks, methods employed, and study partners; handing over control of the learning process (Garrison 1992); and allowing control over environmental conditions such as where to study, the level of noise or quiet in the environment, and when to study (Zimmerman 2001). This improves motivation. However, as Hattie (2009) cautions, one needs to take care not to have irrelevant, less effortful choices and not to have so many choices as to be overwhelming. Appropriately controlled levels of choice with complex tasks need to be provided in a well-organized and well-structured environment.

Competence

Opportunity without competence, and its concomitant successes, however, can lead to disillusionment and abandonment of goals. As shown through this paper, becoming a critical thinker requires competence in the domain; in the use of critical thinking skills, dispositions, strategies, and processes; and in the use of self-regulatory skills, strategies, and behaviors. Meta-analysis of the research suggests that developing competence in critical thinking may be enhanced by explicit instruction *coupled with* opportunity *and* explicit goals (Abrami, Bernard, Borokhovski, Wade, Surkes, Tamim, and Zhang 2008).

Simply providing opportunity is not enough as students can construct understandings incorrectly and reason poorly (Paul 1993). Further, they may not self-regulate well, particularly with complex tasks that take time and perseverance to complete. This is where explicit direct instruction aimed at achieving the desired results of becoming a well-informed self-regulated critical thinker is important. Hattie's (2009) extensive meta-analysis of teaching interventions found that explicit instructional techniques, aimed specifically at improving competence and goal achievement, are significantly more effective in improving student outcomes than approaches in which students find their own way with minimal guidance.

Explicit, direct instruction in critical thinking has been found to be particularly powerful for educational achievement in units of study within the discipline, as it focuses on the skills, behaviors, and attitudes specific to the assessment tasks (Schraw et al. 2011). These can be directly taught by providing students with (1) an explanation of the critical thinking requirements of the tasks; (2) discipline-specific exercises that provide practice in the underlying critical thinking skills and dispositions; and (3) written models showing how sound reasoning was developed and expressed in other similar tasks (Vardi 2013a).

This type of instruction within the discipline improves domain knowledge as well as critical thinking within that domain in a mutually reinforcing cycle. However, as transfer of thinking skills between domains or subject areas is a challenge (Schraw et al. 2011), explicit, direct instruction in general critical thinking, outside of the discipline, may also be necessary to provide students with the tools and practice to apply skills both within and outside of the educational context, thus supporting the broader development of the critical thinker.

Direct instruction on the strategies that students can use to self-regulate has also been found to increase competence, knowledge, interest, self-regulation, and deep processing (Greene and Azevedo 2007). Such strategies can include verbalizing steps and understandings (Schunk 2001), using elaborative questioning, deconstructing problems or issues to identify and focus attention initially on the elementary components (Zimmerman and Schunk 2001), self-recording, positive self-talk, self-rewards (Zimmerman 2001), and regularly reviewing how one is going (Schunk 2001).

This type of instruction lays the basis for providing an appropriate balance of control, guidance, and support for students to take advantage of the opportunities provided to critically think. The literature points to a number of important aspects in the environment that enable and improve students' abilities to take advantage. The first is providing clearly stated goals, learning objectives, or outcomes for critical thinking (Abrami et al. 2008) accompanied by standards or benchmarks for reaching these (Paul and Elder 2001; Sadler 1989; Zimmerman and Schunk 2001). Standards enable self-regulation by providing a basis for comparison and meaningful self-evaluation of performance. To be useful, these standards need to clarify what constitutes "quality," and students need to learn how to objectively compare their performance with the standards and how to improve (Sadler 1989).

The second is the use of feedback. Effective instructional feedback, which explains the strengths and the improvements that need to be made in relation to the standards, is one of the most powerful influences on student achievement (Hattie 2009). Vardi (2013a) shows how powerful feedback can be provided to students that develops both their knowledge and critical thinking in the discipline. First is feedback explaining the grade awarded to the task and its strengths. Next is feedback that addresses and provides direction for conceptual misunderstandings, task-related misunderstandings, and any problems with the reasoning process. Last is the feedback that addresses how the student could improve *in relation to the standards* in both the current task and the next task. This type of feedback is most effective when it is prescriptive and, in the case of written tasks, when it makes links between the context, the content, the thinking, and the structuring of the student's text (Vardi 2009; 2012; 2013b).

Hattie (2009) points out that the power of feedback comes not only from teachers providing feedback to the students, but also from students showing the teacher what they understood, where they are struggling, and where they have misconceptions. Feedback to and from students synchronizes teaching and learning and makes it all the more powerful (Hattie 2009).

The final important component in successfully guiding and supporting students in an environment full of opportunity is the personal mentoring by the teacher. Abrami et al. (in press) found that mentoring in the presence of opportunities to engage in complex authentic tasks *together with* teacher-led critical discussion resulted in a significantly greater effect on students' critical thinking than did either of these opportunities on their own. Research further shows that good mentors and models connect personally with students. They impact on student achievement by believing in the students, caring for and having high expectation for their achievement, and supporting their autonomy (Martin and Dowson 2009).

Conclusion

The development of critical thinkers is an important societal remit for universities. It develops citizens who can (1) use sound reasoning to make decisions and solve problems in their private and professional lives; (2) contribute to and influence societal debate; and (3) progress knowledge and our understanding of the world.

Examining this remit through the lens of self-regulation and personal epistemology provides further direction to universities on how to develop critical thinking for use within and beyond the university context. This perspective highlights the need for universities to look well beyond embedding skills in assessment tasks to developing a person who, in the absence of assessment, grading, and marks, *wants to become* a good critical thinker, and to do so for good purpose.

Addressing the development of critical thinkers through this perspective raises the importance of addressing goals, beliefs, opportunities, *and* competence. It points to the need to support and encourage students to develop personal goals and beliefs that underpin becoming a critical thinker; to provide them with an environment replete with opportunity for using and taking control of the entire process of critical thinking; and to actively and purposefully develop their competence in both self-regulation and critical thinking. This broader approach aims to do more than develop and assess critical thinking skills; it aims to develop the critical thinkers that society needs now and in the future.

References

Abrami, P. C., Bernard, R. M., Borokhovsk, E., Waddington, D. I., Surkes, M. A., Persson, T., and Wade, C. A. (in press). *Teaching Students to Think Critically.* Concordia University, Montreal, QC..

Abrami, P. C., Bernard, R. M., Borokhovski, E., Wade, A., Surkes, M. A., Tamim, R., and Zhang, D. 2008. "Instructional Interventions Affecting Critical Thinking Skills and Dispositions: A Stage 1 Meta-Analysis." *Review of Educational Research* 78 (4): 1102–1134.

Bandura, A. 2001. "Social Cognitive Theory: An Agentic Perspective." *Annual Review of Psychology* 52 (1): 1–26.

Barnett, R. 2009. "Knowing and Becoming in the Higher Education Curriculum." *Studies in Higher Education* 34 (4): 429–440.

Bartholomae, D. 1985. "Inventing the University." In *When a Writer Can't Write*, edited by M. Rose. New York: The Guilford Press. 134–166.

Butler, D. L., and Winne, P. H. 1995. "Feedback and Self-Regulated Learning: A Theoretical Synthesis." *Review of Educational Research* 65 (3): 245–281.

Carver, C. S., and Scheier, M. F. 2000. "On the Structure of Behavioral Self-Regulation." In *Handbook of Self-Regulation*, edited by M. Boekaerts, P. R. Pintrich, and M. Zeidner. San Diego: Academic Press. 41–84.

Ennis, R. H. 1987. "A Taxonomy of Critical Thinking Dispositions and Abilities." In *Teaching Thinking Skills: Theory and Practice*, edited by J. B. Baron and R. J. Sternberg. New York: W. H. Freeman and Company.

Ennis, R. H. 1996. "Critical Thinking Dispositions: Their Nature and Assessability." *Informal Logic* 18 (2 and 3): 165–182.

Eshel, Y., and Kohavi, R. 2003. "Perceived Classroom Control, Self-Regulated Learning Strategies, and Academic Achievement." *Educational Psychology* 23 (3): 249–260.

Facione, P. 1990. *Critical Thinking: A Statement of Expert Consensus for Purposes of Educational Assessment and Instruction: Research Findings and Recommendations.* Newark, DE: American Philosophical Association.

Garrison, D. R. 1992. "Critical Thinking and Self-Directed Learning in Adult Education: An Analysis of Responsibility and Control Issues." *Adult Education Quarterly* 42 (3): 136–148.

Greene, J. A., and Azevedo, R. 2007. "A Theoretical Review of Winne and Hadwin's Model of Self-Regulated Learning: New Perspectives and Directions." *Review of Educational Research* 77 (3): 334–372.

Hattie, J. 2009. *Visible Learning: A Synthesis of Meta-Analyses Relating to Achievement.* New York: Routledge.

Hilbert, M. 2011. "Toward a Synthesis of Cognitive Biases: How Noisy Information Processing Can Bias Human Decision Making." *Psychological Bulletin* 138 (2): 211–237.

Hofer, B. K. 2004. "Exploring the Dimensions of Personal Epistemology in Differing Classroom Contexts: Student Interpretations during the First Year of College." *Contemporary Educational Psychology* 29 (2): 129–163.

King, P. M., and Kitchener, K. S. 2002. "The Reflective Judgement Model: Twenty Years of Research on Epistemic Cognition." In *Personal Epistemology: The Psychology of Beliefs about Knowledge and Knowing*, edited by Barbara K. Hofer and Paul Pintrich. Mahwah, NJ: Lawrence Erlbaum Associates Publishers. 37–61.

Kuhn, D. 1999. "A Developmental Model of Critical Thinking." *Educational Researcher* 28 (2): 16–25, 46.

Maclellan, E., and Soden, R. 2012. "Psychological Knowledge for Teaching Critical Thinking: The Agency of Epistemic Activity, Metacognitive Regulative Behaviour and (Student-Centred) Learning." *Instructional Science* 40 (3): 445–460.

Martin, A. J., and Dowson, M. 2009. "Interpersonal Relationships, Motivation, Engagement, and Achievement: Yields for Theory, Current Issues, and Educational Practice." *Review of Educational Research* 79 (1): 327–365.

Moore, T. 2011. "Critical Thinking and Disciplinary Thinking: A Continuing Debate." *Higher Education Research & Development* 30 (3): 261–274.

Palmer, B., and Marra, R. M. 2004. "College Student Epistemological Perspectives across Knowledge Domains: A Proposed Grounded Theory." *Higher Education* 47 (3): 311–335.

Paul, R. 1993. *Critical Thinking: What Every Person Needs to Survive in a Rapidly Changing World.* (revised third edition). Santa Rosa, CA: The Foundation for Critical Thinking.

Paul, R., and Elder, L. 2001. *Critical Thinking: Tools for Taking Charge of Your Learning and Your Life.* London: Prentice-Hall International (UK) Limited.

Pintrich, P. R., and Zusho, A. 2002. "The Development of Academic Self-Regulation: The Role of Cognitive and Motivational Factors." In *Development of Achievement Motivation*, edited by A. Wigfield and J. S. Eccles. San Diego: Academic Press.

Ryan, R. M., and Deci, E. L. 2000. "Self-Determination Theory and the Facilitation of Intrinsic Motivation, Social Development, and Well-Being." *American Psychologist* 55 (5): 68–78.

Sadler, D. R. 1989. "Formative Assessment and the Design of Instructional Systems." *Instructional Science* 18: 119–144.

Schommer-Aikins, M. 2004. "Explaining the Epistemological Belief System: Introducing the Embedded Systemic Model and Coordinated Research Approach." *Educational Psychologist* 39 (1): 19–29.

Schraw, G., McCrudden, M. T., Lehman, S., and Hoffman, B. 2011. "An Overview of Thinking Skills." In *Assessment of Higher Order Thinking Skills*, edited by G. Schraw and D. R. Robinson. Charlotte, NC: Information Age Publishing.

Schunk, D. H. 2001. "Social Cognitive Theory and Self-Regulated Learning." In *Self-Regulated Learning and Academic Achievement: Theoretical Perspectives*, edited by B. J. Zimmerman and Dale H. Schunk (second edition). London: Lawrence Erlbaum Associates Publishers. 125–151.

Simons, J., Dewitte, S., and Lens, W. 2004. "The Role of Different Types of Instrumentality in Motivation, Study Strategies, and Performance: Know Why You Learn, So You'll Know What You Learn!" *British Journal of Educational Psychology* 74 (3): 343–360.

Tsui, L. 2000. "Effects of Campus Culture on Students' Critical Thinking." *The Review of Higher Education* 23 (4): 421–441.

Vardi, I. 2009. "The Relationship between Feedback and Change in Tertiary Student Writing in the Disciplines." *International Journal of Teaching and Learning in Higher Education* 20 (3): 350–361.

Vardi, I. 2012. "The Impact of Iterative Writing and Feedback on the Characteristics of Tertiary Students' Written Texts." *Teaching in Higher Education* 17 (2): 167–179.

Vardi, I. 2013a. *Developing Students' Critical Thinking in the Higher Education Class*. Milperra, NSW: Higher Education Research and Development Society of Australasia.

Vardi, I. 2013b. "Effectively Feeding Forward from One Assessment Task to the Next." *Assessment & Evaluation in Higher Education* 38 (5): 599–610.

Vrieling, E. M., Bastiaens, T. J., and Stijnen, S. 2012. "Effects of Increased Self-Regulated Learning Opportunities on Student Teachers' Metacognitive and Motivational Development." *International Journal of Educational Research* 53: 251–263.

Zimmerman, B. J. 1990. "Self-Regulated Learning and Academic Achievement: An Overview." *Educational Psychologist* 25 (1): 3–17.

Zimmerman, B. J. 2000. "Attainment of Self-Regulation: A Social Cognitive Perspective." In *Handbook of Self-Regulation*, edited by M. Boekaerts, P. R. Pintrich, and M. Zeidner. San Diego: Academic Press. 13–39.

Zimmerman, B. J. 2001. "Theories of Self-Regulated Learning and Academic Achievement: An Overview and Analysis." In *Self- Regulated Learning and Academic Achievement: Theoretical Perspectives*, edited by Barry J. Zimmerman and Dale H. Schunk (second edition). London: Lawrence Erlbaum Associates Publishers. 1–37.

Zimmerman, B. J. 2002. "Becoming a Self-Regulated Learner: An Overview." *Theory into Practice* 41 (2): 64–70.

Zimmerman, B. J., and Schunk, D. H. 2001. "Reflections on Self-Regulated Learning and Academic Achievement." In *Self- Regulated Learning and Academic Achievement: Theoretical Perspectives*, edited by B. J. Zimmerman and Dale H. Schunk (second edition). London: Lawrence Erlbaum Associates Publishers. 289–307.

13
Using Argument Diagramming to Teach Critical Thinking in a First-Year Writing Course

Maralee Harrell and Danielle Wetzel

Introduction

The importance of teaching critical thinking skills at the college level cannot be overemphasized. Teaching a subcategory of these skills—argument analysis—we believe is especially important for first-year students with their college careers, as well as their lives, ahead of them. The struggle, however, is *how* to effectively teach argument analysis skills that will serve students in a broad range of disciplines.

Why is it so hard to teach argument analysis skills? Martin Davies articulates a good answer:

> In addition to the complexities of distinguishing different parts of the argument, students must also deal with the complexities of academic language. The student must, in addition, be able to:
> (1) Succinctly paraphrase claims;
> (2) Distinguish premises from conclusions;
> (3) Locate crucial hidden premises;
> (4) Put the claims into the appropriate logical order;
> (5) Show the inferential link(s) from premises to conclusions. (Davies 2009, 802–803)

The teaching method we want to advocate here is argument diagramming. There are, however, several different models of argument diagramming from which to choose. One of the most popular models was promoted by Stephen Toulmin in *The Uses of Argument* in 1958 (Toulmin 1958). Over the past several decades, for example, the Toulmin model has been adopted by English, rhetoric, and composition departments all over the United States. An alternative model for diagramming arguments, however, has recently gained some traction with teachers of critical thinking and informal logic. This model originated with Monroe Beardsley in 1950, was refined by James Freeman in the 1980s and 1990s (Freeman 1991), and is now known as the Beardsley-Freeman model.

This kind of argument diagram is a visual representation of the content and structure of an argument. For illustration, consider the following argument:

> The ability to think critically is more important now than it has ever been. People have always had to make important decisions in their daily lives, but now, more than ever, these decisions can affect millions of others around the word, as well as many more millions in future generations. When we vote for particular criminal or national health policies, these decisions resonate through our communities. When we vote for candidates for particular political offices, these decisions can impact other people around the world who are affected by our foreign policy. And, when we vote for particular environmental policies, we are making decisions that will determine the kind of world our child and grandchildren will inherit. Since these decisions are so important, it stands to reason that we need these decisions to be the product of careful research and thoughtful reasoning, which are the hallmarks of critical thinking.

For diagramming using a modified Beardsley-Freeman model, the claims are put into boxes, the inferential connections are represented by arrows, and all the excess verbiage is removed (see figure 13.1).

In what follows, we argue that teaching argument analysis skills in a first-year composition course using a modified version of the Beardsley-Freeman model of diagramming is better than doing so using the Toulmin model. To make this case, we first explore the nature and importance of critical thinking skills in the twenty-first century. We then explore the mounting evidence that teaching argument diagramming is a good way to improve students' critical thinking skills. The method one uses for diagramming arguments, however, depends on one's theory of argumentation, so we analyze Toulmin's theory and its conceptual and pedagogical problems. We then describe the development of a modified Beardsley-Freeman method of argument diagramming, as well as the results of a study we conducted to test the difference between teaching using the Toulmin method of argument diagramming and using the modified Beardsley-Freeman method.

The importance of critical thinking skills

Completion of at least one critical thinking course is a requirement at many colleges and universities in the United States (e.g., California State University, State University of New York, San Francisco State University, Pomona College)

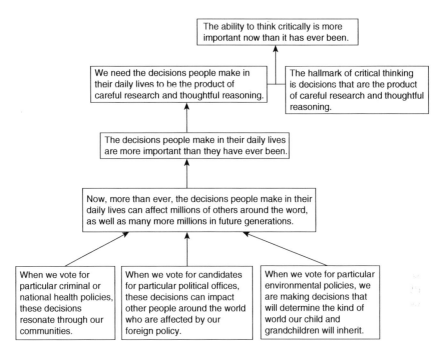

Figure 13.1 An argument diagram representing an argument for the importance of critical thinking.

and in other countries around the world (e.g., Athabasca University, Canada, McMaster University, Canada, Atkinson College, Canada, Universiteit Maastricht Faculty of Law, The Netherlands, Al Akhawayn University, Morocco, Edith Cowan University, Australia, Australian Catholic University, Australia, Charles Sturt University, Australia). In addition, many more colleges and universities have "development of critical thinking skills" or similar language as a part of the mission statement of the institution (e.g., Washburn University, Iowa State University).

In the United States, these requirements may have been based on the mounting evidence that, as a nation, our efforts to impart these skills are woefully inadequate. In 2008, only 13% of American 13-year-olds and 39% of American 17-year-olds could "understand complicated information" and "learn from specialized reading materials" (Rampey, Dion, and Donahue 2009), and in 2011, only 27% of American eighth graders and 27% of American twelfth graders performed at or above "proficient" on a nationally administered writing test (National Center for Education Statistics 2012). College presidents note a rising tide of freshmen unprepared for the intellectual rigors of college, and employers observe that college graduates are seldom prepared for the rigors of the twenty-first-century workplace.

Philosophers of education have said that the development of these skills should be the first priority of any educational activity (Siegel 1980; 1988), and a number of national advisory councils in the past two decades have cited the development of critical thinking skills of our nation's youth during secondary and postsecondary education as one of our most important challenges (Honey, Fasca, Gersick, Mandinach, and Sinha 2005; National Commission on Excellence in Education 1985; Rothman, Slattery, Vranek, and Resnick 2002). Indeed, the development of critical thinking skills is a part of the educational objectives of most universities and colleges, and the possession of these skills is one of the most sought-after qualities in a job candidate in many fields (Bieda 2011; Casserly 2012; Graves 2013).

Although there is no generally accepted, well-defined list of skills that constitutes the set we call "critical thinking skills," there seems to be fair agreement on the types of skills to which educators are referring when they speak about teaching critical thinking to their students. Many of these skills have been identified broadly as a global package of knowledge and behaviors (Brookfield 1987; Ennis 1987; Nickerson 1987; Resnick 1987), or more specifically as a deliberative activity (Carey 2000; Kurfiss 1988).

Even though there are a few generally accepted measures of these skills (e.g., the California Critical Thinking Skills Test and the Watson Glaser Critical Thinking Appraisal, but see also Halpern [1996] and Paul, Binker, Jensen, and Kreklau [1990]), there is surprisingly little research on effective methods for improving the critical thinking skills of college students. The research that has been done shows that the population in general has very poor skills (Kuhn 1991; Means and Voss 1996; Perkins, Jay, and Tishman 1992), and that very few college courses actually improve these skills (Annis and Annis 1979; Pascarella 1989; Resnick 1987; Stenning, Cox, and Oberlander 1995).

Most philosophers and educators agree that one aspect of critical thinking involves the ability to reconstruct, understand, and evaluate an argument— tasks we may call, for the sake of brevity, "argument analysis." For example, Kuhn (1991) says that "argumentative reasoning skills are in fact fundamental to what educators call 'critical' thinking" (5), and Ennis (1987) says that "analyzing arguments" is one of the critical thinking abilities. This covers identifying the stated and unstated premises and the conclusion, and "seeing the structure of an argument" (12).

Teaching critical thinking skills

Although critical thinking courses are required at several universities, many, if not most, undergraduate students never take a critical thinking course in their time in college. There may be several reasons for this: the classes are too hard to get into, the classes are not required, the classes do not exist, students' schedules are already overloaded, etc. It is difficult to understand, though, why any of these would be the case since the development of critical thinking skills is a part of the educational objectives of most universities and colleges, and since the possession of these skills is one of the most sought-after qualities in a job candidate in many fields.

Perhaps, though, both the colleges and employers believe that the ability to reason well is the kind of skill that is taught not intensively in any one course, but rather across the curriculum, in a way that would ensure that students acquired these skills no matter what major they chose. The research seems to show, however, that this is not the case; on tests of general critical thinking skills, students average a gain of less than one standard deviation during their entire time in college, while most of this gain comes just in the first year (Arum and Roska 2011; Pascarella 1989; Pascarella and Terenzini 2005).

Recent research, however, suggests that students' critical thinking skills do improve substantially if they are taught how to construct argument diagrams to aid in the understanding and evaluation of arguments. Some of these studies have shown that instruction that includes the students' critical thinking skills over the course of a semester (Dwyer, Hogan, and Stewart 2012; Harrell 2008; 2011; 2012). Other studies have shown the advantages of using argument diagrams to enhance comprehension and recall (Dwyer, Hogan, and Stewart 2010; Dwyer, Hogan, and Stewart 2013). In addition, studies specifically on *computer-supported* argument visualization have shown that the use of software specifically designed to help students construct argument diagrams significantly improves critical thinking abilities over the course of a semester undergraduate course (Davies 2012; Kirschner, Shum, and Carr 2003; Twardy 2004; van Gelder, Bissett, and Cumming 2004), or a semester of graduate-level work (Carrington, Chen, Davies, Kaur, and Neville 2011; Pinkwart, Ashley, Lynch, and Aleven 2009). Additionally, research in this area has shown that student's critical thinking about specific topics is improved if students collaborate on argument diagram instruction instead of working alone (Scheuer, McLaren, Harrell, and Weinberger 2011a; 2011b).

Two models for argument analysis

Most scholars, if they have any experience with argument diagrams at all associate these diagrams with Stephen Toulmin's seminal work *The Uses of Argument* (1958) (see figure 13.2). But the history of argument diagramming begins much earlier, in the previous century (for an overview of the historical development of argument diagramming see Reed, Walton, and Macagno 2007).

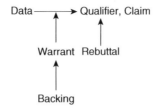

Figure 13.2   Toulmin's example of his model of argument diagramming (1958, 97).

What was revolutionary about Toulmin's approach is the emphasis he placed on needing a method of understanding and evaluating ordinary human rational discourse. Toulmin criticizes the philosophical discipline for focusing nearly exclusively on formal logic and the relationship between logic and mathematics. It had lost sight, he says, of the original motivation: studying actual human reasoning. Toulmin exhorted philosophy to focus more on real (messy) arguments and less on the artificial valid argument forms with which they were concerned.

Thus, Toulmin ushered in a new focus on what we now call "informal logic." However, Toulmin was not the only philosopher interested in emphasizing informal logic. In the 1980s and 1990s, James Freeman developed an alternative method for diagramming arguments. These different methods of argument representation are ultimately based on different theories of basic argument structure. Any theory of argument structure identifies the fundamental elements of arguments and describes how these elements fit together. Both Toulmin and Freeman take an argument to be made up of statements, in which one is the conclusion and others are offered in support of that conclusion. From there, however, their theories diverge.

Toulmin

In Toulmin's theory, there is only one pattern of argument made up of six different elements. Not all of these elements, though, describe kinds of statements, so Toulmin believes that an argument is more than just a collection of statements. There are three basic elements in an argument. The first is the *claim*, which is the original statement controversially asserted to be true. The second is the *data*, which are the reasons offered to support the claim. Finally, there is the *warrant*, which is offered as the link between the data and the claim, the fact that makes it so the data really do support the claim.

There are also three additional elements in an argument, according to Toulmin. There is the *backing*, which is offered as a reason for accepting the warrant. There is the *qualifier*, which indicates whether the data are supposed to be conclusive proof of the claim. And there is the *rebuttal*, which lays out the conditions under which the data do not, in fact, support the claim.

In Toulmin's theory, claims, data, and backing are clearly statements that appear in the argument. And, just as clearly, qualifiers and rebuttals are not statements ("Qualifiers" are words like "usually," "mostly," or "probably," while "rebuttals" are words like "unless" or "except."). Warrants, on the other hand, are not as well defined. Toulmin describes warrants as statements, but often (though not always) says or implies that they are always implicit.

Every argument, then, according to Toulmin, consists of at least a claim that is explicitly supported by data, plus a warrant that implicitly sanctions the inference from the data to the claim. Some arguments have, in addition, backing, qualifier, and/or rebuttal, depending on the nature of the data and the claim.

In representing an argument diagrammatically, Toulmin introduces arrows to his set of six elements. Arrows can begin with either data or backing and end in a qualifier, claim, or warrant. Or, arrows can begin with warrant or rebuttal and end in another arrow. Thus, on the one hand, Toulmin employs a complicated ontology in his theory of argument: six different elements, only some of which are different kinds of statements, and at least two different kinds of connections only some of which are between different elements.

On the other hand, Toulmin's argument pattern is quite simple. On this model, different, independent reasons to believe a claim can only be treated as separate arguments, and objections to the claim, or to the data or backing can also only be treated as separate arguments. (While it may sound like an objection, a rebuttal on Toulmin's model is actually just another kind of qualifier, representing circumstances in which it is acknowledged that the data do not support the claim.)

Beardsley-Freeman

In his *Dialectics and the Macrostructure of Arguments: A Theory of Argument,* James Freeman (1991) credits Monroe Beardsley's *Practical Logic* (Beardsley 1950, subsequently published as *Thinking Straight*; 1966), with what he calls the "standard method" of argument diagramming. Freeman combines this method with the argument patterns identified by Stephen Thomas in *Practical Reasoning in Natural Language* (1986) to offer an alternative theory of argument to Toulmin's. In so doing, he created what we now call the Beardsley-Freeman (B-F) method of argument diagramming.

The first important departure from Toulmin's theory is Freeman's insistence on a simpler ontology. In Freeman's theory, there are only two basic elements that constitute an argument—premises and conclusions—and they are both different kinds of statements. There are however, five basic argument patterns that Freeman acknowledges, as opposed to only one acknowledged by Toulmin. The first is the simplest: one premise that supports one conclusion. The other four patterns require at least three statements. The first is a *divergent* argument, in which one premise supports two different conclusions. The second is a *serial* argument, in which one premise supports another premise, which in turn supports the conclusion. Third is a *convergent* argument, in which two premises each offer independent support for the conclusion. And the last is a *linked* argument, in which two premises must work together to provide support for the conclusion.

On the B-F model of diagramming, then, all the statements in an argument are numbered. The statements are then represented by their corresponding numbers in circles, and the inferential connections between statements are indicated by arrows. Here, there is only one kind of arrow, one that begins at a premise (or premises) and ends in a conclusion. The four patterns involving more than two statements above are represented as diagrams in figure 13.3.

According to Freeman, these patterns can accurately represent any argument, no matter how complicated. Serial arguments can be made up of chains of premises of an arbitrary length, convergent arguments may have many more than just two premises supporting the conclusion, and one can link many premises together if necessary to support a conclusion. In addition, all of these elements may be combined in any way necessary to represent a single complex argument, as shown in figure 13.4.

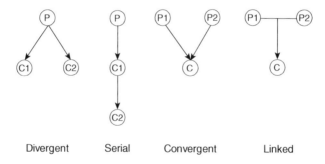

Divergent   Serial   Convergent   Linked

Figure 13.3 Freeman's example of the B-F method of diagramming the four basic argument patterns involving more than two statements (1991, 2).

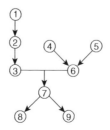

Figure 13.4 Freeman's example of a diagram of a complex argument that combines the basic elements for an accurate representation (1991, 2).

Using Toulmin's method of argument analysis in writing courses

Many colleges and universities across the United States require a general purpose writing or composition course for all students. For example, University of Michigan, University of Minnesota, North Carolina State University, Kent State University, George Washington University, Clemson University, Wellesley College, Swarthmore College, and Amherst College all have versions of this course required of every first-year student. These courses have slightly different course objectives, but a common thread is that students will learn how to analyze arguments in academic texts and create their own arguments—developing thesis statements, marshaling evidence, and synthesizing research—in addition to learning to write in a variety of genres and communicating through various media.

One of the perennial challenges of these kinds of courses is the wide variation in the background knowledge and skills each student brings to the class. Most students in American high schools learn how to write the standard "five paragraph essay," but the emphasis on rigorous argumentation can be variable. (The "five paragraph essay" is a one-paragraph introduction with a thesis statement, a one-paragraph conclusion that rephrases the introduction, and three body paragraphs. The body paragraphs each contain one argument, example, or illustration, and the template calls for the strongest to come in the first body paragraph, and the weakest to come in the third.) In addition, high school students in countries outside the United States can have wildly differing experiences of reading and writing in an academic context. Thus, most students have little experience analyzing arguments when they first arrive on a college campus. The composition teacher, then, has her work cut out for her.

For this reason alone, Toulmin's impact in various academic fields cannot be overstated. Almost immediately after the publication of *The Uses of Argument* in1958, there were calls for Toulmin's model of argument representation to be incorporated into the pedagogical practices of teachers in many disciplines. For example, the Toulmin model was widely adopted by speech departments across the United States, partly due to Wayne Brockriede and Douglas Ehninger urging rhetoricians to pay attention to this landmark book. "Toulmin's analysis and terminology are important to the rhetorician . . . they provide an appropriate structural model by means of which rhetorical arguments may be laid out for analysis and criticism" (Brockriede and Ehninger 1960, 46).

By two decades after its publication, Toulmin's *Uses of Argument* was exerting a huge influence. In urging those in composition studies to adopt the Toulmin model, Charles Kneupper claims that "speech instruction has largely abandoned the syllogistic paradigm, and most recent texts in public speaking, argumentation, and persuasion are now using a model of argument developed by the philosopher, Stephen Toulmin" (Kneupper 1978, 237). In the same article, Kneupper explains that teaching students to read for an argument is notoriously difficult, and he pushes Toulmin's method as a way to make this task easier (Kneupper 1978, 237–240).

The advice of Brockriede and Ehninger, as well as Kneupper, seems to have been heeded. In a citation analysis, Ronald Loui found that "everyone associated with scholarship in rhetoric, dialectic, or informal logic seems to have read Toulmin's 'Uses of Argument'" (Loui 2005, 266). One of the reasons offered for this broad impact is that Toulmin's model is simpler and less cumbersome than formal logic (Fulkerson 1986; Gross 1984).

Developing argument analysis skills can be difficult, especially for first-year students who may have to unlearn bad habits developed in high school. For this reason, more than a decade ago, we followed this trend in using the Toulmin method to teach argumentation at Carnegie Mellon University (CMU), in our First-Year Writing (FYW) Program. The broad purpose of this program is to develop academic reading and writing skills each student needs to be successful in his or her college career. Each student at CMU must take the course "Interpretation and Argument," which is the core of this writing program.

Thus, though not titled "Critical Thinking," the FYW course taken during the first year is generally one of the student's first introductions to important aspects of thinking critically at a college level. Among other goals, the specific learning objectives for the FYW Program is for students to be able to:

1. *Analyze a written argument*: identify the conclusion and the premises (both implicit and explicit) and describe how the premises support the conclusion.
2. *Evaluate a written argument*: determine whether the premises do in fact support the conclusion, and determine whether the premises are reasonable.
3. *Write an essay*: analyze and evaluate one or more arguments.

The goal for the FYW course is to provide foundational reading and writing skills that will enable students to develop advanced literacy in their own disciplines.

Recently, however, we have realized that most of the teachers of the FYW course encounter some serious difficulties when teaching the Toulmin method of argument analysis to their students. And we are not alone. There is mounting evidence that teachers in all disciplines are facing hurdles using the Toulmin model. In particular, the notion and identification of "warrants" in an argumentative text is notoriously difficult to teach (Brunk-Chavez 2004; Fulkerson 1996a; Rex, Thomas, and Engel 2010; Warren 2010). As Warren says, "Teaching students to identify warrants can be so difficult that many teachers simply omit them when teaching the Toulmin model" (2010, 41). In

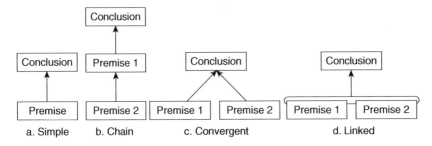

Figure 13.5   The CMU method of diagramming the four basic argument patterns.

fact, this difficulty with warrants is not restricted to students; composition instructors also find warrant identification difficult (Fulkerson 1996b, 62; Warren 2010, 43).

Thus, in the past few years, we have incorporated a new argument diagramming curriculum into the FYW course, based on the B-F model articulated above. Like the B-F model, in the CMU model, there are only two main categories of statements: premises and conclusions. In detailed analysis, there can be different kinds of premises: subconclusions, implicit premises, objections, etc. Instead of numbering the statements, though, they are rewritten and put into boxes. The boxes are then connected by arrows representing the inference(s) from premise(s) to (sub-)conclusion. In the CMU model, there are four basic argument structures. The simplest is one premise supporting one conclusion (figure 13.5a). Like the B-F method, the other three structures are chain (serial), convergent, and linked (figures 13.5b–d), and these structures can be combined in any way necessary to accurately represent any argument.

The reader may notice that we have inverted the diagrams compared to the B-F method outlined above. This is deliberate. The B-F method was developed to mirror traditional formal logical proofs that start with the given premises at the top and work their way down to the conclusion at the bottom. Our thinking, however, is to use the metaphor of a premise supporting a conclusion like a foundation supports a house to make the direction of the arrows easier for students to understand.

Using the CMU argument diagramming model

We have developed an argument diagramming curriculum that is used in many of our classes. The curriculum includes an online tutorial (citation) and a series of in- and out-of-class practice activities with diagramming throughout the semester. The online tutorial defines an argument as a set of *statements*, one of which is the *conclusion*, and the others are *premises*, where the premises are supposed to provide support for the conclusion. Most importantly, the exercises in the tutorial teach the students how to

- identify premises and conclusions using indicator words and phrases in the text,
- identify linked, convergent, and chain arguments using indicators,
- identify subarguments,
- identify implicit premises and conclusions,
- diagram arguments using a specialized software program,
- interpret an argument according to the principles of fairness and charity, and
- diagram objections and replies.

Our move to use this method has enabled us to dodge some problems that we had previously encountered when using the Toulmin model of argument. As reading and writing teachers, our interest in using any kind of heuristic for reading and writing is to facilitate rhetorical invention for analysis and production. The CMU diagramming method and vocabulary, because of its simplicity, has allowed us to apply it across a variety of texts from the wild (i.e., a variety of texts from a variety of contexts), has allowed us to both identify and name parts of the arguments, also visualize those parts, and how they connect together.

Specifically, the method has helped us remedy two particular areas where teachers found themselves becoming "stuck" with the students in the course. The first problem related to the argument vocabulary that we had been using. The second problem stemmed from the fact that our methods, to that point, were insufficient for representing a whole argument at one glance. Together, these two problems merged into an overarching issue: How could students identify pieces of an argument within their reading and represent that argument visually, if the vocabulary for identification was so complicated that they became bogged down in the process of identification?

We had been using Toulmin to teach analytical reading in our FYW course because the students needed a vocabulary for discussing parts of an argument. What the teachers enjoyed from the Toulmin model was the concept of *warrant*. For Toulmin, a *warrant* connects the data or "facts" with claims within an argument. Toulmin, Rieke, and Janik describe warrants as implied generalizations that individuals consider to be trustworthy (Toulmin, Rieke, and Janik 1984, 45). The authors go on to explain that constructing effective arguments depends upon the kinds of "general ways of arguing we are going to rely on, and employ, in this particular case" (Toulmin, Rieke, and Janik 1984, 48). These warrants, according to the authors, are situated, context dependent, and field- or discipline specific.

Prior to using the CMU diagramming method, however, students found that Toulmin's concept of *warrant* (as explained above) was something that they struggled with early in their argument analyses processes. Some students struggled with distinguishing between "data" and "warrant," especially whenever a warrant was an explicit one within the text. In no way do we mean to imply that our students could not eventually understand the concept of warrant. Rather, the nuances of the concept became nearly impossible to navigate within the context of an academic writing course simply because there is a limited amount of time to spend on these terms in light of the course objectives. By the end of the course, the students are not only analyzing individual arguments but are building research syntheses and their own inquiry-driven, written contributions. In order to get students from writing an analysis of one text, to a synthesis and analysis of a field of texts, and then to authoring their own research questions and contributions, teachers could not spend much time problematizing the argumentative heuristic vocabulary. The objective of the course is to give students resources, or flexible tools, that they can apply to text analysis and production. If the students could not trust these tools, they would not be able to (or even wish to) use them. We found that students enjoyed analyzing and discussing warrants but that they could not use the concept of warrant for analyzing their own writing or that of their peers to help them with revision.

The argument vocabulary, according to the CMU diagramming method, is a simple one. All elements in an argument can be identified as statements. Those statements can be further classified as conclusions, subconclusions, premises, and implied premises. (There are other vocabulary items as well, but these are the items relevant for this discussion.) Most usefully, though, the CMU method has simplified the use of the term *warrant* through its use of the term *implied premise*. An implied premise can be a statement of any sort, not just the kind of "if, then" statements that Toulmin, in various places, seems to require. Representing an argument as a whole is much easier with a more streamlined vocabulary because the students spend less time struggling with their comprehension of "data," "warrant," "backing," and so forth. Therefore, students can move toward seeing connections between the "chunks" of arguments and visualizing the argument as a whole.

Students' cognitive burden has been reduced for representing a whole argument visually. The vocabulary and structure of conclusion-subconclusion-premise enables us to build, rather quickly, visual representations of texts because the number of textual categories is not an overwhelming one. The categories themselves reduce the "noise" that a reader might experience when identifying and classifying statements for the purpose of diagramming their relationships. Each item, whether a statement about data, explanation, or reasoning, fits under another statement. Because the structure allows students to think in terms of how the pieces fit together, students can more quickly move toward evaluating the logical they have diagrammed.

We should note that this more streamlined vocabulary and structure does not preclude the teacher from introducing the concept of *warrant* later in the course—or other terms in argument theory that might highlight the rhetorical situation of a text, its purpose and audience, and its overall social context. However, students will have already learned how to identify structures within the arguments they read before grappling with other argument terms. Some examples of other terms that teachers have used to discuss types of statements within the argument diagramming structure include not only Toulmin's terms (e.g., claims that function as warrants), but also stasis terms (e.g., definition or causal claims), rhetorical proofs (e.g., ethical claims), and policy concepts (e.g., problem statements). In these ways, teachers have been able to address a variety of texts through the diagramming heuristic.

Effectiveness of the CMU argument diagramming method

Recently, we presented the data from a study we conducted testing the hypothesis that students in the FYW courses who learn the CMU model of argument diagramming will improve in performance on argumentative writing tasks over the course of a semester-long composition class significantly more than students in the FYW courses who do not (Harrell and Wetzel 2013).

In the Fall of 2009, and the Spring and Fall of 2010, we administered pre- and post-tests to 81 students in the FYW program. Each test consisted in reading some text and completing two tasks. In Task 1, the student was asked to write an essay analyzing the argument presented by the author in the text. This analysis was to consist in identifying both the content and the structure of the argument. In Task 2, the student was asked to write an essay evaluating the argument presented by the author in the same text. The evaluation was to consist a claim about the quality of the argument and reasons to support that claim.

We recognize that text features alone do not constitute "good writing" and that there is no "right way" to read or write a text. We also recognize that privileging some text features over others might ignore other significant features. The features that we chose will help us locate change in demonstrable critical thinking between the pre-test and post-test. We analyzed the texts for markers of text development and text coherence. We were interested in seeing to what extent there would be any kind of change in how many different ideas students could generate—about someone else's argument and about their own arguments. Within this category of "development," we identified the following for both Tasks 1 and 2 of the pre- and post-tests: the number of *different* reasons or premises offered for the argument conclusion, and the number of *counterargu-ments* considered within the text.

For Task 1, we wanted to determine how much the students could understand the argument in the text and what statements they would prioritize in their representations of it. For Task 2 alone, we also considered whether students provided evidence or elaboration of their reasons. We wanted to distinguish between reasons that were supported with evidence and those that were not. Our concern was instances when students produced a lot of different ideas but failed to support them; we did not want to report "growth" in development without attempting to represent to what extent students were actually support-ing their claims.

Because the number of ideas alone does not necessarily equate with good writing, and, in fact, one could argue that too many different ideas within an argument will result in chaos for a reader, we also looked for features that sig-naled an overall coherence in a written text. Vande Kopple has defined coher-ence as "prose in which nearly all the sentences have meaningful connections to sentences that appear both before and after them" (1989, 2). We also draw upon Enkvist's definition of coherence, "the quality that makes a text conform to a consistent world picture and is therefore summarizable and interpretable" (1990, 49). So, by *coherence*, we mean those features that enable a reader to make particular kinds of connections within the text. In coding Task 1, we considered the following as coherence markers: logical connections between premises and the argument conclusion and logical connections between different premises.

In coding Task 2, we looked at the following as markers of coherence: logical connections between premises and the argument conclusion, logical connec-tions between different premises, and metacommentary (or "metadiscourse"). *Metacommentary* is language that writers use, according to Hyland (2003), to compose a text that is clear to a reader.

By providing linguistic "signposts" to readers, writers can create the effect that a text is coherent and holds together in an intentional way. Because these bits of language give clues for making sense of the text, their presence in a text can indicate that a writer is aware of a reader's needs for navigating the text successfully. These bits of language can also show that a writer understands his or her own text in particular ways and can point to a writer's strategic view of his or her writing. For this study, we were only interested in the effect that metacommentary has upon the readers—we were not interested in counting the different types. Therefore, coders scored Task 2 holistically for effective use of metacommentary.

The results from Task 1 (given in figure 13.6) show that, when reading an argument, students who were taught argument diagramming were signifi-cantly more likely than those who were not to identify more of the relevant

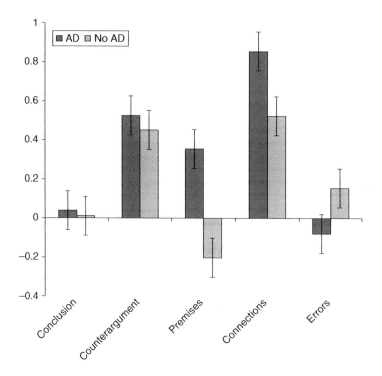

Figure 13.6 Comparisons of gains in each category of Task 1 from pre-test to post-test for students who were and were not taught argument diagramming.

premises offered that support the author's conclusion, and explain more explicitly how the premises are supposed to work together to support the conclusion. In addition, these students were much less likely to make any errors in their analysis.

The results from Task 2 (given in figure 13.7) show that, when evaluating the argument in a text, students who were taught argument diagramming improved significantly more than those who were not in their ability to (1) provide more premises to support their own thesis, (2) offer more evidence in support of each premise, (3) have fewer mismatches between premises and evidence, (4) explain more explicitly how the premises are supposed to work together to support the conclusion, (5) offer possible counterarguments, and (6) provide metacommentary on their response.

Thus, it seems that students who were taught argument diagramming are developing new schema for reading arguments and learning how to effectively translate this into their own writing. This is reflected most noticeably in the improvement of the metacommentary from pre-test to post-test. We conclude that incorporating argument diagramming into the curriculum of *Interpretation and Argument* is positively beneficial to realizing several of our course objectives.

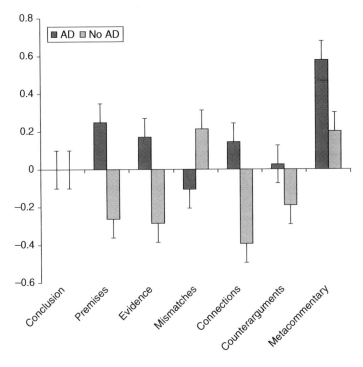

Figure 13.7 Comparisons of gains in each category of Task 2 from pre-test to post-test for students who were and were not taught argument diagramming.

Conclusion

Because of our personal experience teaching in the FYW program, talking to others who teach in the program, and the studies we have conducted, we have concluded that teaching argument analysis skills using our modified B-F argument diagramming method is superior to using the Toulmin method.

These results are important because the creation of one's own arguments as well as the analysis of others' arguments occur in nearly every discipline, from Philosophy and Logic to English and History to Mathematics and Engineering. We believe that the use of argument diagrams, and the CMU diagramming method in particular, would be helpful in any of these areas, both in developing general critical thinking skills and in developing discipline-specific analytic abilities.

Acknowledgments

We would like to thank David Danks, Martin Davies, and three anonymous reviewers for their helpful comments. This study was funded by a joint grant from the Spencer Foundation and the Teagle Foundation, under the title "Systematic Improvement of Undergraduate Education in Research Universities."

References

Annis, D., and Annis, L. 1979. "Does Philosophy Improve Critical Thinking?" *Teaching Philosophy* 3: 145–152.

Arum, R., and Roska, J. 2011. *Academically Adrift: Limited Learning on College Campuses.* Chicago and London: University of Chicago Press.

Beardsley, M. C. 1950. *Practical Logic.* (third edition). Englewood Cliffs, NJ: Prentice Hall.

Beardsley, M. C. 1966. *Thinking Straight.* (third edition). Englewood Cliffs, NJ: Prentice Hall.

Bieda, A. S. 2011. *Closing the Gap between Career Education & Employer Expectations: Implications for America's Unemployment Rate.* Washington, DC: Accrediting Council for Independent Colleges and Schools.

Brockriede, W., and Ehninger, D. 1960. "Toulmin on Argument: An Interpretation and Application." *Quarterly Journal of Speech* 46: 44–53.

Brookfield, S. 1987. *Developing Critical Thinkers: Challenging Adults to Explore Alternative Ways of Thinking and Acting.* San Francisco: Jossey-Bass.

Brunk-Chavez, B. L. 2004. "What's So Funny about Stephen Toulmin? Using Cartoons to Teach the Toulmin Analysis." *Teaching English in the Two Year College* 32 (2): 178–185.

Carey, S. S. 2000. *The Uses and Abuses of Argument.* Mountain View, CA: Mayfield.

Carrington, M., Chen, R., Davies, M., Kaur, J., and Neville, B. 2011. "The Effectiveness of a Single Intervention of Computer-Aided Argument Mapping in a Marketing and a Financial Accounting Subject." *Higher Education Research and Development* 30 (3): 387–403.

Casserly, M. 2012. *The 10 Skills That Will Get You Hired in 2013.* December 12, 2013. Available from http://www.forbes.com/sites/meghancasserly/2012/12/10/the-10-skills -that-will-get-you-a-job-in-2013/.

Davies, M. 2009. "Computer-Assisted Argument Mapping: A Rationale Approach." *Higher Education* 58 (6): 799–820.

Davies, M. 2012. "Computer-Aided Argument Mapping and the Teaching of Critical Thinking (Part 2)." *Inquiry: Critical Thinking across the Disciplines* 27 (3): 16–28.

Dwyer, C. P., Hogan, M. J., and Stewart, I. 2010. "The Evaluation of Argument Mapping as a Learning Tool: Comparing the Effects of Map Reading versus Text Reading on Comprehension and Recall of Arguments." *Thinking Skills and Creativity* 5: 16–22.

Dwyer, C. P., Hogan, M. J., and Stewart, I. 2012. "An Evaluation of Argument Mapping as a Method of Enhancing Critical Thinking Performance in E-Learning Environments." *Metacognition Learning* 7: 219–244.

Dwyer, C. P., Hogan, M. J., and Stewart, I. 2013. "An Examination of the Effects of Argument Mapping on Students' Memory and Comprehension Performance." *Thinking Skills and Creativity* 8: 11–24.

Enkvist, N. E. 1990. "Seven Problems in the Study of Coherence and Interpretability." In *Coherence in Writing: Research and Pedagogical Perspectives*, edited by Ulla Connor and Ann M. Johns. Alexandria, VA: Teachers of English to Speakers of Other Languages, Inc.

Ennis, R. H. 1987. "A Taxonomy of Critical Thinking Dispositions and Abilities." In *Teaching Thinking Skills: Theory and Practice*, edited by J. B. Baron and R. J. Sternberg. New York: W. H. Freeman and Company. 9–26.

Freeman, J. B. 1991. *Dialectics and the Macrostructure of Arguments: A Theory of Argument Structure.* New York: Foris Publications.

Fulkerson, R. 1986. "Logic and Teachers of English?" *Rhetoric Review* 4 (2): 198–209.

Fulkerson, R. 1996a. "Stephen Toulmin." In *Encyclopedia of Rhetoric and Composition.* New York: Garland 726–727.

Fulkerson, R. 1996b. "The Toulmin Model of Argument and the Teaching of Composition." In *Argument Revisited; Argument Redefined; Negotiating Meaning in the Composition Classroom*, edited by Barbara Emmel, Paula Resch and Deborah Tenney. Thousand Oaks, CA: American Forensic Association, Sage Publications. 45–72.

Graves, J. A. 2013. *7 Key Skills You Need to Get Hired Right Now: In-Demand Skills Job Candidates Must Cultivate.* February 24, 2013. Available from http://money.usnews .com/money/careers/articles/2013/02/21/7-key-skills-you-need-to-get-hired-right-now.

Gross, A. G. 1984. "A Comment on the Uses of Toulmin." *College English* 46 (3): 310–314. http://www.jstor.org/stable/377043.

Halpern, D. 1996. *Thought and Knowledge: An Introduction to Critical Thinking.* (third edition) Mahwah, NJ.: Lawrence Erlbaum Associates.

Harrell, M. 2008. "No Computer Program Required: Even Pencil-and-Paper Argument Mapping Improves Critical Thinking Skills." *Teaching Philosophy* 31: 351–374.

Harrell, M. 2011. "Argument Diagramming and Critical Thinking in Introductory Philosophy." *Higher Education Research and Development* 30 (3): 371–385.

Harrell, M. 2012. "Assessing the Efficacy of Argument Diagramming to Teach Critical Thinking Skills in Introduction to Philosophy." *Inquiry* 27 (2): 31–38.

Harrell, M., and Wetzel, D. 2013. "Improving First-Year Writing Using Argument Diagramming." In M. Knauff, M. Pauen, N. Sebanz, and I. Wachsmuth, *Proceedings of the 35th Annual Conference of the Cognitive Science Society* (pp. 2488–2493). Austin, TX: Cognitive Science Society.

Honey, M., Fasca, C., Gersick, A., Mandinach, E., and Sinha, S. 2005. *Assessment of 21st Century Skills: The Current Landscape.* New York: Partnership for 21st Century Skills.

Hyland, K. 2003. *Second Language Writing.* Cambridge: Cambridge University Press.

Kirschner, P. A., Shum, S. J. B., and Carr, C. S. 2003. *Visualizing Argumentation: Software Tools for Collaborative and Educational Sense-Making.* New York: Springer.

Kneupper, C. W. 1978. "Teaching Argument: An Introduction to the Toulmin Model." *College Composition and Communication* 29 (3): 237–241.

Kuhn, D. 1991. *The Skills of Argument.* Cambridge: Cambridge University Press.

Kurfiss, J. 1988. *Critical Thinking: Theory, Research, Practice and Possibilities.* Washington: ASHE-Eric Higher Education Report No. 2, Associate for the Study of Higher Education.

Loui, R. P. 2005. "A Citation-Based Reflection on Toulmin and Argument." *Argumentation* 19 (3): 259–266.

Means, M. L., and Voss, J. F. 1996. "Who Reasons Well? Two Studies of Informal Reasoning among Children of Different Grade, Ability, and Knowledge Levels." *Cognition and Instruction* 14: 139–178.

National Center for Education Statistics. 2012. "The Nation's Report Card: Writing 2011" NCES 2012–470. Washington, DC.

National Commission on Excellence in Education. 1985. "A Nation at Risk: The Imperative for Educational Reform: A Report to the Nation and the Secretary of Education." Washington, DC.

Nickerson, R. S. 1987. "Why Teach Thinking?" In *Teaching Thinking Skills: Theory and Practice*, edited by J. B. Baron and R. J. Sternberg. New York: Freeman. 27–37.

Pascarella, E. 1989. "The Development of Critical Thinking: Does College Make a Difference?" *Journal of College Student Development* 30: 19–26.

Pascarella, E., and Terenzini, P. 2005. *How College Affects Students: Findings and Insights from Twenty Years of Research.* Vol. 2: A Third Decade of Research. San Francisco: Jossey Bass.

Paul, R., Binker, A., K., J., and Kreklau, H. 1990. *Critical Thinking Handbook: A Guide for Remodeling Lesson Plans in Language Arts, Social Studies and Science.* Rohnert Park, CA: Foundation for Critical Thinking.

Perkins, D., Jay, E., and Tishman, S. 1992. *Assessing Thinking: A Framework for Measuring Critical Thinking and Problem Solving Skills at the College Level.* Washington, DC: The National Center for Educational Statistics Workshop on the Assessment of Higher Order Thinking and Communication Skills of College Graduates: Preliminary listing of Skills and Levels of Proficiency.

Pinkwart, N., Ashley, K., Lynch, C., and Aleven, V. 2009. "Evaluating an Intelligent Tutoring System for Making Legal Arguments with Hypotheticals." *International Journal of Artificial Intelligence in Education* 19: 401–424.

Rampey, B. D., Dion, G. S., and Donahue, P. L. 2009. "Naep 2008 Trends in Academic Progress " NCES 2009–479. Washington, DC.

Reed, C., Walton, D., and Macagno, F. 2007. "Argument Diagramming in Logic, Law and Artificial Intelligence." *The Knowledge Engineering Review* 22: 87–109.

Resnick, L. B. 1987. *Education and Learning to Think.* Washington, DC: National Academy Press.

Rex, L. A., Thomas, E. E., and Engel, S. 2010. "Applying Toulmin: Teaching Logical Reasoning and Argumentative Writing." 99 (6): 56–62.

Rothman, R., Slattery, J. B., Vranek, J. L., and Resnick, L. B. 2002. "Benchmarking and Alignment of Standards and Testing." CSE Tech. Rep. 566. Los Angeles, CA.

Scheuer, O., McLaren, B., Harrell, M., and Weinberger, A. 2011a. "Scripting Collaboration: What Affects Does It Have on Student Argumentation?" In *Proceedings of the 19th International Conference on Computers in Education: Icce 2011*, edited by T. Hirashima, G. Biswas, T. Supnithi and F. Yu. Chiang Mai, Thailand: National Electronics and Computer Technology Center.

Scheuer, O., McLaren, B., Harrell, M., and Weinberger, A. 2011b. "Will Structuring the Collaboration of Students Improve Their Argumentation?" In *Lecture Notes in Computer Science: Artificial Intelligence in Education—15th International Conference*, edited by G. Biswas, S. Bull, J. Kay and A. Mitrovic. Berlin: Springer-Verlag, 6738. 544–546.

Siegel, H. 1980. "Critical Thinking as an Educational Ideal." *Educational Forum* 45 (1): 7–23.

Siegel, H. 1988. *Educating Reason: Rationality, Critical Thinking, and Education.* New York: Routledge.

Stenning, K., Cox, R., and Oberlander, J. 1995. "Contrasting the Cognitive Effects of Graphical and Sentential Logic Teaching: Reasoning, Representation and Individual Differences." *Language and Cognitive Processes* 10 (3–4): 333–354.

Thomas, S. N. 1986. *Practical Reasoning in Natural Language.* (third edition). Englewood Cliffs, NJ: Prentice Hall.

Toulmin, S. E. 1958. *The Uses of Argument* (first edition). Cambridge, England: Cambridge University Press.

Toulmin, S. E., Rieke, R., and Janik, A. 1984. *An Introduction to Reasoning.* (second edition). New York, London: Macmillan Publishing Co., Inc. Collier Macmillan Publishers.

Twardy, C. R. 2004. "Argument Maps Improve Critical Thinking." *Teaching Philosophy* 27: 95–116.

van Gelder, T., Bissett, M., and Cumming, G. 2004. "Cultivating Expertise in Informal Reasoning." *Canadian Journal of Experimental Psychology* 58: 142–152.

Vande Kopple, W. 1989. *Clear and Coherent Prose: A Functional Approach.* Glenview, IL: Scott, Foresman, & Company.

Warren, J. E. 2010. "Taming the Warrant in Toulmin's Model of Argument." *English Journal* 99 (6): 41–46.

14
Virtue and Inquiry: Bridging the Transfer Gap

Tracy Bowell and Justine Kingsbury

Introduction

When the benefits of tertiary education are listed, the development of critical thinking is often near the top of the list (Bok 2006). The critical thinker can, among other things, assess evidence, judge the relevance of new information to existing beliefs, and break down a complex problem into less complex parts and work through them in an orderly way. Abilities like these are useful in myriad contexts beyond the classroom. It is easy to see why critical thinking is seized upon as an important part of higher education's contribution to transforming students into lifelong learners.

Critical thinking and the ability to argue well are part of inquiry and intellectual life more generally, and they are undoubtedly of great value both to individuals and to the communities to which they belong. Good argument can lead to cognitive and practical gains. It can rationally persuade us toward true beliefs and a better understanding of the world and of each other. It can help us to make good choices about how to act. *Lack* of critical thinking can lead to undesirable outcomes in, for instance, purchasing decisions, public debate, the conduct of personal relationships and, above all, science.

Critical thinking also serves epistemic justice. It helps us gain a better understanding of who counts as a reliable source of evidence, who should be respected as authoritative with respect to the question in hand, who can reasonably be deemed an expert, and how we should not reject someone's position solely on the basis of who they are or what other beliefs they may hold. When the views of certain people are excluded from consideration for irrelevant reasons, it is a failure of inquiry, because evidence or truths those people could bring to the table are neglected (Fricker 2007).

In sum, critical thinking has the potential to contribute to our development as conscientious enquirers, that is, as citizens who

- value significant truths and seek to avoid holding inconsistent beliefs,
- want to be able to enquire effectively as to the right course of action to take in a particular situation, both personally and at the community or even the global level, and
- are motivated to act on the conclusions of their inquiries.

The standard way of explicitly teaching generic critical thinking at tertiary level is in a first-year undergraduate course aimed at all students, no matter what their major. Such courses are meant to develop skills that are very generally applicable: in particular, the ability to evaluate arguments, which also requires being able to identify an argument in a written or oral text and being able to figure out its structure. However, numerous studies suggest that such courses are not very effective in producing a lasting improvement in students' reasoning outside of the classroom, or even in classrooms other than the critical thinking classroom. (See Behar-Horenstein and Niu [2011] for an overview.) This is the problem of transfer: the reasoning abilities developed in a critical thinking course do not appear to transfer well to other environments.

In this paper we suggest that a virtues-oriented approach to teaching critical thinking has the potential to help bridge this transfer gap. If critical thinking skills are not sticking, perhaps that is at least in part because students lack certain intellectual virtues or dispositions toward conscientious inquiry. We conclude with some suggestions about how these virtues might be fostered in the context of a first-year undergraduate critical thinking course.

Virtuous inquiry

Virtue epistemologists distinguish two kinds of intellectual virtues: reliabilist virtues and responsibilist virtues. The *reliabilist* virtues are so called because they involve the reliable functioning of cognitive and perceptual processes: they include perceptual abilities, introspection, memory, observational skills, and the ability to reason deductively and inductively (see for example Sosa [1991]). Reliabilist virtues might be thought of as cognitive and perceptual abilities. The *responsibilist* virtues, on the other hand, are intellectual virtues the exercise of which is a matter of choice: consequently it makes sense to think of a person as *responsible* for exercising or not exercising them. Linda Zagzebski has developed a comprehensive list of intellectual virtues that she considers responsibilist (similar lists can be found in Paul [2000], Aberdein [2010] and Cohen [2009]). It includes "the ability to recognize the salient facts [to which we would add 'and properly to weigh them up as evidence']; sensitivity to detail; open-mindedness; fairness; epistemic humility; perseverance; diligence, care and thoroughness; the ability to recognise reliable authority; intellectual candour; intellectual courage, autonomy, boldness, creativity and inventiveness" (Zagzebski 1996). Several of these are cognitive capacities that one might consider fall under the reliabilist heading, though they plausibly count as responsibilist insofar as an agent might choose whether or not to exercise her ability to recognize the salient facts or to recognize reliable authority.

One might think of the reliabilist virtues as abilities with which we are naturally endowed (or not), whereas the responsibilist virtues are acquired. But even if perception, memory, introspection, and the ability to reason deductively and inductively are native abilities, training and motivation are still required in order that they be properly developed and appropriately employed. Perceptual abilities can be honed to take heed of features unavailable to those with standard perceptual ability—we can learn to listen properly to music, to appreciate the subtleties of fine wines, or to identify different types of aircraft from the shape of their wings or tails.

We think that what really distinguishes the responsibilist virtues from the reliabilist ones is the role they play in inquiry, rather than whether they are innate or acquired or whether or not one can choose to exercise them. Responsibilist virtues regulate the use of our cognitive abilities, and they tend to be closely related to moral virtues. They often serve to constrain our cognitive and communicative exchanges with others, enabling us to treat other enquirers properly and to take proper account of the contributions that they make to inquiry. Henceforth (following Lepock [2011]) we will call them *regulatory virtues*.

There is one further kind of intellectual virtue that plays an important role in inquiry. Practice is required in order to hone cognitive and perceptual abilities, and practice requires motivation. Motivation may also be required in order to exercise some such abilities even after they have been thoroughly developed. It might be that once a person has honed their ability to appreciate the subtleties of fine wines, they cannot taste a wine without exercising that ability: on the other hand, even someone with finely honed reasoning abilities might sometimes read a complex argument and think, "There's something wrong with that but I can't be bothered figuring out what." Thus, in order to hone reliabilist virtues and (at least in some cases) in order to exercise them even after they are thoroughly honed, one needs also to possess certain *motivational* virtues. Lorraine Code describes the underlying motivations of the intellectually virtuous person as follows:

> The intellectually virtuous person...is one who finds value in knowing and understanding how things really are. S/he resists the temptation to live with partial explanations where fuller ones are attainable, the temptation to live

in fantasy or in a world of dream or illusion, considering it better to know, despite the tempting comfort and complacency that a life of fantasy or illusion (or well-tinged with fantasy or illusion) can offer. (Code 1984)

Once acquired, the *regulatory* virtues are employed in the service of these broader aspirations toward conscientious inquiry.

The regulatory virtues have an analogue in many activities in which we employ skills that have been honed by training and by practice. Consider, for example, the tennis player who, taking her natural cognitive and physical faculties (perception, the ability to run and jump, fine motor skills, and so on) as a starting point, has, with the help and support of her coach, developed and honed her skills at serving, volleying, hitting deep and accurately from the baseline, hitting a winner off a short ball, chasing down a lob, and so on. Impressive though they may be, these skills are insufficient to enable her to be as successful a player as she could be. She needs persistence to stay in rallies, resisting the urge to go for a winner too early in the piece. Conversely, she needs courage to go for riskier shots at tight moments in a match. And, when things go awry or she is simply bettered by her opponent, she needs to display humility and grace in the face of defeat.

There is a clear affinity between the critical thinking habits we are describing as virtues and what are often referred to as the *thinking dispositions*. Virtues are dispositions to behave in certain ways under certain conditions all other things being equal. However, the notion of a virtue includes an important normative element that is lacking in the more general notion of a disposition. A virtue is something that, all things considered, one ought to try to acquire and ought to be motivated to exercise when the circumstances are appropriate. Not exercising a virtue in an appropriate situation is a failing. Take, for example, the disposition to seek and offer reasons (Ennis 1994). When Ennis calls this a *disposition*, he is talking merely about a tendency to think and act in a certain way in certain conditions: when we call it a *virtue*, we are also saying that seeking and offering reasons is a good thing to do, something one *ought* to do. In the rich literature on thinking dispositions, they are characterized in a variety of ways. One characterization—the triadic conception, offered by Tishman, Jay, and Perkins (Tishman 1993) —maps onto our triad of virtues of inquiry. *Sensitivity*—the ability to perceive the appropriateness of a particular thinking behavior—is an analogue of our *regulatory virtues*. *Inclination*—the impetus to think and act in a certain way—is an analogue of our *motivational virtues*. *Ability*—the capacity actually to perform the behavior—is an analogue of our *reliabilist virtues*.

In the following scenario we illustrate how each of these types of trait—motivational virtues, regulatory virtues, and reliabilist virtues—might come into play in the process of inquiry.

Consider a situation that might arise on a Monday morning in a large first-year class. Imogen is sitting near the middle of the lecture theater and paying careful and thoughtful attention to everything the lecturer says. Sebastian is sitting next to her, and he is trying to pay attention too, but he is impeded by a hangover. Robert is sitting in the front row writing it all down word for word so he can study it later for the exam. James is reading the student newspaper in the back row. Further along in the back row, Jenny is telling Megan in a whisper about her disastrous weekend, in the hope of being given a comforting hug, and Megan is responding with a detailed whispered analysis of exactly what Jenny did to cause things to go so wrong and how she can best get out of the resulting situation.

The lecturer asserts something that Imogen, on careful reflection, thinks is mistaken. Sebastian has the vague sense that there is a problem with it too, but he can't muster the will to figure out exactly what it is. None of the others have noticed the problem; the three in the back row aren't listening, and Robert, although he is listening, is not thinking about what he hears. Imogen raises her hand and puts her point and the lecturer readily engages, reconsidering his claim. He adjusts his claim in the light of Imogen's objection, and the discussion goes back and forth a few times until a version of the claim is reached that both parties think is defensible. Sebastian is following the moves, and is persuaded that the revised claim is better than the original one. He is pleased that his inchoate unease with the original claim has been vindicated, and impressed with Imogen's clarity of mind and boldness in speaking up. Robert is irritated—Imogen is wasting time that the lecturer needs if he is to cover all the points that the course outline says are to be covered in this lecture. James is amused by the comics page. Jenny feels miserable. Megan is very satisfied with her own analysis of Jenny's problem.

In this scenario, everyone who is paying attention—Imogen, the lecturer, and those members of the class who are listening—is likely to benefit from the interaction between Imogen and the lecturer. First of all, the participants and the listeners are likely to get closer to the truth about the topic under discussion, which may be a topic of significance to the students or the lecturer or both. In addition, the rest of the class (those who are listening) will have had habits of good inquiry modeled. Perhaps Robert will notice, reading over his notes, that Imogen's speaking up actually advanced and clarified the lecturer's exposition of the topic of the lecture, and change his view of what it is to be a good student in this class. Furthermore, those students who are properly motivated as enquirers have had an opportunity to exercise intellectual diligence, open-mindedness, and the ability to recognize trustworthy authority. None of these advantages would have been gained if Imogen had not spoken up, or if the lecturer had defensively stuck to his initial position.

In the process of noticing the lecturer's mistake, Imogen has exercised reliabilist virtues such as perceptiveness, the capacity for effective deductive or inductive reasoning, and memory (perhaps of something she has already learned). To be in a position to exercise these capacities, she must be motivated: she is paying attention to what is said, and thinking about it. Perhaps she is motivated by a generalized desire to get closer to the truth, or a specific desire to get closer to the truth about the point at hand; perhaps she is generally diligent, and sees that what this class requires is careful and critical thinking; perhaps she is motivated by the desire to impress Sebastian or the lecturer. There are any number of motivations she might have, but she must have one. Paying careful and critical attention is not something that just happens: it takes effort.

Not all of the possible motivations we have suggested for Imogen seem equally virtuous. If she is trying to impress, that seems a less good motivation than if she is trying to get to the bottom of the question at hand because she thinks it is genuinely interesting and important. Nevertheless, all of the benefits (for Imogen, the lecturer, and the rest of the class) that we suggested would follow from her behavior follow from it independently of what motivated it. Some motivations, however, will tend in general to lead to critical thinking in the situations in which critical thinking is required. The desire to impress is not one of them; the desire to get to the bottom of things is. Valuing significant truths, and inquiry and discussion as a means of getting to them; wanting to have, communicate, and act on true beliefs: these are the kinds of characteristics we are calling motivational virtues.

As well as reliabilist and perhaps motivational virtues, in speaking up and putting her point Imogen displays the regulatory virtues of intellectual courage and autonomy. She also displays a good eye for a situation in which critical thinking is called for (unlike Robert, who thinks that students should stay quiet and take notes).

The lecturer too displays all three kinds of virtues. Deploying his own reliabilist virtues, he evaluates Imogen's point and recognizes its force. He displays the regulatory virtues of open-mindedness and epistemic humility in being prepared to reconsider and to adjust his claims, and to engage with points raised by someone who might be considered to have less knowledge of the matter in hand, and over whom he has power. And he has the motivational virtues, one might suppose, of desiring to get closer to the truth, and also of modeling and encouraging attitudes and ways of thinking that will help his students to get there too.

This scenario shows that being a truly critical thinker involves more than simply being good at evaluating arguments and weighing evidence. You might have those abilities and fail to deploy them in a situation in which they would be appropriate. This might happen because for some reason you cannot be bothered—as when Sebastian thinks something is amiss but doesn't manage to put in the effort to figure out exactly what, or when James reads the newspaper and Jenny tells Megan her troubles in the back row of the lecture theater—a failure of motivation. Or it might happen because you fail to identify your situation as one in which critical thinking is called for—as Robert does when, having the view that lecturers have absolute epistemic authority, he occupies himself with writing down every word the lecturer utters rather than thinking critically about what is said. You might also deploy critical thinking skills where they are not appropriate—as Megan does when Jenny confides in her seeking a sympathetic ear rather than a solution, and Megan gives a detailed analysis of precisely what her problem is and how it can be solved. The genuinely critical thinker has the reliabilist epistemic virtues—in particular, the ability to reason inductively and deductively. She also has the motivation to deploy those abilities in appropriate circumstances. And she has the regulatory virtues: she can judge when the situation is one in which she should think critically and how she should go about it.

In our scenario, the protagonists exercise these characteristics in a balanced way. The good student strikes a balance between autonomy and the recognition of appropriate authority, recognizing the authority of the lecturer but not according it too much weight, thereby remaining prepared critically to evaluate the points the lecturer raises. The lecturer displays intellectual humility, but not too generously. If he was too humble (e.g., if he capitulated immediately without having properly considered Imogen's point) then he might mistakenly revise his own position, or revise it correctly but on the strength of little or no evidence. The lecturer displays open-mindedness, but it would be a vice to be entirely open-minded. If it were immediately clear that Imogen's point was irrelevant to the matter at hand, then there would be no need to consider revision of his own position in the light of it. The intellectually conscientious move would be to explain why it is irrelevant.

The scenario also illustrates that the ways in which intellectual capacities come into play and the degree to which they need to be employed is in part determined by the role an enquirer plays within a process of inquiry. On the one hand, while both arguers and their audiences need to exercise their perceptual capacities, their capacities to reason well, their intellectual diligence, boldness, and ability to recognize and heed evidence, the person putting forth an argument may be required to exercise intellectual courage in greater degree than the person who receives it, especially if the conclusion is controversial. The receiver of an argument, on the other hand, is more likely to be required to display open-mindedness, especially if the argument requires them to revise or relinquish their existing beliefs.

Traditional critical thinking teaching tends to focus on enabling students to gain the skills associated with identifying and appraising weak arguments rather than on constructing strong arguments. Thus students are often positioned as an audience for the arguments with which they are presented. Regardless of any disadvantages of this approach, which of the regulatory virtues would it be most important to inculcate and nurture in our students in addition to the skills of argument appraisal upon which we already concentrate? The virtues helpful to good inquiry in this context include the courage to interrogate an argument, even when it is presented by someone who is generally taken to be authoritative or influential (so students also need the ability to recognize appropriate authority), and open-mindedness (to be prepared to revise or relinquish beliefs on the strength of a good argument that gives good reason to doubt those beliefs). Students also need to learn to exercise fairness in their approach to an argument, and this will include avoiding *ad hominem* responses that reject an argument out of hand on the basis of who or what the arguer supposedly is. First and foremost, however, before coming to be able appropriately to manage their own inquiry and their responses to arguments, students need to be able to recognize situations in which they ought to use their reasoning skills and to be motivated to use them in these situations. They need to have come to the view that the ability to reason well, and to act on the results, is a good.

The standard critical thinking course, the virtues, and the problem of transfer

What we are calling "the standard critical thinking course" is a first-year undergraduate course intended to develop cognitive abilities that any undergraduate will find useful, both in their studies and outside the university. (Searching university websites in the United States, Canada, the United Kingdom, Australia, and New Zealand confirms that this is by far the most common model of explicit critical thinking instruction.) The ability particularly emphasized is the ability to evaluate arguments. In the interests of teaching this general skill and of keeping the attention of students with a variety of interests and backgrounds, such courses tend to use examples from a wide range of different areas to illustrate argument types and argument-evaluation methods, rather than being embedded in some particular discipline.

There is a plethora of textbooks designed to be used in such courses (Bowell and Kemp [2014], Waller [2011], and Salmon [2012] are examples). The textbooks and the courses that use them follow a fairly standard pattern. They begin with a brief discussion of the value of critical thinking and go on to talk about what arguments are and how to identify them in a written or oral text. They present at least one method for setting out an argument so that it is clear what the premises and conclusion are and what the structure is: often a "tree diagram" method is presented as a way of visually laying out the structure of an argument. Typically they draw a distinction between deductive and inductive arguments and talk about the criteria of evaluation for these different kinds of argument. They go on to provide a list of common kinds of arguments that are bad and yet persuasive (fallacies), how they can be recognized, and what is wrong with them. At the end of each chapter, exercises are provided to enable the student to practice their skills (argument diagramming, argument evaluation, recognizing and labeling examples of fallacies, etc.). The emphasis is very much on the acquisition of abilities, or the development and refinement of abilities already possessed to some degree, as opposed to the learning of facts.

It appears that the standard critical thinking course does not live up to its billing. While we may be reasonably successful at helping students to become adept at identifying, diagramming, and evaluating arguments in the critical thinking classroom and in the end of semester exam, evidence suggests that fewer of them use these abilities in environments beyond the classroom and the immediate end of gaining the required course credits: these abilities are not necessarily becoming second nature to our students (Behar-Horenstein and Niu 2011; Goldberg 2014).

We suggest that teaching with the aim of developing not just the relevant (reliabilist) abilities but also the kinds of motivational and regulatory virtues discussed in the previous section might have benefits, among them the mitigation of the problem of transfer. Developing these higher-level virtues in students is worthwhile in its own right—we have suggested above that they are characteristics that we want citizens to have, and a critical thinking course seems a likely venue in which to try to develop them. But independently of this, we suggest that a focus on developing motivational and regulatory virtues might increase the likelihood that students will deploy their reasoning skills outside the classroom.

Why might a student who did well in the end of semester critical thinking exam nevertheless not think critically when they are considering whether or not to become a vegetarian, or deciding who to vote for? One possibility is that although they can solve a critical thinking problem when it is presented to them as such, they are not good at recognizing critical thinking problems in the wild—they do not see that *this* is the kind of occasion those classes were supposed to prepare them for. Another is that although they have the skills and can see that this is the sort of situation in which they are applicable, they can't be bothered applying them; they lack the right kind of motivating virtues. Yet another is that they have not properly acquired the skills—either they had them but they did not stick, or what they acquired were not critical thinking skills in any broad sense but only the ability to do well in a critical thinking exam. It seems likely that all three of these (and no doubt others as well) are part of the explanation of the transfer problem. If we could instil regulatory virtues (which are abilities to judge when and to what extent to employ which reliabilist virtues) in our students, that would help with the first kind of situation. If we could instil motivational virtues (get the students to see the importance of employing the reliabilist virtues in those contexts), that would help with the second.

Consider now the third possibility: the student has not properly or lastingly acquired the abilities involved in reasoning well and evaluating other people's reasoning. The results of research on what improves transfer are very mixed, but one feature that consistently appears to make a difference is the length of the critical thinking course. Students in courses lasting five months or more are more likely to show gains (Behar-Horenstein and Niu 2011). Most critical thinking courses, however, are one-semester courses, and there are not in general follow-up courses that give the student the opportunity to further develop and practice the skills taught in the first-year course. It would be unsurprising if the abilities developed do not have time to become deeply engrained, and the exercise of them habitual, in the brief period covered by the course.

Compare the one-semester critical thinking intervention to, for example, how people learn to do long division. Every year, for years and years, children do long division in math class, the exercises gradually becoming more complicated as the years go by. The same procedure keeps getting used, over years of practice exercises, until it is thoroughly engrained. For many adults who have been through this process, long division is second nature—they can recognize a situation in which long division is called for and do it, even if years have passed since the last such situation arose.

It might well be a good thing if critical thinking were taught in a sustained way in primary school. But since most of us are, at least for the moment, stuck with the one-semester critical thinking course at tertiary level, it may be worth thinking about how one might try to get students to act as though they were engaged in a longer course of study that continued after the formal critical thinking course had ended. Everyday life throws up many occasions for critical thinking, and the more of it students do, the more likely they will be to continue to do it, and to do it in the situations in which it is important. Again, the moral is that motivational and regulatory virtues not only need to be instilled in the student during the critical thinking course, but also, they need to be explicitly told what kind of skills are involved in critical thinking and that finding opportunities for practicing them both during and after the course will help to develop and maintain them.

How might one teach in order to develop the motivational and regulatory virtues?

We conclude with some thoughts about how this might in practice be achieved.

Critical thinking courses often begin with a spiel about the value of critical thinking, which gives examples of real-life situations in which it is important. For example, one might ask students to consider the situation of the juror (a situation in which some of them will already have been, and in which many of the others can expect to find themselves at some point in their lives). Jurors have to make decisions that may make an immense difference to the life of the defendant. Different parties are trying to persuade them of different conclusions, and a lot hangs on their being able properly to weigh evidence and evaluate arguments. Furthermore, having come to a decision, they may have to persuade other jurors that it is the right one, and also be prepared to listen to and evaluate the arguments of jurors who think otherwise. The message is that critical thinking can matter, a lot, in real-life situations. Students should conclude that the course is worth taking seriously, and that the skills it teaches them are ones that they can and should use in their everyday lives and not just in an academic context.

In the standard critical thinking course, the emphasis on important real-life situations diminishes rapidly as the course goes on. It takes much longer to situate an argument in a realistic-seeming situation in which evaluating the argument correctly has significant consequences than it does to present and evaluate a simple example of, say, *modus tollens*. There are a lot of different argument types to be got through, and furthermore students need to be exposed to and work on exercises involving a lot of different examples of each kind. It is easy to see why examples tend to be oversimplified and to lack context. A further problem is that the students enrolled in a critical thinking course have a very wide range of interests and backgrounds: what is a realistic scenario to one of them may be, for another, a situation in which they cannot imagine ever finding themselves.

If the critical thinking course were conceived of and presented as merely the *beginning* of a process of developing critical thinking abilities, it might be possible to cover less ground but in more detail, and to thoroughly contextualize all examples. Perhaps, for example, not every one of the traditional long list of fallacies needs to be covered in class, since they are only more examples on which to practice the very general skill of evaluating whether or not a set of premises provides good reason to believe some conclusion. It might also be possible (especially if the class is small or if there are small-group discussion sessions) to elicit suggestions from students about situations they have encountered in which critical thinking was or would have been useful, and to use these as examples. Ideally, perhaps, *all* examples used, in set exercises as well as in class, would be contextualized and would be ones that would seem like real-life examples to some member of the class.

In addition, if we want students to practice their critical thinking skills in contexts where the outcome actually matters after the course is over, we had better get them in the habit of doing so outside the classroom while the course is in progress. This might be achieved by setting assignments involving constructing arguments about topics that are of practical importance to them, and finding and evaluating other people's arguments about such topics. One might also require a critical thinking journal as part of the assessment, in which the student reflects on occasions during each week on which they have used (or should have used) their critical thinking skills: this would both motivate them to find occasions on which to use the skills, and perhaps begin to make them more self-aware and reflective about their critical thinking practices. (See for example Garner [1990, 519] on the importance of self-monitoring.)

Conclusion: The critical thinking teacher as coach

Recall our tennis player, who has with long hours of thoughtful practice honed her serve, her volley, and her ability to chase down a lob. Had she not had the motivation to become an excellent tennis player, she would never have put in the hours. Had she not developed such character traits as persistence, patience, and the right degree of calmness under pressure, she would not be winning nearly so many games. And perhaps if she had not had a good coach providing encouragement and feedback, her motivation would have flagged, or she would have failed to notice that she was often losing crucial points because she went for the winner too soon, or that there was an easily remediable problem with her second serve.

In the critical thinking context, the abilities that need to be developed are those involved in argument evaluation, and one should expect to have to put in long hours of practice in order to develop them. Becoming an excellent critical thinker takes hard work, and hard work needs motivation. The job of the critical thinking teacher might usefully be seen as similar to the job of the coach. Perhaps the most important part of the job is to motivate the student to practice, by keeping both the value of critical thinking and the fact that it requires practice at the forefront of the student's mind.

References

Aberdein, A. 2010. "Virtue in Argument." *Argumentation* 24: 165–179.
Behar-Horenstein, L., and Niu, L. 2011. "Teaching Critical Thinking Skills in Higher Education: A Review of the Literature." *Journal of College Teaching and Learning* 8 (2): 25–41.
Bok, D. 2006. *Our Underachieving Colleges: A Candid Look at How Much Students Learn and Why They Should Be Learning More.* Princeton: Princeton University Press.
Bowell, T., and Kemp, G. 2014. *Critical Thinking: A Concise Guide.* (third fourth edition). London/New York: Routledge.
Code, L. 1984. "Towards a 'Responsibilist' Epistemology." *Philosophy and Phenomenological Research* 45: 29–50.

Cohen, D. 2009. "Keeping an Open Mind and Having a Sense of Proportion as Virtues in Argumentation." *Cogency* 1 (2): 49–64.

Ennis, R. 1994. "Critical Thinking Dispostions: Their Nature and Assessability." *Informal Logic* 18 (2): 165–182.

Fricker, M. 2007. *Epistemic Injustice: Power and the Ethics of Knowing.* New York: Oxford University Press.

Garner, R. 1990. "When Children and Adults Do Not Use Learning Strategies: Toward a Theory of Settings." *Review of Educational Research* 60 (4): 517–529.

Goldberg, I., Kingsbury, J., and Bowell, T. 2014. Measuring Critical Thinking about Deeply Held Beliefs. In *Virtues of Argumentation: Proceedings of the 10th International Conference of the Ontario Society for Argumentation,* edited by Dima Mohammed and Marcin Lewinski. Windsor, ON: OSSA.

Lepock, C. 2011. "Unifying the Intellectual Virtues." *Philosophy and Phenomenological Research* 83: 106–128.

Paul, R. 2000. "Critical Thinking, Moral Integrity, and Citizenship: Teaching for the Intellectual Virtues." In *Knowledge, Belief and Character,* edited by G. Axtell. Lanham, MD: Rowman and Littlefield.

Salmon, M. 2012. *Introduction to Logic and Critical Thinking.* (sixth edition). Stamford, CT: Wadsworth.

Sosa, E. 1991. *Knowledge in Perspective.* Cambridge: Cambridge University Press.

Tishman, S., Jay, E. and Perkins, D. 1993. "Teaching Thinking Dispositions: From Transmission to Enculturation." *Theory into Practice* 32 (3): 147–153.

Waller, B. 2011. *Critical Thinking: Consider the Verdict* (sixth edition). Upper Saddle River, NJ: Pearson.

Zagzebski, L. T. 1996. *Virtues of the Mind : An Inquiry into the Nature of Virtue and the Ethical Foundations of Knowledge.* New York: Cambridge University Press.

15

Proposition Testing: A Strategy to Develop Critical Thinking for Essay Writing

Sara Hammer and Phil Griffiths

Introduction

One of the proudest claims of the humanities and social sciences is that our disciplines develop the capacity of students for critical thinking. This is not a claim that is supported by research, and there are good reasons to be skeptical of it. Davies (2011) has suggested that universities are actually teaching very little critical thinking. In particular, there are serious doubts about the value of the most important student learning activity in most of our subjects, and especially those in second year and above: the student essay. Complaints about the shallowness of student essays are widespread.

One of the first questions to ask is whether we should be even asking students to write essays. Well-known British higher-education scholar Phil Race reminds us that essay writing is inherently difficult. He also believes that it tends to reward students who are more capable of adopting the rules of essay writing as a genre over those who are better at other forms of writing and presentation (Race 2009). This was the position of the first author of this chapter, Sara Hammer, at the beginning of our collaboration.

As well as outlining an alternative approach to teaching essay writing and using it to develop critical thinking, this chapter will report on a qualitative case study led by Sara that explored two separate but interrelated issues: the first was students' preconceived ideas about essay writing and critical thinking as part of essay writing, and the potential of these to shape the way they believed essay writing should be approached. The second was the effectiveness of the second author, Phil Griffiths's, assessment process redesign in developing his students' ability to test and evaluate author propositions. This research was conducted by someone other than the teacher and the teaching team. The assessment of student performance was undertaken by Sara, who had access to all student assignments, and did this independently of the official marking of those assignments. It was she who selected students for interview or forum post analysis, and conducted the subsequent data analysis.

Essays and critical thinking

What critical thinking consists of and how it should be taught to students is a continued subject for debate. One dimension of this debate relates to critical thinking as a personal disposition; another relates to whether or not critical thinking can be learned as a generic skill or must be embedded within an academic discipline (Davies 2006; 2013; Moore 2004; 2011). Mummery and Morten-Allen (2009) define critical thinking as the learner's development of "effective reasoning, interpretation, analysis, inference, evaluation and the monitoring/adjustment of one's own reasoning processes." At its heart, critical thinking in the humanities and social sciences has to begin with an understanding that knowledge is contested and that social phenomena are explained in radically different ways by different theories and approaches that have not been definitively found to be right or wrong. This means that any argument made by any writer is inherently contested by a rival school of thought. Among other things, critical thinking involves subjecting these rival theories and approaches to intellectual examination. This can include identifying and challenging their assumptions, testing their theories against real-world experience and empirical research, and examining the structure of their logic.

For university academics, a key outcome of many essay-writing tasks is for students to show evidence of critical or higher-order thinking (Andrews 2003). However, our experience of reading and marking student essays over many years suggests that they fail in this. As academics in the humanities, we have also routinely found that the majority were poorly researched and contained little real critical evaluation. Our experience is echoed by writers such as Clarsen (2009, 83) who describes student essays as "depressingly simplistic." In our experience, all but a small minority of essay texts are disorganized, the work of other authors cobbled together with little real consideration for its meaning. We have found it not uncommon for students unwittingly to copy a fragment of text in support of an argument that the author of that text actually opposed. We found little in the way of research related to essay writing and critical thinking that would provide us with solutions to these problems. This shallowness of student performance may be a result of what Norton (1990) and Vardi (2000) argue are conflicting and confusing instructions from lecturers and tutors about what essays should consist of, what is important, and what is prioritized by markers.

Research by Hounsell (1997, 112–114) found that students held three different conceptions of essay writing. These were:

1. the essay as "arrangement...an ordered presentation embracing facts and ideas";
2. the essay as "viewpoint...the ordered presentation of a distinctive viewpoint on a problem or issue"; and the most sophisticated,
3. the essay as "argument...an ordered presentation of an argument well-supported by evidence."

The last was the only conception involving some intellectual priority for finding data and interpreting it. Yet even students who held this view of essay writing appeared, when interviewed by Hounsell, to focus unreflectively on finding information to support their own argument (Hounsell 1997, 115).

Our perception was that the essay as an argument supported by evidence is widely taught in Australian universities. To confirm this we examined twenty-six essay writing guides from the full range of institution types. The guides we found were easily accessible to students on websites, and were provided by individual disciplines and by centralized student learning centers. Only four conceived of essay writing as involving research and participation in a debate that sits within a specific knowledge domain. Many offered a decontextualized process that took students through a series of essay-writing stages, such as the following:

1. analyzing the question and defining key terms,
2. establishing possible argument or thesis,
3. researching the topic, and taking notes from readings,
4. developing an essay plan, then writing the first draft,
5. editing and redrafting, and
6. completing final draft, including references and citations.

The most common type of guide we found made explicit the requirement for research or reading as well as writing, but made no reference to a knowledge domain or a debate. Only a few gave students any sense of how they might develop a credible argument, or that developing a position on an academic question is often a vexed intellectual process. Their silence on such matters sends an implicit message that the argument a student makes is not as important as other tasks such as developing a logical structure and finding supporting evidence.

In such guides, divergent opinion is often referred to briefly, or not at all. At best, students are asked to assess the argument of an author without reference to other research, or to be critical by evaluating evidence used to justify conclusions without reference to either authors or a debate. Certainly there is no advice on how to deal with debate among writers. For this most common type of essay guide, critical thinking techniques were not outlined. The end result is that students using such guides are expected to carry out a task for which they are largely unprepared.

Learner conceptions, assessment activities, and learning outcomes

How assessment tasks are conceived matters because this conception directs learner activities as they complete the final product. Biggs's well-known "3P" (Presage, Process, and Product) model is based on the idea that effective curriculum design should account for:

1. Presage—or, contributing factors that impact on what and how students learn, including their prior learning as well as what the teacher intends for them to learn.
2. Process—whether learning activities result in the outcomes desired by the teacher.
3. Product—the desired outcomes themselves (Biggs 2003, 19).

He argues that the interaction between these three domains can produce outcomes that do not align with the teacher's intentions (2003, 19). According to this model, a typical learning and teaching scenario related to argument-focused essay writing and the development of critical evaluation skills might be represented as in table 15.1.

Our discussion in this chapter reflects each component of this model, focusing particularly on student preconceptions of an essay and the assessment process Phil designed to more closely align student activities with his intention to develop critical evaluation skills.

Table 15.1   A typical learning and teaching scenario

Presage	• Students' prior conceptions of an essay • Students' prior learning in the target discipline • Teacher intention to develop critical-evaluation skills
Process	• Student activities to complete assessment include analyzing essay question and finding research evidence to support their argument
Product	• Student essays, structured arguments with evidence used to support their own position; occasional instances of evaluation, which are of broad positions rather than author claims and evidence

Source: Based on J. Biggs's (2003) 3P model.

Background to this study

Phil teaches three politics courses at a regional university, while Sara works in a centrally located learning and teaching support unit at the same university. Phil's position was that developing in students the ability to write a meaningful essay was one of the fundamental objectives of a liberal arts education. The challenge was to work out how to develop students' critical thinking skills as part of an essay-writing assessment.

He began by rejecting the idea that students should be required to write an essay as an argument, because it directed students away from the task of critically examining their own and others' positions. Phil's alternative approach involved four core elements:

1. He conceived an essay as being "a contribution to a debate," not an argument.
2. He argued that the first step in writing such an essay should be to develop a little understanding of the debate around an essay question. He therefore developed a supporting assignment in which students were required to outline the arguments, on the essay topic, of two authors representing rival positions.
3. After considerable experimentation, he found that the critical thinking tasks most achievable by disciplinary novices were proposition testing and evaluation. These involve taking a major component of an author's argument, doing research to test the argument, and evaluating the argument's validity in the light of that research. Validity, here, is meant in the sense of consistency with evidence and the results of research, rather than logical validity.
4. Students were then given a strategy for essay research (called a "research program"), which involved taking a small number of propositions from rival authors, doing research to test these arguments, then evaluating them in the light of that research. This material would form the body of a student's essay.

Subsequent subjects in a disciplinary major could develop students' ability to undertake more sophisticated critical thinking tasks as part of essay writing and research, such as structuring an essay around issues in debate, the examination of rival arguments using theories that challenged their own theoretical basis, analysis of the logic of rival positions, and analysis of theoretical polemics. Our institutional context, where there is no politics major, has meant that we have not been able to implement and test such an approach; the opportunity to do so, and to research its impact, would be of great benefit. It was not Phil's aim to produce in students a rounded or complete ability to think critically; rather it was to focus on developing a few elementary critical thinking skills that could be deepened and broadened in other courses.

Student conceptions of critical thinking and essay writing revealed in this study

As part of what was initially a wider project, Sara first looked at student conceptions of essay writing and critical thinking by analyzing interview and forum data from two of Phil's politics units, including the unit in this study and a first-year unit. Students in both units were disciplinary novices since these were isolated offerings within a business faculty. She interviewed students using open-ended questions and analyzed their online forum posts to discover their preconceptions of essay writing and critical thinking. Sara's analysis of the data was focused on interpretations of essay writing and critical thinking as constructed by the students (Schwandt 1994).

Data were collected from a total of eighteen students over a three-year period in two separate stages, using different collection methods. In the first stage, interview data were collected from students in both units and explicitly raised critical thinking in connection with essay writing. In the second stage, data were collected from students in the second-year unit only, using unsolicited, unprompted posts from online, unit forums. As all students were required to post to forums as part of a participation-type assignment, we were able to analyze a representative range of posts from students enrolled in the unit.

The reason for narrowing data collection to one unit was the removal of the essay as an assessment task in the first-year unit. For the interviews, students were asked open-ended questions, the most relevant of which included:

- What do believe an essay assignment is for? What does it require you to do?
- What is your technique for essay research and writing?
- How do you see the role of an opinion in your essay? Is expressing it easy or hard?
- Lecturers talk about critical thinking. What do you think this is?
- How do you show critical thinking in your essay?

Sara used a "naturalistic inquiry" approach to the collection and interpretation of student data. From this perspective, "data are, so to speak, the *constructions* offered by or in the sources" (italics in the original, Lincoln and Guba 1985). She observed that students were providing valuable data on their learning experience within unit online forums with no direct intrusion from her as a researcher. So, in the third year she instead chose to purposively sample online forum exchanges posted by volunteer students before their assessment submission(s). Their exchanges were posted in response to a forum question that asked them to share strategies they were using to complete the assignments. Sara analyzed posts that dealt with conceptions of and approaches to critical thinking and essay writing. Interview and forum data were analyzed to identify any general themes in student responses over the three-year period.

While data were collected from the same students, before and after their completion on assessments, Sara found that there were too many other factors that may have influenced students' practices and conceptions to draw any firm conclusions from the interview data about the impact of Phil's novel essay-writing strategy on their conceptions of either critical thinking or essay writing. For this reason only analyses of first round interviews and forum posts will be discussed here. In any case, the main reason for obtaining the interview and forum data were to examine these participating students' prior conceptions of critical thinking and essay writing to see what impact it might have on the way students normally approached an essay-writing task.

Conceptions of critical thinking

Analysis of student conceptions of critical thinking revealed four broad themes that were categorized as

1. weighing up different author arguments;
2. making up your own mind;
3. critical thinking as analysis; and
4. assessing the validity of author claims, or testing hypotheses.

Three students claimed not to understand the meaning of critical thinking and three of the six students using the online unit forums did not address the issue of critical thinking in their responses. There was no apparent difference to be found between students in first- and second-year units.

A common conception of critical thinking cited by four different student participants was aligned with the idea of weighing up different author arguments to come to an opinion. One student explains: "I suppose you get given the left and the right of the issue or both sides of the topic and you go, well, I think that's side is right or that side is right, or I think both" (cs2011). A similar theme cited by two students was the idea of making up your own mind and showing that you have thought about a topic. Two students defined critical thinking in terms of analysis, either with reference to breaking down a topic or analysis of theory and practice.

Four other students did refer to activities such as assessing the validity of author claims, or testing hypotheses. For example, one participant focused on: "Verifying their [authors'] sources and opinions because most writers will look to express some sort of opinion or slant" (tk2010): of these two had prior experience in another of the politics units.

Conceptions of an essay

We categorized the most common prior student conceptions of an essay that emerged from these data as: "essay-as-understanding" and "essay-as-argument." We found no discernible difference between learners in first- and second-year units.

Six of the eighteen participants outlined the purpose of an essay in a way that focused on their demonstrating understanding or acquisition of knowledge, as expressed here: "It is for whoever's examining the course to get an idea of my understanding of the particular question and how well I understand it by how well I write it" (cs2010).

The second most popular theme that emerged among student responses and posts was the concept of an "essay as argument" with five of the eighteen participants conceiving of an argument as central to an essay writing task. However, responses to a different question about essay-writing technique indicate that this conception may have been more widely held. For a few there was also an awareness that counterarguments needed to be dealt with, although this student's goal was to produce a balanced essay:

> Once all the information is collected, I choose the ones that I think provide the point [I am making] and the ones that rebut the point to show the differences of opinion for a balanced essay. (sb2011)

One of the less commonly expressed ideas included the role of essays in developing skills, such as communication and information literacy. Another was the idea expressed by two students that essays deal with the "for and against" of disciplinary debate, with a further two who had previously enrolled in another politics unit appearing to differentiate between "politics essays" and those elsewhere:

> I don't think they're [politics essays] really essays. They're more like, here's a question: do you agree or don't you agree and give me some evidence why that's your opinion. (mk2011)

The interview and forum data also indicated that the majority of student participants believed they should direct their essay-preparation activity toward question analysis and research based on key terms, as in this example: "It's very random. I kind of just research—well, I just do a lot of searching on databases for key words and stuff and spend a lot of time reading through journals" (mk2011).

While the number of interviews and forum posts analyzed were small, the association between student conceptions of essays and the approach to essay writing they describe is supported by similar findings in other research (Hounsell 1997; McCune 2004; Norton 1990; Prosser and Webb 1994).

Connections between conceptions of critical thinking and understanding of essays

It became clear to Sara that students' conceptions of essay writing might negatively affect their development of critical-evaluation skills. Overall, her analysis of the data showed no discernible relationship between a students' conception of critical thinking and their proclaimed essay-writing approach. Indeed, for these students there appeared to be no clear connection between essay writing and critical thinking. However, exploring how their approach to essay writing might impact on the development of critical thinking proved more fruitful.

These findings, together, strongly suggest that student preconceptions of an essay-writing task may direct them toward activities that work against their development of critical-evaluation skills. For this reason, Phil had good cause to purposefully reconceptualize his essay task as taking part in a debate, and redesign his assessment scheme to refocus student activity on reading and research as part of identifying, testing, and evaluating author arguments.

The new assessment process

Students in Phil's second-year politics subject were given two major written assignments: a formative assignment titled "Argument Analysis," and a major essay. For the argument analysis, students were to choose an essay topic and were given two short texts arguing rival positions on it. They were told to

1. outline each author's argument on the essay topic, as brief dot points,
2. choose one of these dot-point arguments—it can be from either reading—and search for concrete evidence and other research material to test its validity (proposition-testing), and
3. evaluate the validity of the argument chosen in the light of the evidence found.

When it came to writing their essay, students were told that the body of their essay would consist of three, four, or five proposition-testing exercises, testing other key arguments they had found in their reading. The aim of this assessment structure was to oblige students to engage in critical evaluation, and to do so in a context where it was clear that there was no right/wrong position.

For the argument analysis assignment, students were given detailed instructions on how to carry out each step, and a range of support material, including "how to" videos prepared by library staff. This support material provided clear guidance as to what constitutes good and poor evaluation. Assessment criteria included the accuracy and clarity of their two argument outlines, the relevance of the argument they chose to test, the quality and quantity of their research, and the quality of their evaluation of the author's argument in light of the evidence found. Students found each of these tasks extremely challenging. Feedback on these assignments focused on what students would need to do to write a good essay.

When it came to writing the major essay, students were told that good essays involved:

1. *Engagement.* A politics essay is a contribution to a debate, so we expect you to discuss a range of different opinions that people have on the essay topic, and their reasons. Two of those opinions are represented in the readings you analyzed in your first assignment, but a good essay will also discuss additional viewpoints.
2. *Critical thinking and research.* We don't just accept something as true because it's in writing; that's why we expect you to do extensive research to test the validity of what various authors have said. Your research needs to be critical—trying to find the truth.

Phil found that he had to explicitly differentiate his approach from that of the more common essay-as-argument to get students to go beyond the techniques they had adopted in previous courses. He presented the essay-writing task as building on the argument analysis. From it they had learned how to identify an author's opinion, and they were told they needed to use this technique to deepen their understanding of the debate around the essay topic, reading and outlining the arguments in a few more relevant texts. Their essay's introduction would need to include a brief survey of the different positions they found on the essay topic. When it came to testing and evaluating between three and five arguments for their essay, they would need to think about which arguments to test, and this was described to them as developing their "research program." The aim of this was to show students a way to plan their essay research and break it down into a series of discrete tasks; and to do this in a way that maximized critical analysis and intellectual depth.

Finally, he instructed them to synthesize the evaluations they made of the specific arguments they had tested to write an overall conclusion to the essay question. They would then have a first draft of their essay. Students would then need to consider any gaps in their research, any other issues that needed investigation, etc., and the text would need to be edited to become a coherent whole.

Assessment criteria for marking these essays included the quality of their understanding of debates surrounding the essay topic, the degree to which their essay is focused on the essay question and engages with issues relevant to it, the quality of their research program, the quality and depth of their research, the quality of their evaluation of rival arguments in light of the evidence, and the degree of their understanding of relevant course material. The second-last criterion specifically refers to the type of critical thinking his assessment task was seeking to promote. Students' attainment of this, on a sliding scale related to grade-level performance, would demonstrate whether Phil had been successful or not.

Determining the effectiveness of the new assessment process

To evaluate the effectiveness of Phil's assessment scheme, student performance in essay writing for the second-year politics unit was investigated by Sara, independently of the teaching staff, and with full access to all student assignments. She analyzed scripts of interviewees and forum volunteers, plus an additional sample of eight scripts each year from successful students (thirty-six overall), ranging from those attaining a pass to a high-distinction grade. Eight scripts from passing students in the second-year unit before Phil's assessment innovation were also analyzed so that Sara could compare the effectiveness of a different assessment scheme culminating in an essay to Phil's new assessment process in developing students' critical evaluation skills.

Scripts were analyzed and categorized using,

1. relevant assessment criteria developed and used by Phil and his tutors and
2. Bloom's (Krathwohl 2002) and Biggs's (2003) Structure of the Observed Learning Outcomes (SOLO) taxonomies.

Sara herself had taught similar courses and checked her understanding by discussing the relevant criteria with Phil. To this she added another analytical layer by using educational taxonomies. These are based on theories about how we learn and describe developmental stages of student learning, from the basic to the more complex.

She first analyzed some student work that was a product of the previous essay-based assessment process, which included an article review and an essay. For the article review students were asked to outline three different author positions from set texts on an essay topic and offer a conclusion, giving the relative merits of each. In this version of the assessment, students were guided to identify competing author propositions but were offered no technique for evaluating them, beyond the use of their own judgment as disciplinary novices.

Sara's analysis of this set of student scripts highlighted only one successful example of a student attempting to find evidence about author propositions within the essay. Some students compared and contrasted broad positions on a topic or evaluated broad positions based on their own reasoning, but with no exploration of evidence. Case studies or examples were occasionally used by the students to evaluate a general proposition. There was no clear relationship between grade performance and the ability to evaluate broad positions on a topic, but higher graded papers tended to show better coherence of structure and purpose, and more evidence of research. The relevant criterion for the assessment prior to Phil's innovation was "critical treatment of sources and evidence" (an earlier version of the criterion he would develop as part of the new assessment scheme). Reference was made within the marking sheet to "critical examination of…major arguments and evidence," but the language he used also reflected some lack of precision about what his expectations were of students in relation to their evaluation of author propositions and evidence.

Sara's analysis of student script samples over three years after the implementation of Phil's new assessment process showed that it was more successful at developing students' critical-evaluation skills. All students who attained a passing grade and above had attempted to find evidence about the proposition/s they were testing. She assessed their work using the relevant criterion developed by Phil for the new assessment: "quality of evaluation of rival positions on essay topic and of materials found in research." This included a more precise requirement for students to obtain a minimum passing grade: "evaluation of at least one position on the essay topic in the light of research done." She also applied Bloom's definition of evaluation as: "making judgments based on criteria and standards" (Krathwohl 2002, 215). In this case students were to judge the validity of author propositions based on available evidence.

To determine the overall standard (Stake 2004) of student work, Sara used Biggs's SOLO taxonomy, which describes different levels of student mastery of an area of knowledge. Student performance can be categorized from the lowest level of "incompetence" to the highest level, "extended abstract," which means that students can, among other things, comfortably theorize and generalize about knowledge in their domain of study (Biggs 2003, 19). Sara categorized less sophisticated student work as "multi-structural" (according to SOLO taxonomy). This means that students can successfully complete separate components of an assessment task but they are not able to well integrate them as a coherent "whole." More sophisticated work was categorized as "relational," which meant that students were able to do all the different tasks required and integrate all the material they found into a coherent assignment.

Pass

At the passing-grade level students in the second-year unit demonstrated one or two instances of proposition testing, as in this excerpt, which examines the claim that water privatization leads to higher prices and poorer water quality:

> Saal and Parker (2000) examined the impact of privatization and regulation of the water in England and Wales...the study included 5 years before privatization and 10 years after privatization. The study concluded that privatization led to increase in water prices and showed little variation in efficiency compared to the public sector. Other studies [that] supported this include Shiva (2000). (p22012)

The overall level of work for this student could be categorized as multistructural because of its weaker coherence as an essay. There was little logical connection between separately tested arguments and an overreliance on just a few resources.

Credit

Student performance for a credit generally demonstrated a greater ability to pull material together coherently than students attaining a passing grade, placing them somewhere between a multistructural and relational level, according to the SOLO taxonomy. Either testing of claims and discussion of concrete evidence were brief, or there was a more general discussion of broad-brush evidence as in the following excerpt. The student is testing the proposition that political freedom is a prerequisite for prosperity:

> Another example is the small state, Singapore, which was able to generate a lot of wealth even though it is only a one-party state (Cohen, 2007). Also, Russia has been able to seize economic growth without being democratic. (c12012)

Quality, nonpartisan references were still comparatively few.

Distinction

Student performance for a distinction demonstrated greater overall coherence in their essays with a clearer relationship between the relevant debate, propositions tested, and conclusions reached and, generally, a better use of the essay structure, including introduction, paragraphing, and conclusions; these were categorized as relational, according to the SOLO taxonomy. Distinction essays also demonstrated a more systematic approach to the testing of propositions and the evaluation of evidence. Some of the evidence found by students was still too general, or too secondary. More quality references were used.

High distinction

Student performance for a high distinction demonstrated a clear relationship between the context of the debate, the debate, propositions tested, and conclusions reached. The overall work attained a relational level, with a relatively seamless integration between introduction body and conclusion. Evidence was systematically weighed and propositions tested at a greater level of sophistication, as in this excerpt below:

> The evidence evaluated suggests there are both advantages and limits to private management of water. In some cases such as in Hamilton, private companies failed in efficient management which produced numerous externalities to the community while others such as Thames in England succeeded in efficient and effective managements. Thus it appears that water privatization [*sic*]...is a case by case issue...with correct regulations and government involvement the private sector can manage water efficiently. (Student HD22011).

The experience of Phil and his markers was supported by Sara's analysis of student scripts. Over the three years following the implementation of the new assessment scheme, she consistently found that students who attained a passing grade could reach the benchmark of testing at least one author proposition as part their overall essay discussion. Students attaining higher grades tested these more systematically and with greater levels of sophistication. However, students' capacity to develop logical and coherent essays as a whole was unaffected, with higher-graded student work continuing to show a greater capacity in this area.

Discussion

We both conclude that Phil's new process for essay writing and research has been effective in developing students' critical-evaluation skills because it situated their essay writing within a debate with multiple positions on a topic and no right/wrong answers. This new approach also helped students to develop an investigative research technique for critically engaging with rival arguments based on a research program, rather than an essay plan. The research program provided a structure to assist students in developing the capacity for sustained intellectual work. Studies have shown the value of integrating research and research skills into undergraduate subjects (Wass, Harland, and Mercer 2011;

Willison 2012), and more fundamental skills such critical reading (Clarsen 2009; Hammer and Green 2011; Wilson, Devereaux, Machen-Horarik, and Trimingham-Jack 2004).

Supporting Phil's students in this way was important because the preconceptions they enrolled with may have led them to pursue activities to complete an essay-writing task that worked against the development of their critical-evaluation skills. The majority of the students interviewed interpreted an essay task as demonstrating what they have learned, or developing an argument about a topic supported by evidence. Some students understood the need for balance, impartiality, and engagement with a range of views but the learning activities described by the majority were focused on unpacking an essay question to find key words and conduct their research on the basis of these. While the number of participating students was too small for generalization, other studies have found students to hold similar conceptions and have shown, likewise, a connection between these conceptions and their performance (McCune 2004; Prosser and Webb 1994).

Phil believes that the improvement in student performance has been most marked among "average" students, those achieving grades ranging from high passes to high credits. Sara's evaluation of the new assessment process showed that it was effective in developing the specific critical evaluation skills targeted by Phil. However, it did not appear to improve students' capacity to synthesize the results of their research. Only students achieving distinction and high-distinction grades could convincingly synthesize their work into a coherent "whole." This is in no way to minimize what has been achieved, but instead suggests the level of complexity and range of skills and intellectual development required to successfully write a convincing essay. Students' weakness in synthesizing their work may in part be a product of one of the strengths of the approach, the breakdown of the essay task into a series of proposition-testing activities. Lack of time to read widely in their course as a whole and/or to complete the assignment thoroughly may also be factors. Another issue may be the lack of research evidenced by pass- and credit-level essays and the related failure of students to situate their work within a broader debate related to the topic. Deitering and Jameson (2008) suggest that students need to read widely on a topic to familiarize themselves with the relevant debates within the literature. Most of Phil's students were disciplinary novices, so another explanation may simply be that some of them had not fully understood the task (Biggs 2003), or were unable to complete it. More work is needed to examine in greater detail students' experiences of critical reading and research as part of essay writing.

Students and markers alike were greatly assisted by increased clarity and precision of teacher expectations of their performance as it related to critical evaluation, as reflected in the new assessment criteria and descriptors. This affirms the importance of clear conceptualization and explicit identification of desired forms of learning as part of teaching and assessment practice. Educational taxonomies such as Bloom's (Krathwohl 2002) and Biggs's (2003) SOLO may be useful tools to help academic teachers conceptualize and make explicit both the required threshold for key learning outcomes and expected levels of overall performance for each grade.

Conclusion

Essay writing is a complex writing genre that continues to evolve over time. Because of this complexity, the essay is often the first choice of assessment for those of us who practice in the humanities and social sciences. However, our experience has suggested the need to critically examine our assumptions as they relate to essay writing and the development of students' critical-evaluation skills. In particular we found that teaching students using a model where the essay must be built around an argument is an obstacle to critical thinking and engagement with disciplinary debates. By contrast, reconceptualizing essay writing as contributing to a debate, and teaching students the skills to outline arguments, test, and evaluate propositions, led our novices to practice elementary critical thinking. Providing precise, guided activities and techniques to develop these skills also contributed to the success of the assessment scheme.

References

Andrews, R. 2003. "The End of the Essay?" *Teaching in Higher Education* 8 (1): 117–128.
Biggs, J. 2003. *Teaching for Quality Learning at University.* (second edition). Buckingham: The Society for Research into Higher Education and Open University Press.
Clarsen, G. 2009. "Challenges of the Large Survey Subject: Teaching and Learning How to Read History." Paper read at *The Student Experience: Proceedings of the 32nd HERDSA Annual Conference,* July 6–9, 2009, at Darwin.
Davies, M. 2006. "An 'Infusion' Approach to Critical Thinking: Moore on the Critical Thinking Debate." *Higher Education Research and Development* 25 (2): 179–193.
Davies, M. 2011. "Introduction to the Special Issue on Critical Thinking in Higher Education." *Higher Education Research and Development* 30 (3): 255–260.
Davies, M. 2013. "Critical Thinking and the Disciplines Reconsidered." *Higher Education Research and Development* 32 (4): 529–544.
Deitering, A.-M., and Jameson, S. 2008. "Step by Step through the Scholarly Conversation: A Collaborative Library/Writing Faculty Project to Embed Information Literacy and Promote Critical Thinking in First Year Composition at Oregon State University." *College and Undergraduate Libraries* 15 (1): 57–79.
Hammer, S., and Green, W. 2011. "Critical Thinking in a First Year Management Unit: The Relationship between Disciplinary Learning, Academic Literacy and Learning Progression." *Higher Education Research and Development* 30 (3): 303–315.
Hounsell, D. 1997. "Contrasting Conceptions of Essay-Writing." In *The Experience of Learning: Implications for Teaching and Studying in Higher Education,* edited by D. Hounsell and Noel Entwhistle F. Marton. Edinburgh: Scottish Academic Press. 106–125.

Krathwohl, D. R. 2002. "A Revision of Bloom's Taxonomy: An Overview." *Theory into Practice* 41 (4): 212–218.

Lincoln, Y. S., and Guba, E. G. 1985. *Naturalistic Inquiry.* Newbury Park, CA: Sage.

McCune, V. 2004. "Development of First-Year Students' Conceptions of Essay Writing." *Higher Education* 47: 257–282.

Moore, T. 2004. "The Critical Thinking Debate: How General Are General Thinking Skills?" *Higher Education Research and Development* 23 (1): 3–18.

Moore, T. 2011. "Critical Thinking and Disciplinary Thinking: A Continuing Debate." *Higher Education Research and Development* 30 (3): 261–274.

Mummery, J., and Morton-Allen, E. 2009. "The Development of Critical Thinkers: Do Our Efforts Coincide with Students' Beliefs?" In *The Student Experience: Proceedings of the 32nd Higher Education Research and Development Society of Australasia Annual Conference,* Darwin: 306–313.

Norton, L. S. 1990. "Essay-Writing: What Really Counts." *Higher Education* 20: 411–442.

Prosser, M., and Webb, C. 1994. "Relating the Process of Undergraduate Essay Writing to the Finished Product." *Studies in Higher Education* 19 (2): 125–138.

Race, P. 2009. "Designing Assessment to Improve Physical Sciences Learning." In *A Physical Sciences Practical Guide.* Hull: Higher Education Academy.

Schwandt, P. 1994. "Constructivist, Interpretivist Approaches to Human Inquiry." In *Handbook of Qualitative Research,* edited by N. Denzin and Y. Lincoln. Thousand Oaks, CA: Sage.

Stake, R. E. 2004. *Standards-Based and Responsive Evaluation.* Thousand Oaks, CA: Sage.

Vardi, I. 2000. "What Lecturers Want: An Investigation of Lecturer's Expectations in First Year Essay Writing Tasks." In *First Year Experience Conference,* Eds. Brisbane, QLD. December 18, 2012. http://.www.fyhe.com.au/past_papers/papers/VardiPaper.doc.

Wass, R., Harland, T., and Mercer, A. 2011. "Scaffolding Critical Thinking in the Zone of Proximal Development." *Higher Education Research and Development* 30 (3): 317–328.

Willison, J. W. 2012. "When Academics Integrate Research Skill Development in the Curriculum." *Higher Education Research and Development* 31 (6): 905–919.

Wilson, K., Devereaux, L., Machen-Horarik, M., and Trimingham-Jack, C. 2004. "Reading Readings: How Students Learn to (Dis)Engage with Critical Reading." Paper read at Transforming Knowledge into Wisdom: Holistic Approaches to Teaching and Learning. *Proceedings of the 2004 Annual Conference of the Higher Education Research and Development Society of Australasia,* at Miri, Sarawak.

16

Conditions for Criticality in Doctoral Education: A Creative Concern

Eva M. Brodin

> One of the supervisors said: "Yes, we're going to reshape you here." It seems as if *that* is what they do in doctoral education, they make us into researchers...I mean, make us think in another way. And you could say that they've succeeded with that in my case: I mean with the critical thinking. What you learn is that you need to be critical of *everything*— *everything* you read. And I suppose that's the point with these seminars; it has to be, that you discuss the texts and critically reflect upon them. (Interview with doctoral student in pedagogical work, Pe1)

The demand for developing profound critical thinking in doctoral education is a serious concern since today's doctoral students are the academics and societal leaders of tomorrow. Thus they need to be well prepared for handling the rapid changes of academia, and society at large, in deliberate, transformative, and responsible ways. Such a concern extends beyond the traditional understanding of critical thinking in terms of critical reasoning. It also involves critical self-reflection and critical action (Barnett 1997). Underpinned by a range of scholars who argue for a close relationship between critical and creative thinking (Baer and Kaufman 2006), I shall in this chapter argue that criticality of this all-embracing kind involves an ample amount of creativity.

Walters (1994, 11) states that the critical thinker recognizes when it is necessary for "creatively suspending strict rules of inference and evidence in order to envision new possibilities, innovative procedures, and fresh, potentially fecund, problems." Unfortunately, there seem to be many obstacles along the path of assisting doctoral students in their development toward becoming critical beings of this powerful and creative kind. Doctoral education is a practice with "actors, actions, settings, tools and artefacts, rules, roles and relationships" (Lee and Boud 2009 13), all of which have a pivotal impact on the students' development. Disciplinary traditions, controlling supervisors, gate-keeping senior researchers, funding stakeholders' interests, and limited time frames can therefore be confining factors for the critical and creative development of doctoral students.

Against this background, the educational conditions for developing doctoral students into critical and creative scholars will be discussed in relation to a composite theoretical framework based upon Barnett's (1997) notion of criticality in higher education and Arendt's (1958) political philosophy on the human condition. This picture reflects the sum of my own experiences from conducting research and developmental work in doctoral education across all faculties. Through the chapter, illustrative examples will be used from one of my interview studies with doctoral students from four disciplines (musical performance, pedagogical work, psychiatry, theoretical philosophy) at four universities in Sweden, which has been presented in detail elsewhere (Brodin 2014).

Why creativity matters in critical thinking

Even though an impressive amount of literature exists on critical and creative thinking, these concepts are seldom explicitly defined by academics in practice. However, in his interview study with tenured academics, Moore (2013) found that critical thinking was understood as: judgment, skepticism, a simple originality, sensitive readings, rationality, an activist engagement with knowledge, and self-reflexivity. As regards creativity, Jackson and Shaw (2006) concluded that academics generally conceptualize creative thinking in terms of originality, imagination, and problem working. Creative research contributions, on the other hand, are recognized for being not only original, but relevant as well (Bennich-Björkman 1997). Even though these studies are valuable for understanding the meanings of scholarly critical and creative thinking, they do not address the vital connection between these phenomena in scholarship.

Brookfield (1987) states that critical thinkers are aware of and reinterpret the habitual actions, values, beliefs, and moral codes within their own context in order to liberate themselves from uncongenial ways of living. A similar notion is also provided by Barnett (1997), who suggests that profound criticality implies a process of personal and societal emancipation, in which the individual realizes that the world can be other than it is. "Reflection and critical evaluation, therefore, have to contain moments of the creation of imaginary alternatives" (Barnett 1997, 6). Releasing and creative approaches of this kind are certainly desirable in critical scholars for avoiding scientific dogmatism and for giving birth to new knowledge paradigms (Kuhn 1962). In a wider scholarly perspective, such approaches are also imperative for developing doctoral students into "scholars who not only skillfully explore the frontiers of knowledge, but also integrate ideas, connect thought to action, and inspire students" with deep concerns for the future (Boyer 1990, 77).

Due to the integrative relationship between critical and creative thinking, it is occasionally difficult to distinguish between the two phenomena in the literature. For instance, *open-mindedness* and *wisdom* appear to bridge the gap between creative thinking and reasoning in conceptual research (Fasko 2006), while the two concepts of *originality* and *independence* are frequently connected to both critical and creative thinking in research on doctoral work (Brodin 2014). Furthermore, the converging notion of *critical creativity* has recently been introduced as a methodological concept in action research for developing practices in emancipatory directions (e.g., see McCormack and Titchen 2006; Ragsdell 1998; 2001; Titchen and McCormack 2010). The concept has also been used for describing doctoral students' transformative learning processes toward becoming responsible scholars who express their critical creativity in cognition, action, and speech (Brodin and Frick 2011).

Since critical and creative thinking seem to be two sides of the same coin, Bailin (1993) claims that a distinction between the two ways of thinking is mis-leading in practice. Instead she recommends using the all-embracing concept of "good thinking," which is nurtured through teaching disciplines as modes of inquiry. Critically, this implies teaching the disciplinary norms for knowl-edge development as regards disciplinary problem fields, research methods, and criteria for quality assessment. Creatively, it involves demonstrating the open-ended and dynamic nature of disciplinary knowledge as well. However, with respect to doctoral education, where students' socialization into disciplin-ary traditions is essential (Golde 2010; Weidman 2010), it is valuable to keep a *theoretical* distinction between critical and creative thinking.

In my interview study with fourteen doctoral students from four disciplines (Brodin 2014), the students primarily related their critical thinking to "obtain-ing scholarly legitimacy" within scholarly traditions, while creative thinking was rather associated with "being open-minded" in new knowledge develop-ment. Both critical and creative thinking were connected to academic author-ship in the sense that the students had to "write in a publishable form." The students also associated both phenomena with "being independent" in prag-matic action.

Nevertheless, along with their reflections upon the meanings of critical and creative thinking in doctoral educational practice, it appeared that the creative features of "good thinking" were not always encouraged in their research envi-ronments. This fact laid the foundation for the current chapter. Thus, we need to ask: What kinds of assumptions are challenged in doctoral education? What kinds of alternatives are doctoral students encouraged to imagine? The answer to these questions is found in the current educational conditions for criticality.

Framing the educational conditions for criticality

The notion of criticality

Barnett (1997) suggests that the purpose of critical thinking can be sum-marized in three forms of criticality, which, taken together, constitute the full-grown critical being: *critical reason* (directed to formal knowledge), *critical self-reflection* (directed to the self), and *critical action* (directed to the world). In Barnett's view, the problem in higher education is that students are primarily trained in critical reason, while critical self-reflection and critical action are not emphasized. Consequently, students learn how to critically develop formal knowledge to a larger extent rather than reflect upon themselves and their worldview.

With respect to the present chapter it is important to notice that academia *is* the immediate world for most doctoral students. This is where they think, act, speak, eat, laugh, cry, and sometimes even fall asleep among their colleagues and community members through the years of studying. Academia is also the place where most students develop their scholarly self. Thus critical self-reflection and critical action in doctoral education need to be as much directed to the world of academia (with all the knowledge issues that that conveys) as it is directed to society at large. Unless doctoral students are impelled to critically reflect upon their scholarly selves and the academic context into which they are thrown, one cannot expect them to be prepared for critical self-reflection and critical action in any other context either.

Furthermore, Barnett (1997) claims that students in higher education seldom reach the highest levels of criticality in which they start questioning the frames themselves through transformatory critique that implies existential, social, and personal transformation in the individual. Instead, criticality has been captured within the frame's own instrumental and pragmatic process with no ambitions to challenge the given worldview in itself: Critical reasoning contributes to rationally strengthening existing knowledge paradigms, while critical self-reflection befits a process of conformation of the self to existing norms. Critical action, however, becomes a kind of problem solving where the actions are worked out in a taken-for-granted world.

Even though nearly two decades have passed since Barnett (1997) propounded his critical view of higher education, I daresay that this picture has not changed much in reality—at least not in doctoral education. Criticality is still predominantly constrained to critical reason, and critical thinking seldom reaches its peak in transformatory critique in any of the domains for criticality. You may wonder why. Assuredly, as Barnett points out, this circumstance can be explained by the fact that the university delivers what society asks for: effective operators who serve instrumental and pragmatic agendas. However, another way of explaining this circumstance is to take a closer look at the educational conditions for realizing the full range of criticality in doctoral educational practice.

Educational conditions through the lens of Hannah Arendt

In her comprehensive philosophical work, Hannah Arendt (1958) depicts the most elementary social activities of human beings in terms of three conditions: *labor, work,* and *(political) action,* which will be further explained in the chapter. Presuming that these conditions are all necessary dimensions in civilized human existence, Arendt sheds light on the superior position of labor in contemporary modern society, which leads to the human being becoming captured in her own instrumentality.

As we shall see, this circumstance is in many respects a depiction of the situation for doctoral students as well. Accordingly, doctoral studies are, in this case, basically understood as social learning activities that can be characterized as labor, work, and political action. For the purpose of this chapter, Arendt's original meanings of these concepts will be further developed in relation to the specific context of academia. Thus, I shall suggest how *academic* labor, work, and political action can be understood, and display how these activities are embodied (or not) in doctoral education. Such a context-sensitive framework necessarily includes a thoroughly conceptual foundation in criticality as the critical and creative spirit constitutes the heart of scholarship. The crucial point is that the activities of academic labor, work, and political action reveal diverse educational conditions for creative thinking, which in turn has a deep impact on the potential scope of criticality.

Criticality in labor: a questioning approach

A couple of years ago, a renowned researcher within the field came out with a book about the musician I am researching. And there it said the musician studied in a particular town between this and that year. Oh my, I thought, this is completely new information! Because the information I had obtained was between two other years. And it never crossed my mind that this established researcher could be wrong, so I wrote to him: "What is your source? This is completely new information for me." Then I got the answer: "That was an error in writing." Really??? He can write something wrong? Actually that was a milestone for me. After that I started to understand that: Well, I have to be a little bit critical here. (Doctoral student in musical performance, Mu2)

According to Arendt's (1958) conceptualization of labor, it has no clear beginning or end. Instead, labor consists of an endless process of drudgery, which is necessary for survival. This is how human beings reproduce themselves for the next generation, for example, through housework. Hence, the results of labor are soon consumed, and there is nothing concretely apprehensible left. Therefore, the value of labor is not found in the products but in the mere process of productivity. Since labor is a reproducing activity it entails hardly any creative features, although it constitutes the indispensable foundation for higher levels of both critical and creative thinking. For instance, without the accurate use of references it would be inappropriate to assess a piece as critical and creative in a scholarly sense. With respect to doctoral education, such laboring activities can have a deep impact on the doctoral students' initial experiences of scholarly critical thinking as is evident in the quotation above.

Labor is found in a number of other scholarly activities as well, such as the undertakings of data collection and writing, which also contribute to the reproduction of academia. What makes these laboring activities critical is their foundation in a *questioning approach*. It appears that the questioning approach of doctoral students is generally directed toward theories, methods, and conclusions, both in their own work and in that of others; theories and methods have to be understood and used correctly, and the conclusions need to be well-founded:

> Not so long ago, I read some articles and then I could actually review these articles myself: In this part there weren't a great deal of arguments for the method—since I have tried to be more stringent myself. Then I asked myself: How can they state that it is this method? (Doctoral student in psychiatry, Ps2)

Doctoral students certainly need to acquire sufficient research skills and understanding within their research field, or else they cannot critically participate in any scholarly discourse. This side of labor is not problematic, but the problems arise when the students' individual creativity is not encouraged. Then the student becomes captured within the frames of labor, yet deprived of its critical features. Without a scope for creativity, there is not much left for the student to put into question either:

> Often, as a doctoral student, you are the person who does things. And when it comes to trying experiments, then what I am supposed to do is already given. In that process I do not feel creative, because then I conduct a simple concrete thing that's already been planned. But I feel we have a creative environment and that my supervisors are creative... We've talked a lot back and forth about research questions and how we should do things, and about ideas and stuff like that. But I don't feel like I'm the one who's hatched the ideas. (Doctoral student in psychiatry, Ps3)

Strictly directing supervisors are probably one of the most decisive factors for confining doctoral studies into mere labor. In some cases, the supervisors' directedness is due to the limited time frames (as there is no time for making mistakes) or to the interests of the funding institution (as there is no space for further ideas). In other cases, the supervisors' directedness has its roots in the disciplinary norms or in personal views on learning. However, some of the firmly directed students still manage to find their own creative paths out of the grip of labor:

> For about the last year, I've started reading a lot of books that aren't philosophical works, but rather books in social science or history. When I was admitted to doctoral education I was more like this: Leaning back and thinking that everything would be served. I needed to do everything they told me to do and I needed to publish, but not much more than that. However, then I realized that it becomes very boring doing things that way and it doesn't lead to my development either. So, in that sense, I think that, at least there, I've become more creative because of all my reading. (Doctoral student in theoretical philosophy, Th1)

Criticality in work: an expressive approach

> My metaphor of the dissertation work is that it is a patchwork quilt. Many people asked me: "How far have you come in your work now?" Then I answered that I've cut out many pieces and that I'm busy laying out the pattern...And that vision and that pattern are beginning to emerge, the big patterns—but each piece that belongs with each other piece, I'm busy figuring that out. I think it's creative. (Doctoral student in pedagogical work, Pe2)

In Arendt's (1958) conceptualization, work has a defined beginning and end, with connotations to craftsmanship based on competent and professional knowledge. While labor is captured in its own instrumentality, productivity is now only a means for creating constant products that exist independently of their original creators once they are completed. Hence in contrast to labor, work embraces many opportunities for both individuality and creativity.

In doctoral education, the product of work is displayed in the final dissertation. Listening to the doctoral students' stories, two of their general activities can be related to their work in progress: making analyses and writing the results for new knowledge development. Thus, their critical questioning approach is now combined with an *expressive approach* that conveys potential for their individual and creative voice to be heard. According to doctoral students' understanding of creativity, it implies somehow making yourself interesting to the readers. For instance, it could be a matter of providing a new perspective or combining theories and methods. However, when it comes to expressing their creative and individual voice in their writings, many students feel they have to restrain their creativity as it would otherwise impair the critical quality of their work. They often experience that their individual voice is hampered by the predetermined style and form of scholarly texts:

> The dissertation is not a piece of art…I mean, in principal they [the dissertations] are meant to have the same form and preferably be written with the same ink. Of course you could have a *style* in your writing. I've tried that, but in this context I see it more as being the *thoughts* that I'm trying to express rather than the *writing style* that has an intrinsic value. Because as soon as I have tried that [a writing style] it becomes incredibly chatty and difficult to express my lines of thought. As a composer you're used to the *style* you have—that's what *must* put its stamp on the work. But in this case [the dissertation], there's no sense in putting my stamp on it. Rather, it makes things worse. (Doctoral student in musical performance, Mu1)

Against this background, it seems that doctoral students are less aware of how to express their own voice in their writings, that is, writing texts that not only contribute to original knowledge development, but also reveal the person behind the text. In this process, critical self-reflection can be a powerful tool for developing and expressing the individual voice in creative directions. However, doctoral students do not primarily use this creative kind of self-reflection in their dissertation work. Rather, they are concerned with checking if they are representing their theoretical and methodological frameworks correctly, if they have done justice to their data, and if their language is fluent and clear enough. At its most explicit level, the students also use critical self-reflection for acknowledging the weaknesses of their own study to the reader.

It would seem that the common component in the students' critical self-reflections is that of avoiding critique from the scholarly community. Through avoiding critique, the students' critical self-reflections aim at *survival* rather than creative development. This is how the students' dissertation work tends to be transformed into critical labor. With respect to the critical seminar culture in academia, I am inclined to say that the students' self-critical carefulness is justified. Yet the crucial point is that students need support in how to articulate their creative voice in a way that the scholarly community would find acceptable. Without such support, the students will continue to be cautious within quite narrow frames in their dissertation work. They then weaken their most powerful tool for criticality, namely their individual expression.

Criticality in political action: a relational approach

> You have to show how you have thought on the whole. You have to show that in every article, and in the choice of method—and defend your results and discussions…I think that it's embedded in every moment, that you need to defend what you've done. (Doctoral student in psychiatry, Ps2)

According to Arendt (1958), political action has a clear beginning, in a communal sense of dissatisfaction about the current conditions, but lacks a predetermined end. With its conceptual roots in ancient Greece, political action has barely anything in common with our everyday understanding of contemporary politics in general. It is not a matter of heated discussions with camps of politicians who are struggling to gain ground in favor of their party's interests. Instead, in this context, political action implies a social process in which individuals act and speak *with* each other rather than *for* or *against* each other. This presupposes that the human beings involved are considered as *equals* in the sense that they can all make correct statements about themselves and their environment. They are also *individuals*, contributing with their unique perspectives through their speech. Criticality in political action thus involves a *relational approach* in which responsibility, care, and respect for the individual voice are key words.

New knowledge development stemming from political action aims not at maintaining the status quo of the disciplines, but rather at opening up new paths beyond existing paradigms. Hence, political action provides promising conditions for embodying the most constructive and creative features of criticality. However, considering the quotation above, it seems as if political action is conspicuous by its absence in doctoral education. Instead, doctoral students need to be prepared for the battlefield at the seminar, equipped with a defensive approach that is rather unconstructive in terms of learning and knowledge development. It leads only to strengthening one side in the debate, where the doctoral students are seldom the winners. This can have serious consequences for the individual student if he or she proposes a new idea beyond the scholarly frames of the department:

> I've internalized some norms because at the seminar here, I was severely cut down by another philosopher...Yes, I was actually hurt in that situation. Although the person had concrete arguments which were good to some extent...but you know, there're different ways of giving critique. (Doctoral student in theoretical philosophy, Th1)

As a consequence, this student had progressively conformed to the scholarly traditions of the department through "accepting a lot of things that are not self-evident at all" without explicitly questioning the conditional frames for knowledge development. For the purpose of this chapter, the following lesson can be learned from this incident: Through negative experiences of this kind, students learn fight and flight through conforming to the scholarly traditions of their research environment. Accordingly, the academic community can be a powerful force in choking the individual voice of a doctoral student. This is how political action is transformed into the reproductive process of labor, in which doctoral students may become more or less face-less while serving the hegemonic powers in their research environment.

It should be mentioned that supervisors have an essential role in support-ing their doctoral students both before and after the students present their progressing dissertation work at seminars or conferences. These moments are frequently charged with the student's nervousness as he or she is often uncer-tain if the work is up to standard. If then, the reactions of the public turn out to be negative, the student's feelings of dejection are not far away. However, with a supportive supervisor (that is, someone who *listens* to the student) the student can be strengthened in his or her position even though the critique is harsh:

> One year ago, I was quite severely attacked by the professor at a seminar. But it was so funny, my supervisor supported me. It was incredible, my supervi-sor just thought afterwards that: "How interesting! Not even the professor understood this!"...So I was strengthened by this in fact. And, after that, I also met the professor, because he suggested that we could meet. Then I was a little bit better prepared verbally, and I think he understood what I meant. He backed down. (Doctoral student in musical performance, Mu2)

Certainly, there are some doctoral students who choose to go against the stream and continue to oppose themselves to the conventional frames of mind at their department. These students become highly independent in their approaches, but they nevertheless run the risk of becoming lonely scholars with no audi-ence for their critical and creative voice. It does not matter how substantial the student's viewpoints are, they will not have any transformative effects on the community unless someone is listening to them. Thus, irrespective of whether the doctoral student is conforming to the norms or not, a lack of political action in the educational environment will unavoidably lead to a diminishing of the critical power of the individual voice.

Grasping the conditions and thresholds for criticality in doctoral education

Conditions for criticality

So far, I have illuminated how the educational conditions of labor, work, and political action are connected to different attitudes of criticality in terms of a questioning approach (in labor), an expressive approach (in work), and a relational approach (in political action). Each of these attitudes, in turn, has the potential to embrace all three forms of criticality (critical reason, criti-cal self-reflection, critical action) if both critical and creative thinking are encouraged.

The ideal image would be an educational practice in which there is a *delicate balance* between labor, work, and political action, on the one hand, while critical reasoning, critical self-reflection, and critical action are also balanced, on the other hand. The outcome of such an educational practice would be doctoral students who are critical and creative, and who embrace knowledge, themselves, and the world with their questioning, expressive, and relational approach in an integrative manner. Presuming that all these components need to be realized for comprehensive criticality in doctoral students, a balanced and holistic image of this notion is illustrated in figure 16.1.

Unfortunately, such a balanced picture is quite far from how criticality is embodied in doctoral educational practice. This is due to the fact that creativity is seldom educationally encouraged in students at deeper levels.

Without doubt, this state of affairs can partially be explained by a range of external impeding factors for nurturing doctoral students' creativity. One such factor is the increasing standardization of doctoral education. No matter what subject or problem the doctoral student is investigating, the dissertation and courses should be completed within three or four years of full-time studies. Such rigid time frames require carefully prepared project plans before the student is admitted to doctoral education, which thereafter need to be strictly followed for financial reasons. Transformative transitions aligned with the student's development may thus be inhibited or made impossible.

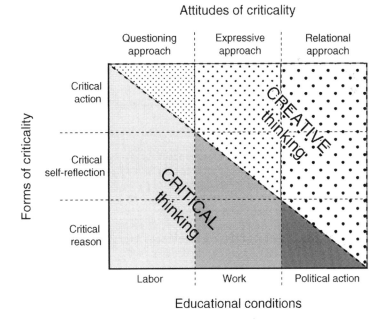

Figure 16.1 The full scope of criticality in doctoral education.

Furthermore, the monograph will soon be but a memory, at least in Sweden. At faculties covering the hard sciences, almost all doctoral writings consist of theses by publication. In the social sciences about a third of the dissertations are of this kind today, while the number of theses by publications is still increasing (Dellgran and Höjer 2011; Vetenskapsrådet 2006). Even in disciplines well suited for monographic writings (e.g., within the humanities), the thesis by publication is beginning to take root (Sjöstedt Landén 2012). This, too, can be a constraining factor for creativity as it precludes project ideas that cannot be cut into the smaller pieces of articles. Moreover, the financing industry can be another compelling factor restricting students' creative thinking; the scope of criticality is now seriously confined:

> There was a study I made earlier, and I even wrote a manuscript to it, but in that case we got an explicit "no" from the industry as regards publishing the results because it put the drug in a bad light. I found the results very interesting, although they could absolutely not be published…That's critical. No, it isn't easy. But, when you want to get articles published, you want to defend your dissertation, you try to produce something that can be published and presented in broad daylight. (Doctoral student in psychiatry, Ps3)

Even though these circumstances certainly restrict doctoral students' creativity, and hence their potential scope of criticality, I assert that we have a bigger problem to deal with. Doctoral students are not only hampered by external forces but they are occasionally let down by the academic community as well, as a supervisor pointed out:

> I would say that some discussions and problematizing are not encouraged, and the doctoral students have probably experienced some censure, from time to time. It's almost here we have the most dogmatism; you cannot relativize certain aspects of our research field for instance, and I'm very concerned about such limitations. You may not even suggest that there might be more urgent global problems to take care of, before we deal with the specific research problems of this doctoral school. (Interview with supervisor, Brodin and Avery 2014, 287)

The supervisor interview above was conducted in an interdisciplinary doctoral school, where the students were allowed to design their own research projects, choose their own theories and methods, and preferably combine knowledge from different disciplines. One might think that such an intellectual freedom would create favorable conditions for doctoral students' criticality at transformative levels. Apparently, this is not always the case.

Thus, it is not surprising that doctoral students are basically molded into questioning laborers rather than being socialized into political actors who embrace their fellowmen with their relational approach. The students learn how to express their critical thinking in their dissertation work, but they find it difficult to express their creative voice beyond the fact that they (re)produce new knowledge. Other creative processes, such as reading, take place in the background where the individual student becomes rather invisible. As a consequence, the most creative features of students' criticality will neither be encouraged nor made available to the scholarly community.

Thresholds for criticality

Against this backdrop it is possible to delineate the thresholds for criticality in doctoral education. In the literature on "threshold concepts," this notion represents fundamental pieces of understanding that students need to acquire for advanced learning, but where they may experience great difficulty crossing the threshold leading to the development of deeper levels of understanding. While this framework was originally developed to detect disciplinary threshold concepts in undergraduate education, such as the concept of *gravity* in physics (Land et al. 2008; Meyer and Land 2006), it has recently been adopted in research on doctoral education for identifying crucial components of doctoral students' understanding with respect to their development as researchers both in context-specific (Humphrey and Simpson 2012; Wisker & Robinson 2009) and general terms (Kiley 2009; Kiley and Wisker 2009; Trafford and Leshem 2009). For instance, based upon supervisors' reports across a range of disciplines, Kiley and Wisker (2009) found six generic threshold concepts at a doctoral level: *argument; theorizing; framework; knowledge creation; analysis and interpretation;* and *research paradigm.*

For our further reasoning, the threshold concept of *knowledge creation* is of special interest since experienced supervisors often mention that their doctoral students have difficulties grasping, developing, and articulating the quality of scholarly creativity in their doctoral work (Kiley and Wisker 2009). Therefore, combined with the doctoral students' stories presented in this chapter, creative thinking appears to be a threshold concept in developing criticality at transformative levels.

Yet the crucial point is not that scholarly creative thinking appears to be a threshold concept in doctoral students' learning, as this is already known through the work of Kiley and Wisker (2009). Rather, the urgent matter is that many students do not genuinely get through this developmental phase during their doctoral studies. In other words, they continue to be "stuck" as Kiley (2009) defines doctoral students as those who are struggling to cross a liminal stage without success. Kiley suggests that one way out of the liminal stage goes via interaction with supervisors and peers who are more experienced. Students who

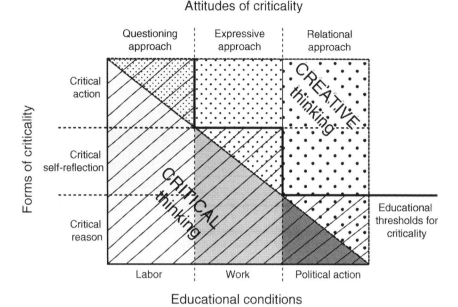

Figure 16.2   Educational thresholds for criticality in doctoral education.

are stuck can then initially mimic their discourse while gradually developing their own conceptual understanding: "fake it until you make it," as a supervisor expressed in Kiley's (2009, 296) study. As regards overcoming the threshold of creative thinking, however, the conditions seem to be far more challenging.

Very likely, initial mimicking could facilitate development of the necessary skills and knowledge needed for scholarly creative thinking. Nevertheless, such proceedings do not evoke creative thinking per se, but doctoral students also need to release themselves from the prevailing frames of knowledge in order to attain creative and hence transformative levels of criticality. *This* is the point where the students get stuck—not primarily owing to insufficient individual understanding, but rather because they get trapped into a liminal stage of dogmatism where the educational conditions firmly circumscribe the frames for students' creative thinking. Thus doctoral education itself needs to be transformed in order to nurture the creative components in the students' expressive and relational approach, which is necessary for embracing the full scope of criticality. The educational thresholds for criticality are illustrated in figure 16.2.

A call for strengthening the individual and creative voice

There is a widespread conception that doctoral studies of a laboring kind are primarily found in the hard sciences. Considering the general conditions for conducting doctoral studies today, however, it appears that this state of affairs can more or less be found in any research environment. Thus doctoral students' labor is no longer a disciplinary issue, but it has become a structural phenomenon with roots in the overall changed meaning of attaining a doctorate.

While the dissertation was previously considered to be an individual prestigious achievement with a scholarly value in its own right, it has now rather become a driver's license for furthering a career both within and outside of academia (e.g., see Lovitts 2007, 29–30; Mullins and Kiley 2002). Thus, the dissertation has become a necessary means for making other more valuable activities possible beyond the doctoral degree. Sometimes, these activities are not primarily connected to the doctoral students' future career, but rather to the supervisors' career (e.g., in externally financed projects). The doctoral student then runs the risk of becoming alienated from his or her own research activities, which paves the way for increased control in research supervision and hence may transform doctoral studies into labor.

Along with this unfortunate development, another serious problem emerges: The overall educational neglect of the individual voice. Having acknowledged that doctoral education is the most individual form of schooling in the entire educational system, I know that this is a controversial statement indeed. Still, the gist of doctoral students' stories generally provides a picture of reconciled acceptance in which most of the students learn to subordinate themselves to the hegemonic powers in their research environment. Doctoral students who raise their critical voice against the given agenda of their scholarly community are, figuratively speaking, suppressed to silence. In silence, no transformative criticality takes form.

Of course, exceptions exist in the doctoral students' stories, although too many students are framed within the somewhat dark picture of this chapter. Exceptional cases are also evident in that outstanding dissertations exist. Based upon her comprehensive focus group study with two-hundred-and-seventy-six faculties in seventy-four departments across ten disciplines, Lovitts (2007, 38) concludes that:

> Outstanding dissertations are characterized by originality, high-quality writing, and compelling consequences. They display a richness of thought and insight, and make an important breakthrough. The body of work in outstanding dissertations is deep and thorough. Each individual component of the dissertation is outstanding and the components are integrated throughout the dissertation in a seamless way. The writing is clear and persuasive and provides a glimpse into the mind of the author—you can see how the student is thinking.

Assuredly, the performance behind an outstanding dissertation is partially conditioned by the student's individual capacity. Based upon the thinking styles and dissertation quality of doctoral students, Lovitts (2008) found that "distinguished completers" possess high levels of practical and creative intelligence and produce a high-quality original dissertation. With references to Sternberg and Lubart (1995), Lovitts (2008) describes practical intelligence as "the ability to solve problems and use ideas and their analyses in effective ways, present them effectively to an audience, and react properly to criticism so that the ideas gain acceptance" (302), and creative intelligence as "the ability to formulate good problems and good ideas" (304).

Nevertheless, the individual outcome of the doctoral journey is also, to a large extent, an educational concern. Lovitts (2005) highlights that being a good course taker is not enough for doctoral students to become independent researchers in which they make an original contribution to knowledge. Besides individual resources such as intelligence, knowledge, thinking styles, personality, and motivation, there is a range of environmental factors influencing the outcome as well. Hence, the relationship between the *individual student* and his or her *educational context* is central for understanding the conditions for developing critical and creative scholars.

Against this background, I claim that it is not enough to train doctoral students in critical thinking. Students have to be supported in how to express their creative thinking as well. In addition to practical and creative intelligence, this involves a personal dimension. Frick (2010) clarifies that achieving a doctorate implies a complex learning process in which the students' personal view of what it means to become a scholar, their approaches to different forms of knowledge and methodologies, and their disciplinary values and ethics need to be aligned.

Unfortunately, it seems as if doctoral students are generally not encouraged to critically reflect upon their scholarly selves and the intellectual frames of their own research environments. Unless the students participate in contexts across disciplines and faculties (e.g., in joint PhD courses), such critical self-reflections seem to be far away. Accordingly, this is a developmental area in doctoral education. Strengthening the individual voice opens up for creativity, and hence criticality. Combined with a supportive scholarly community that is imbued with the educational conditions for political action, the doctoral student then receives the necessary qualities for becoming a critical and responsible scholar who:

> ...holds out his or her hands to the world, saying: "I care for the world and all life in it. There is too much that could be better, although it does not have to be this way. Thus, by opening my mind, better ways of being will be revealed. All I need to do is to use my will to search beyond what is already given. I am willing to challenge the limits of thought, forcing myself to action." (Brodin 2008, 222)

References

Arendt, H. 1958. *The Human Condition*. Chicago: University of Chicago Press.

Baer, J., and Kaufman, J. C. 2006."Conclusions." In *Creativity and Reason in Cognitive Development*, edited by J. C. Kaufman and J. Baer Cambridge: Cambridge University Press. 351–355.

Bailin, S. 1993. "Epilogue: Problems in Conceptualizing Good Thinking." *American Behavioral Scientist*, 37 (1): 156–164.

Barnett, R. 1997. *Higher Education: A Critical Business*. Buckingham: Society for Research into Higher Education & the Open University Press.

Bennich-Björkman, L. 1997. *Organising Innovative Research: The Inner Life of University Departments*. Oxford : Published for the IAU Press [by] Pergamon.

Boyer, E. L. 1990. *Scholarship Reconsidered: Priorities of the Professoriate*. San Francisco: Jossey-Bass.

Brodin, E. M. 2008. *Critical Thinking in Scholarship: Meanings, Conditions and Development*. (Revised edition). Saarbrücken: VDM Verlag Dr Müller.

Brodin, E. M. 2014. "Critical and Creative Thinking Nexus: Learning Experiences of Doctoral Students." *Studies in Higher Education*, DOI: 10.1080/03075079.2014.943656.

Brodin, E. M., and Avery, H. 2014. "Conditions for Scholarly Creativity in Interdisciplinary Doctoral Education through an Aristotelian Lens." In *Creativity Research: An Inter-Disciplinary and Multi-Disciplinary Research Handbook*, edited by E. Shiu. London: Routledge. 273–294.

Brodin, E. M., and Frick, B. L. 2011. "Conceptualizing and Encouraging Critical Creativity in Doctoral Education." *International Journal for Researcher Development* 2 (2): 133–151.

Brookfield, S. D. 1987. *Developing Critical Thinkers: Challenging Adults to Explore Alternative Ways of Thinking and Acting*. San Francisco: Jossey-Bass.

Dellgran, P., and Höjer, S. 2011. "Nya trender och gamla mönster: Doktorsavhandlingarna i socialt arbete 1980–2009." [New trends and old patterns: The doctoral theses in social work 1980–2009.] *Socialvetenskaplig tidskrift* 2: 85–105.

Fasko, D. Jr. 2006. "Creative Thinking and Reasoning: Can You Have the One without the Other?" In *Creativity and Reason in Cognitive Development*, edited by J. C. B. Kaufman. Cambridge: Cambridge University Press. 159–176.

Frick, B. L. 2010. "Creativity in Doctoral Education: Conceptualising the Original Contribution." In *Teaching Creativity—Creativity in Teaching*, edited by C. Nygaard, N. Courtney, and C. Holtham. Faringdon: Libri Publishing. 15–32.

Golde, C. 2010. "Entering Different Worlds: Socialization into Disciplinary Communities." In *On Becoming a Scholar: Socialization and Development in Doctoral Education*, edited by S. K. Gardner and P. Mendoza. Sterling, VA: Stylus. 79–95.

Humphrey, R. and Simpson, B. 2012. "Writes of Passage: Writing up Qualitative Data as a Threshold Concept in Doctoral Research." *Teaching in Higher Education* 17 (6): 735–746.

Jackson, N., and Shaw, M. 2006. "Developing Subject Perspectives on Creativity in Higher Education. In *Developing Creativity in Higher Education: An Imaginative Curriculum*, edited by N. Jackson, M. Oliver, M. Shaw, and J. Wisdom. London: Routledge. 89–108.

Kiley, M. 2009. "Identifying Threshold Concepts and Proposing Strategies to Support Doctoral Candidates." *Innovations in Education and Teaching International* 46 (3): 293–304.

Kiley, M. and Wisker, G. 2009. "Threshold Concepts in Research Education and Evidence of Threshold Crossing." *Higher Education Research and Development* 28 (4): 431–441.

Kuhn, T. S. (1962). *The Structure of Scientific Revolutions*. Chicago: University of Chicago Press.

Land, R., Meyer, J.H.F., and Smith, J. (Eds.). 2008. *Threshold Concepts within the Disciplines*. Rotterdam: Sense.

Lee, A., and Boud, D. 2009. "Framing Doctoral Education as Practice." In *Changing Practices of Doctoral Education,* edited by D. Boud and A. Lee. Abingdon, Oxon, New York: Routledge. 10–25.

Lovitts, B. E. 2005. "Being a Good Course-Taker Is Not Enough: A Theoretical Perspective on the Transition to Independent Research." *Studies in Higher Education* 30 (2): 137–154.

Lovitts, B. E. 2007. *Making the Implicit Explicit: Creating Performance Expectations for the Dissertation.* Sterling, VA: Stylus.

Lovitts, B. E. 2008. "The Transition to Independent Research: Who Makes It, Who Doesn't, and Why." *Journal of Higher Education* 79 (3): 296–325.

McCormack, B., and Titchen, A. 2006. "Critical Creativity: Melding, Exploding, Blending." *Educational Action Research* 14 (2): 239–266.

Meyer, J.H.F., and Land, R. (Eds.). 2006. *Overcoming Barriers to Student Understanding: Threshold Concepts and Troublesome Knowledge.* Abingdon, Oxon: Routledge.

Moore, T. 2013. "Critical Thinking: Seven Definitions in Search of a Concept." *Studies in Higher Education* 38 (4): 506–522.

Mullins, G., and Kiley, M. 2002. "'It's a PhD, Not a Nobel Prize': How Experienced Examiners Assess Research Theses." *Studies in Higher Education* 27 (4): 369–386.

Ragsdell, G. 1998. "Participatory Action Research and the Development of Critical Creativity: A 'Natural' Combination?" *Systemic Practice and Action Research* 11 (5): 503–515.

Ragsdell, G. 2001. "From Creative Thinking to Organisational Learning via Systems Thinking? An Illustration of Critical Creativity." *Creativity and Innovation Management* 10 (2): 102–109.

Sjöstedt Landén, A. (2012). Sammanläggningsavhandling i monografisk tradition? Reflektioner kring metod och etik. [Thesis by publication in monographic tradition? Reflections upon methods and ethics.] In *Mitt i metoden: Kulturvetenskapliga reflektioner,* edited by B. Nilsson and A. S. Lundgren. Umeå: Institutionen för kultur- och medievetenskaper, Umeå universitet. 71–82.

Sternberg, R. J., and Lubart, T. I. 1995. *Defying the Crowd: Cultivating Creativity in a Culture of Conformity.* New York: The Free Press.

Titchen, A., and McCormack, B. 2010. "Dancing with Stones: Critical Creativity as Methodology for Human Flourishing." *Educational Action Research* 18 (4): 531–554.

Trafford, V. and Leshem, S. 2009. "Doctorateness as a threshold concept." *Innovations in Education and Teaching International* 46 (3): 305–316.

Vetenskapsrådet. (2006). *Svenska avhandlingars kvalitét och struktur: Har den ökade volymen på forskarutbildningen påverkat kvalitén på svensk forskning? En bibliometrisk analys.* [Quality and structure of Swedish dissertations: Has the increased volume of doctoral education influenced the quality of Swedish research? A bibliometric analysis]. Stockholm: Vetenskapsrådet.

Walters, K. S. 1994. "Introduction: Beyond Logicism in Critical Thinking." In *Re-Thinking Reason: New Perspectives in Critical Thinking,* edited by K. S. Walters. Albany: State University of New York Press.

Weidman, J. C. 2010. "Doctoral Student Socialization for Research." In *On Becoming a Scholar: Socialization and Development in Doctoral Education,* edited by S. K. Gardner and P. Mendoza. Sterling, VA: Stylus. 45–55.

Wisker, G., and Robinson, G. 2009. "Encouraging Postgraduate Students of Literature and Art to Cross Conceptual Thresholds." *Innovations in Education and Teaching International* 46 (3): 317–330.

17

The Effectiveness of Instruction in Critical Thinking

David Hitchcock

Undergraduate instruction in critical thinking is supposed to improve skills in critical thinking and to foster the dispositions (i.e., behavioral tendencies) of an ideal critical thinker. Students receiving such instruction already have these skills and dispositions to some extent, and their manifestation does not require specialized technical knowledge. Hence it is not obvious that the instruction actually does what it is supposed to do.

In this respect, critical thinking instruction differs from teaching specialized subject matter not previously known to the students, for example, organic chemistry or ancient Greek philosophy or eastern European politics. In those subjects, performance on a final examination can be taken as a good measure of how much a student has learned.

A good examination of critical thinking skills, on the other hand, will not be a test of specialized subject matter. Rather, it will ask students to analyze and evaluate, in a way that the uninitiated will understand, arguments and other presentations of the sort they will encounter in everyday life and in academic or professional contexts. Performance on such a test may thus reflect the student's skills at the start of the course rather than anything learned in the course. If there is improvement, it may be due generally to a semester of engagement in undergraduate education rather than specifically to instruction in critical thinking. There may even be a deterioration in performance from what the student would have shown at the beginning of the semester.

Measuring instructional effectiveness

We therefore need well-designed studies of the effectiveness of undergraduate instruction in critical thinking. An ideal design would take a representative sample of undergraduate students. It would then divide them randomly into two groups, an intervention group and a control group. The intervention group would receive the critical thinking instruction. The control group would receive a substitute that is assumed to have no effect on the outcomes of interest, an educational placebo. Otherwise the groups would be treated the same way. Each group would be tested before and after the instructional period by a validated test of the outcomes of interest. If the intervention group on average improves more than the control group, and the difference is "statistically significant," then the critical thinking instruction has in all probability achieved the desired effect, to roughly the degree indicated by the difference in average gains.

A similar design could be used to investigate whether one method of teaching critical thinking is more effective than another. The two groups would both receive instruction in critical thinking, but by different methods. If one group has higher mean gains than another, and the difference is statistically significant, then its method is probably more effective than the method used with the other group.

Practical constraints make such ideal designs impossible. Students register in the courses they choose. They cannot be allocated randomly to an intervention group and to a control group getting an educational placebo. Even random allocation to two groups exposed to different methods of instruction is difficult.

A standard design therefore administers to a group of students receiving critical thinking instruction a pre-test and a post-test using a validated instrument for testing critical thinking skills. For comparison, one can use a non-randomized control group, such as a class of undergraduate students who are not receiving the critical thinking instruction but who are generally similar in other respects. With such a purpose-built control group, one can compensate for the likelihood that the control group does not perfectly match the intervention group at pre-test by controlling statistically for known differences that exist then. This approach allows for more robust inferences of causation than a simple pre-post design with no control group. An example of such a study is Facione (1990a), where the intervention group consisted of students in 39 sections of courses approved as meeting a critical thinking requirement and the control group consisted of students in six sections of an introductory philosophy course.

A simpler design tests critical thinking skills before and after an instructional intervention, with no control group. In the absence of a control group, reported gains should be reduced by the best available estimate of the gains that the students would have made without the critical thinking instruction. Such gains would presumably be due to such factors as full-time university study, maturation, and familiarity with the test.

Whatever the study design, statistically significant differences are not necessarily educationally meaningful. With large groups, even slight differences will be statistically significant, but they will not reflect much difference in educational outcome. Judgment is required to determine how much of a difference is educationally meaningful or important.

A rough estimate of educational significance can be provided by a statistic known as Cohen's *d* (Cohen 1988, 24–27). To calculate this statistic, one needs an estimate of the standard deviation of scores on the test one is using in the "population" or "universe" to which one wishes to project one's results. (The standard deviation [SD] is a measure of the spread of scores around the mean, or average. A high SD means that the scores are widely spread, a low SD that they are bunched closely around the mean. One's "universe" should be the group from which one's "sample" has been selected. The sample should be representative of the universe in relevant respects.)

Cohen's *d* is a simple comparison of a difference (such as a difference in mean scores) to this standard deviation. One divides the difference by the standard deviation to get its quantity as a fraction of a single standard deviation. This fraction is commonly called the "effect size."

In a simple pre-test, post-test design with no control group, if SD_t is the standard deviation on the test used and μ_{pre} and μ_{post} are the mean scores on the pre-test and post-test respectively, then Cohen's *d* is given by the formula $(\mu_{post} - \mu_{pre})/SD_t$. For example, if the mean score on the post-test is 19, the mean score on the pre-test is 17, and the estimated standard deviation in the population is 4, then the effect size is (19–17)/4, or 0.5 SD–half a standard deviation.

As a rule of thumb, a difference of half a standard deviation (0.5 SD) is a medium effect size. Norman, Sloan, and Wyrwich (2003) report that minimally detectable differences in health studies using a variety of measurement instruments average half a standard deviation. They explain this figure by the fact, established in psychological research, that over a wide range of tasks the limit of people's ability to discriminate is about 1 part in 7, which is very close to half a standard deviation.

Roughly speaking, a difference of 0.8 SD is a large effect size.

Besides giving a rough sense of educational significance of an intervention, Cohen's *d* has the advantage of allowing comparison of effect sizes in studies using different tests. The scoring system of a particular test drops out of the picture and is replaced by an effect size expressed as a fraction of a standard deviation.

Effectiveness of computer-assisted instruction in critical thinking

With the widespread diffusion of the personal computer, and financial pressures on institutions of higher education, instructors are relying more and more on drill-and-practice software, some of which has built-in tutorial help. This software can reduce the labor required to instruct the students; at the same time, it provides immediate feedback and necessary correction in the context of quality practice, which some writers (e.g., van Gelder 2000; 2001) identify as the key to getting substantial improvement in critical thinking skills. In addition, well-designed software can enhance the intrinsic motivation that tends to promote learning more than external motivation (Lepper and Greene 1978). It does so by giving users optimal degrees of control, challenge, and stimulation of curiosity (Larkin and Chabay 1989).

Does the use of such software result in greater skill development, less, or about the same? Can such software completely replace the traditional labor-intensive format of working through examples in small groups and getting feedback from an expert group discussion leader? Or is it better to combine the two approaches?

Computational assistance can also reduce the labor of marking students' work. Can machine-scored testing, in multiple-choice or other formats, completely or partially replace human grading of written answers to open-ended questions?

Answers to such questions can help instructors and academic administrators make wise decisions about formats and resources for undergraduate critical thinking instruction.

An opportunity to answer some of these questions came when face-to-face tutorials in a critical thinking course at McMaster University in Hamilton, Canada, were replaced with computer-assisted instruction with built-in tutorial help. The grade depended entirely on multiple-choice testing. To judge the effectiveness of the new design, the students' critical thinking skills were tested at the beginning and at the end of one offering of the course.

At the first meeting the course outline was reviewed and a pre-test announced, to be administered in the second class. Students were told not to do any preparation for this test. In the second class students wrote as a pre-test either Form A or Form B of the California Critical Thinking Skills Test (CCTST). There followed 19 lectures of 50 minutes each, that is 15.8 hours of critical thinking instruction. In the second-last class, students wrote as a post-test either Form A or Form B of the CCTST. The last class reviewed the course and explained the final exam format.

There were no tutorials. Two graduate teaching assistants and the instructor were available for consultation by e-mail (monitored daily) or during office hours. These opportunities were used very little, except just before term tests. The course could have been (and subsequently was) run just as effectively with one assistant. Review sessions before the mid-term and final examination were attended by about 10% of the students. Two assignments, the mid-term and the final examination were all in machine-scored multiple-choice format. There was no written graded work.

Students used as their textbook Jill LeBlanc's *Critical Thinking* (LeBlanc 1998b), along with its accompanying software LEMUR (LeBlanc 1998a), an acronym for Logical Evaluation Makes Understanding Real. The course covered nine of the textbook's ten chapters, with the following topics: identifying arguments, standardizing arguments, necessary and sufficient conditions, language (definitions and fallacies of language), accepting premises, relevance, arguments from analogy, arguments from experience, causal arguments. There were two multiple-choice assignments, one on distinguishing arguments from causal explanations and standardizing arguments, the other on arguments from analogy. The mid-term covered the listed topics up to and including accepting premises. The final exam covered all the listed topics.

The software LEMUR consists of multiple-choice exercises and quizzes tied to the book's chapters, with tutorial help in the form of explanations and hints if the user chooses an incorrect answer. If the user answers an item correctly, there is often an explanation why that answer is correct. As readers can confirm for themselves, working one's way through the exercises provides immediate feedback that shapes one's future answers. One can observe oneself correcting one's misunderstandings and improving one's performance as one goes along.

LEMUR's argument standardization exercises have pre-structured box-arrow diagrams into whose boxes students can drag the component sentences of an argumentative text so as to exhibit its argument structure graphically. It is possible to construct original diagrams in more sophisticated software, such as Athena Standard (Bertil 2005), Araucaria (Reed and Rowe 2012), and Rationale (Donohue, van Gelder, Cumming, and Bissett 2002; van Gelder 2013).

There was a Web site for the course, on which answers to the textbook exercises were posted, as well as past multiple-choice assignments, tests, and exams with answers, along with other help. There was no monitoring of the extent to which a given student used the software or the Web site.

To encourage students to do their best on both the pre-test and the post-test, 5% of the final grade was given for the better of the two marks received. If one of the two tests was not written the score on the other test was used, and if neither test was written the final exam counted for an additional 5%. In accordance with the test manual, students were not told anything in advance about the test, except that it was a multiple-choice test. A few students who asked what they should do to study for the post-test were told simply to review the material for the entire course. Students had about 55 minutes on each administration to answer the items, slightly more than the 45 minutes recommended in the manual.

The original intention was to use a simple crossover design, with half the students writing Form A as the pre-test and Form B as the post-test, and the other half writing Form B as the pre-test and Form A as the post-test. This design automatically corrects for any differences in difficulty between the two forms. As it turned out, far more students wrote Form A as the pre-test than Form B, and there were not enough copies of Form B to administer it as a post-test to those who wrote Form A as the pre-test. Hence the Form A pre-test group was divided into two for the post-test, with roughly half of them writing Form B and the rest writing Form A again. This design made it possible to determine whether it makes any difference to administer the same form of the test as pre-test and post-test, as opposed to administering a different form.

Of the 402 students who completed the course, 278 wrote both the pre-test and the post-test. Their average score went from 17.03 out of 34 on the pre-test to 19.22 on the post-test, an increase of 2.19 points, which corresponded to half a standard deviation (.49 SD, to be precise). Thus the course had a moderate "effect size." More detailed information about the results can be found in (Hitchcock 2004), on which the present chapter is based.

It made no difference to the gain in average score whether students wrote the same form at post-test as at pre-test. Form B was slightly harder than Form A: the students who wrote Form B first and Form A second (the "BA" group) had a somewhat bigger average gain than those who wrote Form A first and Form B second (the "AB" group). (Jacobs [1995, 94; 1999, 214] also found that students did somewhat worse on Form B than on Form A.) The gain in average score among students who wrote Form A both times (the "AA" group) fell squarely in between the mean gains among the AB and BA students (see figure 17.1). Thus there was no difference between writing the same form of the test twice and writing a different form in the post-test. As the test manual reports, "We have repeatedly found no test effect when using a single version of the CCTST for both pre-testing and post-testing. This is to say that a group will not do better

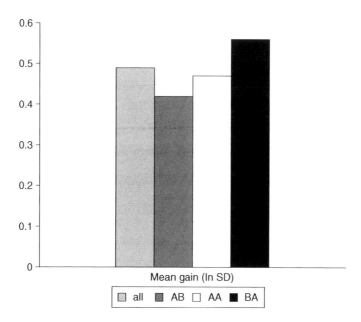

Figure 17.1   Mean gain (in SD) by group in the McMaster study.

on the test simply because they have taken it before" (Facione, Facione, Blohm, Howard, and Giancarlo 1998, 14).

These results raise two main interpretive questions. First, how much of the improvement in test scores can be attributed specifically to the critical thinking instruction? Second, how does the improvement compare to the improvement after other ways of teaching critical thinking?

First, to determine how much of the improvement can be attributed to the critical thinking course, we need to subtract the improvement that the students would have shown if they had been taking some other course instead. Pascarella and Terenzini (2005) estimate, on the basis of a synthesis of studies done in the 1990s, that the first three years in college provide an improvement in critical thinking skills of about .55 of a standard deviation (SD). Most of the gains occur in a student's first year of college. They estimate the sophomore advantage over freshmen at .34 SD, the junior advantage over freshmen at .45 SD, and the senior advantage over freshmen at .54 SD. If we assume that in each year gains are distributed evenly between the two semesters, we can estimate that on average college students gain .17 SD in each semester of their first year in college and .05 SD in each subsequent semester of undergraduate studies. Hitchcock (2004) reports other evidence consistent with this estimate.

Almost all the students in the present study were registered in Level 2 or above. Thus they would be expected to improve their scores on a critical thinking test by .05 SD with a semester of full-time study that did not include a critical thinking course. So almost all their gain of .49 SD can be attributed to their computer-assisted critical thinking instruction–.44 SD, to be exact. This is still close to a moderate effect size.

Second, similar studies, all of which used the CCTST, have found mean gains following a one-semester critical thinking course ranging from .32 SD to .89 SD (Hitchcock 2004). These studies investigated three different methods of critical thinking instruction.

Traditional design: An instructor teaches a small group (25 to 30 students) for a semester and marks assigned exercises. There are no tutorials and no computer-assisted instruction or marking. Assignments, tests, and exams require written answers, marked manually. Studies of this type of instruction show gains ranging from .28 SD (Twardy 2004) to .32 SD (Facione 1990a). Since the students in these studies are mostly beyond their first year, and so would be expected otherwise to show a gain of .05 SD, the gain attributable to the critical thinking instruction is about a quarter of a standard deviation, a small effect.

Full-year freshman course combining critical thinking and writing instruction: An instructor teaches critical thinking to a group of 20 students for seven weeks. Subsequently the group receives instruction in writing skills and writes a series of five essays. Studies of this type of instruction show gains ranging from .46 to .75 SD (Hatcher 1999; 2001 personal communication). Since freshmen can

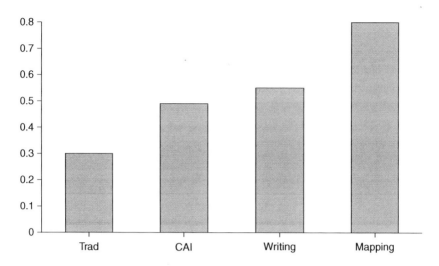

Figure 17.2   Mean gain (in SD) by design.

be expected to improve their scores by .34 SD during an academic year even without specific critical thinking instruction, the contribution of the critical thinking instruction in this design should be estimated at .12 to .41 SD, a small to moderately small effect.

One-semester freshman course using computer-assisted argument mapping: An instructor teaches methods of analyzing arguments to a large class (135 students in one case), with the students meeting in small tutorial groups (15 students on average) once a week. Almost the entire semester is devoted to argument mapping using computer assistance to produce box-arrow diagrams of argument structure. Studies of this type of instruction show gains ranging from .73 SD to .89 SD. Allowing for the expected gain otherwise of .17 SD in one semester of first-year undergraduate education, we can attribute to the critical thinking instruction in this design an effect of .56 SD to .72 SD, which is moderately large.

If we amalgamate the results of these studies, as displayed in Table 1 of (Hitchcock 2004, 188), we get the pattern displayed in figure 17.2. The highly efficient computer-assisted instructional design of the McMaster course is more effective than traditional critical thinking courses, about as effective as a freshman course combining critical thinking and writing, and less effective than computer-assisted instruction focused on argument mapping.

Comparative effectiveness of different methods

Despite the optimism of such titles as "Why Critical Thinking Should Be Combined with Written Composition" (Hatcher 1999) and "Argument Maps

Improve Critical Thinking" (Twardy 2004), the studies just mentioned do not establish conclusively what instructional methods are most effective at improving the critical thinking skills of undergraduate students.

For one thing, the groups studied differ in many ways: the instructor's academic background and experience, the topics, the textbook, the feedback to students, the incentives for taking the pre-test and post-test seriously, the students' majors and levels of registration, their facility with multiple-choice tests, their academic ability, their critical thinking skills at pre-test, and so on.

For another, one can raise questions about the validity of the CCTST (Facione 1990a; b; c; d; Facione et al. 1998), that is, whether it really measures critical thinking skills. The CCTST is based on an expert consensus statement of the critical thinking skills that might be expected of college freshmen and sophomores (American Philosophical Association 1990). Its 34 items, however, test only some of the skills mentioned in this statement. There are also legitimate questions about the soundness of some of its items. Further, other conceptualizations of critical thinking, such as those of Robert Ennis (1962; 1987; 1991) or of Alec Fisher and Michael Scriven (1997), imply a somewhat different set of critical thinking skills.

To address these concerns, which are described in detail in Hitchcock (2004), we need better validated tests of critical thinking skills. And we need studies like those described in this chapter with different groups of students receiving critical thinking instruction with different content from different instructors using different methods. Such studies should use a nonrandomized control group that permits a covariance analysis to control for differences at pre-test with the experimental group. They should report on the topics covered, the textbook used, the types of work used to determine the students' grade in the course (in particular, the balance between essay-type questions, short-answer questions, and multiple-choice items), class size, the instructor's relevant training and experience, the students' level of registration, the students' verbal and mathematical aptitude, the percentage of students whose mother tongue is not English, the instrument used at pre-test and post-test, incentives to do well on the pre-test and the post-test, and the stage of the course at which the post-test was given.

Useful guidance on exploring different instructional designs can come from a systematic meta-analysis by Philip Abrami and his colleagues (Abrami, Bernard, Borokhovski, Waddington, Wade, and Persson 2014) of the effect of instructional interventions on generic critical thinking skills. As a measure of effect size, they modified Cohen's d to correct for bias in small samples. In 684 studies, with 867 effect sizes, they found an average effect size of .39. This low-to-moderate effect size suggests that it is possible to teach generic critical thinking skills.

For more detailed analysis, Abrami and his colleagues confined their attention to true experiments or quasi experiments where the particularities of the intervention could be confidently identified and standardized outcome measures with determinate reliability and validity were used. Also, studies were removed where the intervention was over a semester long. This collection of more methodologically sound studies included 341 effect sizes with a somewhat lower average of .30, with high heterogeneity. Effect sizes in these studies did not differ significantly by educational level, subject-matter, or duration of the intervention.

In an earlier preliminary meta-analysis of fewer studies, Abrami, Bernard, Borokhovski, Wade, Surkes, Tamim, and Zhang (2008) were able to account for 32% of the variance in effect sizes by two features of the intervention: its pedagogical grounding and its type. The most effective type of pedagogical grounding, with an average effect size of 1.00, was special training of the instructor for teaching critical thinking. The most effective type of intervention, with an average effect size of .94, was a "mixed intervention" (Ennis 1989) combining subject-matter instruction with a unit devoted specifically to critical thinking. This type of intervention was also the most effective in the more recent meta-analysis (Abrami et al. 2014), but the average effect size of .38 did not differ significantly from the effect size with other types of intervention.

The more recent meta-analysis (Abrami et al. 2014) did not analyze their data by pedagogical grounding. Instead, they used a second set of instructional variables: dialogue, anchored instruction, and coaching. Dialogue involves learning through discussion. Anchored instruction (also called "authentic instruction") presents students with problems that make sense to them, engage them, and stimulate them to inquire. In coaching (also called "mentoring" or "tutoring"), someone with more expertise models a task to a "novice" with less expertise and then corrects the novice's errors based on critical analysis. A combination of all three of these strategies produced the highest effect size: .57, compared to .32 for a combination of anchored instruction and dialogue, .25 for anchored instruction alone, and .23 for dialogue alone.

Since these meta-analyses combined studies of interventions at various educational levels, their implications for undergraduate instruction in critical thinking are not straightforward. In particular, neither meta-analysis analyzed the results by whether the instructional design included argument mapping, which some studies of critical thinking instruction of undergraduates have shown to be particularly effective. But the meta-analyses suggest exploration of mixed designs with explicit teaching of critical thinking, in the context of subject-matter instruction, by an instructor specially trained for teaching critical thinking, where students engage in dialogue, apply the skills being taught to problems that engage them, and have some individual coaching. The substantial unexplained heterogeneity in both meta-analyses reinforces the need for further well-designed experimental and quasi-experimental studies in which the instructional strategies and instructor and student characteristics in the intervention and control groups are fully described.

Summary

One way to measure the effectiveness of an instructional intervention in improving critical thinking skills is to compare the mean gain of its recipients, on a validated test of critical thinking skills, to the mean gain of a control group. Studies of this kind have shown that traditional stand-alone undergraduate critical thinking courses tend to produce only a small improvement. There tends to be moderate improvement when such courses involve computer-assisted tutoring or are combined with writing instruction and practice. The largest improvements have been found mainly in courses that focus on computer-assisted argument mapping. In addition, two recent meta-analyses suggest that the most effective method may be a unit of critical thinking instruction by a purpose-trained instructor in the context of subject-matter instruction with student discussion, engagement with a problem, and coaching.

References

Abrami, P. C., Bernard, R. M., Borokhovski, E., Wade, A., Surkes, M. A., Tamim, R., and Zhang, D. 2008. "Instructional Interventions Affecting Critical Thinking Skills and Dispositions: A Stage 1 Meta-Analysis." *Review of Educational Research* 78 (4): 1102–1134.

Abrami, P. C., Bernard, R. M., Borokhovski, E., Waddington, D. I., Wade, C. A., and Persson, T. 2014. "Strategies for Teaching Students to Think Critically: A Meta-Analysis." *Review of Educational Research*. Advance online publication retrieved from http://rer.sagepub.com. doi:10.3102/0034654314551063.

American Philosophical Association. 1990. "Critical Thinking: A Statement of Expert Consensus for Purposes of Educational Assessment and Instruction." ED 315423. ERIC Document.

Bertil, R. *Athena Standard*, Software accessed August 16, 2013 2005. Available from http://theguide.ntic.org/display_lo.php?oai_id=oai%3Aeureka.ntic.org%3A4c9917 65e30281.68868667.

Cohen, J. 1988. *Statistical Power Analysis for the Behavioral Sciences*. (second edition). Hillsdale, NJ: Lawrence Erlbaum.

Donohue, A., van Gelder, T. J., Cumming, G., and Bissett, M. 2002. "Reason! Project Studies 1999–2002 (Reason! Project Technical Report 2002/1)." Melbourne. August 18, 2013. http://www.kritischdenken.nl/Rationale/Tutorials/Overview/ReasonProject Report2002_1.pdf.

Ennis, R. H. 1962. "A Concept of Critical Thinking: A Proposed Basis for Research in the Teaching and Evaluation of Critical Thinking Ability." *Harvard Educational Review* 32: 81–111.

Ennis, R. H. 1987. "A Taxonomy of Critical Thinking Dispositions and Abilities." In *Teaching Thinking Skills: Theory and Practice*, edited by J. B. Baron and R. J. Sternberg. New York: W. H. Freeman and Company. 9–26.

Ennis, R. H. 1989. "Critical Thinking and Subject Specificity: Clarification and Needed Research." *Educational Researcher* 18 (3): 4–10.

Ennis, R. H. 1991. "Critical Thinking: A Streamlined Conception." *Teaching Philosophy* 14 (1): 5–25.

Facione, P. 1990a. "The California Critical Thinking Skills Test: College Level, Technical Report #1: Experimental Validation and Content Validity." ED 327549. ERIC Document.

Facione, P. 1990b. "The California Critical Thinking Skills Test: College Level, Technical Report #2: Factors Predictive of CT Skills." ED 327550. ERIC Document.

Facione, P. 1990c. "The California Critical Thinking Skills Test: College Level, Technical Report #3: Gender, Ethnicity, Major, CT Self-Esteem and the CCTST." ED 326584. ERIC Document.

Facione, P. 1990d. "The California Critical Thinking Skills Test: College Level, Technical Report #4: Interpreting the CCTST, Group Norms and Sub-Scores." ED 327566. ERIC Document.

Facione, P. A., Facione, N. C., Blohm, S. W., Howard, K., and Giancarlo, C. F. 1998. *Test Manual: The California Critical Thinking Skills Test* (revised edition). Millbrae: California Academic Press.

Fisher, A., and Scriven, D. 1997. *Critical Thinking: Its Definition and Assessment.* Point Reyes, CA: Edgepress.

Hatcher, D. 1999. "Why Critical Thinking Should Be Combined with Written Composition." *Informal Logic* 19: 171–183.

Hatcher, D. 2001. "Why Percy Can't Think: A Response to Bailin." *Informal Logic* 21: 171–181.

Hitchcock, D. 2004. "The Effectiveness of Computer Assisted Instruction in Critical Thinking." *Informal Logic* 24 (3): 183–217.

Jacobs, S. S. 1995. "Technical Characteristics and Some Correlates of the California Critical Thinking Skills Test, Forms A and B." *Research in Higher Education* 36: 89–108.

Jacobs, S. S. 1999. "The Equivalence of Forms A and B of the California Critical Thinking Skills Test." *Measurement and Evaluation in Counseling and Development* 31: 211–222.

Larkin, J. H., and Chabay, R. W. 1989. "Research on Teaching Scientific Thinking: Implications for Computer-Based Instruction." In *Toward the Thinking Curriculum: Current Cognitive Research*, edited by L. B. Resnick and L. E. Klopfer. Alexandria, VA: Association for Supervision and Curriculum Development. 150–172.

LeBlanc, J. 1998a. *LEMUR (Logical Evaluation Makes Understanding Real) to Accompany Thinking Clearly.* New York: W. W. Norton.

LeBlanc, J. 1998b. *Thinking Clearly: A Guide to Critical Reasoning.* New York: W. W. Norton.

Lepper, M. R., and Greene, J. A. 1978. *The Hidden Costs of Reward: New Perspectives on the Psychology of Human Motivation.* Hillsdale, NJ: Lawrence Erlbaum.

Norman, G. R., Sloan, J. A., and Wyrwich, K. W. 2003. "Interpretation of Changes in Health-Related Quality of Life: The Remarkable Universality of Half a Standard Deviation." *Medical Care* 41: 582–592.

Pascarella, E., and Terenzini, P. 2005. *How College Affects Students: Findings and Insights from Twenty Years of Research.* Vol. 2: A Third Decade of Research. San Francisco: Jossey Bass.

Reed, C., and Rowe, G. 2012. *Araucaria 3.1* August 16. Available from http://araucaria. computing.dundee.ac.uk/doku.php.

Twardy, C. R. 2004. "Argument Maps Improve Critical Thinking." *Teaching Philosophy* 27: 95–116.

van Gelder, T. 2001. "How to Improve Critical Thinking Using Educational Technology." In *Meeting at the Crossroads: Proceedings of the 18th Annual Conference of the Australasian Society for Computers in Learning in Tertiary Education,*edited by G. Kennedy, M. Keppell, C. McNaught and T. Petrovic. University of Melbourne. 539–548.

van Gelder, T. J. 2000. "Learning to Reason: A Reason! Able Approach." In *Cognitive Science in Australia 2000: Proceedings of the Fifth Australasian Cognitive Science Society,* edited by C. Davis, T. J. van Gelder, and R. Wales. Adelaide: Causal.

van Gelder, T. J. 2013. *Rationale 2.0.10.* August 16. Available from http://rationale .austhink.com/download.

Part IV
Critical Thinking and Culture

Does critical thinking remain culturally appropriate to the higher education curriculum in the twenty-first century?

The changing face of international higher education is increasingly becoming Asia-focused, and non-Anglo-centric. Can and should critical thinking retain its historic place as a fundamental aspect of a Western-style university education? Even if it does, will this require compromise? What does "critical thinking" mean in the developing regions of the world such as the East and the Middle East? No account of critical thinking in higher education can progress without grappling with some of these key issues.

The difficulties here are partly pedagogical, and partly political. In the past, there has been a hesitance on the part of the Chinese government to adopt Western-style cultural norms (e.g., an unwillingness to foster a culture of debate and free exchange of ideas). This is now changing. However, old habits die hard, and it may no longer be possible for critical thinking to be taught in traditional "Western" ways, even if critical thinking remains a part of the curriculum. Asia, and Asian concerns, will increasingly dominate our educational future.

What do we make of countries in the Middle East in this respect? Countries of this region have undergone much turmoil and upheaval since the Arab Spring. Is it too much to hope for critical thinking to have a renaissance in this region of the world? And what is to be made of criticality in relation to indigenous populations in countries like Australia, Africa, and Canada? This section of the book offers four stimulating papers on the topic of critical thinking and culture.

Manalo, Kusumi, Koyasu, Michita, and Tanaka raise the vexed issue of whether critical thinking is understood differently across cultures. There is much literature on this, but few data-based studies. Using a qualitative methodology, rich in commentary, they survey the views of twenty-three undergraduate students in New Zealand, and from similar institutions in different cities in Japan. Contrary to prevailing intuitions and attitudes about Asian students being "passive" and uncritical learners, they found no evidence of an East-West difference in the importance, use, and definitions of thinking skills, including that of critical thinking. Does it follow from this that such assumptions are misguided and uninformed?

Bali looks at critical thinking in Egypt, and Islamic societies in general. She argues that critical thinking is not anathema to Islamic scholarship, nor unique to Western countries, though she agrees that there are certainly issues that inhibit the teaching of critical thinking in the Muslim world. These include the notion of "cultural capital," the extent to which there has been previous exposure to critical thinking in domestic, social, or pre-tertiary educational contexts. Variability in cultural capital is a factor that impinges on confidence in critical thinking, and the ability to "speak-out." A second issue is pedagogies that are not culturally neutral. In many cases, teachers revert to norms of teaching that are culturally acceptable. It takes additional effort for teachers in Islamic countries to break the mold. Linguistic competence is a third issue as this prevents the development of criticality. None of these issues are unique to Egypt, of course, but there is more. The fourth issue is the sociopolitical environment, which in the case of many countries in the Middle East is increasingly destabilized and fraught with conflict and danger. In such circumstances, teaching critical thinking by means of fostering skepticism and antagonistic questioning is likely to be viewed with deep suspicion. A better approach, Bali argues, might be to approach critical thinking by means of invoking human empathy and a sense of social justice.

Huijser and Chirgwin outline a "both-ways" approach to critical thinking acknowledging the values of indigenous cultures and belief systems. They attempt to construct a criticality of "hybridity" that places Western-style criticality on a similar footing with deeply held intergenerational systems of thinking accepted by these ancient cultures. Making mention of the suspicion with which Western education is viewed, as a "residual colonial legacy," they argue that learning critical thinking using a "both-ways" approach can foster cultural code-switching. This enables participants to lower their defenses, appreciate the value of what Western critical thinking offers, and allows the student to function effectively both in Western and indigenous worlds, which are essential skills for the twenty-first century.

Dong investigates critical thinking in China. He draws attention to an educational paradox: China has made great strides in introducing critical thinking in the curriculum along with a degree of political openness, massive investment in education, material development, and increasing prosperity. Despite this, the spread of critical thinking has been slow, and the quality of instruction is patchy. Why is this? Dong claims that the introduction of critical thinking is beset by what he calls "Chinese characteristics." These include traditional exam-orientated goals, irrelevance of content to practice, rote-learning pedagogies, and inadequate critical content. Rejecting the common view that "collectivist" attitudes are responsible, Dong suggests that the inhibiting factor is the concept of 'truth as dogma.' This normative Confucian view, he suggests, is "not subject to falsification...is not a reality-based truth statement." It undergirds a distinctively Chinese approach to education, that is, teachers as omniscient sages who dispense "truths." Changing this is difficult as it requires more than improved inquiry-based instruction in critical thinking. Instead, it requires wholesale cultural transformation.

Biggs (1997) noted that a "conceptual colonialist"' view has prevailed in many Western tertiary institutions in relation to the teaching of international students. The assumption was that non-Western cultures are bereft of criticality. This has led to work on how best to "assimilate" international students into the culture of critical thinking via various kinds of academic preparation programs. However, given the dynamics of universities in the twenty-first century, and the pivot to Asia, this approach may need reconsideration. Political expedience—not to mention funding considerations—might result in the much-lauded skill of "critical thinking" being downplayed and deemphasized in the modern university. Some argue that this is already happening (see Cowden and Singh, section 7). What impact will this have on the role of the university in society? Will this mean a reorientation of "Western"-style education away from approaches emphasizing critical thinking as a desirable skill? What will this mean for higher education? What will it mean for teaching critical thinking skills and dispositions? What will it mean for teaching *for* criticality? Or teaching for critical pedagogy? Critical thinking and culture is a vital topic, hence its importance to this volume.

Reference

Biggs, J. (1997) "Teaching Across and Within Cultures: The Issue of International Students." In *Learning and Teaching in Higher Education: Advancing International Perspectives*, edited by R. Murray-Harvey and H. C. Silins. Proceedings of the Higher Education Research & Development Society of Australasia Conference, Adelaide: Flinders.

18

Do Students from Different Cultures Think Differently about Critical and Other Thinking Skills?

Emmanuel Manalo, Takashi Kusumi, Masuo Koyasu,
Yasushi Michita, and Yuko Tanaka

In this chapter, we explore what students from different cultural backgrounds think "good" thinking skills are, including the skills they perceive as being necessary in their studies. We report on findings from focus group interviews we conducted with undergraduate university students from Kyoto and Okinawa in Japan, and from Auckland in New Zealand. What the students said during the interviews shows important similarities in views about what "good thinkers" possess, including many qualities associated with critical thinking such as consideration of different or alternative perspectives. However, when we specifically asked about the meaning of "critical thinking," many of the students from Okinawa indicated uncertainty in their responses, and the students from Auckland and Okinawa also referred to thinking approaches that are not commonly associated with critical thinking such as intuition and positive thinking. The findings from our investigation suggest that students need more explicit instruction to promote critical thinking skills development, and that they should be provided clearer, more transparent explanations of the thinking skills they are expected to demonstrate in their courses of study.

The need for critical thinking

In most educational environments worldwide, it is considered as very important for students to develop critical thinking skills. Cultivating students' abilities to critically assess the soundness of knowledge claims and arguments is one of the most important objectives of education (e.g., Glassner, Weinstock, and Neuman 2005). Documents about tertiary education aspirations, like the Association of American Colleges and University's "College Learning for the New Global Century" (2007), reflect this view. This document, for example, places critical thinking alongside writing and quantitative reasoning (math) as one of the *essential* intellectual skills. Such a view is also held in many universities outside the United States. In New Zealand, for example, critical thinking is listed in the University of Auckland's "graduate profile"—where the University describes the personal qualities, skills, and attributes that students who graduate from there are expected to develop (University of Auckland 2003). Another example is the Asian University for Women in Bangladesh: on its Web site, it declares its intention to equip students with "strong skills in critical thinking, analysis, and communication" (Asian University for Women 2008–2013).

But what exactly does critical thinking mean? One very simple and often-cited definition is it involves "correct assessing of statements" (Ennis 1962, 81), but there are more detailed definitions that describe the thinking processes it entails, such as "skilled and active interpretation of observations and communications, information and argumentation" (Fisher and Scriven 1997, 21). Among researchers, there is some disagreement about the exact meaning of critical thinking, including debates about the nature of the skill itself and the personal qualities that people who exercise critical thinking tend to possess (see, e.g., Halpern 1998; Mason 2007). From an educational perspective, however, the objective of equipping students with a skill they can use both in their studies as well as in society at large would be the most important consideration. Hence, from such a perspective, critical thinking should be viewed in terms of its practical, teachable, and measurable aspects (see, e.g., Ennis 1962; Fisher and Scriven 1997). As various authors have explained (Scharrer, Bromme, Britt, and Stadtler 2012; Thomm and Bromme 2011), one of the biggest challenges in modern society is developing people's skills in evaluating and making sensible judgments about the overwhelming amount of information that is now available through the Internet, television, and other mass media sources. Formal education clearly has a crucial role to play in meeting that challenge.

What students think about thinking skills

Although most tertiary institutions worldwide share the view that developing students' critical thinking skills is important, little is known about the extent to which student share that view. We are not aware, for example, of any study that has in any systematic way examined what students think about critical and other thinking skills they are expected to develop during their years in tertiary studies. One study, which Tapper (2004) conducted, examined undergraduate students' perceptions about the critical thinking components of a course they had taken in science and communication. However, a more general attempt at understanding students' knowledge and views about such thinking skills appear not to have been previously carried out.

Understanding students' views about an educational issue such as the thinking skills they should use is actually quite important. There is much research evidence to show that what students think they are *expected to do* has a huge influence on what they *actually do* in their studies (Jussim and Eccles 1992; Miller and Turnbull 1986). Thus, it is important to know whether students think the same about *thinking skills they need to develop* as expected by the institutions. If there are mismatches between these expectations, appropriate educational measures need to be put in place.

Do cultural differences impact student thinking?

One hotly debated question in this area of research is whether students from different cultural backgrounds might differ in their views and perceptions about thinking skills that are required in their studies. Some authors have argued that critical thinking is largely a Western concept that can be very difficult for students from Asian and other non-Western cultures to learn and use (Atkinson 1997; Fox 1994). Other authors strongly oppose such a view, pointing out that Asian and other non-Western students are equally capable of grasping the requirements of critical thought and demonstrating those in the work they produce (e.g., Paton 2005; Stapleton 2002).

More recent investigations into student critical thinking performance in relation to cultural factors have revealed some differences. However, those differences could have been due to students' language proficiency rather than culture per se. For example, Lun, Fischer, and Ward (2010) found that New Zealand European students performed better than their Asian counterparts on some measures of critical thinking skills, but they pointed out that those differences could have arisen because of the Asian students being less proficient in the English language they had to use in the critical thinking tasks. Floyd (2011) also reported evidence that, for Chinese students, critical thinking was more difficult in a second language (English) compared to their first language (Chinese).

Apart from language proficiency, educational experiences could also affect how students approach thinking tasks. Because of likely differences between cultures in such experiences, it would be helpful to find out whether—as a consequence—students who come from non-Western educational backgrounds might differ in some of their views about the thinking skills needed for success in tertiary studies. In Japan, for example, development of students' thinking skills is emphasized in the education system as much as in other countries. However, critical thinking is not specified as a skill that students need to develop. For instance, the Central Council for Education of Japan's MEXT (Ministry of Education, Culture, Sports, Science and Technology) submitted a report titled "Towards the Enhancement of Undergraduate Education," which described the competencies that students should acquire through a university bachelor's degree. The competencies are based on generic skills (communication skills, logical thinking, and problem-solving skills), knowledge/understanding, and comprehensive learning and its application—but *critical thinking* is never mentioned. Also, the Global Human Resource Development Committee of the Industry-Academia Partnership for Human Resource Development of Japan's METI (Ministry of Economy, Trade and Industry) proposed a set of fundamental competencies that working persons should possess. These include: thinking skills, basic social skills, and execution skills (METI 2010). Again, however, critical thinking was not specified. It is possible that these kinds of differences could result in students forming different views about the kinds of thinking skills they are expected to use to succeed in their studies.

Investigating student perceptions about thinking skills

With these questions in mind, we set out to explore the extent to which students from different cultural backgrounds might hold the same—or different—perceptions and views about thinking skills that are useful in tertiary studies. We conducted focus group interviews with university students in Japan and New Zealand. We decided on this method of data collection (i.e., focus group interviews) because it is generally considered an effective method for understanding the experiences, interests, attitudes, perspectives, and assumptions of a group of people on a specific topic (Wilkinson and Birmingham 2003).

We chose New Zealand as a country that has a Western culture, and Japan as a country that has an Asian culture, but we were aware of the limitations involved in using such classifications. New Zealand, for example, has a multi-ethnic population, comprising not only Europeans and the Maori (the indigenous people of New Zealand) but also various Asian and Polynesian ethnic groups. The students who took part in this study were solicited from a university in Auckland, which is the biggest and most ethnically diverse of the cities in New Zealand. Likewise, although the Japanese student population does not share the same multiethnic characteristics as the student population in New Zealand, Japan is a country with considerable variations in cultural environments according to location and other factors. It is for this reason that we decided to gather data from two locations in Japan—Kyoto and Okinawa—to find out the extent to which student views might be similar or different across "subgroups" within the same overarching "cultural group." Of the two locations, Kyoto is the bigger and more modern city, but at the same time it is one of the best traditionally preserved cities in Japan. In contrast, Okinawa is part of the Ryukyu Islands group in the southern part of Japan with a more ethnically diverse population, being closer to China and other East Asian countries. Since the end of World War II, the United States has also had prominent military bases in Okinawa.

How we conducted our study

In total we had twenty-three undergraduate university students voluntarily taking part in our study. Eight of the students were from a university in Kyoto (two of whom were female, and six were male), seven were from a university in Okinawa (three females, four males), and eight from a university in Auckland (same number of females and males). The students came from a range of subject disciplines: in Kyoto, they came from education, engineering, law, and science; in Okinawa, they were from education, engineering, law and literature, and science; and in Auckland, they were from arts and humanities, engineering, law, and science. The students came from all levels (first year to final year) of their undergraduate degrees. All were in the eighteen to twenty-two age range, except for two students (one male and one female) in Auckland who were already in their thirties.

We held the focus group interviews in a quiet seminar room in the students' home universities. In Kyoto and Okinawa, we interviewed all the students together in one group at each location. However, in Auckland, because of scheduling difficulties, we interviewed the students in two groups, with four students in each group. In Kyoto and Okinawa, the interviews lasted approximately 80 minutes, while in Auckland they lasted approximately 45 minutes. (Note that in Japanese universities, regular class periods usually last 90 minutes, while in New Zealand universities they last 50–55 minutes.)

With the students' permission, we audio-recorded the interviews we conducted in Japanese in Japan, and in English in New Zealand. At the beginning of the interviews, we provided a brief explanation of the study. As a warm-up, we also asked the students some general questions about their interest and participation in courses they were enrolled in, and about the aspects of their courses they were finding interesting. After that, we asked the students the same basic questions in each of the interview sessions. Those questions were:

1. In the courses you are taking, what kinds of thinking are important?
2. In those courses, what kinds of thinking do you think your lecturers expect?
3. In everyday life, what kind of people do you think can be described as good thinkers?
4. To what extent do you think you can use those kinds of thinking skills in university classes and learning activities?
5. When do you think those thinking skills cannot be used?
6. In what ways do you think you can improve those thinking skills?
7. Have you heard of the term "critical thinking"? What do you think it means?

Although we were focused on soliciting the students' answers to these questions, we allowed sufficient flexibility for the students to bring up other comments and ideas as long as they were at least loosely related to these questions. We also made every effort to encourage all the students to contribute, without placing any particular pressure on any of them to contribute if they did not want to.

We transcribed and then thematically analyzed (see, e.g., Boyatzis 1998) the audio recordings to identify patterns of responses to each of the questions in each group. We then compared the patterns of responses across the three groups.

Our results: students' thoughts on thinking skills

In table 18.1 we provide a summary of the key points that the students made in response to each of the questions. We describe and explain these responses in more detail in the following subsections.

What kinds of thinking are important at university?

In response to the first question, all three groups referred to the importance of understanding what is taught in their courses. They expressed this in a variety of ways such as "comprehension of ideas," "considering what the lecturer is saying," and "[paying] attention to the detail and the concepts of your subject." All three groups also referred to the importance of understanding other people's points of view. For example, a student from Kyoto mentioned the value of "thinking from a third person's point of view," while a student from Okinawa indicated "thinking flexibly based on the opposite opinion" as being beneficial. A student from Auckland described basically the same idea as approaching things in an "open-minded" way.

Additionally, students from Kyoto referred to various thinking strategies that involve deeper processing such as abstraction and linking of ideas (e.g., relating data to existing thinking). They also referred to the importance of thinking in a rational manner and observing reality, and of finding things out for oneself and challenging ideas when necessary. One student, for example, described the important aspects of thinking as, "think rationally...think logically using a step-by-step method...develop your own ideas for yourself."

In contrast, students from Okinawa also pointed out the value of being able to express one's own ideas about the subject matter being dealt with in lectures, and of the ability to express such ideas clearly when writing. These students also noted the importance of considering the cultural meaning of ideas taught in courses, and of appreciating the relevance of that cultural meaning to oneself. For instance, one student expressed this point as "considering the cultural meaning based on history, life, and values without prejudice."

Table 18.1 Summary of student responses[a]

Issues	Common responses	Location-specific responses		
		Kyoto	*Okinawa*	*Auckland*
1. Thinking skills important at university	Comprehension of course content Understanding others' perspectives	Deeper thinking processes	Expression of ideas Appreciation of cultural meaning and relevance	Management of course demands Instructor management Study planning
2. Instructor expectations about thinking skills	Development of multiple perspectives Thinking for oneself Questioning	Flexibility in thinking Discovery learning Development of one's own thinking skills	Social responsibilities in thinking	Application of research skills Integration of information learned
3. Characteristics of good thinkers	Logical and systematic Considers different perspectives Reflective Adaptable Able to see the "big picture"	Plans and thinks ahead Considers and listens to others Possesses direction in thinking Understands the rules that apply	Plans and thinks ahead Considers and listens to others	Can generate and organize ideas Metacognitive
4. Thinking skills' application to studies		In classes: Logic Thinking independently Planning In group work: Understanding the way people think	In classes where own thoughts are asked for	Existence of considerable variation in instructor expectations Variations according to subject matter
5. When thinking skills might not apply		In some subjects When time is constrained When deciding emotionally Activities that depend on skill In sports	When things have already been decided When time is constrained When being conciliatory to others	In some lectures, assignments and tests Some instructors do not elicit their use

continued

Table 18.1   Continued

Issues	Common responses	Location-specific responses		
		Kyoto	*Okinawa*	*Auckland*
6. How thinking skills might be improved	Practice Learning more about the skills	Remembering rules and prerequisites	Articulating ideas Being exposed to different perspectives Meeting challenges	Working in interdisciplinary teams
7. Meaning of critical thinking	Reflective and questioning Deliberate and structured thinking Avoids blind acceptance of ideas	Based on principle Being a skeptic Avoidance of stereotyping and prejudice Searching possible options	Generation of ideas Fixing problems in thinking Returning to the origin of an idea	Avoidance of bias Possession of a balanced view Comprehension Being intuitive Positive thinking Negative thinking Being constructive

[a]These responses are based on the issues that were raised by students when answering the questions during the focus group interviews.

Students from Auckland, on the other hand, additionally referred to the more pragmatic aspects of thinking—such as the thinking that one needs to do to manage course demands and lecturer expectations. For example, one student noted that it is important in producing academic work to "try and do your best to make it easy on the [assignment or test] marker as well," while another student referred to the value of "using the lecturer's 'speak'…to try and get inside the lecturer's head and think with the same sort of strategy." Several of the students also noted the importance of, as well as the challenges associated with, planning and executing study-related plans.

What do instructors expect?

The students from the three groups were very similar in the kinds of thinking they believed their instructors expected them to use in their studies. They mentioned the expectation to view things from different perspectives. For example, an Auckland student explained that lecturers expect students to be able to "stand in different points and see different perspectives of the topic or question." A Kyoto student mentioned that instructors expect students to "compare the merits and demerit of ideas." And an Okinawa student explained that their instructors expect students to view things "not from one perspective, but from multiple perspectives."

Another thinking quality that appeared to be commonly expected by instructors across the three locations was "thinking for oneself." For example, an Okinawa student noted that "lecturers expect students to think thoroughly about their own ideas," and a Kyoto student noted that lecturers expect students to be able to reach their "*own* conclusions" about information they get presented. One Auckland student explained,

> You have to know your stuff, basically... What do I mean by "know your stuff"? [You have to] understand the question; do research on the question to find out the positive side, the negative side, and then *your* argument.

Self-questioning was another thinking characteristic that students from all three locations reported that their instructors expected them to utilize. An Auckland student, for example, explained, "They always want us to question what they're teaching us... to ask questions like, does this actually work in real life?" A similar point was raised by an Okinawa student who noted that students are expected to question the applications of the information they are learning, while a Kyoto student expressed the same idea as an expectation to be able to "create questions" about what they are learning.

Students also referred to subject-specific skills, such as those that might apply to chemistry, law, or psychology. Apart from these, however, there were also some expectations mentioned that were specific to each group. For example, the Kyoto students also mentioned that their lecturers expected them to think flexibly, to "discover reality," and to take steps in promoting their own thinking development. The Okinawa students, on the other hand, referred to expectations concerning their relationships with others. For example, they mentioned that their lecturers expected them to think about how to effectively communicate their own ideas to others.

Likewise, there were some expectations that only the Auckland students mentioned. These included applications of research skills: as one student put it, they were expected to "know how to use the library system,...[to] understand all the statistics, how to do it and analyse it...[to] do a literature review." The connection of different ideas was also mentioned: one student, for example, explained that "my lecturer seems to like us to connect the dots...to be able to draw all the threads and make [a] more coherent, big picture."

What characteristics do good thinkers possess?

When asked about what they themselves considered as qualities of "good thinkers," the students from all three locations referred to mostly similar, overlapping qualities. Many of these qualities pertained to applications of skills commonly associated with critical thinking, such as considering different perspectives and taking a logical approach. For example, a Kyoto student suggested the value of "reaching logical conclusions by continuous thinking." An Okinawa student expressed the view that it is useful to "generate new ideas from different aspects [of the subject or topic]." One Auckland student described a related point in the following way, "In approaching an idea…not to just assume what it's saying but…to sort of stand back and suspend judgment and have a capacity to start looking at it from all different angles." Students from all groups also referred to the importance of being able to see "the big picture" or, as one Kyoto student put it, "to grasp things in their entirety."

Students from all three groups also referred to the importance of adaptability in thinking. Suggestions about this point from the Kyoto students included: "those who adapt to requirements," "those who think dynamically," and "those who seek improvement." The point about being able to "improve things" was also mentioned by an Okinawa student, while an Auckland student noted the benefits of being able to "change your way of thinking to match your situation…that's quite important because not all situations demand the same type of thinking." Another Auckland student referred to the value of being able to use both "deductive" and "inductive" approaches to thinking.

Students from Kyoto and Okinawa also noted the importance of planning-related thinking. One Kyoto student, for example, described good thinkers are "those who can structure their time," while another suggested "those who think proactively." An Okinawa student expressed the view that good thinkers are "those who think ahead and act efficiently." Additionally, the Kyoto and Okinawa students pointed out the importance of considering others in the way they think. A Kyoto student, for example, suggested "those who are popular and understand people's thinking," while an Okinawa student suggested "those who really ask and listen to the opinion of others." However, listening to others did not necessarily mean changing one's views according to what others think: as an Okinawa student pointed out, good thinkers can think for themselves and "aren't swayed by others."

The Kyoto students also put forward two additional qualities of good thinkers that were not mentioned by students from the other groups. These were the exhibition of direction in the way one thinks, and understanding the "rules and prerequisite conditions" that apply to the situation in which thinking is called for.

Likewise, the Auckland students suggested several unique attributes of good thinkers. These included being able to generate and organize ideas through the use of techniques like brainstorming and mind mapping. One student also referred to the importance of metacognition, describing it in the following way:

> To be able to assess the usefulness of thinking...being aware of your thinking...thinking about thinking...

How much can those thinking skills be used in university studies?

When asked about the extent to which the good thinking qualities they described might be applicable to university classes and learning activities, the Kyoto students explained that logic, thinking independently, and planning can all be used in study and research activities. They also pointed out and agreed that "understanding the way people think is effective for group work."

When asked the same question, one student in Okinawa responded by saying, "I can use them in classes where my thoughts are asked for." Another student noted that such thinking skills are useful in deciding how to study.

The students from Auckland were more explicit in voicing their perceptions of considerable variations in instructor expectations about the use of such thinking skills. They pointed out that they can apply the skills to their studies, but that some instructors simply require students to demonstrate an "ability to regurgitate information." One student pointed out that "some [instructors] just seem to get up and talk at you and not be too concerned with thinking on that deeper level at all, but others...stimulate and challenge you to think on that deeper level." A few students also referred to differences in thinking expectations according to the subject matter—comparing a medical science course, for example, to a course in statistics.

When might such skills not be applicable?

On being asked about when these skills may not apply or could not be used, the Auckland students had plenty to say. They pointed out that in some lectures, assignments, and tests, these thinking skills were not applicable. One student provided an example of tests where "you're just basically asked to...give definitions of something, just to know that you have been to your lectures, you have been doing your readings, and you can repeat back—remember stuff." Another student explained,

> It's not required in the lecture if you just listen, you don't think...you do the thinking before and after...When you're in there, you watch, basically the lecturer, you listen to what he says and try to relay back...Maybe it's thinking too—relaying back?

Like their response to the previous question, the Auckland students stressed the considerable variation in teaching approaches between instructors, pointing out for example that "some lecturers elicit from us those thinking skills" more so than others.

The Auckland students also noted that some of their instructors were not good role models in the use of those thinking skills. One student, for example explained:

> Sometimes they are not actually explaining things properly—they're not giving evidence or giving examples and explaining what's expected of us. When they themselves are not using good thinking skills, it breaks the capacity for you to enter deeply into what they are saying.

In response to the same question, the Okinawa students provided cases when these good thinking skills could not be put to practical use: instances when things have already been decided to some extent and cannot be changed; when time is limited; and when it is necessary to be conciliatory to others. The following two comments exemplify the views expressed:

> When I don't have time, for example when I'm preparing for an exam, or doing a lot of tasks in class, I concentrate on finishing the work without thinking too deeply.

> When I am playing team sports or working as part of a team, I cannot be inflexible and hold on too strongly to my opinions. If I become too rigid about my opinions, our team won't work well.

The Kyoto students' responses overlapped in some cases with the responses from the Auckland and the Okinawa students. For example, they pointed out that in some subjects—such as "in the arts"—these thinking skills might not be required. They also noted that these skills cannot be used in many cases when "speed is required." However, they also mentioned additional cases, such as in everyday life, when making emotional decisions, while taking part in sports, or in activities where skill is necessary. As two students explained: "When people decide in a hurry or during an emotional confrontation, they cannot think logically in those situations," and "sports is based on physical skills, which are not controlled by just thoughts."

How can we improve our thinking skills?

When asked their opinions about how they might be able to improve their thinking skills, students in all the groups suggested various forms of practice as well as learning more about these thinking skills. Suggestions from the Kyoto students included "becoming conscious of them [the thinking skills] in everyday life," "practicing," and "reading books." Suggestions from the Okinawa students included "routinely thinking about things," and "generating questions even about trivial matters." Similarly, the Auckland students suggested "practicing good thinking skills will develop my thinking skills," "reading some books on thinking," and "taking some courses about how to learn."

In addition, the Kyoto students mentioned "remembering rules and prerequisites"—suggesting a more systematic approach to the development of thinking skills. One of the Auckland students also suggested the helpfulness of working in interdisciplinary teams, explaining the benefits as follows:

> I think what actually facilitates it [development of thinking skills] is working in inter-disciplinary teams where you've got people from a whole lot of different fields, and you're having to...I'm from arts...[and] trying to talk to a person who's in engineering is very difficult. But that makes you think...being challenged.

Students from Okinawa likewise suggested the value of articulating ideas ("verbalizing"), being exposed to a wide range of perspectives ("listening to a range of opinions"), and being challenged ("be pressed or feel the urgency to do something").

What does critical thinking mean?

The students were asked if they had heard of, or knew, the term "critical thinking," and what they thought it meant. All the students in the Kyoto group had heard of it, and they put forward various characteristics of it such as "not accepting information passively but doing so critically," "independence in thinking," "being aware of bias," "to distrust," and "to be reflective." In contrast, most of the students in Okinawa indicated that they had not heard of it, or they had heard of it before but had forgotten what it meant. One person who had heard of it put forward the following as its meaning: "to think more deeply and derive a good direction." The other students, when pressed for what they thought, suggested the following meanings: "to judge one idea and generate new ideas", "to consider [something] from many facets," and "to think for oneself rather than simply swallowing a variety of ideas." They also put forward the following as possible meanings of the term: "to submit different ideas," "to fix problems in your own thinking," and "to go back to the origin of an idea."

Like the Kyoto students, the students in Auckland indicated familiarity with the idea of critical thinking. When asked what it meant, they put forward definitions and descriptions like "being aware of the thinking processes and biases that creep into ideas," "looking at things from different perspectives to get a balanced view," "using objective rather than subjective arguments," "being structured in thinking," "asking the four Ws and one H—why, where, when, what, and how," and "being able to absorb new information, [and] trying to find a connection between that and some of the information that you already know." However, they also suggested that critical thinking is "being able to learn, being able to understand and then being able to apply it" and "understanding the subject matter"—suggesting that at least some of them were not able to distinguish the concept from comprehension and application. Furthermore, other definitions and characteristics that are not normally associated with critical thinking were proposed such as "intuitive thinking," "positive thinking," "negative thinking," and "being more constructive." One student even suggested that it is "a bit like playing a game...you know that game, but you sometimes wonder when you go out into the world if it's the same game."

Similarities and differences between the groups: some implications

The findings of our investigation suggest that, irrespective of where they come from, students share many common views about good thinking skills that are useful in tertiary studies. We found no evidence of an East-West difference in knowledge and awareness about useful thinking skills, which supports views that have earlier been expressed by authors like Paton (2005) and Stapleton (2002) about cultural equivalence as far as the possession of such skills is concerned. We also found it quite encouraging to see that many qualities commonly associated with critical thinking—such as viewing things from multiple perspectives, thinking for oneself, questioning, and using systematic, logical approaches—were mentioned by students in all three groups. Perhaps this similarity between the three groups indicates the extent to which many tertiary institutions globally are increasingly becoming more alike in the student competencies and values they are promoting.

The Kyoto and Okinawa students did not appear to be any more similar to each other in their views and perceptions compared to the students from Auckland. In fact, in some aspects, either the Kyoto or the Okinawa group was more similar to the Auckland group. For example, where improving thinking skills was concerned, the Okinawa and Auckland students suggested similar strategies involving exposure to others' views, articulation of knowledge, and rising to meet challenges, which were not suggested by the students from Kyoto. This again perhaps indicates that educational environments, more than cultural factors, influence students' views about how they should apply themselves to their studies.

We did, however, find one similarity between the Kyoto and Okinawa students that appears important, and that is their common assertion that consideration of others is one of the qualities of good thinkers. The consideration they were referring to did not appear to be necessarily for self-benefit—unlike, for example, instances when the Auckland students mentioned "getting inside the lecturer's head" so that they could produce work accordingly (and hence obtain better grades). Rather, the suggestion about consideration of others from the Kyoto and Okinawa students appeared more concerned with a genuine desire to understand and establish good relationships with other people. This finding supports earlier observations by Markus and Kitayama (1991) that people from many Asian and non-Western cultures possess a more *interdependent* self-construal. According to their theory, interdependent people think of themselves as being part of a bigger social relationship with significant others, and this view affects not only their behavior but also their ways of thinking.

Other differences between the groups, however, suggest the influence of the students' educational and social environments rather than their cultures per se. In discussing thinking skills that are important at university and instructor expectations about those skills, there were many similarities between the three groups, but also some interesting differences. The Kyoto students referred to the importance of deeper thinking processes and instructor expectations about responsibility for the development of one's own thinking skills. However, the Kyoto students who took part in our study were enrolled in one of Japan's top-ranked universities. As a group, these students could be considered as high achievers, and so perhaps these responses from them were not surprising. These students would likely have been better aware of the range of thinking processes required for effective learning, and instructor expectations of them would likely have been higher compared to expectations in many other universities.

In contrast, when discussing the same issues, the students from Okinawa referred to the importance of appreciating cultural meanings and relevance, and to instructor expectations about social responsibilities. Again, one could argue that, with the multicultural mix of people in Okinawa and the social and political issues surrounding the US military presence there, such views and perceptions about thinking would likely be impressed upon university students who live there.

Finally, the Auckland students responded to the same questions by mentioning course and instructor management strategies and by noting their lecturers' expectations about research skills applications. In New Zealand universities, including the one where the participants for the current study came from, undergraduate student attrition is high: entry into universities is comparatively easier (i.e., compared to many Asian countries, for instance) but, once in a university, students are expected to work very hard to keep up with course requirements—and significant proportions of students fail and/

or drop out. Study management strategies are therefore heavily emphasized in New Zealand tertiary institutions, including the need to effectively employ skills learned to meet coursework demands. This situation may well have influenced the views and perceptions expressed by the Auckland students.

Knowledge and expectations about critical thinking: some implications

As we noted earlier, students from all three groups referred to numerous characteristics of critical thinking when discussing study-related thinking requirements. The students also appeared well aware of situational factors that may limit or make the use of such thinking skills unwise. These findings are encouraging indicators that students are developing not only knowledge about, but also practical understanding of, various thinking skills—including aspects of critical thinking—that, ideally, formal education should inculcate (see, e.g., Glassner, Weinstock, and Neuman 2005; Halpern 1998; Thomm and Bromme 2011).

What is not as encouraging, however, is the finding that, when explicitly asked what they thought critical thinking meant, some misconceptions about it also surfaced in the students' responses. As we noted earlier, the Auckland and Okinawa students mentioned other thinking characteristics that would not normally be considered as aspects of critical thinking, like creativity, positivity, and intuition (cf. critical thinking definitions provided by Ennis 1962; Fisher and Scriven 1997). The Kyoto students put forward definitions and qualities that were appropriate, but they did not say many of these, suggesting that perhaps they too were not so certain about the exact meaning of critical thinking. This finding about students' misconceptions and uncertainties about the meaning of critical thinking suggests that there is a need to provide more explicit, systematic, and comprehensive education about this thinking approach. If the capacity to think critically is as important to develop in students as many tertiary institutional documents worldwide suggest, then critical thinking should be taught more explicitly and its development incorporated more systematically into course curricula.

An issue related to student knowledge and skills development is *teacher* knowledge and skills development. We found ample evidence from the comments provided by students that thinking skills expectations vary considerably across courses and instructors, with some apparently requiring only superficial thinking approaches such as rote memorization. This indicates the need to ensure that tertiary-level instructors are knowledgeable about the thinking skills that students need to develop and are sufficiently skilled in the facilitation of their development through the courses they teach. Hence, appropriate means for teacher professional skills development in these areas (e.g., training and resources) ought to be made available to all tertiary education instructors.

Where to from here?

Our investigation was carried out with only 23 undergraduate students and we used no systematic method to ensure that the students who participated possessed profiles that reflected the demographic profiles of the student populations in the corresponding institutions and locations. There are therefore obvious limitations in the extent to which our findings can be generalized. In future research, it would be useful to examine students' views and perceptions about the same thinking skills issues, but with the use of other groups of students, in other institutions, and in other locations. Other methods of data collection could also be employed, such as the use of questionnaires, which would enable larger samples of students to convey their opinions on these issues.

It would also be helpful in future research to examine not just self-reports but also students' actual use of the target thinking skills in the work they produce. The effects of providing instruction on these thinking skills, particularly on student competencies as may be observed through work they produce, would also be an important future direction to take in research. It would be important to identify any variations in student performance that may relate to cultural factors—including, for example, language—so that they could be addressed as may be deemed appropriate.

References

Asian University for Women. 2013. *Academic Programs* 2008–2013. February 3. Available from http://asian-university.org/academic_programs.htm.

Association of American Colleges and Universities. 2013. *College Learning for the New Global Century* 2007. February 2. Available from http://www.aacu.org/leap/documents/GlobalCentury_final.pdf.

Atkinson, D. 1997. "A Critical Approach to Critical Thinking in Tesol." *TESOL Quarterly* 31 (1): 71–94.

Boyatzis, R. E. 1998. *Transforming Qualitative Information: Thematic Analysis and Code Development.* Thousand Oaks, CA: Sage.

Ennis, R. H. 1962. "A Concept of Critical Thinking." *Harvard Educational Review* 32: 81–111.

Fisher, A., and Scriven, M. 1997. *Critical Thinking: Its Definition and Assessment.* Norwich, UK: Centre for Research in Critical Thinking.

Floyd, C. B. 2011. "Critical Thinking in a Second Language." *Higher Education Research and Development* 30 (3): 289–302.

Fox, H. 1994. *Listening to the World.* Urbana, IL: National Council of Teachers of English.

Glassner, A., Weinstock, M., and Neuman, Y. 2005. "Pupils' Evaluation and Generation of Evidence and Explanation in Argumentation." *British Journal of Educational Psychology* 75 (1): 105–118.

Halpern, D. F. 1998. "Teaching Critical Thinking for Transfer across Domains: Dispositions, Skills, Structure Training, and Metacognitive Monitoring." *American Psychologist* 53 (4): 449–455.

Jussim, L., and Eccles, J. S. 1992. "Teacher Expectations II: Construction and Reflection of Student Achievement." *Journal of Personality and Social Psychology* 63 (6): 947–961.

Lun, V. M. C., Fischer, R., and Ward, C. 2010. "Exploring Cultural Differences in Critical Thinking: Is It about My Thinking Style or the Language I Speak?" *Learning and Individual Differences* 20 (6): 604–616.

Markus, H. R., and Kitayama, S. 1991. "Culture and the Self: Implications for Cognition, Emotion, and Motivation." *Psychological Review* 98 (2): 224–253.

Mason, M. 2007. "Critical Thinking and Learning." *Educational Philosophy and Theory* 39: 339–349.

Miller, D. T., and Turnbull, W. 1986. "Expectancies and Interpersonal Processes." *Annual Review of Psychology* 37 (1): 233–256.

Ministry of Economy, Trade and Industry (METI), Japan. 2013. *Developing Global Human Resources through Industry-Academia-Government Collaboration* 2010. February 7. Available from http://www.meti.go.jp/english/press/data/20100423_02.html.

Ministry of Education, Culture, Sports, Science and Technology (MEXT), Japan. 2013. *Higher Education Bureau. Higher Education in Japan* 2009. February 7. Available from http://www.mext.go.jp/english1koutou/detail/_icsFiles/afieldfile/2009/12/0311 287370_1_1.pdf

Paton, M. 2005. "Is Critical Analysis Foreign to Chinese Students?" In *Communication Skills in University Education: The International Dimension*, edited by E. Manalo and G. Wong-Toi. Auckland, New Zealand: Pearson Education. 1–11.

Scharrer, L., Bromme, R., Britt, M. A., and Stadtler, M. 2012. "The Seduction of Easiness: How Science Depictions Influence Laypeople's Reliance on Their Own Evaluation of Scientific Information." *Learning and Instruction* 22 (3): 231–243.

Stapleton, P. 2002. "Critical Thinking in Japanese L2 Writing: Rethinking Tired Constructs." *ELT Journal: English Language Teachers Journal* 56 (3): 250–257.

Tapper, J. 2004. "Student Perceptions of How Critical Thinking Is Embedded in a Degree Program." *Higher Education Research and Development* 23 (2): 199–222.

Thomm, E., and Bromme, R. 2011. "'It Should at Least Seem Scientific!' Textual Features of 'Scientificness' and Their Impact on Lay Assessments of Online Information." *Science Education* 96 (2): 187–211.

University of Auckland. 2013. *Graduate Profile* 2003. February 2. Available from http://www.auckland.ac.nz/uoa/home/for/current-students/cs-academic-information /cs-regulations-policies-and-guidelines/cs-graduate-profile

Wilkinson, D., and Birmingham, P. 2003. *Using Research Instruments: A Guide for Researchers*. Edited by Peter Birmingham. London, UK: Routledge Falmer.

19
Critical Thinking through a Multicultural Lens: Cultural Challenges of Teaching Critical Thinking

Maha Bali

Introduction: critical thinking in non-Western cultures

When I first started doing research about critical thinking (CT), I had not realized how contested a notion (Atkinson 1997) it was, even though it is widely accepted as an important goal of (at least Western) higher education (Barnett 1997; Norris 1995). Most people agree that critical thinking is essential for citizens to participate actively in a democracy (Brookfield 1987; Johnson and Morris 2010; ten Dam and Volman 2004). There is also a need for citizens to criticize the systems and hierarchies in which they live, whether academic or corporate, whether in a democratic state or not, questioning and challenging even the structures within which we conduct critical thinking. Some scholars claim that critical thinking is culturally biased (e.g., Atkinson 1997; Fox 1994; Norris 1995), which is a view I initially dismissed, because the ideas and practices of CT exist in my own Egyptian Islamic culture. Some scholars argue that viewing critical thinking as distant from non-Anglo (i.e., non-English speaking) cultures is a symptom of misunderstandings (Ennis 1998), and even ignorance of, and condescension toward non-Westerners' capacities for rational thinking (Nussbaum 1997). Claims of cultural distance or difference are often presented under the guise of cultural sensitivity, while hiding reductionist and deficit-oriented tendencies (Zamel 1997).

We cannot, however, dismiss the evidence of practical difficulties in teaching critical thinking to international (particularly Asian) students in Anglo universities (e.g., Egege and Kutieleh 2004; Vandermensbrugghe 2004). Some have suggested that this is due to linguistic difficulties (Floyd 2011) or pedagogical biases in how academics tend to teach critical thinking (Ennis 1998). It has also been suggested that critical thinking is valued differently in various cultures (Egege and Kutieleh 2004).

Much of the research on cultural bias in CT has been conducted with reference to Asian students. In what follows, I use evidence from the American University in Cairo (AUC) to examine cultural issues of developing critical thinking in the Egyptian context, but many of the issues raised can be recontextualized to other cultures.

I present my argument in two parts. First, I argue that critical thinking is not an exclusively Western notion, but one ingrained in Islamic scholarship and informal Egyptian culture. In spite of that, the teaching of CT to students in the Arab and Islamic world continues to pose challenges for educators. The second part of the chapter highlights some of the cultural and contextual challenges faced by those teaching critical thinking to non-Western university students, using research conducted at AUC. On the classroom level, these include variability in students' incoming cultural capital, previous pedagogical exposure, and linguistic ability. On a macrolevel, these include the ways in which the sociopolitical environment can hinder students' capacity to practice critical thinking outside the classroom.

I focus on the particular context of AUC teaching CT to mostly Egyptian students because I have in-depth experience with this institution and have conducted research there. However, many of the arguments in the second part of the chapter apply to other contexts in the region and beyond. I shed light on the challenges of teaching critical thinking in any Western institution (wherever located) to an international audience (of whatever background) because the emphasis is on understanding the *diversity* of the learners and their backgrounds, and how this poses a challenge for teaching CT, rather than on the *particular* characteristics of the learners' backgrounds. Because the AUC setting is one of mostly non-Western students living in the same country, the perspective is different from that commonly used when a Western institution has small numbers of international students from, for example, Asia. Rather than generalize about Egyptian students, I focus on the elements of diversity among them—as is mostly likely the case among Asian students, about whom generalizations are often done.

North American critical thinking

When I first started researching conceptions of CT, I came across what I thought was the agreed-upon definition: the traditional North American/ Western conception of critical thinking as it appears in Facione's (1990) *Expert Consensus*. The report articulates critical thinking based upon the collaboration of a panel of experts (predominantly North American, including the well-known Robert Ennis, Richard Paul, and Stephen Norris). In the report, CT is defined as (Facione, 1990, 2)

> purposeful, self-regulatory judgment which results in interpretation, analysis, evaluation, and inference, as well as explanation of the evidential, conceptual,

methodological, criteriological, or contextual considerations upon which that judgment is based.

CT is then broken into a set of cognitive skills, namely interpretation, analysis, evaluation, inference, evaluation and self-regulation (6), as well as dispositions, including fair-mindedness, inquisitiveness, open-mindedness to divergent worldviews, and diligence in seeking relevant information, among others.

This is the approach to critical thinking closest to that used in North American universities, but it is by no means the only way CT is conceptualized in the literature (Bali 2013b; Brodin 2007; 2008). It is also the conceptualization of CT closest to that used at the American University in Cairo (AUC), which I use as the context for this paper.

Critical thinking in Islam and in Egypt

Critical thinking as "Ijtihad" (Islamic scholarship)

Some consider critical thinking to be a Western-influenced educational ideal and suggest that it opposes some of the values inherent in other cultures. For example, Cook (1999) suggests that although Islam is tolerant to variety in perspectives, it still does claim that there is a universal truth and would not comfortably accommodate a notion of equally viable perspectives. However, the notion of "ijtihad," the approach of Islamic scholarship fundamental to interpreting Islamic law (shari'a), overlaps strongly with modern-day Western notions of critical thinking in terms of the hermeneutic process of evaluating credibility of sources, examining multiple meanings behind a text, making contextual connections, and using logic to arrive at what are usually multiple divergent but equally valid interpretations (Nurullah 2006; Said 2004). Islam's primary sources (Quran and Sunnah) strongly encourage critical reflection, rationality, and scientific study, and consider such deliberation one of the highest forms of worship. In a recent lecture by Bradley Cook at AUC, there was a discussion on how some Quranic text can be interpreted in very orthodox ways that prohibit questioning, while some progressive and spiritual scholars of Islam would consider the exact same text as an invitation to reflect, explore, and question. In practice, Arab and Muslim societies nowadays are less likely to apply critical and questioning approaches to Islamic scholarship (Nurullah 2006). This may be related to oppressive political regimes and educational curricula, which I discuss next.

Critical thinking in Egypt

School curricula across the Arab region are known to "encourage submission, obedience, subordination and compliance, rather than free critical thinking" (UNDP 2003, iv). Egyptian educational curricula at school *and* university levels are notorious for emphasizing memorization and shunning criticism (Aboulghar 2006). However, lack of critical thinking in formal educational contexts in some cultures can contrast with critical thinking in informal contexts such as everyday discussions of politics (Fox 1994), and Egypt is one such culture. However, because of oppressive regimes, such as those that have existed in the region for years after independence from colonial rule, public expressions of criticism toward power pose risks many people are often unwilling to take (Asgharzadeh 2008).

The January 2011 uprising proved that Egyptians who have various levels of literacy and education were able to think critically despite the educational system and willing to take critical action in spite of the oppressive state. However, the lack of significant reform today, three years after the uprising, implies that more is needed in order to promote a critical citizenry capable of moving beyond protest/opposition toward construction and democratic reform (Bali 2013a; Beinin 2013). The masses in Egypt were able to protest and be critical in terms of objection, but this did not relate directly to being critical in ways we traditionally mean in the academic setting, which involve evaluating credibility of evidence, questioning assumptions, and arriving at conclusions logically using criteria that incorporate different perspectives.

Critical thinking is therefore not an exclusively Western notion: it exists in Islamic scholarship and in Egyptian informal contexts, as is the case in other cultures (Fox 1994). However, the lack of it in formal educational contexts can and does pose contextual challenges for educators. I support my argument by using examples from the American University in Cairo, which aims to promote critical thinking to Egyptian undergraduate students via a liberal arts education.

To support my argument, I refer to student and faculty views at AUC, using pseudonyms to protect individual identities. I use quotes taken from two studies I conducted previously at the American University in Cairo. The first is my (2013b) thesis, which included semistructured interviews with students and faculty on critical thinking development. The second is a survey for faculty in the disciplines, asking about persistent language-related issues (Bali and Carpenter 2009). The latter was conducted as part of a needs assessment exercise conducted by the English Language Institute (a unit that teaches remedial English to students who enter AUC since English is the language of study at AUC).

Contextual considerations in promoting critical thinking

AUC faculty and students are not a monolithic group. There are American/ Western and Egyptian/Arab professors, but among these are Egyptians who have been educated in the West, American professors who have lived in the

Arab world for long periods of time, and there are those who are bicultural/ dual-nationality. Among the mostly Egyptian students, there is diversity in terms of degree of "Westernization": some are scholarship (i.e., less affluent) recipients, students from traditional Egyptian families and traditional Egyptian Arabic–language public schooling, whereas others have always had a Western education and have lived abroad, and other students fall within a continuum of "cultural hybridity" (Lash 2001), to use Bhabha's term.

Therefore, it would be inaccurate to use any blanket generalizations about the cultures of "Egyptian students" at AUC. Rather, it is important to look at individual students in a nondeterministic manner, to see how their agency works within the constraints and opportunities posed by their cultural and social backgrounds. It is also important to note that although AUC is an American institution aiming to teach critical thinking via a liberal arts curriculum, the faculty who teach at AUC themselves have varying degrees of familiarity with liberal arts philosophy (e.g., part-time faculty who have never been exposed to American education) on the one hand, and Egyptian culture on the other hand (e.g., young Western faculty coming to Egypt for the first time).

Cultural capital

The first difficulty AUC educators face in promoting critical thinking in Egypt is the variability in students' cultural capital in terms of exposure and familiarity with critical thinking in previous schooling and family backgrounds. There is a belief among scholars that deferring the teaching of CT until college is unlikely to be effective (Facione 1990). In the Arab region, it is difficult to use the 4–5 years of college to promote a capacity that has been suppressed or at least not developed over many years of schooling (Rivard, 2006, cited in Hall 2011). This becomes even more difficult when it also contrasts with what is encouraged in the home environment and public discourse. Some educators have suggested that Arab students may resist questioning certain cultural taboos and struggle to do so without strong support from teachers (Raddawi 2011). This means that a household may encourage youth to question certain topics but suppress questioning of others such as cultural taboos, and teachers in higher education have difficulty reversing those unsaid rules.

My research found that variability in students' schooling affected their capacity and comfort with critical thinking as they entered college. It is important to note these differences, because the more practice a student has in critical thinking, the better he or she will become at it (van Gelder 2005), as is the case for many skills.

On the one hand there are AUC students who have studied in what are called "International schools" (teaching using the American, British, German, or French systems), for example, and who repeatedly mentioned ways in which their schooling encouraged critical thinking via in-class discussions of controversial topics, conducting small-scale research-writing projects, and extracurricular activities that promote critical thinking such as the Model United Nations. On the other hand, we have the example of one student, Noha, who came from traditional Egyptian schooling, was unused to questioning authority, and felt uncomfortable participating in discussions and debates, even though such discussion was encouraged at home. She said:

> I guess we were taught to always think that the teacher is right as opposed to college...you don't question authority...as a freshman, it wasn't the norm for me to challenge professors at least intellectually and I'd take their word for granted and I learned from people around me...that everything they [professors] say is not necessarily true.

Her ability to question grew slowly:

> I guess my self-esteem or my confidence in my own intelligence was limited because of the education I had as a younger person. But I mean that's improving, but it is taking me a very very long time to adapt to.

Although her confidence (which Bourdieu 1973 considers an expression of social capital) grew, the transition was not easy. She says it has been a "tough transition" from her previous schooling and remains so:

> I was so used to having one right answer growing up, so it's still very frustrating at times to not have a right answer.

Professors who teach at AUC notice this issue among students. One attributes this to a lack of confidence/comfort:

> Some students are not comfortable expressing themselves—some want to simply repeat what's been given to them, are comfortable staying close to the text, don't want to venture on their own...no one made them feel confident enough to say their view even if others may not agree.

Another American professor compares this to teaching in the West:

> In a Western setting, students have just been brought up that way...it becomes more natural for students...to think independently. Here [in Egypt], a different type of culture, where you [student, say], "I'm not the authority so who am I to speak, to offer my independent observation about this? The professor is the authority or the author of the book is the authority."

This professor is referring to the lack of cultural capital in students' backgrounds that did not promote critical thinking and questioning of authority while students were growing up. These ideas are supported by research done by Nelson, El Bakary, and Fathi (1996) showing that Egyptian students generally show higher discomfort with uncertainty than US students.

It is important for educators to recognize these differences in students' cultural capital, their familiarity and comfort with critical thinking, when thinking about pedagogies to use in class to promote it.

Pedagogical considerations

We often assume that good pedagogy transfers well across contexts and cultures. We as educators refer to dialogue and inquiry-based learning, and often practice this pedagogy with our students, regardless of their background. But pedagogy is not necessarily culturally neutral. It is better conceived of with a lens of "diversimilarity" that recognizes the importance of differences without ignoring cultural similarities (Skelton 2005). Discussion/dialogue has often been used to promote critical thinking, as it helps students reflect on their own assumptions and worldview in dialogue with peers (Brookfield 1987). While Egyptian society is comfortable with criticism in informal social contexts, discussion and criticism are absent from traditional Egyptian schooling.

Arab cultures are considered to be oral cultures (Hall 2011), and Egypt is no exception. Students at AUC are generally more comfortable with in-class discussions in disciplines where they do not feel they need to be content experts, and where they feel they can question authority: these are often the humanities and social sciences; but they feel less comfortable doing so in sciences, engineering and some professional disciplines (Bali 2013b). A Rhetoric & Composition administrator at AUC said that instructors create "the environment for [students] to express themselves," because the students are "often more willing to be critical verbally than in [formal] writing." Since students are comfortable having informal debates outside of class, instructors use classroom discussions to bring those debates into the classroom context. While this strategy is culturally sensitive, one often sees students transferring informal debate into the classroom by stating unsupported opinions rather than critical, thoughtful ones.

Moreover, educators at AUC who come in with the intention of reducing their own authority in class and conducting in-class discussion with peers are sometimes faced with students unwilling to participate (possibly because of lack of confidence as in Noha's case above). This can create a situation as Burbules (1986, 109) describes:

> To the extent that students enter the classroom with preexisting antipathy to, or ignorance of, consensual relations (based on their family experiences,

friendships, or their socialization via the media), they often act in ways which interfere with even the best teacher intentions, thereby "justifying" [teacher] authoritarianism.

By opening the floor for discussion, the teacher cannot assume all students will be able to participate equally. The use of dialogue/discussion as a pedagogy for promoting critical thinking automatically privileges students who are more comfortable and familiar with this pedagogy, as well as those more confident. It took Noha, discussed in the previous section, a couple of years to build up confidence to start actively participating. Moreover, Noha said that she eventually realized that some of the students who were comfortable and confident speaking up were not necessarily intellectually better than her, but merely more confident and eloquent.

Dialogue as an ideal falsely assumes all students have equal power (Ellsworth 1989), even though it actually

> ...privileges students comfortable with spontaneous and oral, rather than reflective and written communication. It privileges Western[ized] students generally more familiar with the idea of interactive classrooms, than those unfamiliar with it, such as Arabs schooled in traditional ways that discourage student participation altogether. (Bali 2013c, 4)

Kamal, another student from traditional Egyptian schooling background (and whose English language was weaker than other students), preferred to participate in online discussions:

> The discussion board [in a particular course] was one of the main reasons that made me actually have this broad mind...you can see that 3–4 people are talking, each taking a side, and you read the four and you try to make something to support, or for, or against a specific side, and you try to support it differently than the other four. Clearer than in class—[which is] limited in time, maybe I won't have time to say my point and explain it, maybe I just say the point and shut up; but in the discussion board I am not limited neither [sic] by time or anything so I can say whatever info I want and can support my argument as I want; in class it is otherwise. In class you have to be to the point and concise.

Kamal's preference for online discussion highlights some of the challenges for him in face-to-face discussions. A professor who used online discussions in her classes felt that it allowed more equal participation than face-to-face discussions:

> I find it very difficult [to promote critical thinking]. Especially if there are different levels in the class...In discussion online [there] was a big difference

between those that are mature or have a different background or at a higher level. They bring in something different but at the same time challenge the others...if it happens I have to draw others in so as not to be intimidated by the more [eloquent]...but on the whole in online discussions I didn't feel they were intimidated.

It is also important to note that some students are uncomfortable with conflict (Grundy 1987), and that this can create problems during in-class discussions that center around conflict. For example, all the female students I interviewed said they were uncomfortable with the notion of questioning hidden agendas: they entered AUC not looking for hidden agendas but reluctantly started to see them and look for them, whereas male students were much more comfortable with finding hidden agendas and proud of themselves for doing so. A professor of English Literature said thatstudents "hold back in fear of displeasing the instructor," and a professor of Biology found that students were afraid to "risk" expressing their own ideas and opinions. Furthermore, students at AUC are often unfamiliar with the idea of a professor playing "devil's advocate" in class. Whereas students more familiar with this approach would understand why the professor is doing it, some Egyptian students misunderstand this as a professor's criticism of their arguments, and think that they should instead stop expressing their own views that dissent from the professor's own (Bali 2013b). It therefore becomes imperative for the teacher to clarify and emphasize to students that conflict is in fact encouraged in the classroom (Browne and Freeman 2000), and to ensure the instructor's behavior supports students in expressing dissenting views.

Critical thinking is often also taught via courses in writing, as is the case at AUC, and I discuss this approach next, within the discussion of linguistic considerations.

Linguistic considerations

Floyd's (2011) study found that thinking critically may be easier in one's native language, and also that reading in a non-native language that is orthographically different from English (as is the case for Chinese and also Arabic, both of which use non-Latin alphabets) can hinder critical reading ability (Floyd 2011 citing Koda, 1996; 2005). She also suggests that when students write in a second language, instructors may end up focusing on correcting grammar and thus deemphasize the focus on criticality. Research also suggests that students more familiar with Western education (Nelson 1992) or use of English in their everyday lives (Fox 1994) participate more easily in writing classes in American universities versus their peers. In my research (Bali 2013b), students from American international schools had had exposure to writing and research, whereas others were learning to do this for the first time in college in introductory Rhetoric & Composition courses.

Students enter AUC with variable linguistic competence, as well as familiarity with research writing. Despite remedial courses in English, administrators suggest that differences in linguistic competence persist throughout the college years. One engineering professor reported that the majority of third/fourth year students continue to use Arabic in oral classroom discussion, regardless of teachers' efforts to enforce English. A political science professor takes this point further in suggesting linguistic difficulties result in students being unable to understand undergraduate textbooks well, which further makes it difficult for them to read critically. Linguistic difficulties can also affect students' abilities to understand subtleties in readings (Chandler 1995; Kaplan 1966). A social science professor said:

> One of the biggest problems I have is that students don't understand the straw man concept; they frequently critique the author for making exactly the opposite argument that the author was making, based on the author's straw man. This occurs at all levels, but especially the lower ones.

When teaching critical thinking relies heavily on language (and how can it not?) and that language is not the students' native language, it poses an additional challenge to the teacher. In order to think critically about a text, a student needs to first comprehend the surface meanings of the text, then be able to read some of its hidden meanings, and then also be able to pick out the rhetorical devices used in the language. When a teacher is faced with a class of mixed linguistic abilities, it is more difficult to cater to the needs of those who cannot grasp those subtleties of language that are more familiar to students who had grown up immersed in the language. In some of my own classes for adult learners (whose linguistic abilities were usually weaker than undergraduate AUC students), I found myself unable to use readings from UK newspapers because students inevitably misunderstood the arguments in them because of the unfamiliar rhetoric used.

Furthermore, for students to be able to express themselves critically in writing is another layer of complexity, particularly for students unfamiliar with writing in the English language, unfamiliar with conducting research, and unfamiliar with integrating evidence from outside sources. I have found that many students in nonwriting intensive disciplines (e.g., as evidenced by their graduation theses on Turnitin.com) reach their final years of college without having captured the basics of plagiarism and correct integration of sources. Informal discussions with faculty in writing-intensive social science disciplines also bring forth complaints of poor integration of sources. This is a relatively mechanical skill, and students are unable to learn it and use it in four years of college. To be able to write critically in one's non-native language is a much more complex ability that is unlikely to develop and become internalized any faster.

Awareness of these three challenges may help educators find ways to support students' CT development. While I believe that there should be some teaching of CT in students' native languages, most English-language universities such as AUC do not commonly offer that option, nor would they be expected to if they have a diverse student body of different native languages. However, one option would be to allow students to learn the CT process by critiquing materials in their own native language (whether or not the instructor understands that language). Some AUC faculty incorporate elements of Egyptian popular media in their classes—whether the material is in Arabic or English, it is culturally closer to students than material about other regions, and students' familiarity with the topic might facilitate comprehension for them in order to set the stage for critical reflection.

I have highlighted three differences among students that may make teaching critical thinking challenging: their cultural capital and exposure to critical thinking before college; their exposure to pedagogies that promote critical thinking before college; and their linguistic ability, which impacts their ability to read/write critically. All of these factors influence students' openness and ability to develop critical thinking throughout college. But beyond individual students in classrooms, there remain factors outside the influence of the classroom or educational institution that may hinder students' capacities to think critically.

Combined capability: influence of the sociopolitical environment

There is a need to account for the influence of the sociopolitical context when teaching critical thinking: How is critical thinking received socially? How critical are media outlets in their presentation of events and viewpoints? Are there examples of critical thinking in public discourse? In what ways does the political culture encourage or discourage dissent? Which topics are culturally acceptable to critique, and which are not? What happens to the person who dissents from the mainstream discourse? These questions highlight how context influences students' capacity to apply critical thinking outside the classroom, but more importantly, it influences our understanding as educators of *why* we need to develop our students' critical thinking, and *how* to do so in this environment, because it highlights the kind of *barriers to criticality* our students have grown up with and often continue to face every day.

Promoting critical thinking as an instrumental skill (a means to a defined end), as is often done in higher education, tends to ignore the importance of also promoting an individual's capacity to exercise "will" and "judgment" in using what has been learned appropriately in context (Barnett and Coate 2005). Moreover, we need to recognize the ways in which the external environment can limit an individual's capacity and agency to exercise a learned capability, a notion Nussbaum calls "combined capability" (2011, 22).

In a country like Egypt, previously oppressive regimes limited the amount and type of public discourse that could oppose or criticize the regime overtly. It is only quite recent that opposition newspapers with some credibility started to emerge, and non-state-run satellite TV channels started to appear and gain audiences. The spread of Internet and social media enabled individuals to access views outside the mainstream, state-run media, and this social media revolution is known to have helped mobilize the youthful masses who initiated the Egyptian and Tunisian revolutions. More recently, after the 2011 uprising, there continued to be risks in public criticism of whichever regime is in power, as well as more subtle resistance from public opinion towards dissenting voices (Fadl 2013). This has been done in both subtle and explicitly violent ways.

It is important to note that informal forms of criticism such as those rife in Egyptian media and social conversations differ from what is generally considered academic critical thinking, in that use of rhetorical devices and sensationalism can often eclipse drives toward evidence-based, balanced, and logical criticism. The discourse is antagonistic and often one-sided, usually also hiding the extreme uncertainty looming on Egypt's future.

This means that students are often willing to engage in opinionated discussions in classrooms, but these are not necessarily "critical" discussions in the academic sense. They often display what Richard Paul (1994) calls "weak sense" critical thinking: "egocentric" or "sociocentric" thinking, aiming at supporting one's own opinion, while criticizing the opposing view, which he suggests is uncritical. What is needed, therefore, is something closer to what Paul (1994) terms "strong sense" critical thinking that incorporates diverse worldviews and is able to turn inwardly to question one's own biases (Brookfield 1987). Currently, the public discourse in Egypt leans toward modeling and promoting weak sense critical thinking rather than strong sense critical thinking.

The lack of reform in Egypt since the 2011 uprising indicates that criticism and protest are insufficient to create reform in Egypt, and that a different conception of "critical citizenship" needs to be advanced in universities, because currently, citizenship that is

> …based on opposition, seems unable to change tactics and work towards reconciliation and reconstruction. It just recreates the protest cycle over and over again. The most recent escalations of violence [summer 2013] further complicate chances for reconciliation. (Bali 2013a)

In the current Egyptian context, developing critical thinking as a mechanical skill would do little to improve the situation. A re-thinking of critical thinking that centers on incorporating empathy and social justice as central to our teaching of critical thinking may be needed.

The traditional North American conception of CT is fragmented into skills and dispositions. It does not capture the essence of what it means to think critically, though it can provide helpful tools for educators to teach students about CT as a series of steps. However, CT is often approached from an antagonistic, confrontational standpoint, meant to promote skepticism rather than understanding of the viewpoints of others (Belenky, Clinchy, Goldberger, and Tarule 1986). I have already described how such an approach to CT can intimidate some students unwilling to question in this manner.

It is possible that the more confrontational, masculine approach to promoting critical thinking used in pedagogies such as debate are unhelpful in Egypt's current circumstances, which require people from plural worldviews to work together constructively for the purpose of democratic reform. Centering CT around skepticism, as is usually done, may stand in the way of helping people trust each other in order to work together toward reconciliation and reconstruction.

A new, socially constructed and contextually driven (as in Thayer-Bacon 1998) conception of CT may be needed: a rethinking of what we believe critical thinking is, why we value it, and how we teach it. I would not suggest that I, as a single individual, can come up with an appropriate reconception of critical thinking, but I can contribute a preliminary approach and open it to discussion (Bali 2013b). A revised conception of critical thinking that is more communal and centers around empathy may be more appropriate to conditions such as those facing Egypt today. Such a conception could incorporate ideas such as Nussbaum's (1997) notion of "Narrative Imagination" (basically, understanding the views and actions of "the other" based on understanding their worldview and context). This notion also exists in Edward Said's "philological hermeneutics" (as described in Nixon 2006, 34) which involves understanding a creator's work first as it was intended by the creator before critiquing it.

These notions are not completely absent from the critical thinking literature, but are often referred to within the context of female preferences for knowing/learning that emphasize community and connection (Belenky et al. 1986). Baxter Magolda (2004) shows these are different from traditionally dominant masculine approaches to developing critical thinking, but that they are equally viable pathways to developing critical thinking, and are preferred by most women and some men. I would suggest that in a caustic and antagonistic sociopolitical environment, such approaches to developing critical thinking and citizenship are more constructive than traditional ones.

Another very important notion is the incorporation of a social justice orientation into pedagogy that develops critical thinking. By social justice orientation, I mean an awareness of how one's thinking and decision making impacts upon social inequalities and injustice, coupled with a desire to challenge this injustice in one's current context, and to take action to influence change for the better. This inevitably contains an element of empathy, so that the privileged who have more power to effect change imagine themselves in the position of the unprivileged, and the oppressed begin to understand the different oppressions of others. It entails using our critical thinking not as a neutral context-free skill, but as a goal- and value-driven capacity to influence society for the better.

While this approach is not the focus of the North American conception of critical thinking (as in Facione, 1990), some educators influenced by the critical pedagogy movement and the Frankfurt School do infuse a social justice orientation within their pedagogy in higher education (Benesch 1999; 2001; Brookfield 1987; 2007). This approach would entail encouraging students to challenge the unjust status quo, and to continue to reflect on how power and authority may be perpetuating injustice, even when one supports those in power for other reasons. Currently, the case in Egypt is that people tend to be critical of the "side" they dislike, but are often unable to think critically about the "side" they support, or think empathetically about the "other." This defeats the purpose of critical thinking in helping individuals make informed decisions upon which to base their actions.

The infusion of empathy and social justice into our teaching of CT can also fit quite comfortably with Quranic notions of coupling "mercy" with "knowledge," such that the person who has knowledge benefits from first learning merciful qualities before using that knowledge.

I presume that emphasizing social justice and empathy into teaching critical thinking could enrich curricula in other contexts as well (there is no shortage of injustice and antagonism in our increasingly uncertain world), but the extreme uncertainty and political instability in Egypt highlights the importance of these elements that are not normally considered essential dimensions of critical thinking in the North American context. Therefore, examining the sociopolitical environment can make educators question what critical thinking is *for* (Barnett, 1997), and reflect on how this would impact our pedagogy.

A practical example

In my own teacher-education classes, I apply culturally relevant pedagogy to introduce the concepts of critical thinking and citation by first prompting students to think of how Islamic scholarship (*ijtihad*) arrives at multiple valid conclusions and how their validity is evaluated. This, because they are mostly Muslim, or have grown up in an Islamic society, is common knowledge, which is then useful to describe critical thinking and citation throughout the course, as well as highlight the questioning of authority and credibility of sources and experts. We discuss this process without directly talking about critical thinking or citation. A light goes on in students' eyes when they later realize that even though we don't learn about critical thinking and citation in Egyptian schooling, the notions are not Western/imported, but exist already in our own culture. More importantly, they then start to see how the Islamic notion of multiple valid viewpoints, each applicable to different contexts, is relevant to the Egyptian situation, where currently tolerance for diverse viewpoints is rare. Students were independently able to compare the current antagonistic media discourse that pushes only one viewpoint while excluding others, with the Islamic notion of multiple valid pathways. They also managed to recognize the relevance of the concept of "mercy" to general ways of doing critical thinking: in the Islamic approach, the multiple views are considered a "mercy" as difference in interpretation is embraced, allowing people to choose after considering their own personal context. This is an example of a local approach that is culturally relevant but would *not* transfer well as students become more diverse, without also including other examples from different student cultures.

Conclusion

I have shown that critical thinking is not a uniquely Western concept: it is inherent in Islamic scholarship, and common in Egyptian informal contexts. I have shown, however, that lack of critical thinking in some formal educational contexts in Egypt can pose problems to educators promoting critical thinking in higher education. Using examples from the American University in Cairo, I have shown four important considerations that Western universities should consider when teaching critical thinking to non-Western students. The first relates to cultural capital: educators should consider variability in students' previous exposure/familiarity with critical thinking in school and at home, and how this may impact their comfort and confidence with uncertainty and critical thinking when encountered in university. The second relates to recognizing that pedagogies we use to promote critical thinking are not necessarily culturally neutral, and that it may take effort and intentionality from an instructor to consider ways of promoting student participation in pedagogies such as in-class discussion, when some may be uncomfortable with questioning authority or dealing with conflict in the class. A third issue relates to linguistic competence and how it can hinder student capacity to read and write critically, as well as their confidence in expressing themselves orally. These three areas represent diversity among students that needs to be considered when teaching critical thinking. These differences may be found in a regular North American classroom, particularly when there are many international students present. One cannot assume that non-Western students face a mono-lithic difficulty in developing critical thinking, even if they all come from the same culture/country. Within the same culture/country, youth have variable exposure to critical thinking in school, have varying degrees of comfort with pedagogies for promoting it, and have different linguistic abilities that affect their capacity to express criticality effectively.

Finally, instructors should consider how the sociopolitical environment may impact students' capacity to apply critical thinking outside the classroom, as well as whether the sociopolitical environment requires a differently conceived contextual approach to critical thinking that helps students respond to the most pressing challenges in their environments.

Reconceiving critical thinking in a way that incorporates elements of empa-thy and social justice may not only better address the needs of a sociopolitical environment riddled with conflict, but also be a better pedagogical approach for students unfamiliar with and intimidated by more well-known approaches that incorporate skepticism, antagonistic questioning, and debate.

While all of the challenges posed throughout this chapter are faced to vary-ing degrees in most educational contexts, they are more exaggerated in Western institutions promoting critical thinking with non-Western students, particu-larly when these students' home environments contain large degrees of socio-political upheaval and uncertainty. Critical thinking as a general concept need not be considered culture specific, but as a notion that can be reconstructed and contextualized, and approaches to developing it do need to consider cul-tural context, with all its nuances.

Acknowledgments

I would like to thank my PhD thesis supervisors at the University of Sheffield, Professor Jon Nixon and Dr. Alan Skelton, for their feedback on earlier thinking about this topic. I would also like to thank the reviewers for their comments.

References

Aboulghar, M. 2006. "Egyptian Educational Decline: Is There a Way Out?" *Al-Ahram Weekly*, February 2–8, 2006.

Asgharzadeh, A. 2008. "The Return of the Subaltern: International Education and Politics of Voice." *Journal of Studies in International Education* 12 (4): 334–336.

Atkinson, D. 1997. "A Critical Approach to Critical Thinking in TESOL." *TESOL Quarterly* 31 (1): 71–94.

Bali, M. 2013a. "Critical Citizenship for Critical Times." *Al-Fanar Media*, August 19, 2013.

Bali, M. 2013b. *Critical Thinking in Context: Practice at an American Liberal Arts University in Egypt*, Education. Thesis. Sheffield, England: University of Sheffield.

Bali, M. 2013c. Why Doesn't This Feel Empowering? The Challenges of Web-Based Intercultural Dialogue. *Teaching in Higher Education*. Advance online publication.

Bali, M., and Carpenter, V. 2009. "English Language Institute Needs Assessment." TESOL, Boston, MA March, 2010. *Report at the American University in Cairo.*

Barnett, R. 1997. *Higher Education: A Critical Business.* Buckingham: Society for Research into Higher Education and the Open University Press.

Barnett, R., and Coate, K. 2005. *Engaging the Curriculum in Higher Education.* Berkshire, England and New York: Society for Research into Higher Education and Open University Press.

Baxter Magolda, M. B. 2004. "Evolution of a Constructivist Conceptualization of Epistemological Reflection." *Educational Psychologist* 39 (1): 31–42.

Beinin, J. 2013. "Was There a January 25 Revolution?" *Jadaliyya*, January 25, 2013.

Belenky, M. F., Clinchy, B. M., Goldberger, N. R., and Tarule, J. M. 1986. *Women's Ways of Knowing: The Development of Self, Voice, and Mind.* New York: Basic Books.

Benesch, S. 1999. "Thinking Critically, Thinking Dialogically." *TESOL Quarterly* 33 (3): 573–580.

Benesch, S. 2001. *Critical English for Academic Purposes: Theory, Politics, and Practice.* Mahwah, NJ: Erlbaum Associates.

Bourdieu, P. 1973. "Cultural Reproduction and Social Reproduction." In *Knowledge, Education, and Cultural Change,* edited by R. Brown. London: British Sociological Association. 56–69.

Brodin, E. 2007. *Critical Thinking in Scholarship: Meanings, Conditions and Development.* Dissertation. Department of Education. Lund University. Linnéuniversitetet, Sweden.

Brodin, E. M. 2008. *Critical Thinking in Scholarship : Meanings, Conditions and Development. Revised Edition.* Saarbrücken: VDM Verlag Dr Müller.

Brookfield, S. 1987. *Developing Critical Thinkers: Challenging Adults to Explore Alternative Ways of Thinking and Acting.* San Francisco: Jossey-Bass.

Brookfield, S. 2007. "Diversifying Curriculum as the Practice of Repressive Tolerance." *Teaching in Higher Education* 12 (5 and 6): 557–568.

Browne, M. N., and Freeman, K. 2000. "Distinguishing Features of Critical Thinking Classrooms." *Teaching in Higher Education* 5 (3): 301–309.

Burbules, N. C. 1986. "A Theory of Power in Education." *Educational Theory* 36 (2): 95–114.

Chandler, D. 1995. *The Act of Writing: A Media Theory Approach.* Prifysgol Cymru Aberystwyth: University of Wales.

Cook, B. J. 1999. "Islamic versus Western Conceptions of Education: Reflections on Egypt." *International Review of Education* 45 (3/4): 339–357.

Egege, S., and Kutieleh, S. 2004. "Critical Thinking: Teaching Foreign Notions to Foreign Students." *International Education Journal* 4 (4): 75–85.

Ellsworth, E. 1989. "Why Doesn't This Feel Empowering? Working through the Repressive Myths of Critical Pedagogy." *Harvard Educational Review* 59 (3): 297–324.

Ennis, R. H. 1998. "Is Critical Thinking Culturally Biased?" *Teaching Philosophy* 21 (1): 15–33.

Facione, P. 1990. *The Delphi Report: Critical Thinking: A Statement of Expert Consensus for Purposes of Educational Assessment and Instruction.* Millbrae: California Academic Press.

Fadl, B. 2013. "حكومة صناعة الإرهاب" (Translation: The Terrorism-Manufacture Government. *Al-Shorouk*, December 28, 2013. http://www.shorouknews.com/mobile/columns/view .aspx?cdate=28122013&id=71e8d7ad-8413-4224-b2f6-760bf82c607a.

Floyd, C. B. 2011. "Critical Thinking in a Second Language." *Higher Education Research and Development* 30 (3): 289–302.

Fox, H. 1994. *Listening to the World.* Urbana, IL: National Council of Teachers of English.

Grundy, S. 1987. *Curriculum: Product or Praxis?* London: Falmer.

Hall, K. L. 2011. "Teaching Composition and Rhetoric to Arab EFL Learners." In *Teaching and Learning in the Arab World,* edited by C. Gitsaki. Berne, Switzerland and New York: Peter Lang. 421–440.

Johnson, L., and Morris, P. 2010. "Towards a Framework for Critical Citizenship Education." *Curriculum Journal* 21 (1): 77–96.

Kaplan, R. B. 1966. "Cultural Thought Patterns in Intercultural Education." *Language and Learning* 16 (1–2): 11–25.

Lash, J. 2001. *Exporting Education: The Case of the American University in Cairo.* Dissertation. Geography, Southwest Texas State University, San Marcos, Texas.

Nelson, G. 1992. "The Relationship between the Use of Personal, Cultural Examples in International Teaching Assistants' Lectures and Uncertainty Reduction, Student Attitude, Student Recall, and Ethnocentrism." *International Journal of Intercultural Relations* 16 (1): 33–52.

Nelson, G., El Bakary, W., and Fathi, M. 1996. "A Cross-Cultural Study of Egyptian and U.S. Education Using Hofstede's Four-Dimensional Model of Cultural Differences." *International Education* 26: 56–76.

Nixon, J. 2006. "Towards a Hermeneutics of Hope: The Legacy of Edward W. Said." *Discourse: Studies in the Cultural Politics of Education* 27 (3): 341–356.

Norris, S. P. 1995. "Sustaining and Responding to Charges of Bias in Critical Thinking." *Educational Theory* 45 (2): 199–211.

Nurullah, A. S. 2006. "Ijtihād and Creative/Critical Thinking: A New Look into Islamic Creativity." *The Islamic Quarterly* 50 (2): 153–173.

Nussbaum, M. 1997. *Cultivating Humanity: A Classical Defense of Reform in Liberal Education.* Cambridge, MA: Harvard University Press.

Nussbaum, M. 2011. *Creating Capabilities.* Cambridge, MA: Harvard University Press.

Paul, R. 1994. "Teaching Critical Thinking in the Strong Sense: A Focus on Self-Deception, World Views, and a Dialectical Mode of Analysis." In *Re-Thinking Reason: New Perspectives on Critical Thinking,* edited by Kerry S. Walters. Albany, NY: SUNY. 181–198.

Raddawi, R. 2011. "Teaching Critical Thinking Skills to Arab University Students." In *Teaching and Learning in the Arab World,* edited by C. Gitsaki. Berne, Switzerland and New York: Peter Lang.

Said, E. W. 2004. *Humanism and Democratic Criticism.* New York: Columbia University Press.

Skelton, A. M. 2005. "Internationalization and Intercultural Learning." In *Understanding Teaching Excellence in Higher Education: Towards a Critical Approach,* edited by Alan M. Skelton. London and New York: Routledge. 102–115.

ten Dam, G., and Volman, M. 2004. "Critical Thinking as a Citizenship Competence: Teaching Strategies." *Learning and Instruction* 14 (4): 359–379.

Thayer-Bacon, B. J. 1998. "Transforming and Redescribing Critical Thinking: Constructive Thinking." *Studies in Philosophy and Education* 17: 123–148.

UNDP. 2003. *Arab Human Development Report: Building a Knowledge Society.* New York: United Nations.

van Gelder, T. 2005. "Teaching Critical Thinking: Some Lessons from Cognitive Science." *College Teaching* 53 (1): 41–46.

Vandermensbrugghe, J. 2004. "The Unbearable Vagueness of Critical Thinking in the Context of the Anglo-Saxonisation of Education." *International Education Journal* 5 (3): 417–422.

Zamel, V. 1997. "Toward a Model of Transculturation." *TESOL Quarterly* 31 (2): 341–352.

20

Cultural Variance, Critical Thinking, and Indigenous Knowledges: Exploring a Both-Ways Approach

Sharon K. Chirgwin and Henk Huijser

Introduction

Critical thinking is generally considered to be one of the most crucial attributes in the Western processes of knowledge assemblage and creation and is therefore carefully nurtured in traditional higher education institutions. In Western knowledge systems, critical thinking is considered to drive not only knowledge production but also innovation and development, while it is intimately linked to a colonial history in which "progress" has been the key focus and driving force. Not coincidentally then the university as a "research and development" institution has played a central role in this colonial history. From an Indigenous point of view, such "progress" has been viewed with ambivalent feelings at best, but more often with suspicion and skepticism for good reasons (Tuhiwai Smith 1999).

Colonialism has created an ambivalent relationship between Indigenous peoples and the university as an institution, and this ambivalence is ongoing and creates tension, especially for Indigenous students who may at times feel they are being co-opted into Western ways of thinking. The concept of critical thinking is central to this ambivalence because it is at the very heart of the university's modus operandi. This degree of ambivalence is of course different for different Indigenous students. Not only are there very different degrees to which Indigenous students have been exposed to, or immersed in, Western ways of thinking but more importantly different Indigenous cultures (both outside of and within Australia) have different culturally specific ways of developing knowledge as well as different ways of expressing and passing on knowledge. Cultural variance is an important and often ignored part of Indigenous engagement with the university. This chapter explores whether a both-ways approach can be applied to critical thinking and whether it can, in the process, overcome some of the potential for conflict and ambivalence.

If we accept that critical thinking is central to a Western approach to knowledge creation and development, then it should come as no surprise that the creation of new knowledge is an essential part of obtaining a masters degree or a doctorate. However this has the potential to create problems for Indigenous students because traditional Indigenous knowledge is produced, owned, and distributed quite differently from the way it is done in Western tradition. Knowledge in many Indigenous cultures is not "open" in the same way as it is in the Western context, but instead is guarded by particular individuals, and the handing over of such knowledge is often safeguarded by strict cultural protocol. This is quite different from the Western academic context, which is fundamentally characterized by the ideas of openness to scrutiny and knowledge as situated in the "public domain."

These differences and the link to colonial values raise the question of whether critical thinking in its Western conceptualization is either relevant or desirable in Indigenous contexts. Furthermore if it is deemed to be desirable, then what is an appropriate way to develop it? In attempting to answer such questions, it is important to reiterate that cultural variance means that there are many different degrees of "tradition," and exposure to a Western context for Indigenous students, despite what a "both ways" approach may imply, should not be seen as a strict binary between Indigenous and Western knowledge but rather as a continuum.

This chapter draws on our experiences in our work with Indigenous Higher Degree by Research (HDR) students who are studying at Batchelor Institute of Indigenous Tertiary Education. Batchelor Institute is a dual-sector tertiary education provider specifically established for Indigenous Australians. It has been forced to address these issues, particularly in the context of providing research training and support to those undertaking masters and doctoral degrees by research. Driven by the wishes of Indigenous community leaders, in 1999 the Batchelor Institute adopted the philosophy of "both ways," which in its earliest form represented the opportunity for simultaneous Aboriginal cultural sustainability and academic success (Harris 1990, xii), two key features that have been pivotal in Batchelor Institute's vision. For example in the most recent strategic plan the vision is that the institute should be "a site of national significance in Indigenous Education—strengthening identity, achieving success and transforming lives" (2012).

In its simplest representation, "both-ways" is a philosophy of education that "brings together Indigenous Australian traditions of knowledge and Western academic disciplinary positions and cultural contexts, and embraces values of respect, tolerance, and diversity (2012) (*Batchelor Institute Strategic Plan 2012–2014* 2012, 6). The "both-ways" philosophy is founded on the metaphor of Ganma used by Marika (1999), and based on Yolngu culture of North East

Arnhem Land in Australia's Northern Territory. The Ganma process occurs in a space where fresh water (Yolngu knowledge) and salt water (non-Aboriginal knowledge) come together in a briny lagoon. This lagoon is a nutrient-rich environment in which some plant and animal life lives that is not found elsewhere. So too with both-ways—there is no need to compromise either epistemological position, but rather a new space can come into being that supports the creation of new understandings and knowledge (Bat, Kilgariff, and Doe 2014). Batchelor Institute acknowledges that this is the metaphor that has formed the basis of development of the both-ways philosophy (Ober and Bat 2007). It is also important to note at this point that Batchelor Institute has adopted the Australian Government approach to Indigenous identity. That is, it accepts all students who identify as being Indigenous Australians. While the title of the institution tends to attract interest from Indigenous peoples from other countries, and the institution has the ability to grant special permission for the enrolment of those who do not identify as Indigenous or are Indigenous but not Australian-born, most commonly the only group to take advantage of this is non-Indigenous staff members. While the institution has strong links to a range of remote Northern Territory communities that still have strong cultural traditions and beliefs, the higher education students tend to be a diverse group from all states of Australia. Many students started their tertiary education journey in other institutions and have been attracted to Batchelor Institute because it offers them the opportunity to explore their identity in a culturally safe and encouraging environment.

Cultural safety in this context means an explicit recognition of cultural variance *within* Indigenous identity, which actively works against the perpetuation of an Indigenous versus non-Indigenous binary. In order to describe the approach Batchelor Institute has taken to overcome what may seem a challenging divergence, we first explore the Western concept of critical thinking and some of the relevant characteristics of "both ways," including how it accommodates cultural variance. Further, some of the relevant features of Indigenous knowledges, Indigenous languages, and the nature of thinking within Indigenous knowledge systems need to be understood before the potential "incommensurability" between Western and Indigenous thinking (Moreton-Robinson 2003) can be analyzed and strategies proposed.

Critical thinking skills: defining the context

Critical thinking has increasingly come to be seen as *the* crucial underlying skill or attitude to be developed in a higher education context, since it is seen as the fundamental skill that will drive research and innovation and by extension grow the knowledge-based economies of the twenty-first century (Davies, Fidler, and Gorbis 2011). Irrespective of whether a student is Indigenous or non-Indigenous, Davies notes that "there is not a university today (in Australia at least) that does not proudly proclaim that their graduates will, as a result of a degree program in their institution, learn to think critically" (2011, 255). He goes on to express serious doubts about how successful universities actually are at teaching critical thinking, a position that is well supported by a large body of research (Arum and Roska 2011; Larson, Britt, and Kurby 2009; Rimer 2011).

This perceived lack of success in teaching critical thinking skills may in the first instance be traced back to the many different definitions of critical thinking, which have led to corresponding differences in how it is taught. These differences have been explored in great depth particularly as part of the long-running "generalist" versus "specificist" debate (Robinson 2011). Some argue that critical thinking is a discipline-specific skill and should therefore be taught in the context of a Western discipline, and importantly, by using Western discipline-specific language (Hammer and Green 2011; McPeck 1981; 1990). For others, such as Ennis (1987), "critical thinking is at heart a universal and generic quality" (cited in Moore 2011). We argue that it is more productive to look at it as potentially both. Still others, such as Davies, argue for an "infusion approach" whereby critical thinking is taught as a generic skill in the context of the disciplines (Davies 2006; Davies 2008). Since this debate has been widely discussed elsewhere the arguments will not be revisited here. However, some of the definitions used in the debate will be, since they form the basis of exploring whether, and/or how, they would apply in Indigenous contexts or with Indigenous students from culturally highly diverse backgrounds.

One of the most common descriptions of a critical thinker is someone who possesses higher-order thinking skills in Bloom's (1956) sense of the word or has the ability to analyze effectively and solve problems in an able manner. Thus it is not surprising to find that these characteristics are embedded in the many definitions that have been suggested over time, albeit in slightly different ways. For example John Dewey (1933, 6) provided a foundational definition of critical thinking: "active, persistent and careful consideration of a belief or supposed form of knowledge in the light of the grounds that support it." Edward Glaser (1941, 5), cocreator of the broadly used Watson-Glaser instrument to test critical thinking, refined Dewey's definition of critical thinking to a process guided by three elements: (1) an attitude of being disposed to consider in a thoughtful way the problems and subjects that come within the range of one's experiences; (2) knowledge of the methods of logical inquiry and reasoning; and (3) some skill in applying those methods. Particularly the latter point is seen as a crucial skill in knowledge-based economies, driven by a research and development ethic.

An alternative cognitive psychology perspective can be traced back to Bloom's (1956) work where he identified critical thinking as a skill on the higher scale of his taxonomy of educational objectives such as analysis, synthesis, and/or evaluation, in contrast to lower-order thinking skills such as knowledge, comprehension, and/or application. Other well-known definitions of critical thinking are: "disciplined, self-directed thinking" (Paul 1990, 52); "reasonable reflective thinking focused on deciding what to believe or do" (Ennis 1993, 180); "self-directed, self-disciplined, self-monitored, and self-corrective thinking" (Paul and Elder 2006, 4); and "purposeful, self-regulatory judgment which results in interpretation, analysis, evaluation, and inference" (Facione 1990, 6).

Finally Davies (2006) endorses four sets of skills as outlined by Ikuenobe (2001, cited in Moore 2011, 263–264):

1. understanding concepts of argument, premise, conclusion, propositions (statements);
2. identifying statements from nonstatements, and isolate(ing) premises and conclusions;
3. understanding the concepts of truth and validity, soundness and fallacy; and
4. identifying fallacies in inferences and explain why they are fallacious.

Out of all of these definitions and skill sets associated with critical thinking, Moore ultimately distils one single point on which he claims almost everyone agrees: "that teaching students to be 'critical' in their studies is an intrinsic good and that it is this, perhaps more than anything else, that should be the goal of a higher education" (Moore 2011) for it ultimately leads to critical thinkers. Furthermore "the possession of a critical outlook is seen not only as an essential part of engaging with knowledge in the academy, but also crucially to being an engaged citizen in the world" (261). These are common statements and they appear "universal" in their application but that is precisely what makes them potentially problematic in Indigenous contexts. For a start, and despite the fact that science is an increasingly collaborative pursuit, all of these definitions implicitly approach critical thinking as an *individual* skill that can be assessed on an individual level.

This is not the approach in some Indigenous contexts, where knowledge is seen as communal, and questioning that knowledge, or sharing it beyond the community, is in some cases considered inappropriate and can lead to sanctions for the individual. In other words, in the culturally variant and diverse contexts of Indigenous Australia, Western assumptions that underlie the concept of critical thinking and its value are not always appropriate, and in some cases raise a number of questions that need to be considered when teaching Indigenous students critical thinking skills.

For example, what does being "an engaged citizen in the world" mean for an Indigenous student from a remote community like Maningrida in Australia's

Northern Territory? Moreover how appropriate is critical thinking in the cultural context from which some Indigenous students come? Appropriate here is not a value judgment on our part but rather an awareness of how knowledge is passed on and communicated in various Indigenous contexts, which is in some cases directly related to the social position of the person who *chooses* both whether to pass the knowledge on and the manner in which to do so. Individuals are merely "guardians" of knowledge, which is considered to be communally owned. Even if critical thinking is deemed appropriate, how should it be assessed? For example, should it be assessed in the traditional format of the academic essay? Or are there oral or visual modes through which to express or demonstrate, and therefore also assess, critical thinking skills?

Language, and especially written language, has a privileged position in an academic context, which in turn means that critical thinking is closely linked to academic literacy in the traditional sense of the word. This has huge implications, both at the level of language and culture, for second-language speakers (for some Indigenous students, English is their third or fourth language). For example in an article about international Chinese students in Australia, Floyd notes that there is a perception that critical thinking is "essentially a Western skill and not valued in Confucian cultures" (Floyd 2011, 290). It is associated with core Western values, such as individualism and scientific exploration, which foreground skills such as analytical thinking, active learning, and abstract reasoning (Zhou and Pedersen 2011, 163). This is then contrasted with "Confucian values" such as collectivism and spiritual perfection, which foreground skills such as practical reasoning, passive learning, and holistic thinking (p. 163). Similarly, Nisbett et. al. (2001, 291, original emphasis) found:

> East Asians to be *holistic*, attending to the entire field and assigning causality to it, making relatively little use of categories and formal logic, and relying on "dialectical" reasoning, whereas Westerners are more *analytic*, paying attention primarily to the object and the categories to which it belongs and using rules, including formal logic, to understand its behaviour.

While their findings are very detailed and backed up by a lot of evidence, it is still important to guard against essentialism, as these kinds of binary oppositions have a long history in Western thought, and are in many respects transferable to the ways in which Indigenous students are often perceived. This not only applies to Australian Indigenous students but is also echoed in other contexts with Indigenous populations, such as Canada, New Zealand, and the United States. Sefa Dei (2000, 111, our emphasis) notes for example that "when located in the Euro-American educational contexts, Indigenous knowledges can be fundamentally experientially based, non-universal, *holistic* and relational knowledges of 'resistance.'" The inevitable outcome is that

Indigenous students will often be perceived as learners with "lacks" or "deficits" (Floyd 2011), because they are being measured against a standard which firstly is not of their making, and secondly may not be appropriate, either culturally or educationally. Again, appropriate here is not a value judgment on critical thinking itself but rather the process of teaching and assessing critical thinking skills.

Ultimately, this raises a number of key questions around critical thinking that can be divided into two streams: (1) how, and the extent to which, critical thinking aligns with different and variant cultural frameworks; and (2) how critical thinking relates to linguistic expression. In the former case it requires a rethink about whether critical thinking (in its Western guise) is appropriate in Indigenous contexts, and if so, how. In the latter case, it requires a rethink about how we assess critical thinking skills, and whether for example the written essay is the only appropriate form to assess such skills, especially for second- or third-language learners. If not, what other forms of expression (e.g., visual or oral forms) may be used to express critical thinking skills? These issues relate to culture and inevitably lead to the need for a more detailed consideration of the position of culture (and its implications of power) in a higher education context. The following description of Batchelor Institute's attempt to ensure cultural sustainability, concurrent with efforts to develop critical thinking skills, aims to explore this along with the two key questions raised at the beginning of this paragraph.

Critical thinking in a "both ways" context

Before exploring these key questions in relation to the development of critical thinking skills in Indigenous Higher Degree by Research (HDR) students, several features of "both ways" need to be considered. In the first instance, to operate effectively in a "both ways" environment, educators must view different frameworks as coexistent without allowing any single framework to dominate pedagogical design (Hooley 2000, 3). This requires Indigenous Australian ontologies and epistemologies to be afforded equal status with dominant Western ontologies and worldviews.

It also requires an engagement with, and learning from, non-Indigenous staff who have been trained within dominant Western worldviews. For example, in many Indigenous contexts, connections to country and connections within kinship systems are a fundamental part of how individuals identify themselves and their place in the world. Knowledge transfer is a fully integrated part of that so that when a Western eye might simply see a painting or an artwork, that same painting may convey important knowledge about country and the essence of their being to an Indigenous person (Martin 2007). In this example, a painting is a culturally appropriate way to convey that particular knowledge, while an academic essay may not be, so the painting needs to be afforded equal and appropriate status.

In the second instance, as Ellis (1997, 6) identifies, it is worldviews, not a single worldview, that must be considered for there is "neither a single Indigenous way nor a single 'mainstream' way." To work in a "both ways" framework, Indigenous students not only have the challenge of understanding Western worldviews but also a range of other Indigenous world views that differ from their own, again foregrounding the significance of being conscious of cultural variance. It follows that Western-trained lecturers and supervisors with little exposure to different worldviews have even greater challenges, if they wish to effectively provide support to their students. They not only need to be aware of these differing worldviews but also must be able to understand them sufficiently to both use them and/or to guide their students to use them appropriately.

A further complication is that the Indigenous Australian students who are attracted to study within the "both ways" framework may also have views that exist along a continuum between purely Indigenous and purely Western, with the occasional influence of Eastern (Ober and Bat 2007). There are also students who identify as Indigenous that have Afghan, Chinese, or Japanese parents or grandparents, so they may have up to three different influences formulating their approach to thinking.

Nonetheless there are aspects of Indigenous Australian worldviews that have a commonality across many groups and therefore need to be appreciated in any exploration of critical thinking. Christie (cited in Hughes and More 1997) summarized the most common Australian Indigenous worldview as one where the world takes on meaning through the qualities, relationships, and laws laid down in "the dreaming." For a worldview where objects are related, the value of things lies in their quality and relatedness. This is a specific worldview or ideology, which Sharifian (2005) believes has arisen through an image schema of thinking that is a circular or spiral pattern of interconnections of ideas, events, beings and places.

Grieves (2008, 369) explains the interconnectedness or relatedness in the following manner:

> When Aboriginal and Torres Strait Islander people say that they have a spiritual connection to the land, sea, landforms, watercourses, the species and plant life, this connection exists through the law developed at the time of creation. Thus each person or specific plant or species is linked to the spirit of creation, and thus to each other.

Since objects are related through their spiritual connectedness rather than their physical properties, the Western concepts of counting and even those of contrast and comparison have little meaning. Accordingly in many Australian Indigenous languages there are few numbers, and frequently few terms, to describe many Western analytical concepts such as contrast and comparison (Hughes and More 1997). These documented differences have caused linguists to argue that systems of knowledge and ways of thinking are embedded in language (McConvell and Thieberger 2001). For example Hamilton (1981, 81) records that "the Burarra language has no single word for 'why' and so the question is so seldom asked by anyone, adult or child, that people have to think carefully before finding an expression for it suitable to the context."

Further she noted that in the early years of childhood much is learnt by imitation and that in the specific group studied, there were appropriate times for acquiring specific types of knowledge. While Harris (1990) also believes that much is learnt by imitation, he also contends that observation and trial and error are important ways of learning, particularly "real life" knowledge. Similarly, Castellano (2000) describes three processes for individual acquisition of traditional Australian Indigenous knowledge: traditional teachings, empirical observations, and spiritual insight. While the latter process might not find a ready match in many recognized Western methods of inquiry, empirical observation involves meticulous observation of natural and cultural phenomena over time, and at first glance it may therefore seem similar to the research processes used by many Western natural scientists. Despite this, Christie (2006, 79) identifies that Indigenous knowledges are responsive, active, continually renewed, and reconfigured, very local and in no way universal. Some of these, and especially the latter two, can be considered features that rarely apply to the knowledge produced by Western means.

It may therefore seem that traditional Indigenous ways of learning about and making sense of the world, and in some instances the lack of the very words and concepts that the Western mind uses to think, create barriers to achieving the simultaneous cultural sustainability and academic success described by Harris (1990, xii). But as indicated, a true "both ways" pedagogic approach allows for reciprocal or two-way learning (Yunupingu 1995, 85) and the acceptance that two or more systems can coexist, at best intertwined but at the very least in parallel. While this has challenging implications for the cultural scholarship and expertise of those trying to teach Indigenous students to think critically, the acquisition of this Western skill will have many wider benefits for Indigenous graduates who wish to participate in, and contribute to, a "both ways" (or truly cross-cultural) world.

Critical thinking in an Indigenous higher education institute

At Batchelor Institute many undergraduates come with a desire to acquire Western knowledge, and to learn about Western ways of thinking. Therefore, to some degree, they may be aligning themselves to Moore's (2011, 261) notion that they need this education to be an engaged citizen of the world. Alternatively, many HDR students have indicated that they have a desire to understand Western academia and how it assembles and disseminates knowledge so that they have the tools to improve the standing of Indigenous knowledges and facilitate cultural continuity and sustainability (Croft, Fredericks, and Robinson 1997). Ironically while critical thinking can be seen as a Western skill, those Indigenous graduates who acquire it will be better placed to challenge any residual colonial legacies that may still dominate the academy and society. It can become a powerful part of their cultural code switching, that is, an ability to select the way of thinking according to the culture engaged in at a particular time. It could further be argued that the ability to culturally code switch is another vital twenty-first-century skill.

As indicated previously, while most universities list critical thinking skills as an outcome or attribute for their graduates, the desired outcome is not always reached (Arum and Roska 2011; Larson, Britt, and Kurby 2009; Rimer 2011). Since Batchelor Institute enrols Indigenous HDR students from all over Australia from a variety of institutions and a wide range of cultural and social backgrounds, the assumption is never made a priori that a graduate with a first or even second degree is able to think critically in the Western sense. It is also accepted that by enrolling in Batchelor Institute's HDR program that must meet specified Western standards they are indicating their willingness to acquire this skill and use it at a high level, but that their many different thinking styles will have an impact on the development of a critical thinking disposition, specific to their particular research award (Zhang 2003).

All HDR students are therefore required to pass a carefully designed research training unit, culminating in a public and panel-assessed research proposal presentation, prior to starting their actual research project. The "both ways" framework dictates that there is an exploration of both Indigenous and Western paradigms and research methodologies and methods, with the program organized so that as specific skills are developed they contribute in a step-by-step manner to the thought and planning that must occur before research can commence. It also requires that the trainee researcher has sufficient in-depth knowledge of both systems to choose the most appropriate approach to complete their research. Indeed, they are consistently required to justify their approach with reference to different knowledge systems and therefore consider the potential impact of such knowledge systems on the acquisition and application of research data.

Critical thinking skills are incorporated into the preparation unit, and indeed form a crucial part of that unit, for both masters and PhD students. For example, students spend time in the library to learn how to effectively access and critically assess the wealth of information available (Mahaffey 2006). A part of the unit, *Analysis and Interpretation,* is specifically designed to provide focused practice in the analysis and interpretation of data, as well as revision in how to think critically. Given that many Indigenous students are contextual learners, there is some criticism of teaching the skill in a stand-alone manner. The approach adopted involves presenting critical thinking as an essential research skill and provides practice by critically analyzing the results of previously collected and relevant research data, acquired by both quantitative and qualitative means. This approach is supported by the findings of a study by Lloyd and Bahr (2010, 14) who found that students need to have designated opportunities to develop their critical thinking skills, and that coursework designed to promote critical thinking tends to achieve its aim.

Under the "both ways" approach, critical learning and other higher-order skills are taught using flexible learning frameworks, taking into account the variation of learner styles that Indigenous students may exhibit (Hughes and More 1997; Nichol 2006). However, three key elements determine what materials are used and how they are introduced. In the first instance, in acknowledgment of the interrelatedness of Indigenous worldviews, holistic or integrated examples are used that are situated in Indigenous contexts and may provide links to, for example, ancestral knowledge and country. Second, the method of learning through observation is respected, by providing examples that are first observed and discussed in a group, before the trainee researcher undertakes the exercise individually. Communal learning has been found to be particularly effective as a first stage. Third, methods or skills are not introduced in an isolated or abstract manner, but by initially using real-life examples that are familiar to the students. Once mastered in an everyday manner, the skill is practiced and reinforced using research examples. Ober and Bat (2007b) provide a further exploration of practical ways to implement a "both ways" approach in the classroom.

For some Indigenous Australian students who speak English as a second or third language, understanding the words and concepts associated with critical thinking and developing confidence in their use can be challenging. While most HDR students should have encountered and used these terms before as undergraduates, their understanding may be superficial and even discipline focused. There is therefore a need to reinforce their use in the context of research in a more general sense. What is particularly useful in the HDR context is the increasing amount of reflective work from prominent Indigenous academics who themselves have battled with Western terms and concepts. Their work can be used to justify why students should develop an understanding and mastery of these words. For example, writing to his sons about research as a form of ceremony, Wilson (2008, 13) states:

> In order to tell this story, it may be necessary for me to use some pretty big and daunting words. I try hard not to use these words in everyday conversations, because I think that too many people use big language as a way of

belittling others. However, some of the ideas that I want you to understand require these words, as they are able to get across a lot of meaning. Our traditional language has words that contain huge amounts of information encoded like a ZIP file within them. The English language also has such words.

It is therefore essential that the non-Indigenous academic who in some cases has the role of supervisor and/or research skills trainer is not only patient and supportive where these words and concepts need to be further developed, but also continually engages in the emerging scholarship from Indigenous academics who may provide important insights or even materials that will facilitate their acquisition.

Nakata, Nakata, and Chin (2008, 138) argue that to better support Indigenous Australian students, not only should there be acknowledgment of the areas of deficit, but most importantly there needs to be appreciation of the assets that they bring to their learning. A well-known asset that Indigenous Australians bring to the classroom is their oral skills (Nakata 2007). Therefore, assessment of the student's ability to critique research-based publications, for example, can occur through individual oral presentations rather than the more usual written forms, such as academic essays or reports. Similarly a painting could potentially serve as assessable evidence if appropriate and if the students wish to use that medium, in combination with an oral explanation. At Batchelor Institute we are only beginning to explore such alternative forms of assessment (O'Sullivan 2009), and there is a need for much more research to evaluate how effective such forms of assessment are, especially when it comes to the assessment of critical thinking skills.

Indigenous Australians who seek to gain any form of higher education live and work in a difficult, complex space, where their traditional knowledge systems can be seen as obstacles to progress along the path to modern civilisation (Nakata 2007, 182). Since their traditional knowledge systems are so intricately linked to their identity and their sense of self, they feel strongly about sustaining them. Nevertheless they must understand Western ways of critical thinking at the same time not only to gain the academic qualification that they desire but also so they can interrogate, unsettle, respond to, reinterpret, and construct alternative theories and reshape knowledge (Nakata 2006).

Indigenous Australians who come from strong traditional backgrounds have different concepts of time, ways of questioning, and protocols, but are also very diverse from group to group (Winch and Hayward 1999). A true "both ways" approach ensures that these differing ways of making sense of the world, the Western way and culturally variant Indigenous ways, must all be accepted as valid, equal alternatives. Trying to "fit" a skill like critical thinking into Indigenous knowledge systems is in a sense a retrograde step into a colonial past where mind-sets other than that of the dominant culture were undervalued. In the interest of promoting open-minded academic thinking it is more desirable that alternatives are explored and mutual understanding is attempted.

The true spirit and productive potential of a "both ways" approach ultimately allows for the arrival at what Bhabha (1994) has called "hybridity" in a "third space," where new knowledge and understanding is created that is neither one nor the other. While Bhabha's concept of hybridity was developed in the context of migration and global diaspora movements, we believe it can be applied to a "both ways" context. Moreover, it is specifically useful for a context in which a high degree of cultural variance runs along a continuum with Western culture one end of the scale and highly traditional, Indigenous cultures on the other. The "third space" in which "both ways" knowledge emerges is a productive space, rather than a reductive space, for it values culturally variant knowledge systems, rather than privileging one over the other, or the various others.

Conclusion

There is no doubt that at the present time, Indigenous research students require critical thinking skills to succeed in what remains a Western-dominated academic world, and in turn to affect change from within that world, paradoxical as that may sound. Specifically, the aim here is to carve a space of power for Indigenous knowledge systems, and thereby an acceptance that Indigenous knowledge systems reflect the ways in which knowledge is created, reshaped (Fatnowna and Pickett 2002), and therefore critically thought about. Whether such critical thinking skills are assessed in written form or in any other form or medium (such as orally, in visual form, or even in performance) is less important than the development of such skills in itself, because a true "both ways" approach allows for a variety of expressions of critical thinking that goes well beyond the traditional Western container for critical thinking: the academic essay.

Questions around who should assess such skills in Indigenous contexts, or who is in the right position, both culturally and in terms of expertise, to do so, are yet to be explored. Once such questions have been addressed, there is much research to be done on what the most effective ways would be to address critical thinking within a "both ways" framework. While an Indigenous tertiary education institute like Batchelor Institute is uniquely positioned to conduct such research, the outcomes would have wide potential applications across the higher education sector. Moreover, if we can link such research with strategies to "dislodge" critical thinking from its privileged Western position, we can begin to approach a truly "both ways" approach to teaching such skills.

References

Arum, R., and Roksa, J. 2011. *Academically Adrift: Limited Learning on College Campuses.* Chicago and London: University of Chicago Press.

Bat, M., Kilgariff, C., and Doe, T. 2014. "Indigenous Tertiary Education—We're All Learning: Both-Ways Pedagogy in the Northern Territory of Australia." *Higher Education Research and Development* 33 (5): 871–886.

Batchelor Institute. 2012. *Batchelor Institute of Indigenous Tertiary Education, Strategic Plan, 2012–2014.* Batchelor: Batchelor Press.

Bhabha, H. K. 1994. *The Location of Culture.* London and New York: Routledge.

Bloom, B., Engelhart, M., Furst, E., Hill, W., and Krathwohl, D. 1956. *Taxonomy of Educational Objectives: Handbook I: Cognitive Domain.* Vol. 19. New York: David McKay.

Castellano, M. B. 2000. "'Updating Aboriginal Traditions of Knowledge.'" In *Indigenous Knowledge in Global Contexts,* edited by George J. Sefa Dei, Budd L. Hall, and Dorothy G. Rosenberg. Toronto: University of Toronto Press. 21–36.

Christie, M. 2006. "Trans-Disciplinary Research and Aboriginal Knowledge." *Australian Journal of Indigenous Education* (35): 78–89.

Croft, P., Fredericks, B., and Robinson, C. M. 1997. "Our Agenda: Indigenous Knowledges and Postgraduate Education." In *Australian Curriculum Studies Association Biennial Conference,* Eds. Sydney.

Davies, A., Fidler, D., and Gorbis, M. 2011. "*Future Work Skills 2020.*" In. Palo Alto, CA: Institute for the Future for the University of Phoenix Research Institute.

Davies, M. 2006. "An 'Infusion' Approach to Critical Thinking: Moore on the Critical Thinking Debate." *Higher Education Research and Development* 25 (2): 179–193.

Davies, M. 2008. "'Not Quite Right': Helping Students to Make Better Arguments." *Teaching in Higher Education* 13 (3): 327–340.

Davies, M. 2011. "Introduction to the Special Issue on Critical Thinking in Higher Education." *Higher Education Research and Development* 30 (3): 255–260.

Dewey, J. 1933. *How We Think: A Restatement of the Relation of Reflective Thinking to the Educative Process* (revised edition). Boston: D. C. Heath and Company.

Ellis, B. 1997. "Pegs and Holes as the Focus of Change: Indigenous Students and Institutions of Higher Education." In *1997 Forum of the Australian Association for Institutional Research,* Eds. Adelaide.

Ennis, R. H. 1987. "A Taxonomy of Critical Thinking Dispositions and Abilities." In *Teaching Thinking Skills: Theory and Practice,* edited by J. B. Baron and R. J. Sternberg. New York: W. H. Freeman and Company. 9–26.

Ennis, R. H. 1993. "Critical Thinking Assessment." *Theory into Practice* 32 (3): 179–186.

Facione, P. 1990. The Delphi Report: Critical Thinking: A Statement of Expert Consensus for Purposes of Educational Assessment and Instruction. Millbrae: California Academic Press.

Fatnowna, S., and Pickett, H. 2002. "Indigenous Contemporary Knowledge Development through Research: The Task for an Indigenous Academy." In *Indigenous Knowledge and the Integration of Knowledge Systems: Towards a Philosophy of Articulation,* edited by Odora Hoppers. Claremont: New Africa Books. 209–236.

Floyd, C. B. 2011. "Critical Thinking in a Second Language." *Higher Education Research and Development* 30 (3): 289–302.

Glaser, E. 1941. *An Experiment in the Development of Critical Thinking.* New York: Teachers College, Columbia University.

Grieves, V. 2008. "Aboriginal Spirituality: A Baseline for Indigenous Knowledge Development in Australia." *The Canadian Journal of Native Studies* 28 (2): 363–398.

Hamilton, A. 1981. *Nature and Nurture: Aboriginal Child-Rearing in North-Central Arnhem Land.* Canberra: Australian Institute of Aboriginal Studies.

Hammer, S., and Green, W. 2011. "Critical Thinking in a First Year Management Unit: The Relationship between Disciplinary Learning, Academic Literacy and Learning Progression." *Higher Education Research and Development* 30 (3): 303–316.

Harris, S. 1990. *Two-Way Aboriginal Schooling: Education and Cultural Survival*. Canberra: Aboriginal Studies Press.

Hooley, N. 2000. "Reconciling Indigenous and Western Knowing." In *Annual Conference of the Australian Association for Research in Education*, Eds. Sydney.

Hughes, P., and More, A. J. 1997. "Aboriginal Ways of Learning and Learning Styles." In *Annual Conference of the Australian Association for Research in Education*, Eds. Brisbane.

Larson, A. A., Britt, M. A., and Kurby, C. A. 2009. "Improving Students' Evaluation of Arguments." *Journal of Experimental Education* 77 (4): 339–365.

Lloyd, M., and Bahr, N. 2010. "Thinking Critically about Critical Thinking in Higher Education." *International Journal for the Scholarship of Teaching and Learning* 4 (2): 1–17.

Mahaffey, M. 2006. "Encouraging Critical Thinking in Student Library Research: An Application of National Standards." *College Teaching* 54 (4): 324–327.

Marika, R. 1999. "Milthun Latju Wanga Romgu Yolgnu: Valuing Knowledge in the Education System." *Ngoonjook: A Journal of Australian Indigenous Issues* 16: 23–39.

Martin, K. 2007. "The Intersection of Aboriginal Knowledges, Aboriginal Literacies and New Learning Pedagogies for Aboriginal Students." In *Multiliteracies and Diversity in Education: New Pedagogies for Expanding Landscapes*, edited by Annah Healy. Melbourne: Oxford University Press. 58–61.

McConvell, P., and Thieberger, N. 2001. "State of Indigenous Languages in Australia-2001." In *Australia State of the Environment. Second Technical Paper Series No. 2 (Natural and Cultural Heritage)* ed. Canberra: Department of the Environment and Heritage.

McPeck, J. 1981. *Critical Thinking and Education*. New York: St. Martin's Press.

McPeck, J. 1990. *Teaching Critical Thinking: Dialogue and Dialectic*. New York: Routledge.

Moore, T. 2011. "Critical Thinking and Disciplinary Thinking: A Continuing Debate." *Higher Education Research and Development* 30 (3): 261–274.

Moreton-Robinson, A. 2003. "'I Still Call Australia Home': Indigenous Belonging and Place in a White Postcolonising Society." In *Uprootings/ Regroundings: Questions of Home and Migration*, edited by S. Ahmed, Castañeda, C., Fortier, A. and Sheller, M. Oxford and New York: Berg. 23–40.

Nakata, M. 2006. "Australian Indigenous Studies: A Question of Discipline." *The Australian Journal of Anthropology* 17 (3): 265–275.

Nakata, M. 2007. *Disciplining the Savages, Savaging the Disciplines*. Canberra: Aboriginal Studies Press.

Nakata, M., Nakata, V., and Chin, M. 2008. "Approaches to the Academic Preparation and Support of Australian Indigenous Students for Tertiary Studies." *Australian Journal of Indigenous Education* (37 [Supplement]): 137–145.

Nichol, R. 2006. "Towards a More Inclusive Indigenous Citizenship, Pedagogy and Education." *Primary and Middle Years Educator* 3 (2): 15–21.

Nisbett, R. E., Peng, K., Choi, I., and Norenzayan, A. 2001. ""Culture and Systems of Thought: Holistic versus Analytic Cognition." *Psychological Review* 108 (2): 291–310.

Ober, R., and Bat, M. 2007. "Paper 1: Both-Ways: The Philosophy." *Ngoonjook: a Journal of Australian Indigenous Issues* (31): 64–86.

Ober, R., and Bat, M. 2007b. ""Paper 2: Both-Ways: Philosophy to Practice." *Ngoonjook: a Journal of Australian Indigenous Issues* (32): 56–79.

O'Sullivan, S. 2009. "Intermedia: Culturally Appropriate Dissemination Tools for Indigenous Postgraduate Research Training." *Journal of Australian Indigenous Issues* 12 (1–4): 155–161.

Paul, R. 1990. "Critical Thinking: What, Why and How." In *Critical Thinking: What Every Person Needs to Survive in a Rapidly Changing World*, edited by A. J. A Binker. Rohnert Park, CA: Center for Critical Thinking and Moral Critique, Sonoma State University.

Paul, R., and Elder, L. 2006. *The Miniature Guide to Critical Thinking: Concepts and Tools.* Dillon Beach, CA: The Foundation for Critical Thinking.

Rimer, S. 2011. *Study: Many College Students Not Learning to Think Critically.* May 28. Available from http://www.mcclatchydc.com/2011/01/18/106949/study-many-college -students-not.html–.UaVBAr-jPTp.

Robinson, S. R. 2011. "Teaching Logic and Teaching Critical Thinking: Revisiting McPeck." *Higher Education Research and Development* 30 (3): 275–287.

Sefa Dei, G. J. 2000. "Rethinking the Role of Indigenous Knowledges in the Academy." *International Journal of Inclusive Education* 4 (2): 111–132.

Sharifian, F. 2005. "Cultural Conceptualizations in English Words: A Study of Aboriginal Children in Perth." *Language and Education* 19 (1): 74–88.

Tuhiwai Smith, L. 1999. *Decolonizing Methodologies: Research and Indigenous Peoples.* London: Zed Books.

Wilson, S. 2008. *Research Is Ceremony: Indigenous Research Methods.* Halifax and Winnipeg: Fernwood Publishing.

Winch, J., and Hayward, K. 1999. "'Doing It Our Way': Can Cultural Traditions Survive in Universities?" *New Doctor* Summer (25–27).

Yunupingu, M. 1995. *National Review of Education for Aboriginal and Torres Strait Islander Peoples.* Edited by Education and Training (DEET) Department of Employment. Canberra: Australian Government Publishing Service.

Zhang, L. F. 2003. "Contributions of Thinking Styles to Critical Thinking Dispositions." *Journal of Psychology* 137 (6): 517–544.

Zhou, P., and Pedersen, C. 2011. "Understanding Cultural Differences between Western and Confucian Teaching and Learning." In *Beyond Binaries in Education Research*, edited by W. Midgley, M. A. Tyler, P. A. Danaher, and A. Mander. New York & London: Routledge. 161–175.

21

Critical Thinking Education with Chinese Characteristics

Yu Dong

Introduction

This chapter considers the development of critical thinking education in China. On the one hand, progress in critical thinking education in China has been made since the late 1990s, including textbooks, courses, articles, projects, conferences, etc. On the other hand, the development in reality is sluggish, difficult, and with undesirable "Chinese characteristics." In our analysis, the most important factors underlying the problems are not traditional Chinese collectivism or difficulties in teaching Chinese students. We argue instead that the main resistance comes from the uncritical cognitive dispositions in the Chinese tradition and the materialistic values of modern Chinese society. Based on this understanding, we outline a strategy for achieving progress and steering critical thinking education to a better and faster track.

Critical thinking education development in China

According to Meixia Li (2012), the first articles introducing ideas of critical thinking education to China appeared in 1986 and 1987. However, attempts to apply the ideas to higher education in China did not begin until the mid-1990s, when a need emerged to redesign the largely impractical logic courses to be more useful. With growing appeals in China to reform its higher education to foster thinking skills, a number of logic instructors began to look to the critical thinking movement in the West for ideas and inspiration. They translated and published critical thinking articles and books useful for educational reform (e.g., Browne and Keeley 1994). In 1997, China established its own national MBA program entrance examination, which contained questions similar to those used to test critical thinking skills in the GRE, GMAT, and LSAT in the United States. With rising demand, logic instructors began to extend the contents of their courses to ordinary language arguments. The course and textbook by Zhenyi Gu *Argumentation and Analysis: Application of Logic* (2000) resulted from this stage of the development.

In the following years, some of these instructors continued to reform their logic courses to be more like a critical thinking course as they understood it. 2003 was a milestone year: the first course with the title "Logic and Critical Thinking" was launched simultaneously in two universities in Beijing, China Youth University for Political Sciences and Beijing University. The instructors' textbook (Gu and Liu 2006) was published three years later and the Ministry of Education decided to make it the recommended textbook for all reformed logic courses. Also in 2003, the Formal Logic Sub-Committee of the Chinese Association of Logic decided that the theme of its annual conference would be informal logic and critical thinking, marking a "turn to critical thinking" of the logic education reform movement (Xiong 2006). Soon thereafter, in early 2004, two other universities in Beijing—Chinese University of Politics and Law and Renmin University of China—also started the "Logic and Critical Thinking" course. Renmin later published its own textbooks (Chen and Yu 2011; Yang 2007). Another logic instructor at the forefront of this critical thinking shift was Hongzhi Wu at Yan'an University, who published a textbook *Critical Thinking: Based on the Tool of Logic of Arguments* (Wu and Liu 2005; revised edition Wu and Zhou 2010).Later, these instructors and others continued to introduce the ideas and value of critical thinking to Chinese people.

Another important development was that, as of 2008, a critical thinking course was taught to students at Qiming College in Huazhong University of Science and Technology (HUST). In contents and pedagogy, this course is in a class of its own, and is being advocated by some educators in promoting it as a model of critical thinking education in China.

Qiming College is an institute for educational reforms. Its students are selected for their good attitudes and grades from science, engineering, management, medical, and other schools in HUST. Class sizes for the critical thinking course at Qiming College are mostly between twenty and forty students, taught by the present author and two other Western-trained Chinese instructors of logic and philosophy.

Its textbook, *Principles and Methods of Critical Thinking* (in Chinese) (Dong 2010), was the first textbook, and is still the only one, to combine Western textbook contents with real examples and exercises from China's culture and society. Its topics were based on the models of the process of critical thinking developed by Hitchcock (Hitchcock 1983; 2012) and Ennis (1996), and covered Ennis's twelve constitutive abilities of critical thinking (see the list below, in section 2).

The Chinese real-life examples in the textbook and teaching covered a wide range of topics, including feng shui, traffic problems, pollution, food safety, officials' corruption, the gap between the rich and the poor, economic materialism, fake products, media objectivity, social stability, university fees, smoking, the death penalty, house prices, egoistic individualism, and, yes, Internet censorship.

Another exceptional aspect of this course was its interactive and practical style of teaching and testing. The teachers encouraged students to engage in active, reasonable, open-minded, and practical thinking exercises. Under the title "inquiry-based critical pedagogy," the following active and critical learning and teaching techniques were practiced: Socratic questioning, just-in-time teaching, problem-based learning, cooperation learning, peer teaching, group projects, guided practice, and scenario-based exercises and tests. The students' final marks were based on a combination of participation in learning, a group project on analyses of real-life examples, critical writing, and a classroom presentation by each group.

The course drew high praise from the students and administrators, who concluded that it should be held up as a model of effective teaching and learning for all subject-matter courses as well (Dong and Liu 2013; Gan 2010; Gu 2011). The "inquiry-based critical pedagogy" later formed the core of critical thinking teacher training programs at Shantou University in 2012 and at HUST in 2013. The teachers applauded it as the most enlightening training in their careers (Chen S. B. 2013; Yu 2013).

Encouraged by this success, in May 2011 HUST hosted the first nationwide conference for developing critical thinking courses, attended by more than 70 instructors from 35 Chinese universities (many sent by their administration). Professor David Hitchcock gave a speech on critical thinking as an educational ideal, providing the audience with clarifications for understanding critical thinking and helpful suggestions for designing and teaching such courses (Hitchcock 2012). In the following two weeks Hitchcock and Yu Dong taught a demonstration course to a class of 68 students, with a content and style of teaching that would foster both critical thinking skills and a critical spirit. This direct and comprehensive demonstration was greatly appreciated by the students and a group of teacher observers, many of whom reported the course to have opened their eyes to the value of critical thinking and the styles of teaching and learning.

As an outgrowth of the conference, a group of the participants decided to establish a cooperative network in China: the Association for Critical and Creative Thinking Education (ACCTE). The association has been publishing a bi-monthly electronic newsletter since June 2011.

Also assisted by the ideas and information from the conference, in the Fall 2011 semester, as part of its thinking skills training program, Shantou University started a critical thinking course that was compulsory for all first-year students—the first compulsory critical thinking course in China. A unique feature of this course is that, with financial support from the Li Kashing Foundation, it hires a team of 10–12 teaching assistants to support large-class lectures with additional small-class tutorials, a system common abroad but still rare in China. The course was also recommended by ACCTE as a large-class model for critical thinking education in China. Also a first in China, in 2012 Shantou University held a critical thinking program to train instructors selected from various faculties to infuse critical thinking into their subject-matter courses.

Over the last several years, awareness of the ideas and value of critical thinking has been spreading at an accelerated pace across China. The number of scholarly articles about critical thinking in the fields of education, psychology, logic, philosophy, English studies, and so on is increasing rapidly (Li 2012). More critical thinking books have been translated and have sold well as general reading (Kirby and Goodpaster 2010; Moore and Parker 2012; Paul and Elder 2010; Ruggiero 2010; Weston 2011). This trend reflects growing demands for critical thinking education in China. In academic journals and media, the term critical thinking is sometimes used to connote an exit strategy needed to depart from China's age-old education tradition of rote learning. In 2010, during the fourth international forum of university presidents in Nanjing, Richard Levin, president of Yale University, declared that a major problem with Chinese students is lack of critical thinking education (Levin 2010). This statement from such a prominent figure in Western education strongly reinforced what many Chinese had been saying. Now even high-level government officials, including former premier Jiabao Wen, have begun to affirm the importance of critical thinking education for China in their public speeches (Wen 2012).

Two types of problems with developing critical thinking education: Chinese characteristics

We have briefly outlined the pioneering efforts of critical thinking education as well as the rapid growth of recognition of its value in Chinese society since the late 1990s. However, when one starts to look more closely at its development in higher education, one is surprised and disappointed. Something unexpected has taken place in most educational institutions: nothing. With regard to critical thinking education, China is still in its infancy, and its growing up has been difficult.

The first evidence of the difficulties is with regard to the quantity. Until 2013, only about 50 of more than 2,100 higher education institutions have opened a course in critical thinking, and only a few of these courses are open to students outside philosophy departments or special programs. For many years, critical thinking education has not been able to expand. (Some universities opened logic courses under such titles as "thinking skills" to include ordinary language argument contents; they are at the pre-2003 stage of the logic course reform process.) Since both demand for genuine critical thinking education and supplies of education materials are plentiful and steadily increasing, the actual progress is oddly sluggish.

The efforts to expand critical thinking education to more institutions through the 2011 conference and demonstration courses have so far met with limited success. For most of the instructors participating in the event, either out of their own interest or at the request of their administrations, hoping to start such a course in their institutions, the goal is still not in sight, for various reasons. Many logic instructors have openly expressed their unwillingness to start such a course, declaring it an unnecessary addition to the logic course or that it is too difficult (Chen J. 2013).

The second evidence of difficulty faced in developing critical thinking education in China is with regard to quality. Even in places where a course with the phrase "critical thinking" in its title is taught, mostly it is at best a half logic and half critical thinking course, as acknowledged by Zhenyi Gu, a pioneer in such education (personal communication). The course title "Logic and Critical Thinking" is symptomatic, but even if the word "logic" is omitted from the title, the content stays the same. The journey of logic course reform toward critical thinking has not been completed even if the title has changed.

The quality issue was related to the early understanding of critical thinking as an application of logic to ordinary language arguments, consisting of only theories of definition, logical analyses of arguments, and fallacies (Chen 2002, 242–282). Over the last several years, with more access to studies of the concept and skills of critical thinking by authors such as Robert Ennis (1991), Peter Facione (1990), and Alec Fisher and Michael Scriven (1997), some Chinese scholars have acknowledged that critical thinking is not another logic course. However, even in the most recent textbooks (Chen and Yu 2011; Wu and Zhou 2010), the logical focus and contents predominate, and many constitutive topics of critical thinking are still missing.

The Chinese textbooks typically contain about half of Ennis's twelve constitutive abilities of critical thinking (Ennis 1991; 1998; 2011), usually the following:

1. Focus on a question.
2. Analyze arguments.
3. Ask and answer questions of clarification and/or challenge.
6. Deduce, and judge deduction.
7a. Induce, and judge induction (generalizations).
9. Define terms and judge definitions.

The rest are either missing or given little attention:

4. Judge the credibility of a source.
5. Observe, and judge observation reports.
7.b. Draw explanatory conclusions (including hypotheses).
8. Make and judge value judgments.
10. Attribute unstated assumptions.

11. Consider and reason from premises, reasons, assumptions, positions, and other propositions with which they disagree or about which they are in doubt.
12. Integrate the other abilities and dispositions in making and defending a decision.

Apparently, the textbooks are within the framework of regarding critical thinking as logical analysis of single arguments. There is no mention of the need for, and skills of, searching for alternative ideas, explanations, and arguments. In addition, critical thinking attitudes and dispositions are not discussed in any depth. Some logic instructors defend their exclusion of these topics by arguing that they are not included in examinations like the GRE or GMAT (Chen M. 2013). The examination-oriented goal is strongly reflected in their teaching of the course.

Another shortcoming of the textbooks and teaching is the lack of real-life Chinese examples and exercises. The textbooks mostly use examples, like abortion or euthanasia, which are translated from Western textbooks and do not impact the everyday lives of Chinese students. Even worse is the problem with the traditional one-way transmission style of pedagogy. Critical thinking is taught in Chinese classrooms very much like a formal logic course, where instructors only lecture, with few attempts to encourage thinking and participation from students. Moreover, tests are just used to ensure that the "thinking rules" are memorized. The curriculum methods rarely allow for questioning, coaching, discussing, writing, peer-instruction, inquiry-based researching/learning, and so on to develop independent, cooperative, and hands-on study habits and skills in the students.

In summary, China faces two problems in the development of critical thinking education: (1) the oddly slow pace of expanding the offering of critical thinking courses in higher education; and (2) deficiencies in quality—many existing courses have these undesirable "Chinese characteristics":

1. examination-oriented goals,
2. insufficient contents,
3. irrelevance to practice, and
4. rote-learning pedagogy.

Thus, the courses cannot reach the goal of training students to build critical thinking skills. Their failure can actually add more obstacles to its development.

What is the root of the resistance?

When thinking about the causes of these problems in China, naturally we would first look into political and ideological factors. On the one hand, it is true that governments at all levels have not matched their promises of educational reform to develop students' skills with concrete actions. This is certainly one of the causes for the slow development of critical thinking in higher education. On the other hand, governments have not played a noticeable role in obstructing educational reform either. China has been advocating openness and reform for over thirty years. Any ideology, if it is not capitalism, is already too faded to drive anything in China today. Critical thinking courses have been running in some universities without any interference from anyone; textbooks, translations, and articles are continually published. As noted above, many educational administrators were more passionate about starting a critical thinking course than their faculty members. Interestingly, Kember (2000) found a similar situation in Hong Kong universities: both students and department heads showed a higher level of support for the critical learning project than departmental colleagues.

Thus, the key questions are: Why do the faculty members not want to start a critical thinking course? Why, even in places where such a course has already started, are most of them still largely a logic course, and have remained so for years? What prevents instructors from improving their courses? If we ask these questions, we would soon hear this answer: as in many other cases, first and foremost, it is Chinese tradition that plays a powerful and persistent role in resisting its development (Li 2012).

Then, what is this tradition, and how does it play a resisting role?

It has been commonly acknowledged that Chinese traditional culture is generally uncritical. There are many discussions with regard to its negative impact on critical thinking education in China or on Chinese international students. These discussions are often focused on three kinds of cultural factors.

First, the philosophical and social tradition. As is well known, Chinese culture originated from one of the fundamental features of ancient China, clan society. Confucius (551–479 BCE) based his theory on his view of the most stable and orderly society, the patriarchal clan system in the Western Zhou Dynasty (eleventh century BCE). In his view, to have an ideal society was to build a family-like political order between the rulers and their subjects—to love one's children or younger brothers (human-heartedness) and in return to attend to one's parents and follow one's older brothers (filial piety) (Dong 1992, 40–42; Ng 2001, 29; Richmond 2007, 2). Thus, Confucianism shaped a tradition that valued respect for parents and the elderly, the collective good, social order, and harmony. This is in contrast with ancient Greek civilization, which valued independent thought, reason, and ability to debate and argue in public.

Accordingly, Chinese and Asian societies under Confucian influence are often characterized as following "collectivism" or "group thinking," as opposed to Western "individualism" or "individual thinking." The labels refer not only to the political and social rules of obedience to authority in hierarchy, but also to a type of personality characteristic of Asians: a holistic perspective, conforming to and dependent on an in-group, as opposed to the individual perspective and self-reliant personality trait in Western cultures (Richmond 2007, 2).

Second, the Chinese education tradition intertwined perfectly with Confucianism. For centuries, the goal of education was to ensure that students follow Confucian teachings: obey the rules of their roles in the clan and political order to build an ideal society. Only two types of people were allowed in the education system: authoritarian teachers and passive students. Only one learning process was possible: memorizing exactly whatever the classics and teachers say.

It is easy to see why the combination of such philosophical teachings and such an educational system would deter critical thinking. It would ensure that challenging another person's viewpoints or prescribed ideas be avoided as much as possible (Facione, Tiwari, and Yuen 2009; Ng 2001, 56–87). Most Chinese scholars agree that this too was largely responsible for the lack of new ideas and innovations in Chinese history, in contrast to the Renaissance and scientific and industrial revolutions in the West. This is the main reason for these scholars to support critical thinking education in China—to make its society more creative.

Third, differences in languages or thought patterns. Some scholars and educators argue that other cultural factors, such as differences in languages or thought patterns in Chinese and Asian learners, may affect their learning of critical thinking. In particular, there have been experimental studies intended to prove that Westerners' reasoning is more analytic and uses tacit logical rules, while Asians' thinking focuses on relationships and uses direct experiences, such as deciding logical validity by conclusion plausibility (e.g., Peng 1997; Peng and Nisbett 1999). Some also argued that basic laws of logic—including identity, noncontradiction and the excluded middle—are not applied in Chinese thought. For the Chinese, everything contains contradiction, a student in many ways is not a student (as in the case of graduate student instructors) (Peng and Nisbett 1999, 744). The spirit of the Tao or yin-yang principle entails that A can actually imply that not-A is also the case or at any rate soon will be the case (Nisbett 2004, 27).

There have been long debates among theorists about whether critical thinking or ways of reasoning are universal or culture- and context specific. The above discussions provide support for the view that critical thinking is essentially specific to Western cultures. There was even an unspoken "stereotypical view that Asian students cannot think critically" (Vandermensbrugghe 2004). An implication is that obstacles to critical thinking education in Asia reside in the culture of collectivism in which these students are brought up.

Although most educators might not agree with the stereotypical view, many of them still tend to explain the passive behavior of Chinese international students, or the apparent difficulty they face in teaching them critical thinking, as owing to such cultural factors as collectivism and the lack of alternatives to their own intellectual framework or dominant worldview, as well as to the authoritarian education they receive back home (e.g., O'Sullivan and Guo 2010, 55).

However, I would argue that, although the passive demeanor of students in the past could mostly be explained by cultural factors, we should be cautious today about this sort of explanation. China has changed.

First, in many respects, collectivism is no longer a reality of current Chinese society. It is true that a Chinese group in a restaurant still makes a "collective" decision about what to eat—as the dishes are shared. However, as Xu Jilin, a Chinese philosopher, declared, "A society of individualism has come upon us." This individualism, Xu explained, is not the type of individualism in the West, which denotes moral autonomy. Instead it is a type of egoistic individualism: "It aims to satisfy material desires, lacks any sense of public morality, and cleaves onto a self-interested concept and attitude of life" (Xu 2009). Xu just stated a widely known fact (Steele and Lynch 2012). Universities in China are at the forefront of this age of Chinese-style individualism. In the name of individual freedom, the teachers have rejected collectivism to such an extent that their activities are labeled as "individualistic teaching," meaning self-centered and lacking necessary cooperation (Xu 2007). Their students might still be quiet in the classroom, but their silence has little to do with their respect for the teacher as an authority (Wang 2003).

In fact, many studies have already revealed that the passive demeanor of Chinese or Asian international students is not due to collectivism.

Chuah reported in his case studies that the quiet and passive behavior of Asian students in his classrooms in the United Kingdom was largely due to their lack of confidence in conversing in English. He "observed that, back home in their respective countries, these students are not as quiet and passive as they are here" (Chuah 2010). With his own experience in training developing country professionals, Richmond confirmed that Asian students showed eagerness for active learning and ability to learn critical thinking skills (2007, 7–8). Bell reported that his Chinese students at Tsinghua University exhibited a well-developed capacity to critique existing practices in China and expressed thoughtful and informed views about the world (Bell 2008). In addition, Biggs and others have undertaken long and influential studies rejecting the stereotypes of Chinese learners (e.g., Watkins and Biggs 1996; 2001).

Based on 2,307 responses from students in eight Asian countries, Littlewood (2000) concluded that the "obedient listeners" stereotype did not reflect the roles the students would like to adopt. "They do not see the teacher as an authority figure who should not be questioned; they do not want to sit in class passively receiving knowledge. So if Asian students do show passive classroom attitudes, that is more likely to be a consequence of the educational contexts that have been or are now provided for them, than of any inherent dispositions of the students themselves" (33).

Educational contexts for Chinese and Asian international students often combine multiple barriers: linguistic difficulties, lack of critical thinking skills, unfamiliarity with examples and/or contexts, shyness, etc. There is also a cognitive barrier that I will discuss next. Thus, when the students first learn and do critical thinking, they do so quietly. If the teacher has taken steps to ease the environment so that they feel comfortable in speaking, they can impress the teacher with their thinking and keen ability to analyze (Bell 2008; Richmond 2007).

All these findings resonate with our teaching experience in HUST critical thinking classes. The classes were distinctive because we taught in Chinese, used Chinese real-life examples, and engaged the students. Accordingly, difficulties and confusions from linguistic differences were avoided or resolved quickly. The classes were intellectually active. We were frequently amazed by the thoughtfulness of the students' talk and work (Dong and Liu 2013). Though teaching in English, in his HUST demonstration course in 2011, Hitchcock was also pleased with his students' challenging questions, critiques, and discussions on such sensitive topics as Internet censorship (personal communication).

There may be cultural differences in thought patterns, and their implications for inter-cultural education should be studied. However, as Davies (2007) also pointed out, many of them are insignificant in everyday exchanges. The Tao is more a belief of an influential school of philosophy than an inherent disposition of the Chinese, and believing it does not prevent them from following the basic laws of logic in everyday life. A Chinese teacher in the classroom will certainly treat students as students, regardless of what the teacher thinks the students would be in other situations.

More importantly, we do not see any evidence that the differences will block Chinese students from learning critical thinking. The fact that they can do well in mathematics and logic tests shows that they can learn Western inference patterns, some of which could be contrary to their intuition. In the exercise about understanding logical validity, our students easily identified incorrect statements, including those using conclusion plausibility to decide validity.

As Ennis (1998) argued, critical thinking is not biased against particular cultures. A "group thinking" culture, if it exists, can learn to think critically in the sense of seeking reasons and alternatives. In discussing "whether traditional nonliterate people share with us our logical processes" (some argued that they do not have suppositional ability), Ennis pointed out that even if the nonliterate people do not have this ability (as some do not have it either in the West), they still can learn to use the skill for their benefit (Ennis 1998, 21–23).

In conclusion, traditional collectivism is not the cause of problems in developing critical thinking education in China. It is no longer a dominant force in society or in education. Differences in language or thought pattern do not constitute insurmountable obstacles for Chinese students to learn critical thinking. Evidently, they can learn. The real issue is that they mostly do not have an opportunity to learn. The barriers to critical thinking education in China do not reside in the students, but in the teachers, who are reluctant to teach a critical thinking course (Chen J. 2013).

So, again, why? We think that Chinese culture does play an important role in creating this situation, but in a different way from what many might have thought. In the next section, we shall describe this role, followed by a discussion of current social obstacles.

Impediments from the uncritical tradition and the modern materialistic society

There is no doubt that Confucianism influenced China in multiple ways. The first was by its doctrines—the philosophical, social, and moral teachings: obey the rules of your social roles in the hierarchy, etc. Its second, in our view, was by its status as dogma—this was truth, and therefore was incontrovertible, beyond consideration of evidence or the opinions of others. We have argued that the doctrines are not dominant in contemporary China; but the implied dogmatism about truth and knowledge still is. This cognitive tradition fundamentally shapes intellectuals' ideas of knowledge, education, and their profession. Furthermore, we would argue, it is precisely these ideas that constitute the internal and persistent resistance to critical thinking education in China. Thus the resisting role that the cultural tradition plays is more cognitive than social.

As we know, the Chinese concept of truth—which is fundamentally different from the Western concept—was about the right way of doing things (Facione et al. 2009). As stated above, Confucianism itself was a normative theory of an ideal society and a code of behavior within it. Mencius (372–289 BCE), the second greatest soul of Confucianism, argued for this Confucian paradigm by his theory of human nature: a human being must have such feelings, so the paradigm was as universal and eternal as human nature (Dong 1992, 40–42). As a normative paradigm, it was not subject to falsification by the reality of society: the "knowledge" was not a reality-based truth statement. The Chinese were urged to learn and follow this Confucian paradigm to build this ideal society. If there were any setbacks, it must be the fault of reality, not of the normative theory. Any accusation of inconsistency with the facts, a capital crime in Western academia, was not considered a problem for it at all.

Consequently, this type of knowledge determined the rational way to learn it—reading the classical books without looking outside the windows. Any evidence-based critical questioning or appeal for further proof than that of Mencius was considered not only unnecessary, but even immoral or inhuman. It would be considered an insult to the teachers too, since the knowledge was considered infallible, and therefore the teachers were too. Any difficult question that teachers could not answer would be a proof of their ineptness—a "loss of face," an unforgettable shame for the Chinese. This is one of the contrasts between Confucian questioning and Socratic questioning: an answer that a Chinese teacher would give should be the final solution to their students' puzzles, not a clue or guide for the students to find their own answers.

The dogmatism was further reinforced by the historical examination system. Success at civil service examinations depended on memorizing the classics, which remained the same over two thousand years of study. "Even the slightest deviation in thinking from established orthodox thought was likely to result in failure" (Upton 1989, 21).

This cognitive orientation itself is a tradition of irrelevance between theory and practice. Discussing the difficulties of engineering education in China, Peigen Li, a prominent engineer and then president of HUST, pointed out that the primary cause is lack of practical teaching (Li, Xu, and Chen 2012, 7). And the root of this lack is the indifferent or condescending attitude toward practice in the traditional culture, which affects the content and methods of teaching. Li argued: "A main defect of our traditional culture is a tendency to disconnect theory from practice, to focus on book knowledge rather than on practical experience, and to separate thinking from doing. It seems that our engineering students are more willing to 'explain the world'"(Li, Xu, and Chen 2012, 10).

The preference to "explain the world" is a century-old habit of Chinese intellectuals, reflecting on how they view their own role in society: "pure talk." It really means idle talk: endlessly talking about vague and empty theories without considering facts, context, accuracy, logic, or practical application. It is fairly easy to see that the intellectual separation between theory and practice contributes to the lack of innovation, since the "learning" does not reach to a level of understanding that would come from practicing how a concept works.

Obviously this two-thousand-year-old cognitive tradition is in a head-on collision with dispositions of critical thinking like "reason-seeking" and "seeking-and-being-open-to-alternatives" (Ennis 1998). The effects of the conflict are deep and extensive in both ordinary people and those in academic circles. Articles in Chinese social science journals are often full of claims, general principles, emotional stories, or rhetorical skills, but they are short of specific and reliable reasons and valid arguments. The authors neither see a need for the reasons and arguments nor know how to meet it.

We can see that this cognitive tradition constitutes another aspect of the educational context for the students. What differs today is the books the students study, like science books from the West. However, the ways they are required to learn remain basically the same: strive to read and remember the infallible truth in the books (Upton 1989, 21). Instead of the social authority of the teacher, it is the cognitive authority of knowledge—carried by the teacher—that makes them passive.

That is why critical thinking education in such an uncritical culture is fighting against a current that comes primarily from inside the minds and habits of the intellectuals. Obviously, people love to talk about critical thinking education, and hundreds of articles have been produced. But to start instruction in critical thinking and to do it effectively would be a totally different thing. First of all, this would require instructors to be models of open-mindedness and self-criticism in interactive teaching—to face the risk of losing face. This means intellectual, moral, and spiritual changes to what the instructors have comfortably inherited from their cognitive tradition: a notion of knowledge, an ideal of education, a model of teaching and learning, norms of the profession, and so on. Simply speaking, this entails a cultural transformation.

Needless to say, cultural transformation is a difficult undertaking. And worse, it is made even harder by the current Chinese society. As noted above, China is a materialistic and egoistic society, where a change would likely take place only when people can see its direct material benefit.

Pursuing power and wealth by academic study is nothing new in this world, but what differs in China is that "looking for gold and beauty in books" has been the righteous and exclusive endeavor of the vast majority of students. In history, for most Chinese intellectuals, the real purpose of studying Confucian theory was to obtain a seat in the privileged bureaucratic class with power and wealth. This has fundamentally separated Chinese scholars from many Western scholars, whose interest was mainly to discover truth.

Over the last 30 years of reform, as China has been pursuing economic development at full steam, the materialistic goals and values have only become much more dominant and have created serious problems. A *New York Times* article reported, quite accurately, that "corruption is pervasive in every part of Chinese society, and education is no exception." The universities are full of a "culture where cash is king," a "culture of bribery," and so on (Levin 2012).

In addition, as shown by previous changes, market-oriented reform in education could bring new and deeper problems. As Peigen Li pointed out, the practicality of education is lower today, as universities and the business communities have less interest in developing cooperation without direct financial benefit. Conditions for teaching and practice have deteriorated in the last ten years along with reforms to expand hastily the size of higher education without the addition of resources (Li, Xu, and Chen 2012). The quality of education is also degraded because it is overlooked by the research-based system of incentives. Teachers feel that it is not worth spending time on improving teaching (Chen J. 2013; Zhang and Jin 2010).

All of these factors not only add new roadblocks to critical thinking education, but also reduce people's will to overcome them. Existing deficiencies and restrictions in the educational system, such as the lack of a system of teaching assistants, become even harder to change. Spending money for the quality of education is not an attractive idea today.

Thus, in this materialistic and egoistic society, starting critical thinking education is viewed as an overly challenging and complicated business without good returns, or worse, with the risk of getting bad returns. A serious instructor for critical thinking would need to act like an idealist, willing to fight popular values, push more reforming changes to the system, learn the expertise, model the critical spirit, strive to make the course practical and effective, and face possible grievances. Obviously, a pool of such instructors is small in this society where "nearly everything has a price" (Levin 2012).

In summary, the cultural tradition—mainly its cognitive orientation—has formed barriers to critical thinking education in Chinese educators' minds, as it means a cultural transformation to them. The materialistic values of current society make this transformation an unworthy business. With a combination of such cultural and social factors, we can understand the pace and the way in which critical thinking education in China has been developing. A critical thinking course either would not begin, or would be changed to an examination-oriented rote learning course of logic or of thinking rules, taught in a fashion the instructors feel familiar and comfortable with. The difficulties and struggles are thus reduced, and the instructors are not transformed into a role model for critical thinking. In turn, the course is transformed into a product with Chinese characteristics, which cannot serve the purpose of fostering critical thinkers.

Strategy to promote the future development of critical thinking education

Obviously, developing critical thinking education in such a sociocultural restrictive environment is a long and ongoing endeavor. However, with the above understanding of the underlying causes of the difficulties, and with the successfully implemented HUST and Shantou University models of a critical thinking course, we are optimistic. We should be able gradually to steer critical thinking education to a better and faster track by making coordinated efforts in important areas to achieve likely and concrete progress. In turn, this could also help improve the educational environment. For this, here are some tasks we should focus on in the years to come:

1. Continue to help more people understand the broader concept of critical thinking as other than negative thinking, logical application, technical training, theoretical study, or examination preparation. A proper conceptualization is a precondition for critical thinking education to be comprehensive and practical and to win more acceptance and support.
2. Continue to advocate HUST and Shantou University models of a critical thinking course for various class sizes. They display what more helpful content, examples, pedagogy, and tests should be. This would greatly help put critical thinking education on the right track, ensuring that it will not deviate to another logic course or theoretical study of thinking rules. Recently the models have been gaining a wider recognition nationally. This enhances our confidence.
3. Train instructors of critical thinking courses and subject-matter courses. This would be a priority. In the training, emphasize fostering a critical spirit as well as developing skills. Instructors should be urged to learn to be a model of attitudes of reason seeking, open-mindedness, self-regulation, and so on, and abandon traditional ideas and habits, like that of being a preacher transmitting infallible knowledge. Such individual change might also bring about changes to the cultural and social environment for critical thinking education. It is also important to equip the teachers with an awareness of, and ability to use, effective teaching methods as an integral part of the curriculum. We have implemented the training program for university teachers to learn "inquiry-based critical pedagogy"; we shall strive to expand it nationally.
4. Make use of university English courses to teach critical thinking. If the English teachers are well trained in critical thinking attitudes and skills, their courses could be a very effective channel to teach critical thinking to a very large student population, as English is a mandatory two-year course in all Chinese universities. Combining the learning of critical thinking with learning language skills is a very beneficial solution.
5. Encourage university administrations to develop critical thinking education with more resources for teaching, in order to reform the system to introduce teaching assistants, to reward instructors in various ways for their educational work, and to build a risk-taking atmosphere for experiments in reform.
6. Establish a national research and assessment center for critical thinking information, with real-life Chinese examples and tests. Advocate adding

critical thinking questions to those important academic and professional examinations, such as the national college entry examination, the graduate school entry examination, and the teacher qualification examination. The Chinese examination-oriented tradition, where tests determine teaching and learning, can be used to develop and control what content will be taught in critical thinking courses. That is, better designed tests can enrich the education.

7. Promote critical thinking through nongovernmental organizations and networks for communication and cooperation. The two-year-old Association for Critical and Creative Thinking Education (ACCTE) and its newsletter have been quite instrumental in this respect. Further, expand critical thinking education to secondary schools and business sectors, and build partnerships with them to reinforce the efforts.

8. Last but not least, continue to persuade governments to create policies and provide resources to incorporate critical thinking into the education system at all levels, thus ensuring the fastest means to spread such education across China.

Acknowledgments

For many helpful discussions and corrections I thank David Hitchcock, Zhenyi Gu, Haixia Zhong, Filip Borovsky, and Anton Ovrutsky. I am also indebted to the two anonymous reviewers and the editors who read through a draft of this chapter.

References

Bell, D. 2008. "Chinese Students' Constructive Nationalism." *Chronicle of Higher Education.* October 21, 2013.

Browne, M. N., and Keeley, S. M. 1994. *Asking the Right Questions: A Guide to Critical Thinking.* Translated by Xiaohui Zhang and Jinjie Wang. (third edition). Beijing: Central Compilation & Translation Press. Original edition, 1990.

Chen, B. 2002. *What Is Logic?* Beijing: Peking University Press.

Chen, J. 2013. "On the Characteristics of a Critical Thinking Teacher." *Newsletter for Critical and Creative Thinking Education* 2013 (14): 28–30.

Chen, M. 2013. "Critical Thinking: From a Logician's Point of View." *Newsletter for Critical and Creative Thinking Education* 2013 (14): 28–30.

Chen, S. B. 2013. "Reflections on Critical Pedagogy Workshops." *Newsletter for Critical and Creative Thinking Education* 2013 (12): 10–22.

Chen, M., and Yu, J. 2011. *Logic and Critical Thinking.* Beijing: China Renmin University Press.

Chuah, S. 2013. "Teaching East-Asian Students: Some Observations." *The Economics Network 2010.* July 30.

Davies, M. 2007. "Cognitive Contours: Recent Work on Cross-Cultural Psychology and Its Relevance for Education." *Studies in Philosophy and Education* 26 (1): 13–42.

Dong, Y. 1992. "Russell and Chinese Civilization." *Russell* 12 (1): 22–49.

Dong, Y. 2010. *Principles and Methods of Critical Thinking: Toward the New Knowledge and Action.* Beijing: China Higher Education Press.

Dong, Y., and Liu, Y. 2013. "Critical Thinking: Foster New Type of Creative Engineers." *Higher Engineering Education* 2013 (1): 176–180.

Ennis, R. H. 1991. "Critical Thinking: A Streamlined Conception." *Teaching Philosophy* 14 (1): 5–25.

Ennis, R. H. 1996. *Critical Thinking.* Upper Saddle River, NJ: Prentice Hall.

Ennis, R. H. 1998. "Is Critical Thinking Culturally Biased?" *Teaching Philosophy* 21 (1): 15–33.

Ennis, R. H. 2011. "Critical Thinking: Reflection and Perspective, Part I." *Inquiry: Critical Thinking across the Disciplines* 26 (1): 4–18.

Facione, P. 1990. "The Delphi Report: Critical Thinking." *A Statement of Expert Consensus for Purposes of Educational Assessment and Instruction.* Millbrae: California Academic Press.

Facione, P., Facione, N., Tiwari, A., and Yuen, F. 2009. "Chinese and American Perspectives on the Pervasive Human Phenomenon of Critical Thinking." *Journal of Peking University (Philosophy & Social Sciences)* 46 (1): 55–62.

Fisher, A., and Scriven, D. 1997. *Critical Thinking: Its Definition and Assessment.* Point Reyes, CA: Edgepress.

Gan, Y. 2010. "Shaping Critical Thinking Courses with Critical Thinking." *Journal of Southwest University (Social Sciences Edition)* 36 (6): 51–55.

Gu, Z. 2000. *Argumentation and Analysis: Application of Logic.* Beijing: People's Publishing House.

Gu, Z. 2011. "Basic Thinking Skills and Critical Thinking." *Guangming Daily*, May 16.

Gu, Z., and Liu, Z. 2006. *Critical Thinking Text.* Beijing: Peking University Press.

Hitchcock, D. 1983. *Critical Thinking: A Guide to Evaluating Information.* Toronto: Methuen.

Hitchcock, D. 2012. "Critical Thinking as an Education Ideal." *Journal of Higher Education* 2012 (11): 54–63.

Kember, D. 2000. "Misconceptions about the Learning Approaches, Motivation and Study Practices of Asian Students." *Higher Education* 40 (1): 99–121.

Kirby, G., and Goodpaster, J. 2010. *Thinking: An Interdisciplinary Approach to Critical and Creative Thought.* Translated by Guangzhong Han (fourth edition). Beijing: China Renmin University Press. Original edition, 2007.

Levin, D. 2012. "A Chinese Education, for a Price." *The New York Times*, November 22, A6.

Levin, R. 2010. "Role of General Education in China's Education Development." May 5. Accessed November 8, 2012. Available from http://edu.people.com.cn /GB/8216/188867/189175/11524636.html.

Li, M. 2012. "History and Reflection of Researches on Thinking Skills of English Major Students in China." *Newsletter for Critical and Creative Thinking Education* 2012 (9): 6–17.

Li, P., Xu, X., and Chen, G. 2012. "The Causes of the Practical Education Problems in Engineering Education." *Higher Engineering Education* 2012 (3): 7–12.

Littlewood, W. 2000. "Asian Students Really Want to Listen and Obey?" *ELT Journal* 54 (1): 32–35.

Moore, B., and Parker, R. 2012. *Critical Thinking.* Translated by S. Zhu. (ninth edition). Beijing: China Machine Press. Original edition, 2009.

Ng, A. K. 2001. *Why Asians Are Less Creative Than Westerners.* Singapore: Prentice-Hall.

Nisbett, R. E. 2004. *The Geography of Thought: How Asians and Westerners Think Differently...and Why.* New York: The Free Press.

O'Sullivan, M., and Guo, L. 2010. "Critical Thinking and Chinese International Students: An East-West Dialogue." *Journal of Contemporary Issues in Education* 5 (2): 53–73.

Paul, R., and Elder, L. 2010. *Thinking: Tools for Taking Charge of Your Professional and Personal Life*. Translated by Wei Ding. Shanghai: Genzhi Publishing House. Original edition, 2002.

Peng, K. 1997. *Naive Dialecticism and Its Effects on Reasoning and Judgment about Contradiction*. Michigan: University of Michigan Press.

Peng, K., and Nisbett, R. 1999. "Culture, Dialectics, and Reasoning about Contradiction." *American Psychologist* 54 (9): 741–754.

Richmond, J. 2007. "Bring Critical Thinking to the Education of Developing Country Professionals." *International Education Journal* 8 (1): 1–29.

Ruggiero, V. 2010. *Beyond Feelings: A Guide to Critical Thinking*. Translated by Shu Gu and Yurong Dong. (eighth edition). Shanghai: Fudan University Press. Original edition, 2007.

Steele, L., and Lynch, S. 2012. "The Pursuit of Happiness in China: Individualism, Collectivism, and Subjective Well-Being during China's Economic and Social Transformation." *Social Indicators Research* 2012 (September): 1–11.

Upton, T. 1989. "Chinese Students, American Universities and Cultural Confrontation." *MinneTESOL Journal* 7: 9–28.

Vandermensbrugghe, J. 2004. "The Unbearable Vagueness of Critical Thinking in the Context of the Anglo-Saxonisation of Education." *International Education Journal* 5 (3): 417–422.

Wang, L. 2003. "Why Students Do Not Respect Teachers." *Composition King* 2003 (5): 22–23.

Watkins, D., and Biggs, J. 1996. *The Chinese Learner: Cultural, Psychological and Contextual Influences*. Hong Kong: Hong Kong University Press.

Watkins, D., and Biggs, J. 2001. *Teaching the Chinese Learner: Psychological and Pedagogical Perspectives*. Hong Kong: University of Hong Kong, ACER Press.

Wen, J. 2012. "Actively Meet Challenges of New Revolutions of Science and Technology." June 11. November 8, 2012. Available from http://www.gov.cn/ldhd/2012-07/02/content_2175033.htm.

Weston, A. 2011. *A Rulebook for Arguments*. Translated by S. Qin. (fourth edition). Beijing: Xinhua Publishing House. Original edition, 2009.

Wu, H., and Liu, C. 2005. *Critical Thinking: Based on the Tools of the Logic of Arguments*. Taiyuan: Shanxi People's Publishing House.

Wu, H., and Zhou, J. 2010. *Critical Thinking: From Perspective of Logic of Arguments*. (revised edition). Beijing: China Renmin University Press.

Xiong, M. 2006. "On the Relationship between Critical Thinking and Logic." *Modern Philosophy* 2006 (2): 114–119.

Xu, J. C. 2007. "Individualistic Teaching and Its Critique." *Curriculum, Teaching Material and Method* 2007 (8): 20–25.

Xu, J. L. 2009. "The Dissolution of the Big Self: The Transformation of Individualism in Modern China." *Chinese Social Science Quarterly* Spring.

Yang, W. 2007. *Logic and Critical Thinking*. Beijing: Peking University Press.

Yu, H. 2013. "Report on Critical Pedagogy Workshops." *Newsletter for Critical and Creative Thinking Education* 2013 (12): 22–26.

Zhang, A., and Jin, M. 2010. "Respect Scholarship of Teaching, Practical Solution to Improve Quality of Teaching in Higher Education." *China University Teaching* 2010 (11): 20–23.

Part V

Critical Thinking and the Cognitive Sciences

Work being done in the cognitive sciences is important for critical thinking in higher education. This is because critical thinking is, in part at least, a cognitive skill. Cognitive skills, by definition, involve the brain, and therefore research into the cognitive sciences is clearly relevant. But how relevant is it, and how can this research be applied to the work being done by educators?

From a cognitive science perspective, critical thinking is assumed to be a higher-order skill, similar to learning a language or playing a piano. The literature informs us that expertise in these other fields requires, by some estimates, ten years of practice at four hours per day. In this regard, work is being done by cognitive scientists on what is known as the "deliberative practice hypothesis." Evidence from these studies show that concentrated, graduated, and exercise-based practice, with regular feedback, results in greatest learning gains for higher-order skills. Skills in criticality, it seems, need to be made explicit, and deliberate and purposeful steps need to be provided to reaching levels of high educational attainment. At a purely mechanical level, this kind of evidence should inform educational practices. And yet this is not happening. Is it any wonder that critical thinking is an oft-required, yet seldom achieved, educational goal?

There is considerable work also being done in the cognitive sciences on the notion of skill transfer. This also applies to critical thinking in higher education, especially if critical thinking is (partly at least) a generic skill. How much transfer between knowledge domains occurs in critical thinking? This is, in part, an empirical question, and it clearly has application to higher education. According to Halpern (1998), we must "teach for transfer." We cannot simply hope and expect that critical thinking skills, once learned in a particular knowledge domain, will be automatically and seamlessly applied to another. This kind of evidence, again, rarely reaches to the level of pedagogical practice.

Critical thinking and the cognitive sciences really require treatment in a separate book. However, we hint at its importance in this volume.

Lau opens the charge and offers an account of the importance of metacognition, or "thinking about thinking." For Lau, critical thinking requires the ability to reflect on the reasons for beliefs by what he calls "disciplined self-regulation." He claims that to be better at critical thinking naturally requires being better at metacognition. This involves attention to be given to the psychology of learning, reasoning, and creative problem solving. Harking back to Barnett's point about educating for a world of "complexity" and uncertainty (Part I), and Green's notion of critical thinking as "life-long learning" (Part II), Lau notes that it is only through metacognition that students will be able to overcome the challenges of the modern world beset by turmoil and constant change.

He identifies a number of facets of metacognition that might be used as a basis for metacognitive education. These include "meta-conceptions" or "mindsets," that is, core concepts that ground our knowledge about thinking. He gives an example of how being told that "anxiety improves performance" results in better mathematics test results compared to control groups. Another facet of metacognition is general knowledge about cognition and how it is affected by cognitive biases; yet another is meta self-knowledge (an understanding of one's thinking skills and dispositions), and self-regulation—the ability to monitor cognitive processes and resources effectively. One of the interesting points Lau makes is that universities need to rebuild a learning culture that rewards enhancements in critical thinking. He suggests that this can only be done by explicitly addressing the importance of metacognition, not only by making a revitalized attempt to impart soft-skills, but also by embedding creativity and problem solving within discipline-specific courses of study.

Lodge, O'Connor, Shaw, and Burton extend Lau's discussion about the importance of metacognition by focusing on the fallacies, biases, and heuristics responsible for faulty reasoning about ourselves and the world. They outline some the ways in which these errors might be overcome. They note that, at its core, "critical thinking involves addressing our assumptions about how the world works," and yet frequently our cognitive biases let us down in this regard. The cognitive science literature is replete in examples of such faulty reasoning, and yet this literature often fails to percolate down to educational practice, pedagogy, and regimes of educational assessment and attainment. Providing a corrective to this, they suggest that a better understanding of the cognitive science literature offers a way of circumventing human tendencies toward less rational thought. They suggest that this might offer insights on how mental shortcuts compromise thinking in various disciplinary domains, providing a way of side-stepping the generalist-specifist debate.

Taking the discussion further, Ellington provides a model of metacognition that incorporates critical thinking. Defining metacognition as 'the ability to attend to representations of the world such that the representations themselves and their interactions become objects of study," and critical thinking as "the ability to be meta- cognitively evaluative,"he proposes a model that incorporates beliefs, goals, and desires as part of an integrated account of human higher-order processing. He provides a number of examples of how being "metacognitive evaluative" can shed light on better ways of educating for critical thinking.

It is instructive that none of the above suggestions approximate how we presently educate students in critical thinking. Currently, educational practice has nothing to say about metacognition, nor cognitive biases and blindsights—let alone how to become aware of them or correct them. According to van Gelder (2005): "The way we generally go about cultivating critical thinking is to expect that students somehow will pick it all up through some mysterious process of intellectual osmosis." Given the importance of critical thinking in higher education, and in teaching for an unknown future, this is clearly inadequate. Perhaps the time has come to explicitly teach cognitive models and routines and incorporate these in educational regimes of formative, discipline-specific practice.

Cognitive science clearly has lessons for us in relation to the teaching of critical thinking. These insights need to be made explicit to educationalists. Research intersecting education and the cognitive sciences is at the beginning of an exciting new phase.

References

Halpern, D. 1998. "Teaching Critical Thinking for Transfer across Domains: Dispositions, Skills, Structure Training, and Metacognitive Monitoring." *American Psychologist*, 53 (4), 449–455.

van Gelder, T. (2005). "Teaching Critical Thinking: Some Lessons from Cognitive Science." *College Teaching*, 53 (1), 41–48.

22
Metacognitive Education: Going beyond Critical Thinking

Joe Y. F. Lau

Critical thinking is one of the central aims of education, and many schools and universities have courses specifically devoted to critical thinking. Ennis (1989) defines critical thinking as "reasonable reflective thinking focused on deciding what to believe or do." There are of course many other definitions of critical thinking, but most of them emphasize the importance of rationality, clarity, analysis, and independence of thought. In a typical university course on critical thinking, students might study logic, argument analysis, basic scientific methodology, fallacies, and other related topics. They learn how to distinguish between good and bad reasoning, and use this knowledge to improve their own thinking.

As such, critical thinking necessarily involves a certain amount of "metacognition," or "thinking about thinking." The concept of metacognition started to gain prominence in developmental psychology around the 1970s (Flavell 1976). It is usually understood as having two components: knowledge about cognition, and the use of this knowledge in "self-regulation," which is the monitoring and control of cognition. Critical thinking must involve some amount of metacognition, since a critical thinker ought to be able to reflect upon the reasons for her beliefs, and take careful steps to ensure that her reasoning is correct. The main thesis of this paper is that the teaching of critical thinking should be expanded and re-conceptualized as part of a broader educational program for enhancing metacognition. One of the most basic reasons for teaching critical thinking is to help students improve their decisions about what to believe or what to do. This paper argues that in order to better achieve this goal, we need to go beyond critical thinking. It involves teaching more about other aspects of cognition such as the psychology of learning and reasoning and creative problem solving. We also need to help students gain better insight and control over their work habits and personality. This training in metacognition can improve the quality and effectiveness of thinking. It will in turn strengthen the learning of critical thinking and bring about more lasting cognitive gains.

The case for metacognition

A central argument for expanding the critical thinking curriculum has to do with the cognitive skills necessary for success in the modern world. First, globalization and technology have led to social upheavals, economic volatility, and global competition. Technical knowledge can become obsolete quickly. Linear and stable careers are becoming exceptions rather than the norm. The average US citizen born in the latter baby boomer years (1957–64) would have had ten jobs by age forty (US Department of Labor 2012). Critical thinking is of course more important than ever in this environment of accelerating changes. But it has to be supported by the motivation and ability to engage in lifelong learning. Metacognitive education can help students learn how to acquire new skills and expertise quickly and effectively.

Another facet of the modern economy is the high premium placed upon creativity. Nowadays, a good idea can leverage global capital and technology to achieve a worldwide impact never before possible. Consider Facebook, the popular social networking website. It began as an idea of an undergraduate student, but it reached one billion monthly active users in less than ten years. If Facebook were a country, it would be the third largest in the world after China and India. Its phenomenal success is a good reminder that we should never underestimate the power of a good idea. It also means companies and individuals must constantly adapt and innovate in order to deal with new challenges and opportunities. But when it comes to innovation, it is artificial to separate critical thinking and creativity. They work intimately together in solving the complex problems in our personal and professional lives. Without creativity, critical thinking is impotent in changing the world. But creativity in turn requires critical thinking in testing and implementing ideas. Yet the teaching of critical thinking in universities is typically completely divorced from the topic of creativity. If we are serious about helping students become more effective thinkers, there should be a better integration of these two topics. Metacognitive education can help students become more adept at monitoring their own thinking and reasoning, and there is evidence that this will enhance creative problem solving (Hargrove 2012). Our students can increase their awareness of the heuristics for solving problems and try to internalize them. They can also find inspiration in the thinking processes and habits of creative people and reflect upon the conditions that promote creativity.

However, it should be emphasized that the case for metacognition is not solely a response to new economic realities. Nor should it be seen as an attempt to turn universities into job training camps. Whatever projects we choose to engage in, the complexity of the modern world has created tremendous opportunities and challenges. We live in a world beset with deep problems in politics and social justice, and the destruction of the environment is threatening our survival. Social progress depends in part on an informed citizenry being able to think about complicated issues critically and imaginatively and to overcome parochial biases and prejudices. To help our students make better decisions and improve their reasoning, we need to equip them with a more versatile thinking toolkit. This requires taking into account recent research in education, cognitive science, social psychology, behavioral economics, and related disciplines. This paper argues that a converging theme from these diverse fields is that metacognition plays a crucial role in improving thinking skills in the long run.

A key insight of the metacognitive approach is that being a good thinker is not simply a matter of knowing the principles of correct reasoning. It has to be supported by an appropriate system of knowledge, skills, and character traits. Nearly a century ago, John Dewey argued for the importance of teaching "reflective thinking," which is the examination of an idea "in light of the grounds that support it and the further conclusions to which it tends" (Dewey 1933, 7). Reflective thinking is the precursor to what we now call critical thinking, and it includes a fair amount of logic. But interestingly, Dewey emphasized that theoretical knowledge is not sufficient for developing reflective thinking. Other personal qualities, such as being curious and open-minded, are also relevant if not more important:

> If we were compelled to make a choice between these personal attributes and knowledge about the principles of logical reasoning together with some degree of technical skill in manipulating special logical processes, we should decide for the former. Fortunately no such choice has to be made, because there is no opposition between personal attitudes and logical processes. We only need to bear in mind that, with respect to the aims of education, no separation can be made between impersonal, abstract principles of logic and moral qualities of character. What is needed is to weave them into unity. (Dewey 1933, 34)

More recent authors agree with Dewey that critical thinking requires not just knowledge but a range of thinking dispositions, motivations, and attitudes. Langer (1989) argues for the importance of "mindfulness." The philosopher Richard Paul has urged that critical thinkers ought to develop "fair-mindedness" (Paul, Willsen, and Binker 1993). Costa (1991) lists fifteen "habits of mind," while Perkins, Jay, and Tishman (1993) offer seven key thinking dispositions:

1. to be broad and adventurous,
2. toward sustained intellectual activity,
3. to clarify and seek understanding,

4. to be planful and strategic,
5. to be intellectually careful,
6. to seek and evaluate reasons, and
7. to be metacognitive.

Notice that metacognition is explicitly mentioned in this list as a thinking disposition, which is explained as

> the tendency to be aware of and monitor the flow of one's own thinking; alertness to complex thinking situations; the ability to exercise control of mental processes and to be reflective. (Perkins, Jay, and Tishman 1993, 148)

Halpern (1998) also includes metacognition in her four-part model of critical thinking instruction, one of which is the teaching of "metacognitive monitoring." This includes for example checking for accuracy, examining progress, and making appropriate decisions about the allocation of time and mental effort in problem solving.

We agree there is a whole spectrum of attitudes and dispositions that are conducive to critical thinking. We also firmly believe that metacognition enhances critical thinking (see Magno [2010] for a review of the empirical evidence). Our approach builds upon these observations but is different in at least three ways. First, we think metacognition should not be conceived as just one among many thinking dispositions. Rather, it is a set of higher-order cognitive skills and dispositions that help us acquire and regulate other thinking dispositions. Thinking dispositions are often described as intellectual virtues. Like their moral counterparts, putting special effort into one of them might mean less cognitive resources for the rest. Metacognitive self-regulation helps us achieve a better balance between these dispositions. Moreover, as Aristotle has pointed out, virtues lie between excesses and deficiencies. Being careful is a good disposition, but being overcautious can be just as bad as being careless. It is good to have a plan and be reflective, but it is also possible to overdeliberate. Given individual differences, we each have our own pattern of excesses and deficiencies. Higher-order monitoring is needed to correct and fine-tune our cognitive dispositions, and this is precisely a central function of metacognition.

The second distinctive feature of our approach concerns the role of knowledge in metacognition. Perkins, Jay, and Tishman (1993) and Halpern (1998) focus on the self-monitoring and self-regulatory aspect of metacognition, such as paying attention to our reasoning and tracking our progress. These dispositions are of course very important for critical and creative thinking. But we want to emphasize that these dispositions have to be supported by a suitable level of scientific knowledge about cognition. Reasoning itself is a cognitive process.

There is a wealth of information from psychology and cognitive science about how reasoning might fail, and how it can be made more accurate and efficient. Recent research has found that our thinking processes and dispositions are very much affected by the quirks and biases of our cognitive architecture, often in surprising and unexpected ways (Kahneman 2011). For example, we ought to be aware of our own thinking, but we often overestimate our abilities and underestimate our susceptibility to biases. Seeking more alternatives is a good habit of thought, but having too many choices can be counter-effective and leads to decision fatigue (Iyengar and Lepper 2000). Objectivity and fair-mindedness are admirable traits, but in some situations priming a sense of objectivity can actually increase discrimination (Uhlmann and Cohen 2007). What this means is that a careful, critical, and reflective attitude has its limitations. We can achieve a lot more when this attitude is combined with a suitable level of psychological literacy that helps us combat hard-to-detect biases and enhance the accuracy and effectiveness of our thinking.

The third distinctive aspect of our approach is that the teaching of critical thinking is conceptualized as one component of metacognitive education, rather than the other way around. There is widespread agreement that critical thinking ought to be one of the central aims of education. But it is also hard to deny that there are many other cognitive skills that are desirable for our students. We have already argued for the importance of creativity and lifelong learning. Others might add that our students also need to enhance their social and cultural sensibilities, emotional intelligence, and leadership and self-management skills. Again, it is worth emphasizing that this is not just a matter of getting students ready for the workplace. It is simply a recognition that there is a multitude of skills that helps us become successful at our projects, whatever they are. However, the cultivation of these skills is a lifelong process, depending on factors such as intelligence, upbringing, and personal effort. It is unrealistic to think our students can achieve their full cognitive potential with just a few years of university education. Ultimately, our students have to be responsible for their own learning and personal growth, taking into account their own unique circumstances. This implies putting critical thinking within a larger framework of higher-order cognitive skills that helps students embark on their lifelong journey of self-development. Such a framework will include basic competence in critical thinking and problem solving, an enhanced awareness of the importance of self-knowledge and positive personal habits, and the use of empirically validated methods to acquire new expertise and improve one's performance. This is what metacognitive education is all about. As we shall see later in this paper, metacognitive competence not only enhances critical thinking. More generally, it is also linked to many positive outcomes in life, helping us attain achievements that go far beyond our IQ or innate talent. We shall now discuss the nature of metacognitive education in more detail. In particular, we propose that the curriculum should include four main components:

1. *Meta-conceptions.* These are our core concepts about the nature and norms of high-level cognition. These concepts are of special importance because misunderstanding can prevent us from adopting the correct principles of thinking and learning.
2. *General knowledge about cognition.* This refers to more specific principles about cognition that can improve our thinking. They include: (a) knowledge about good thinking skills, such as the principles of critical thinking, heuristics for creative thinking, problem-solving methods, and decision theory; (b) scientific knowledge about psychological processes such as memory and reasoning, and how their performance might be affected by biases and other factors.
3. *Meta self-knowledge.* Having an accurate understanding of one's thinking skills and related dispositions, as opposed to general knowledge about cognition that applies to most people. Accurate self-understanding is important for knowing our strengths and weaknesses and for identifying areas of improvement.
4. *Self-regulation.* How to monitor and control our cognitive processes and resources effectively and develop cognitive dispositions and personality traits conducive to better thinking and learning and other positive life outcomes.

We now discuss each of these four parts in turn.

Meta-conceptions

Misunderstanding the basic nature of thinking and learning can have detrimental trickle-down effects on everyday cognition. Some of these misconceptions pertain to critical thinking itself. For example, some people dislike critical thinking because they mistakenly believe it just means criticizing others all the time, which they regard as too destructive and confrontational; or they believe creativity is incompatible with critical thinking, because they think critical scrutiny will destroy new ideas before they are fully developed. A person with these views is probably less motivated to improve his or her critical thinking. Similarly, misconceptions about truth and values can also hinder reasoning. People who uncritically accept relativism about truth might not care about arguments and evidence. Or sometimes people end up with incoherent moral judgments because they confuse moral relativism with the view that right or wrong depends on the situation. In decision making, it is not uncommon for people to think that a good decision is one that happens to have a favorable outcome. But if they fail to focus on the quality of the decision process itself, they are more likely to make bad decisions or to repeat past mistakes (Russo and Schoemaker 1990).

It is also important to have the right meta-conception about creativity. Creativity is often thought to be a matter of innate talent or a product of mysterious inspiration. But people may fail to realize that creativity in a given domain depends a lot on extensive knowledge and the development of expertise over a long period of time. Case studies and research in psychology have also documented the importance of intrinsic motivation, self-control, and other personality traits (Ericsson and Lehmann 1996; Torrance 2002). Having the right conception about creativity might turn out to be a crucial step in becoming more creative.

Our meta-conceptions about learning can directly affect our actual learning and problem-solving skills. Most teachers are familiar with students who are more interested in knowing the correct answers than the methods used to arrive at those answers. Low-aptitude students in particular often place little value on careful reasoning in problem solving, and are less likely to engage in detailed analysis (Lochhead and Whimbey 1980). Research by Carol Dweck on mind-sets confirms the importance of meta-conception in learning (Dweck 1986; Dweck and Elliott 1983). According to Dweck, individuals with a "growth mindset" are those who think of intelligence as a malleable attribute that can be improved through effort. These individuals are more likely to persist through adversity and achieve success compared with those who adopt a "fixed mindset," seeing intelligence as an inborn and static trait. The latter group is more ready to give up when they encounter setbacks in solving problems. But research suggests that mindsets can be changed. Students can improve their academic performance when they are taught that intellectual skills can be acquired and enhanced through effort and in overcoming challenges. This line of research is also relevant for the teaching of critical thinking. It is very important for students to understand that their intellectual capacities could be improved well beyond the bounds of their IQ or innate endowment.

According to social psychologists, our interpretation of stress and anxiety (e.g., as fear or excitement) can also affect our performance in solving problems. In a recent study, students who were told that anxiety can improve performance ended up with better scores at a mock Graduate Record Examination (GRE) mathematics test. Furthermore, the effect persisted in that these students on average performed about 8% better than controls in the subsequent official GRE mathematics test, and they reported being more confident of their performance and less worried about their anxiety (Jamieson and Harkins 2010). This is a dramatic illustration of the power of meta-conception. Interestingly, the intervention had no significant effect on either the mock or official GRE verbal score. One hypothesis is that the positive construal of anxiety serves to improve executive functions that involve planning and elaborate computation, which are more important for mathematical reasoning than verbal retrieval tasks. If this is correct, it seems plausible that critical thinking under demanding conditions will exhibit a similar response given its heavy reliance on executive functions.

General knowledge about cognition

Metacognitive education stresses the importance of acquiring knowledge about cognition in becoming a better thinker. We have seen how meta-conception affects critical thinking and cognitive performance. Another main component of knowledge concerns the principles governing good reasoning. They include the standard curriculum of critical thinking courses, such as the rules of logic and scientific reasoning. But as we have argued earlier, critical thinking does not work on its own. We need our creative imagination to come up with arguments, alternative explanations, and counterexamples. In trying to solve complex problems, critical analysis and creativity complement each other. Given that metacognitive education is about effective and useful cognition, the curriculum will include not just critical thinking, but also topics such as heuristics in creative thinking and problem-solving methodology.

There is of course no algorithm for creativity. But many creative individuals seem to make use of similar heuristics and thinking habits. So students might conceivably benefit by incorporating them into their own repertoire. For example, using creative problem solving often follows a cyclical process, starting with extensive research and collection of data. This is followed by intensive analysis involving activities such as reframing the problem, finding connections and patterns, and exploring alternatives. At this stage there are heuristics for problem solving that might be applied, such as those discussed in Pólya (1945). A subsequent incubation period of relaxation or sleep might then facilitate the emergence of new ideas. If so, the ideas can be tested and improved upon. If not, the whole process can be repeated until the problem is solved (Young 1975). Of course, this technique does not always deliver results, and some people might benefit from a different working pattern. But it is still worth teaching because it raises awareness that creativity is an extended process involving preparation, effort, and knowledge. Students can fine-tune a routine that suits them best based on the technique.

Another aspect of creativity that might be emphasized concerns the growing trend in modern society toward collective problem solving. For example, academic and industrial R&D processes are increasingly team based. Team-authored papers generally receive more citations, and play an increasing role in high-impact research and the filing of patents. This is not just a trend in the natural sciences. It can also be observed in the social sciences and humanities

(Wuchty, Jones, and Uzzi 2007). But the curricula of many university courses on critical thinking are often "individualist," focusing on the knowledge and skills that a single thinker ought to possess. It is worth reminding our students of the increasingly social dimension of knowledge production. Exploiting the help of social networks and learning from the best people around us can boost our problem-solving ability. However, at the same time we also need to be vigilant against the dangers of conformity and groupthink.

Conformity and groupthink are examples of thinking traps that we should avoid. Thinking traps include fallacies, and examples include overgeneralization, false dilemma, begging the question, or inappropriate appeal to authority, to name just a few. The topic of fallacy is discussed in almost all critical thinking courses. Philosophers often classify fallacious thinking using semantic or logical categories such as ambiguity, inconsistency, or lack of justification. But failures in critical thinking can also come from psychological dispositions and contextual influences. These cognitive biases are usually prevalent and persistent, affecting our minds in subtle and even unconscious ways. The teaching of critical thinking can become richer and more practically relevant if we expand the topic of fallacy to incorporate related research from psychology and cognitive science.

Take for example confirmation bias, the tendency to selectively recall information and interpret evidence in a way that conforms to our preexisting beliefs. It can lead to overconfidence, also a widespread bias. We end up with an inaccurate picture of our capacities, and our opinions become less objective because we do not pay enough attention to consider alternatives and counter evidence. Systematic and deliberate effort is needed to mitigate the effects of these biases, and it is not just a matter of knowing the rules of logic.

It is worth noting that cognitive biases can occur even when no logical fallacy is being committed. For example, racial or gender biases are sometimes unconscious and hard to detect, even among people who sincerely affirm liberal and egalitarian values. Even the price tag on a bottle of wine can influence our subjective evaluation of its taste (Plassmann, O'Doherty, Shiv, and Rangel 2008). There are also framing effects where the choice of words can unconsciously distort our memory and decision making. For example, when asked whether the tallest redwood tree in the world is higher than one thousand feet, subjects tend to give inflated estimates "anchored" around the arbitrary figure mentioned in the question (Kahneman 2011). In another experiment, subjects who participated in a prisoner's dilemma game called "Wall Street Game" behaved much more selfishly than participants in a "Community Game," even though the two games were exactly the same (Liberman, Samuels, and Ross 2004)! These are dramatic examples that illustrate the powerful but subtle effect of language on our minds.

It is of course impossible to be completely immune to these influences. Our susceptibility to many of these biases seems uncorrelated to cognitive ability. But in some cases, knowing more about them can help us become more resistant to their influence (Stanovich and West 2008). Teaching about these biases therefore has pedagogical value and might offer some protection against manipulative attempts in marketing and politics. It also improves strategic thinking where we need to take into account other people's suboptimal decisions. More generally, we can use such knowledge to design better public policies that nudge people toward better decisions (Thaler and Sunstein 2008).

Cognitive biases provide an important source of information about the architecture of the mind. In cognitive science, many authors have proposed some form of dual-process model of higher cognition (Evans 2003; Kahneman 2011). They make a distinction between psychological processes that are fast, automatic, and unconscious, and those that are slow, deliberate, but conscious. The two groups of processes are often known as System 1 and System 2 respectively. System 1 includes innate skills and automatic reactions that we share with other animals, and is crucial for survival. It serves us well most of the time, but the danger is that it can also lead to unreliable intuitions and rash decisions, in situations where careful analysis is required. This is particularly likely to happen when we are not paying attention, or are tired, emotional, or under stress. The engagement of System 2 to override default responses requires deliberate effort and reflection, and is crucial for metacognition. Interesting, a recent study suggests that this readiness for reflection (as measured by what is known as "The Cognitive Reflection Test") is a better predictor of the ability to combat classic cognitive biases compared with measures of cognitive ability, thinking dispositions, and executive functioning (Toplak, West, and Stanovich 2011).

The delineation between the Systems 1 and 2 is not uncontroversial. Some researchers have even argued that in some situations relying on intuitions rather than deliberate reasoning can lead to more satisfactory decision outcomes (Dijksterhuis, Bos, Nordgren, and van Baaren 2006). But being aware of the divergent sources of our judgments can help us find ways to improve their accuracy and become more alert to potential lapses. For example, one might adopt the strategy of not making drastic decisions when being emotional. Or one might decide as a rule not to follow one's intuitions whenever there is a feeling of uncertainty or anxiety. It is also useful to keep a record of our decisions to explore the effectiveness of different thinking strategies. Teaching about the psychology of reasoning and biases offers a more comprehensive picture of rationality, helping students fine-tune and self-correct their thinking.

Meta self-knowledge

Accurate self-knowledge is essential in order to control and improve our thinking. But psychologists have found that in many areas people tend to overestimate their abilities. For example, they think they are more likely than their peers to get a higher salary or have a gifted child, but less likely to divorce or have a drinking problem (Weinstein 1980). Similarly, the vast majority of drivers believe they drive better than average (Svenson 1981). This "above-average effect" also applies to business managers (Larwood and Whittaker 1977) and football players (Felson 1981). It also extends to college students when they are asked to rank their logical reasoning skills, knowledge of English grammar, and ability to recognize humor (Kruger and Dunning 1999). Furthermore, comparative studies suggest that overconfidence is prevalent across many different cultures (Chen, Kim, Nofsinger, and Rui 2007; Yates, Lee, Shinotsuka, Patalano, and Sieck 1998).

Of course, self-confidence helps sustain a positive self-image, and motivates us to overcome obstacles. But overconfidence can hinder self-improvement by blocking insight into our own weaknesses. To deal with this problem, it is crucial to calibrate self-appraisals using objective measurements, accurate comparative information, and corrective feedback. It might also be useful to keep a journal of our successes and failures for periodic review. It is important to emphasize this aspect of metacognitive monitoring in the teaching of critical thinking as well.

People's optimistic perception of themselves can make them evaluate their own actions more favorably, which can lead to disagreements and conflicts. Because we are often unconscious of the biases operating in our own judgments, we tend to see ourselves as objective and fair-minded. But when other people disagree, we judge them to be irrational or motivated by self-interest. This differential recognition of bias in others but not in ourselves is known as "the bias blind spot." It seems to be a particularly irrepressible bias, and one unfortunate consequence is that if we regard our adversaries as irrational, we are more likely to confront them and resort to more aggressive means (see the review by Pronin 2008). An awareness of this problem might help us become more charitable and improve interpersonal understanding.

Self-regulation

In the metacognition literature, self-regulation refers to the capacity to monitor and control our own cognitive processes. Typically it involves setting up goals, applying and reflecting on the strategies for achieving those goals, monitoring our progress, and making necessary adjustments. Self-regulation is surely crucial for critical thinking. A critical thinker understands the importance of clarity and truth and takes careful steps to achieve those objectives. Deliberate effort is needed to analyze ideas systematically and to avoid rash judgments. Furthermore, good critical thinkers will try to obtain better insight into their own thinking and find ways to improve their thinking skills even further.

Beyond critical thinking, self-regulation provides the discipline necessary for acquiring expertise in domains where there are learnable regularities. This requires intensive and deliberate training taking corrective feedback into account. Strong motivation and discipline are essential in order to endure repetitive exercises over a long period of time. In many cases, daily practice for a whole decade is necessary to achieve world-class performance. This seems to be true across diverse domains, whether it is chess, mathematics, dance, sports, or musical performance (Ericsson and Lehmann 1996).

There are of course individual differences in self-regulation, linking to differences in personality. The five-factor model of personality in psychology describes variation across five dimensions: neuroticism, extraversion, openness, agreeableness, and conscientiousness (Costa and McCrae 1992). Conscientiousness involves being responsible, careful, systematic, and hardworking. There is now a huge body of data pointing to the benefits of conscientiousness. Among the five factors, it is the best predictor of academic performance in high school and college, independent of cognitive ability. It is also the best predictor for self-regulation in undergraduate students (Fein and Klein 2011). Outside of academic performance, conscientiousness predicts physical and mental health, longevity, (lack of) criminal convictions, marital stability, income, leadership, job performance, and occupational attainment. Many of these effects can be separated from other variables such as socioeconomic status and education (Moffitt, Arseneault, Belsky, Dickson, Hancox, Harrington, Houts, Poulton, Roberts, Ross, Sears, Thomson, and Caspi 2011; Roberts, Lejuez, Krueger, Richards, and Hill 2012).

Grit is a related personality trait that has also received a lot of attention recently. Compared with conscientiousness, grit places greater emphasis on persistence and effort, resilience in overcoming hardship, and the ability to stick to long-term goals despite setbacks. It seems to be a common trait among exceptionally creative and successful people according to some extensive case studies (Miles 1926). Grit presupposes self-regulation and is highly correlated with conscientiousness (but not IQ). Grit predicts educational attainment, undergraduate grade point average, retention in college, over and above IQ, and conscientiousness. In fact, grit can propel less intelligent individuals to excel and become more successful than their more gifted counterparts (Duckworth, Peterson, Matthews, and Kelly 2007).

It is interesting to note that when Dewey (1933) argued for "reflective thinking" in education, he characterized it partly as thinking that involves "care" and "persistence," which correspond closely to conscientiousness and grit. University education should do more to help students understand the importance of these traits. As discussed earlier, cognitive skills often require extensive, structured, deliberate practice over a long period of time. Self-regulation in the form of conscientiousness and grit can surely help. Many authors are worried about universities failing to help students improve their thinking. One large-scale study in the United States involving more than two thousand students at twenty-four universities showed that 45% of students failed to improve their critical thinking significantly during their first two years of college, while 36% still showed no gains after four years (Arum and Roska 2011). Not surprisingly, the study also found that students who took courses with more reading and writing showed higher rates of learning (Dollinger, Matyja, and Huber 2007). If universities are serious about improving critical thinking, one of the many things they should do is to rebuild a learning culture that values and rewards hard work and persistence. The enhancement of critical thinking is thus intimately related to metacognitive education.

Of course, it is an open question to what extent we are able to change our personality. But there seem to be plenty of strategies to improve self-control, such as preempting or reappraising an undesirable option, or other methods such as distancing and distraction (Goldin, McRae, Ramel, and Gross 2008; McGonigal 2011). By increasing self-control, one might indirectly strengthen conscientiousness. In the academic context, conscientious self-regulated learning has been reported to correlate with academic success (Kitsantas, Winsler, and Huie 2008; Zimmerman and Schunk 2008). Self-regulated learners are self-aware and take responsibility for their own learning processes. They are motivated to seek out the information and skills they need to acquire, and they take active steps to plan and monitor their learning. There is evidence that explicitly teaching students metacognitive learning strategies that include self-regulation can succeed in raising their grade point averages and graduation rates (Tuckman and Kennedy 2011).

Conclusion and further discussion

Different sources of theoretical perspectives and empirical evidence have converged on the importance of metacognition in the form of disciplined self-regulation supported by self-understanding and knowledge of psychology. It enhances learning, critical thinking, creativity, and academic and career success. There are therefore good reasons to expand the teaching of critical thinking in this direction. In this final section we address two potential reservations.

First of all, there might be a worry about mixing critical thinking with psychology. The teaching of critical thinking often includes a fair amount of logic, broadly conceived. This might include the rules of deductive logic, inductive inferences, and scientific confirmation. But these principles are usually taken to be a priori, fundamentally different from the empirical a posteriori theories in much of metacognition, such as the science of cognitive biases and personality traits. It might be thought that these two sets of theories should not be taught together since they belong to different disciplines.

However, as Dewey has observed, thinking well is not exhausted by knowledge about logic. If there are other important factors that contribute to good thinking, they should also be included into our teaching regardless of disciplinary boundaries. Take medical education as an analogy. The discipline of human physiology is indeed distinct from psychology, risk management, counseling, and ethics. But nobody should deny that the latter topics are also important in training doctors. Similarly, critical thinking is only part of what we ought to know to be an effective thinker. We teach logic because we think it helps students avoid errors in reasoning and make better judgments. Nonetheless, from a pedagogical point of view, it is just as useful to know about the psychology of biases and the importance of self-regulation. By explicitly introducing metacognition into the curriculum, our students acquire a broader perspective about the different factors that contribute to good thinking. This interdisciplinary approach gives them a more solid foundation to acquire other cognitive skills and improve themselves in the long run.

This raises a more practical reservation about metacognitive education. Critical thinking already covers a lot of topics. Teachers of critical thinking inevitably have to decide which topics to include and which to leave out because of limited class time. For example, how much of formal logic should be taught? How useful are Venn diagrams and Aristotelian syllogisms? Metacognition is even more wide-ranging, including diverse topics about creativity, cognitive science, social psychology, and so on. One might wonder whether it is realistic or even desirable for a single course to address all these topics. There is the risk of superficial coverage leading to poor results.

This is a legitimate concern, but the present proposal is not that we abandon courses in critical thinking immediately and start teaching metacognition instead. Rather, the suggestion is that the development of metacognitive competence ought to be an explicit aim of education, and the time is ripe to consider how critical thinking fits within this larger framework. But there is no reason why we should shoehorn every topic related to metacognition into one single course. Some of the points discussed, such as the effect of interpreting anxiety in a positive light, or strategies for improving self-control, are perhaps more appropriate in a learning component designed to impart "soft-skills." Creativity and problem-solving heuristics can perhaps be discussed within discipline-specific courses. Certainly we need more research on how best to teach metacognition. But the university curriculum as a whole should convey a clear message of its importance. There should be appropriate coordination to ensure adequate support and incentives to build a learning culture that centers upon metacognition.

In any case, metacognition ought to be given a more prominent place in the teaching of critical thinking itself. As far as teaching methodology is concerned, critical thinking courses can make more extensive use of problem-based learning instead of lectures to familiarize students with self-regulated learning. As for content, students can learn more about the psychological factors that affect our reasoning and find out how to mitigate the effects of cognitive biases. They should understand that thinking skills go beyond raw intelligence, and that conscientious effort and good personal habits can help us maximize our potential. We also need accurate self-understanding, and take corrective actions in response to feedback. It is a cliché that our students need to learn how to think. It is also a commonplace observation that many of them fail to do so. Hopefully, by thinking more about learning, and about the nature of the thinking process itself, our students will end up becoming better thinkers.

References

Arum, R., and Roksa, J. 2011. *Academically Adrift: Limited Learning on College Campuses.* Chicago and London: University of Chicago Press.

Chen G., Kim K.A., Nofsinger J. R., and Rui, O. M. 2007. "Trading Performance, Disposition Effect, Overconfidence, Representativeness Bias, and Experience of Emerging Market Investors." *Journal of Behavioral Decision Making* 20 (4): 425–451.

Costa, A. L. 1991. *The School as a Home for the Mind.* Palatine, IL: Skylight Publishing.

Costa, P. T. Jr., and McCrae, R. R. 1992. *Revised Neo Personality Inventory (NEO-PI-R) and Neo Five-Factor Inventory NEO-FFI) Manual.* Odessa, FL: Psychological Assessment Resources.

Dewey, J. 1933. *How We Think: A Restatement of the Relation of Reflective Thinking to the Educative Process.* (revised edition). Boston: D. C. Heath and Company.

Dijksterhuis, A. P., Bos, M. W., Nordgren, L. F., and van Baaren, R. B. 2006. "On Making the Right Choice: The Deliberation-without-Attention Effect." *Science* 311: 1005–1007.

Dollinger, S. J., Matyja, A. M., and Huber, J. L. 2007. "Which Factors Best Account for Academic Success: Those Which College Students Can Control or Those They Cannot?" *Journal of Research in Personality* 42 (4): 872–885.

Duckworth, A. L., Peterson, C., Matthews, M. D., and Kelly, D. R. 2007. "Grit: Perseverance and Passion for Long-Term Goals." *Journal of Personality and Social Psychology* 92 (6): 1087–1101.

Dweck, C. S. 1986. "Motivational Processes Affecting Learning." *American Psychologist* 41 (10): 1040–1048.

Dweck, C. S., and Elliott, E. S. 1983. "Achievement Motivation." In *Handbook of Child Psychology: Vol. 14, Socialization, Personality, and Social Development,* edited by E. M. Hetherington, (fourth edition). New York: Wiley. 643–691.

Ennis, R. H. 1989. "Critical Thinking and Subject Specificity: Clarification and Needed Research." *Educational Researcher* 18 (3): 4–10.

Ericsson, K. A., and Lehmann, A. C. 1996. "Expert and Exceptional Performance: Evidence of Maximal Adaptation to Task Constraints." *Annual Review of Psychology* 47: 273–305.

Evans, J. S. 2003. "In Two Minds: Dual-Process Accounts of Reasoning." *Trends in Cognitive Sciences* 7 (10): 454–459.

Fein, E. C., and Klein, H. J. 2011. "Personality Predictors of Behavioral Self-Regulation: Linking Behavioral Self-Regulation to Five-Factor Model Factors, Facets, and a Compound Trait." *International Journal of Selection and Assessment* 19 (2): 132–144.

Felson, R. B. 1981. "Ambiguity and Bias in the Self-Concept." *Social Psychology Quarterly* 44: 64–69.

Flavell, J. H. 1976. "Metacognitive Aspects of Problem Solving." In *The Nature of Intelligence*, edited by L. B. Resnick. Hillsdale, NJ: Erlbaum. 231–236.

Goldin, P. R., McRae, K., Ramel, W., and Gross, J. J. 2008. "The Neural Bases of Emotion Regulation: Reappraisal and Suppression of Negative Emotion." *Biological Psychiatry* 63 (6): 577–586.

Halpern, D. 1998. "Teaching Critical Thinking for Transfer across Domains: Dispositions, Skills, Structure Training, and Metacognitive Monitoring." *American Psychologist* 53 (4): 449–455.

Hargrove, R. A. 2012. "Assessing the Long-Term Impact of a Metacognitive Approach to Creative Skill Development." *International Journal of Technology and Design Education* 23 (3): 489–517.

Iyengar, S. S., and Lepper, M. 2000. "When Choice Is Demotivating: Can One Desire Too Much of a Good Thing?" *Journal of Personality and Social Psychology* 79: 995–1006.

Jamieson, J. P., and Harkins, S. G. 2010. "Evaluation Is Necessary to Produce Stereotype Threat Performance Effects." *Social Influence* 5 (2): 1–12.

Kahneman, D. 2011. *Thinking, Fast and Slow*. London: Allen Lane.

Kitsantas, A., Winsler, A., and Huie, F. 2008. "Self-Regulation and Ability Predictors of Academic Success during College: A Predictive Validity Study." *Journal of Advanced Academics* 20 (1): 42–68.

Kruger, J., and Dunning, D. 1999. "Unskilled and Unaware of It: How Difficulties in Recognizing One's Own Incompetence Lead to Inflated Self-Assessments." *Journal of Personality and Social Psychology* 77 (6): 1121–1134.

Langer, E. J. 1989. *Mindfulness*. Reading, MA: Addison-Wesley Pub. Co.

Larwood, L., and Whittaker, W. 1977. "Managerial Myopia: Self-Serving Biases in Organizational Planning." *Journal of Applied Psychology* 62: 194–198.

Liberman, V., Samuels, S. M., and Ross, L. 2004. "The Name of the Game: Predictive Power of Reputations versus Situational Labels in Determining Prisoner's Dilemma Game Moves." *Personality and Social Psychology Bulletin* 30 (9): 1175–1185.

Lochhead, J., and Whimbey, A. 1980. *Problem Solving and Comprehension*. Philadelphia: The Franklin Institute Press.

Magno, C. 2010. "The Role of Metacognitive Skills in Developing Critical Thinking." *Metacognition Learning* 5 (2): 137–156.

McGonigal, K. 2011. *The Willpower Instinct: How Self-Control Works, Why It Matters, and What You Can Do to Get More of It*. New York: Avery.

Miles, C. C. 1926. *The Early Mental Traits of Three Hundred Geniuses, Genetic Studies of Genius*. Stanford: Stanford University Press.

Moffitt, T. E., Arseneault, L., Belsky, D., Dickson, N., Hancox, R. J., Harrington, H., Houts, R., Poulton, R., Roberts, B. W., Ross, S., Sears, M. R., Thomson, W. M., and Caspi, A. 2011. "A Gradient of Childhood Self-Control Predicts Health, Wealth, and Public Safety." *Proceedings of the National Academy of Sciences United States of America* 108 (7): 2693–2698.

Paul, R., Willsen, J., and Binker, A. J. A. 1993. *Critical Thinking: What Every Person Needs to Survive in a Rapidly Changing World*. (revised third edition). Santa Rosa, CA: Foundation for Critical Thinking.

Perkins, D., Jay, E., and Tishman, S. 1993. "Beyond Abilities: A Dispositional Theory of Thinking." *Merrill-Palmer Quarterly* 39 (1): 1–21.

Plassmann, H., O'Doherty, J., Shiv, B., and Rangel, A. 2008. "Marketing Actions Can Modulate Neural Representations of Experienced Pleasantness." *Proceedings of the National Academy of the United States of America* 105 (3): 1050–1054.

Pólya, G. 1945. *How to Solve It: A New Aspect of Mathematical Method*. Princeton, NJ: Princeton University Press.

Pronin, E. 2008. "How We See Ourselves and How We See Others." *Science* 320: 1177–1180.

Roberts, B. W., Lejuez, C., Krueger, R. F., Richards, J. M., and Hill, P. L. 2012. "What Is Conscientiousness and How Can It Be Assessed?" *Developmental Psychology.*

Russo, J. E., and Schoemaker, P. J. H. 1990. *Decision Traps: The Ten Barriers to Decision-Making and How to Overcome Them.* New York: Simon & Schuster.

Stanovich, K. E., and West, R. F. 2008. "On the Relative Independence of Thinking Biases and Cognitive Ability." *Journal of Persoanlity and Social Psychology* 94 (4): 672–695.

Svenson, O. 1981. "Are We Less Risky and More Skillful Than Our Fellow Drivers?" *Acta Psychologica* 43: 147–151.

Thaler, R. H., and Sunstein, C. R. 2008. *Nudge: Improving Decisions about Health, Wealth, and Happiness.* New Haven, CT: Yale University Press.

Toplak, M. E., West, R. F., and Stanovich, K. E. 2011. "The Cognitive Reflection Test as a Predictor of Performance on Heuristics-and-Biases Tasks." *Memory and Cognition* 39: 1275–1289.

Torrance, E. P. 2002. *The Manifesto: A Guide to Developing a Creative Career.* Westport, CT: Ablex Pub.

Tuckman, B. W., and Kennedy, G., J. 2011. "Teaching Learning Strategies to Increase Success of First-Term College Students." *Journal of Experimental Education* 79: 478–504.

Uhlmann, E. L., and Cohen, G. L. 2007. ""I Think It, Therefore It's True": Effects of Self Perceived Objectivity on Hiring Discrimination." *Organizational Behavior and Human Decision Processes* 104 (2): 207–223.

US Department of Labor. 2012. "Number of Jobs, Labor Market Experience, and Earnings Growth: Results from a Longitudinal Survey" [Press release]. Retrieved from http://www.bls.gov/news.release/nlsoy.htm.

Weinstein, N. D. 1980. "Unrealistic Optimism about Future Life Events." *Journal of Personality and Social Psychology* 39: 806–820.

Wuchty, S., Jones, B. F., and Uzzi, B. 2007. "The Increasing Dominance of Teams in Production of Knowledge." *Science* 316 (5827): 1036–1039.

Yates J. F., Lee J. W., Shinotsuka H., Patalano A. L., and Sieck, W. R. 1998. "Cross-Cultural Variations in Probability Judgment Accuracy: Beyond General Knowledge Overconfidence?" *Organizational Behavior and Human Decision Processes* 74 (2): 89–117.

Young, J. W. 1975. *A Technique for Producing Ideas.* Chicago, IL: Crain Commmunication.

Zimmerman, B. J., and Schunk, D. H. 2008. "Motivation: An Essential Dimension of Self-Regulated Learning." In *Motivation and Self-Regulated Learning: Theory, Research, and Applications*, edited by D. H. Schunk and B. J. Zimmermaned. Mahwah, NJ: Lawrence Erlbaum Associates. 1–30.

23

Applying Cognitive Science to Critical Thinking among Higher Education Students

Jason M. Lodge, Erin O'Connor, Rhonda Shaw, and Lorelle Burton

Introduction

One of the main aims of higher education is for students to develop their analytical and critical thinking in order for graduates to function as competent professionals (e.g., Burton, Westen, and Kowalski 2012). The importance of this supposed generic skill is reflected in the ubiquitous inclusion of critical thinking as a graduate capability in universities (Moore 2011). While there exist many ways of defining and understanding critical thinking, at its core, critical thinking involves addressing our assumptions about how the world works. It is, therefore, essential for competent practice as a professional (Moon 2008). Without exposure to effective training in critical thinking, assumptions are more often than not based on the cognitive biases that are either inherent or conditioned through experience. The cognitive and emotional processes underpinning biases in thinking are often difficult to overcome. Our natural tendency to take mental shortcuts has allowed us to effectively navigate our environment and process only those stimuli that are of immediate value to us and to our survival. These shortcuts, however, often make it difficult for students to engage deeply with a complex concept, idea, or discipline in a higher education context. Ensuring that graduates are capable of thinking beyond their tendency to take mental shortcuts therefore poses a significant challenge for teaching critical thinking in higher education institutions.

A long-standing debate continues into whether or not critical thinking is best taught in a general or specific manner (see Davies 2006; 2011; 2013). The debate between the generalist and specifist positions has given rise to a number of potential problems for understanding critical thinking in a higher education context. The aim of this chapter is to provide a fresh perspective to the generalist-specifist debate in order to make progress in the design of interventions for developing critical thinking in university students. The current chapter will examine how such cognitive theories offer ways to circumvent our tendencies toward less rational and less effortful thought. A key focus will be to examine the biases in thinking and to determine how cognitive sciences can provide insight for enhancing higher education teaching and help university students develop their critical thinking skills.

Cognitive science and the generalist/specifist debate

An example of the ways in which the generalist/specifist debate has led to calls for new perspectives on critical thinking in higher education is provided by McPeck (1981). McPeck argues that the study of informal logic for critical thinking focuses too much on fallacies and not enough on the underlying cognitive mechanisms of critical thought. It follows that, in order to inform the development of curriculum design that best develops critical thinking skills in students, it is necessary to examine the mental processes thought to underpin critical thinking. An alternate view is forwarded by Ennis (1989), who, despite significant disagreement with McPeck about the role of informal logic in critical thinking, similarly suggests that the capacity to evaluate arguments, skepticism, problem solving, higher-order thinking and metacognition are all aspects of critical thinking and therefore all deserve further examination. While there are many principles and practices for helping students to engage in deeper, critical thinking in a higher education setting, those that give the most compelling appreciation of the biases that attenuate critical thinking come from an examination of these factors. All are factors that have been examined in the cognitive sciences. While the translation from laboratory-based cognitive science to pedagogical practice is difficult, the learning sciences have the potential to make a meaningful contribution to debates around the best way to enhance critical thinking.

From a cognitive science perspective, the approach taken to further understand the concept of critical thinking is essentially generalist. Cognitive scientists, in agreement with Ennis (1989), take the view that the broader concept can be broken into smaller constituent cognitive processes that are common to all. Breaking critical thinking into constituent components allows for each process to be experimentally investigated under controlled laboratory conditions. Cognitive science therefore provides a valid method of approaching the teaching of critical thinking skills through a deeper and more rigorous examination of the underlying mechanisms. While a number of clear obstacles exist that potentially limit the ability of students in higher education institutions to develop their critical thinking, cognitive science provides a valuable viewpoint on why this is the case and how it can be addressed. Relevant within this context are the ways in which the human mind takes mental shortcuts and has biases that lead to faulty logic and reasoning. Heuristics are rules of thumb or mental shortcuts (Plous 1993) that serve to reduce the cognitive load associated with complex tasks. It is in the understanding of these experimentally examined heuristics that cognitive science can contribute most to the enhancement of methods to develop critical thinking in higher education.

One example of cognitive science research relevant to understanding critical thinking is that of Kahneman (2011), who has examined biases in thinking over several decades and has concluded that humans will generally take the most efficient rather than the most effective path when it comes to expending cognitive effort. Indeed, Kahneman's research indicates that our students' brains are wired to cut as many corners as possible. In order to apply this type of research to the classroom, Halpern's (1998) evidence-based model for the development of critical thinking will be examined. This model involves specific preparation for the cognitive effort required to overcome heuristics and biases. The model also takes into account that transfer of critical thinking skills across contexts is much less automatic than teachers may expect. This model also provides a framework for curriculum-based approaches to develop critical thinking including (1) preparing learners for critical thinking, (2) developing and practicing skills, and (3) exploring learners' capacity to transfer skills to new and unfamiliar contexts. Both Kahneman's and Halpern's work will be discussed in some detail as examples of the ways in which cognitive science can meaningfully inform the enhancement of practice in a higher education setting.

Fallacies, biases, and heuristics

As a first step in understanding the cognitive components in critical thinking, we will examine some of the common fallacies, biases, and heuristics that are responsible for faulty reasoning. The purpose of this discussion is to highlight some of the most relevant roadblocks to critical thinking.

Derived from the division between formal and informal logic, common fallacies can be divided into formal and informal types. "Formal fallacies are those arguments that derive their psychological persuasiveness from their superficial resemblance to valid deductive argument forms" (Zeidler, Lederman, and Taylor 1992, 440). Informal fallacies, on the other hand, use ambiguous or misleading language to deceive. Zeidler and colleagues suggest informal fallacies are more common than formal fallacies. Copi (1986, cited in Zeidler, Lederman, and Taylor 1992) divided informal fallacies into two further subgroups—*fallacies of relevance* and *fallacies of ambiguity*. Fallacies of relevance are arguments that deceive through the inclusion of at least one statement that is irrelevant to the final conclusion (Zeidler, Lederman, and Taylor 1992, 441). Examples of these fallacies include *ad hominen arguments*. Ad hominen arguments include an irrelevant personal attack or claims that some relationships between or special circumstances relating to the alternate position and their position should render their entire argument implausible. Appeals to popularity, appeals to authority, and circular reasoning are also examples of fallacies of relevance.

Fallacies of ambiguity are "informally fallacious arguments that contain an ambiguous word or term the meaning of which renders the argument fallacious" (Zeidler, Lederman, and Taylor 1992, 443). A common example of ambiguity is fallacies of equivocation—the repeated use of a term with the implication that the word is consistently used throughout an argument when the meaning behind each occurrence is not equivalent. For example "Blacks and whites are historically, culturally, and biologically different; the races are not equal. Consequently, the laws have never meant that blacks and whites are equal in our society" (Zeidler, Lederman, and Taylor 1992, 443). In this example the word "equal" is first used to mean something like "identical", whereas the second appearance of "equal" denotes "entitled to the same rights." Fallacies of this type are relatively common and, without the necessary skills to unpack the argument being made in such statements, it can be difficult to see the fault in the argument.

While fallacies are a mainstay in the study of informal logic, they also provide evidence of the psychological aspects of thinking. The tendency of our brains to be susceptible to fallacies suggests that we find it difficult to deal with complex information without specific exposure or training or indeed without conscious cognitive effort. Just as it is impossible for us to form perceptions of all of the information that impinges on our senses, it is equally impossible for us to consider all the information that may be available to us when making a decision or forming a judgment. We therefore have a tendency to be susceptible to communication that is specifically designed to be ambiguous or difficult to interpret such as that in the fallacies discussed above. One useful framework based on laboratory research in cognitive science that has been used to explain how we make decisions based on the information we are presented, especially under uncertainty, is the heuristics and biases framework (Tversky and Kahneman 1974). Heuristics and biases draw on information that is already known, by what we have already experienced, heard, or felt. One does not deliberately set out to use a particular heuristic; rather it is "elicited by the task at hand" (Gilovich and Griffin 2002, 4). In most instances heuristics are useful; however, they can also lead to systematic errors. Tversky and Kahneman describe three heuristics that are used when making judgments and decisions—*representativeness, availability,* and *anchoring and adjustment*—each of which is associated with a set of biases.

The *representativeness heuristic* refers to the tendency to make judgments based on the degree to which one factor resembles another factor. For example, "the degree to which A is representative of B, that is, by the degree to which A resembles B" (Tversky and Kahneman 1974, 1124). In one study Tversky and

Kahneman (1982) asked participants to determine whether Linda was a bank teller or a feminist bank teller based on a brief description of a fictitious character. The majority of the participants determined that she was a feminist bank teller. In this instance, representativeness lead participants to commit the *conjunction fallacy*—the belief that the conjunction of two events is more probable than either event alone. Further biases that can result from representativeness include the belief in the law of small numbers (a belief that random samples resemble the populations from which they are drawn), discounting or ignoring base rate information (the relative frequency of an event), and making nonregressive predictions (e.g., the expectation that good luck will follow a run of bad luck). A more objective examination of the data in these cases clarifies the actual situation. For example, regression to the mean tells us that extreme scores tend to average out over time so there is no reason to believe that any sustained period of bad fortune will end with similarly extreme good fortune.

Availability is the idea that we make judgments based on how easy an event is to think about (Tversky and Kahneman 1974). Events may be more available because they are more recent, they have received media attention, or they are highly emotional. Gigerenzer and colleagues refer to this as the recognition heuristic (Goldstein and Gigerenzer 2002) rather than availability but the focus for both is on accessibility. Much of the research in this area asks participants to make numerical judgments. As an example, McKelvie (2000) found that judgments about the frequency of male and female names depended on how famous the names on a list were. Participants' judgments were not based on actual numbers of male or female names on a list but by the number of famous names on a list because these were easier to retrieve from memory than nonfamous names. Availability can also result in errors caused by biases in the effectiveness of a memory search, for example, words that begin with the letter "r" come to mind much easier than words that have "r" as the third letter even though there are more words that have "r" as the third letter (Tversky and Kahneman 1974). A further bias associated with the availability heuristic is an illusory correlation that can lead to dismissal of data or information that may contradict common beliefs, for example, that there is a strong relationship between suspiciousness and peculiar eyes.

Anchoring and adjustment is the notion that the estimation of quantities that are not known is based on adjusting from a value that is known. The known information is referred to as the anchor (Tversky and Kahneman 1974). Biases in judgments were originally assumed to result from insufficient adjustments from the anchor. That is, participants tend to estimate close to an anchor. For instance, if asked to estimate the date when past president John F. Kennedy started in office, participants may start from the date Kennedy was assassinated and work backwards from there. According to Epley and Gilovich (2006), adjustment requires effort. Adjustment-based anchoring effects can be reduced if participants are encouraged to carefully consider their adjustments and the point of reference from which they make their judgments. In the standard paradigm in which the anchor is provided by the experimenter, incentives or warnings to avoid satisficing, or the tendency to accept an arbitrary and often inadequate threshold of information, do not prevent participants from making insufficient adjustments.

Epley and Gilovich's (2006) work suggests that, in some instances, the influence of heuristics and their concomitant biases can be reduced. The challenge is to be aware that heuristics, and the associated biases, influence thinking. Critical thinking is about the inhibition of this influence (West, Toplak, and Stanovich 2008). Another way to conceptualize errors and biases is to understand that they, more than cognitive shortcuts, represent the failure of intuitive and deliberate reasoning processes (Kahneman and Frederick 2005). From this conceptualization, it can be derived that there are two processing systems working side by side, a fast system relying on efficient processing strategies and a more deliberate, slow system that engages when more taxing cognitive processing is required. The idea that reasoning and thinking involves two separate but complementary processes is explicit in dual-process theories of judgment and decision making (Evans 2003).

Based on the accumulation of research into fallacies, heuristics, and biases such as those presented here, Kahneman (2011) argues that there are two distinct processing systems in the brain. The first system, "System 1" relies on heuristics and processes information quickly to allow for an efficient response. The second system, "System 2" is slower, more deliberate, and requires greater effort in order to effectively process more complex cognitive tasks. Kahneman has also found over a large number of studies that we tend to rely on the fast, intuitive, error-prone System 1. In other words, we tend to overrely on heuristics and are prone to biases and fallacies because we are prone to cutting cognitive corners. Therein lies the challenge of creating learning experiences for students that push them to overcome this inherent tendency. The research on fallacies, heuristics, and biases has helped to understand the processes responsible for cognitive errors and has helped appreciate why critical thinking is difficult. However, cognitive science also provides possible avenues for addressing these issues.

Ways of overcoming faulty thinking

It is believed that, in some cases, cognitive ability affects the capacity to overcome bias. According to Macpherson and Stanovich (2007), while the ability to overcome biases in knowledge in formal reasoning tasks appears to be related to cognitive ability, whereas the ability to overcome myside (or confirmatory) bias or the tendency to believe only information that confirms existing beliefs appears to be independent of cognitive ability. This claim suggests that there are a number of ways in which biases and heuristics are triggered. It is therefore difficult to establish ways of overcoming all heuristics and biases when teaching future professionals. Based on a wealth of cognitive science research, Halpern (1998) argues that most importantly "in critical thinking instruction, the goal is to promote the learning of transcontextual thinking skills and the awareness of an ability to direct one's own thinking and learning" (451). Halpern suggests that the development of meta-level thinking skills provides the best point for intervention in the thinking process. This view also aligns with the view of that of the generalists (e.g., Ennis 1989) who suggest that formal training and generic intervention on higher-level thinking processes is the most effective approach for enhancing critical thinking. How to go about creating the conditions where such an intervention would be effective is therefore the next issue.

There are a number of ways that have been suggested as possible avenues for breaking critical thinking down into factors that will allow for intervention. Scott-Smith (2006) suggests that critical thinking strategies should include the following essential features:

1. step-by-step analysis (slower inductive thought process combining information toward a solution);
2. analogy (comparison with a similar previous event and using this to interpret new information);
3. visualization;
4. free production of ideas (e.g., "brainstorming," but more appropriate to multidisciplinary group work); and
5. a combination of the above (120).

According to Scott-Smith, the inclusion of these strategies as features of an intervention for critical thinking will enable hypothetico-deductive reasoning (testing hypotheses derived from observations and moving toward a conclusion/diagnosis), and an enhanced capability for pattern recognition or categorization (direct automatic retrieval of information based upon previous events). Furthermore prototypes (problems that resemble each other in a number of common ways) can be developed and participants can begin to develop instance-based recognition ("seen one before"). Each of these outcomes is important for overcoming biases and heuristics that can lead to faulty conclusions when faced with new information.

Scott-Smith's (2006) argument here is also based on developing interventions from an understanding of the underlying cognitive processes. While Scott-Smith focuses on information processing and Halpern more on higher-level thinking or metacognition, other researchers have examined ways of using learning theory as a means of enhancing critical thinking. For example, Novak (2010, 21) used Ausubel's assimilation theory (1963) to show "the distinction between learning by rote versus learning meaningfully." Meaningful learning is based on the premise of building a hierarchically displayed pattern of knowledge. Initially concept maps were developed as a way of testing the knowledge of students, but they were later developed into a method of learning by capturing the knowledge required and displaying the links between that information, which resulted in meaningful learning (Novak 2010). According to Novak, "meaningful learning" is learning that incorporates new information into hierarchical knowledge structures and enables critical thinking. Given the advances in technology, it is now possible to create and maintain concept maps electronically but they should still maintain the basics of a hierarchical setup and cross-links (Novak, 1998, as cited in Vacek 2009). Within this organizational structure, novel problem solving, a characteristic of critical thinking, occurs (Vacek 2009). This perspective suggests that there are multiple avenues for using the findings from cognitive science to enhance critical thinking in students. While each of these theorists has developed effective frameworks for attempting to target specific processes to enhance critical thinking, there still remains the issue of determining what an intervention would look like in practice.

Some early research (Lyle, 1958, as cited in Williams, Oliver, Allin, Winn, and Booher 2003) showed that the efficacy of critical thinking instruction in psychology might be related in part to the aptitude of students. Lyle (1958) compared the effects of a problem-based format and a lecture format on critical thinking in a general psychology course. The problem-based format required students to find valid evidence to address problems assigned by the instructor; the lecture format involved the instructor periodically mentioning problems similar to those introduced under the problem format and then giving possible solutions to them. Lyle found that students with high academic aptitude improved their critical thinking skills under the problem-based format, but the low aptitude students improved their critical thinking skills more under the lecture format. Thus, instructional format and student aptitude may interactively affect critical thinking (Williams et al. 2003, 220). This research demonstrated a clear improvement in critical thinking across the duration of the program through opportunities to practice critical thinking. For example, a course emphasizing project development could integrate critical thinking tasks into project-related activities. "The cardinal guideline is to make the use of critical thinking explicit within course activities rather than expecting critical thinking to accrue as an indirect benefit of those activities" (Chance 1986; Harris and Clemmons, 1996, as cited in Williams et al. 2003, 222).

The teaching of academic writing also provides a specific point of intervention based on principles gleaned from cognitive science. For example, Parameswaram (2007) described the use of inclusive writing in a teaching curriculum to incorporate critical thinking skills via group-based learning. It involves a semester-long process of interactive/problem-based learning following a series of specific steps:

1. Students explore the issues relating to the initial topic and decide cooperatively on the important issues and are encouraged to do some free writing before the next class.
2. The students then break into groups (which they keep for the rest of the semester) and discuss the issues further.
3. Students are given the opportunity to share their free writing with the class. They are not forced to do so and the work is not graded.
4. The class discusses the important themes and develops a cognitive map.
5. The students are instructed to think and reflect on the theme developed and again this is discussed as a group.
6. At this point the students complete a graded paper and are provided feedback.
7. The next step involves the introduction of references and includes a workshop on the library and discussion of the key words and terms used in the previous assignment. Students are also introduced to the concept of a bibliography.
8. Students are directed to use references sourced in the previous step to develop their own questions or narratives and inquire or develop thought on the topic.
9. Students are encouraged to develop the presentation of this material creatively (songs, skits, films, role play) either by themselves or in a group.
10. Again, the students are encouraged to maintain narratives on the topic and process.
11. The final assessment piece brings together all the components of the topic introduced during the semester.

Parameswaram (2007) found that this inclusive writing process allayed students' fears about assessment. As part of the process, students purportedly developed critical thinking skills and achieved a deeper understanding of an important topic through the development of skills in concept mapping, research, and collaboration—three areas essential to the development of good critical thinking skills. Again, this is a sound demonstration of the ways in which cognitive science can be used to enhance critical thinking, in this case by using a concept-mapping approach.

Halpern's model for developing critical thinking

Of the approaches and models for improving critical thinking based on cognitive science, the body of work that has had the most impact is that of Halpern (2003). Halpern (1998) initially proposed an evidence-based, multifaceted model of critical thinking to inform critical thinking curricula. The model depicted not only the skills and methods of critical thinking but also the inclination to apply these skills, the ability to identify appropriate opportunities for critical thinking, and the ability to monitor progress and quality of thinking.

Components of Halpern's model

Halpern (1998) argued that to engage appropriately in critical thinking, first, requires a certain attitude toward critical thinking and analysis. Halpern includes here *disposition*, a positive view of critical thinking and a willingness to engage in and commit some effort to the work required to develop this skill. She suggests that a critical thinker displays these characteristics:

1. a willingness to engage in and persist at a complex task;
2. habitual use of plans and the suppression of impulsive activity;
3. flexibility or open-mindedness;
4. willingness to abandon nonproductive strategies in an attempt to self-correct; and
5. an awareness of the social realities that need to be overcome (Halpern 1998, 452).

Assuming a student has these characteristics, he or she is then capable of developing the skills and abilities of critical thinking. These include an ability to analyze or argue, decision-making and problem-solving skills, an ability to generate alternative explanations or points of view and then sensibly judge these, and an ability to test hypotheses. *Training in structure* is the ability to recognize or notice when to apply the skills and to detect appropriate contexts for application of critical thinking (453). A key aspect of structure is also the ability to transfer knowledge, skills, patterns, or analogies appropriately from the representations of past experience stored in memory to the current problem. *Metacognitive monitoring*, Halpern's last component, is the "executive or boss" (454). Halpern describes metacognitive monitoring as the ability to check the progress of any critical thinking activity, to evaluate the thinking against its goal, and to adjust the efforts accordingly. She suggests that this component can be enhanced through "well structured questions" to encourage reflection on learning (454).

Halpern also noted that many educators do not take into account these foundations of cognitive science when developing critical thinking lessons and curricula. She argued that undergraduate students demonstrate poor transferral of critical thinking skills and abilities, citing an example in which 99% of college students endorsed at least one paranormal phenomenon (see Messer and Griggs 1989). Halpern also suggested that many adults are not able to extend critical thinking skills to contexts outside of formal education settings. For example, the majority of adults read horoscopes and most of these believe that they are written for their own personal situation (Lister, 1992, cited in Halpern, 1998).

Applying Halpern's model to practice

A number of factors must be considered in order to apply Halpern's (1998) model in practice. Among these are student characteristics such as open-mindedness and cultivating a critical disposition. Halpern and others (e.g., Zeidler, Lederman, and Taylor 1992) have identified the importance of holding a positive attitude toward critical thinking. An important part of this disposition is being open to other ideas, having beliefs and biases challenged, and seeing these challenges as a positive thing. A well-cultivated disposition may also assist in identifying how to apply critical thinking appropriately.

Critical thinkers should also have the ability to examine argument structures and to critique their relevance and validity. Critical to this type of examination is some knowledge of argument structures and the skills to distinguish between established, demonstrable facts and hypothetical suggestions or predictions. A critical thinker may look for markers in the language of the argument that may indicate that something is being presented as evidence when it is really speculation. These skills should be developed through regular and sustained practice. Practice, ideally at every available opportunity, leads to higher familiarity with the structure of arguments and improvements in skill (Ericsson and Charness 1994). Members of a general population may choose to seek out debates in their community, media, or government to use as opportunities to practice these skills. For example, McKendree Small, Stenning, and Conlon (2002) suggest that the commonly asked question "What should the government do to stem the tide of immigration?" should be further examined. "Questioning the question" (64) may lead a critical thinker to some appropriate alternative questions including "Is there a tide of immigration?" and "What problems does a tide of immigration pose?"

It is also important, in both general and academic environments, to have some awareness of the context of the argument. An understanding of how bias can influence others and the possible biases of each party in debate or discussion can help to evaluate the context of the evidence presented. To facilitate this knowledge, critical thinkers should also have an awareness of the stakeholders involved in a certain argument (including opponents to a particular point of view) and the incentives they may have to hold a particular view. While this is not an argument unique to or originating from cognitive science, an understanding of the processes beneath the tendency not to give sufficient consideration, such as the work on myside bias, has been important in understanding how best to develop these skills.

Critical thinking instruction should also assist learners to overcome their own biases or "blind spots" and rely less on the heuristics that make up Kahneman's (2011) System 1, discussed earlier. While it is easy to turn automatically to a heuristic, metacognitive processes can be taught to help learners look more carefully at problems. For example, encouragement to "look for a problem's deep structure" or to "consider both sides of an issue"' (Willingham 2007) are reminders that may improve the learner's analysis of a problem (Novick and Holyoak 1991).

Critical thinkers regularly use reflective strategies that allow them to review and monitor the progress in and quality of their analysis of new information. A metacognitive prompt may therefore encourage questioning about the quality of the analysis, any self-arising bias, and questioning the adequacy of information collected. In this regard, activating Halpern's (1998) metacognitive monitoring may encourage more critical thinking.

Based on Halpern's model, a number of other authors have also suggested methods for teaching critical thinking. For example, van Gelder (2005) suggests a series of lessons for teaching critical thinking. He suggests that educators and teachers of critical thinking use cognitive science to inform their teaching practice. First, van Gelder reminds us that critical thinking is not a natural position for humans to acquire and exhibit. A strength of human cognition is that we seek patterns and use past experience and storytelling to organize our thinking (Shermer 2002). Critical thinking is an effortful undertaking and students must be encouraged to develop the disposition that Halpern (1998) argues is required for critical thinking. van Gelder suggests that we need to be ready, willing, and able to put effort into critical thinking.

van Gelder also aligns well with Halpern's concept of skills to suggest that critical thinking takes practice. To ensure that one has honed the skills of critical thinking, students need ample opportunities to practice these skills. Rather than rote-learning activities, this practice should be what van Gelder and others refer to as "deliberate." Borrowed from Ericsson and Charness's (1994) findings in the area of skill acquisition, deliberate practice requires that students are aware that the activity is aimed at improving their skills, hence triggering reflective metacognitive processes. The activity is focused at improving performance of the skill, tasks are graduated and allow repetition of simple activities before moving to more complex or difficult ones, which allows for graduated improvement in metacognition. This process of gradual improvement also relies on close guidance and supportive feedback.

Halpern and van Gelder included transfer in their guides to critical thinking but stressed that this ability is at the center of why appropriate and adaptable critical thinking skills are so difficult to develop. As an illustration of this difficulty, in a study of transfer ability (Novick and Holyoak 1991), subjects were exposed to the solutions and underlying working for four mathematics problems all based on a similar reasoning process. One of these problems was about rows of plants. Later, the students were asked to solve a problem about rows of marching band musicians. Solving this problem relied on the same reasoning process used in the original problems. Only 19% of the students identified the similarities and were able to appropriately apply the correct reasoning process to the new problem.

McKendree and colleagues (2002) suggest that while most formal education focuses on how to apply cognitive representations (symbolic "stand-ins" for ideas or items) to problems, not enough focus is given to the question of how to select different representations for different problems. They suggest that critical thinking involves not just selecting appropriate symbolism but knowing how to evaluate the utility of different representations and to explain why one may be more appropriate than others. This "representational fluency" allows for transfer, and university students who are able to demonstrate this type of ability also score better on tests of reasoning ability (Stenning, Cox, and Oberlander 1995) and problem solving (Monaghan, Stenning, Oberlander, and Sönströd 1999).

Cognitive science, critical thinking, and higher education

Research into various tools to support critical thinking has reaped mixed findings. Some rankings of student-developed concept maps (as opposed to teacher-developed maps) indicate more sophisticated thinking from learners (Daley, Shaw, Balistrieri, Glasenapp, and Piacentine 1999). A meta-analysis of eighteen published reports (divided into nineteen individual studies—fourteen of which included student-developed maps) examining the use of concept mapping in classrooms demonstrated a positive impact on student achievement and attitudes (Horton, Mcconney, Gallo, Woods, Senn, and Hamelin 1993). However, the generalizability of concept maps to applied tasks is not clear. Horton et al. (1993) reported that the subject area and the study location both had an impact on the level of improvement. For example, when used in a clinical reasoning task in the place of traditional tools, concept maps did not result in any significant improvements across a range of critical thinking indicators (Wheeler and Collins 2003).

Some studies of cognition have also indicated that the way that feedback and reinforcement are provided may influence skill development in learners. Dweck and colleagues (Dweck 1975; Dweck and Reppucci 1973) identified different groups of child learners. They found that some children are *incremental theorists* who believe that their intellectual competence consists of a set of skills that may be enhanced through effort. The outcome of effortful behavior is increased intelligence. These children seek out tasks that allow for learning opportunities. Dweck and colleagues identified a separate and opposing group they called *entity theorists*. These children attribute performance outcomes to ability. They see intelligence as a global and stable trait that cannot be increased through effort. Because they equate the need to expel effort with lower intelligence, they do not seek out challenging tasks but instead seek opportunities that guarantee success and minimize the chance of making an error. Without taking into account these different approaches students will take, it is difficult to apply a single feedback or reinforcement strategy based on the experimental evidence. What these studies suggest is that the path from cognitive science to the classroom is far from smooth. Laboratory-based research is difficult to translate to real-life teaching situations and therein lies the challenge for the application of cognitive science to the teaching of critical thinking.

The above discussion traces several ways in which the findings from cognitive science and particularly research on heuristics and biases can be used to create better learning for critical thinking through concept mapping, metacognition, and so on. While there is potential in applying laboratory-based research in this way, caution must be taken when attempting to apply cognitive science or neuroscience to classroom practice (see also Bruer 1997). While cognitive science can indeed shed light on the underlying thought processes, it is also important to take into account the specific contextual features of the institution, the students concerned, and the discipline, which gives some weight to the context-specific view of critical thinking. The journey from the laboratory to the classroom is long and fraught with difficulty. In particular, it is difficult to control all the factors in a classroom to provide any level of certainty about any specific intervention. The combination of these issues has meant that the application of the sciences of learning, including cognitive science, to any level of education is a multistep process that involves an ongoing conversation between researchers and practitioners (Lodge and Bosanquet, 2014). It would appear that this conversation is still at an early stage.

Conclusion

Despite the difficulties in applying cognitive science to higher education, the examples provided here demonstrate that there may be substantial gains to be made by incorporating more of the research from cognitive science into higher education teaching practice. The potential for using these findings more broadly has been demonstrated through the application of several phenomena observed under experimental conditions. For example, Bjork's (1994) *desirable difficulties*, where more challenging learning situations create better memory consolidation, and Roediger's (e.g., Roediger and Karpicke 2006) *testing effect*, where repeated testing of knowledge as it is acquired also leads to better memory consolidation, have shown to be robust when applied to educational settings. Dunlosky, Rawson, Marsh, Nathan, and Willingham, (2013) recently reviewed all of the existing evidence from cognitive science that might be used for enhancing teaching practice more broadly. While cognitive science might be useful in enhancing teaching practices for critical thinking for example, results from these studies have also helped to debunk a number of persistent myths in learning and teaching such as the idea that each student has a "learning style" (e.g., Hansen and Cottrell 2013). The potential for using findings from cognitive science to improve teaching for critical thinking in higher education as in other areas of teaching practice is thus clear but yet to be completely realized.

Perhaps the way forward for developing curriculum design for enhancing critical thinking is to move away from the generalist/specifist debate. Educators could consider a more deliberate adaptation and transfer of findings from research examining the mechanisms known to undermine rational thought. In this way, cognitive science can assist in understanding the efficient but not necessarily effective heuristics and biases that lead to faulty thinking within different disciplinary or professional contexts. Cognitive science helps us to understand where thinking goes wrong so that, rather than simply provide formal education in informal logic, we can directly address situations where students' logic is compromised by mental shortcuts through enhanced, evidence-based curriculum design.

References

Bjork, R. A. 1994. "Memory and Metamemory Considerations in the Training of Human Beings." In *Metacognition: Knowing about Knowing*, edited by J. Metcalfe and A. Shimamura. Cambridge: MIT Press.

Bruer, J. 1997. "Education and the Brain: A Bridge Too Far." *Educational Researcher* 26 (8): 4–16.

Burton, L. J., Westen, D., and Kowalski, R. 2012. *Psychology*. (third Australian edition). Brisbane: John Wiley & Sons.

Chance, P. 1986. *Thinking in the Classroom: A Survey of Programs*. New York: Teachers' College, Columbia University.

Daley, B. J., Shaw, C. R., Balistrieri, T., Glasenapp, K., and Piacentine, L. 1999. "Concept Maps: A Strategy to Teach and Evaluate Critical Thinking." *Journal of Nursing Education* 38 (1): 42–47.

Davies, M. 2006. "An 'Infusion' Approach to Critical Thinking: Moore on the Critical Thinking Debate." *Higher Education Research and Development* 25 (2): 179–193.

Davies, M. 2011. "Introduction to the Special Issue on Critical Thinking in Higher Education." *Higher Education Research and Development* 30 (3): 255–260.

Davies, M. 2013. "Critical Thinking and the Disciplines Reconsidered." *Higher Education Research and Development* 32 (4): 529–544.

Dunlosky, J., Rawson, K. A., Marsh, E. J., Nathan, M. J., and Willingham, D. T. 2013. "Improving Students' Learning with Effective Learning Techniques: Promising Directions from Cognitive and Educational Psychology." *Psychological Science in the Public Interest* 14 (1): 4–58.

Dweck, C. S. 1975. "The Role of Expectations and Attributions in the Alleviation of Learned Helplessness." *Journal of Personality and Social Psychology* 31: 674–685.

Dweck, C. S., and Reppucci, N. D. 1973. "Learned Helplessness and Reinforcement Responsibility in Children." *Journal of Personality and Social Psychology* 25: 109–116.

Ennis, R. H. 1989. "Critical Thinking and Subject Specificity: Clarification and Needed Research." *Educational Researcher* 18 (3): 4–10.

Epley, N., and Gilovich, T. 2006. "The Anchoring-and-Adjustment Heuristic." *Psychological Science* 17 (4): 311–318.

Ericsson, K. A., and Charness, N. 1994. "Expert Performance—Its Structure and Acquisition." *American Psychologist* 49 (8): 725–747.

Evans, J. S. 2003. "In Two Minds: Dual-Process Accounts of Reasoning." *Trends in Cognitive Sciences* 7 (10): 454–459.

Gilovich, T., and Griffin, D. 2002. "Introduction—Heuristics and Biases: Then and Now." In *Heuristics and Biases: The Psychology of Intuitive Judgment*, edited by T. Gilovich, D. Griffin and D. Kahneman. Cambridge: Cambridge University Press. 1–18.

Goldstein, D. G., and Gigerenzer, G. 2002. "Models of Ecological Rationality: The Recognition Heuristic." *Psychological Review* 109 (1): 75–90.

Halpern, D. 1998. "Teaching Critical Thinking for Transfer across Domains: Dispositions, Skills, Structure Training, and Metacognitive Monitoring." *American Psychologist* 53 (4): 449–455.

Halpern, D. 2003. *Thought and Knowledge: An Introduction to Critical Thinking*. London: Lawrence Erlbaum Associates.

Hansen, L., and Cottrell, D. 2013. "An Evaluation of Modality Preference Using a 'Morse Code' Recall Task." *Journal of Experimental Education* 81 (1): 123–137.

Horton, P. B., McConney, A. A., Gallo, M., Woods, A. L., Senn, G. J., and Hamelin, D. 1993. "An Investigation of the Effectiveness of Concept Mapping as an Instructional Tool." *Science Education* 77 (1): 95–111.

Kahneman, D. 2011. *Thinking, Fast and Slow*. London: Allen Lane.

Kahneman, D., and Frederick, S. 2005. "A Model of Heuristic Judgement." In *The Cambridge Handbook of Thinking and Reasoning*, edited by K. J. Holyoak and R. G. Morrison. Cambridge: Cambridge University Press.

Lodge, J. M., and Bosanquet, A. (2014). "Evaluating Quality Learning in Higher Education: Re-Examining the Evidence." *Quality in Higher Eduction* 20 (1): 3–23.

Macpherson, R., and Stanovich, K. E. 2007. "Cognitive Ability, Thinking Dispositions, and Instructional Set as Predictors of Critical Thinking." *Learning and Individual Differences* 17: 115–127.

McKelvie, S. J. 2000. "Quantifying the Availability Heuristic with Famous Names." *North American Journal of Psychology* 2: 347–357.

McKendree, J., Small, C., Stenning, K., and Conlon, T. 2002. "The Role of Representation in Teaching and Learning Critical Thinking." *Educational Review* 54 (1): 57–67.

McPeck, J. 1981. *Critical Thinking and Education*. New York: St. Martin's Press.

Messer, W. S., and Griggs, R. A. 1989. "Student Belief and Involvement in the Paranormal and Performance in Introductory Psychology." *Teaching of Psychology* 16 (4): 187–191.

Monaghan, P., Stenning, K., Oberlander, J., and Sönströd, C. 1999. "Integrating Psychometric and Computational Approaches to Individual Differences in Multimodal Reasoning." In *21st Annual Cognitive Science Society Conference*, Eds: 405–410. http://citeseerx.ist.psu.edu/viewdoc/download?doi=10.1.1.32.2687&rep=rep1&type=pdf.

Moon, J. 2008. *Critical Thinking. An Exploration of Theory and Practice*. Abingdon: Routledge.

Moore, T. 2011. "Critical Thinking and Disciplinary Thinking: A Continuing Debate." *Higher Education Research and Development* 30 (3): 261–274.

Novak, J. D. 2010. "Learning, Creating, and Using Knowledge: Concept Maps as Facilitative Tools in Schools and Corporations." *Journal of e-Learning and Knowledge Society* 6 (3): 21–30.

Novick, L. R., and Holyoak, K. J. 1991. "Mathematical Problem Solving by Analogy." *Journal of Experimental Psychology: Learning, Memory, and Cognition* 17 (3): 398–415.

Parameswaram, G. 2007. "Inclusive Writing in a Psychology Class." *Journal of Instructional Psychology* 34 (3): 172–175.

Plous, S. 1993. *The Psychology of Judgment and Decision Making*. New York: McGraw Hill.

Roediger, H. L. I., and Karpicke, J. D. 2006. "Test-Enhanced Learning: Taking Memory Tests Improves Long-Term Retention." *Psychological Science* 17: 249–255.

Scott-Smith, W. 2006. "The Development of Reasoning Skills and Expertise in Primary Care." *Education for Primary Care* 17: 117–129.

Shermer, M. 2002. *Why People Believe Weird Things: Pseudoscience, Superstition, and Other Confusions of Our Time*. London: Souvenir Press.

Stenning, K., Cox, R., and Oberlander, J. 1995. "Contrasting the Cognitive Effects of Graphical and Sentential Logic Teaching: Reasoning, Representation and Individual Differences." *Language and Cognitive Processes* 10 (3–4): 333–354.

Tversky, A., and Kahneman, D. 1974. "Judgment under Uncertainty: Heuristics and Biases." *Science* 211: 1124–1130.

Tversky, A., and Kahneman, D. 1982. "Judgments of and by Representativeness." In *Judgment under Uncertainty: Heuristics and Biases*, edited by D. Kahneman, P. Slovic, and A. Tversky. Cambridge: Cambridge University Press. 84–98.

Vacek, J. E. 2009. "Using a Conceptual Approach with Concept Mapping to Promote Critical Thinking." *Journal of Nursing Education* 48 (1): 45–48.

van Gelder, T. 2005. "Teaching Critical Thinking: Some Lessons from Cognitive Science." *College Teaching* 53 (1): 41–46.

West, R. F., Toplak, M. E., and Stanovich, K. E. 2008. "Heuristics and Biases as Measures of Critical Thinking: Associations with Cognitive Ability and Thinking Dispositions." *Journal of Educational Psychology* 100: 930–941.

Wheeler, L. A., and Collins, S. A. 2003. "The Influence of Concept Mapping on Critical Thinking in Baccalaureate Nursing Students." *Journal of Professional Nursing* 19 (6): 339–346.

Williams, R. L., Oliver, R., Allin, J. L., Winn, B., and Booher, C. S. 2003. "Psychological Critical Thinking as a Course Predictor and Outcome Variable." *Teaching of Psychology* 30 (3): 220–223.

Willingham, D. T. 2007. "Critical Thinking: Why Is It So Hard to Teach?" *American Educator* 31 (2): 8–19.

Zeidler, D. L., Lederman, N. G., and Taylor, S. C. 1992. "Fallacies and Student Discourse: Conceptualizing the Role of Critical Thinking in Science Education." *Science Education* 76 (4): 437–450.

24

Metacognition and Critical Thinking: Some Pedagogical Imperatives

Peter Ellerton

Introduction

Critical thinking is apparently universally desirable as an educational outcome. It is rare to find an educational institution that does not mention some critical skills in the list of its graduate attributes. Too often, however, critical thinking permeates the talk and spirit of syllabi but the substance of it fails to materialize. It has become the Cheshire Cat of curricula, in that it seems to be in all places, owned by all disciplines, but it does not appear, fully developed, in any of these.

Most, but not all, attempts to understand critical thinking focus on compiling a set of cognitive skills, and perhaps affective dispositions, that together are not so much definitive of critical thinking as they are descriptive of how a paradigmatic critical thinker might be said to operate. Broad-brush definitions of a critical thinker include, taking a very small sample, someone who is able to correctly assess statements (Ennis 1964, 599), test their own thinking using criteria and standards (Paul 1993), think effectively with concepts (Elder and Paul 2001), and uncover "evidential relations that hold between statements" (Mulnix 2010, 467).

It is hard to imagine discounting any of these ideas as not relating to common conceptions of critical thinking, and none of these or other researchers would likely dispute the value of each other's categorizations. Nor do they fail to elaborate on their descriptive summaries of what critical thinking involves. Hence, the current understanding of critical thinking is very broad. This does not mean that it lacks precision, as many researchers have articulated skills and affective dispositions in great detail (Facione 1990), but there is a sense in which the net is cast so widely that our definitions become too diffuse to provide a sharp educational focus. The lack of a deeper and more unified understanding of critical thinking also makes the creation of a pedagogical approach to producing critical thinkers problematic.

In this paper I shall focus less on the set of cognitive skills often thought of as being constitutive of critical thinking (inferring, analyzing, evaluating, justifying, etc.) and more on the role of metacognition in thinking critically. Metacognition, as it is routinely elucidated, is thinking about thinking. This simplistic definition does not, however, shine a bright enough light on what we shall see is a cognitively complex phenomenon. I shall therefore provide a model of metacognition, via an explanation given at a functional level, that I hope will be productive in two ways: first, it will improve our understanding of critical thinking; second, it will provide some clear pedagogical principles that can guide the construction of learning experiences and assessment design. I shall use this model of metacognition to propose a model of critical thinking in which metacognition is a *necessary*, unifying element, but in which it is not sufficient—it is, rather, the element that allows any cognitive skill set to be used most effectively, and provides the experience of thinking critically that can be recalled and applied across disciplines and situations. I shall also focus on the evaluative aspects of critical thinking, an emphasis that moves me to name this the *metacognitively evaluative* (ME) model of critical thinking, with full appreciation of the "me" acronym.

Having developed an understanding of critical thinking involving both metacognition and cognitive skills, and using this to generate useful pedagogical principles, I shall then give examples of the application of these principles that can be applied in most discipline contexts.

Metacognition

Metacognition as a concept suffers from the same lack of categorical precision as critical thinking, so it may be a risky strategy to promote it from its role on the periphery to the center of attention as a critical thinking attribute. Let me therefore provide a brief overview of how metacognition may be understood, along with how it is said to be beneficial as an educational strategy.

Understanding metacognition is not just an issue of constitution (i.e., what makes it up); it is an issue of framing. Are we to understand it as a feeling, as a cognitive skill set, as a clutch of strategies, or as intellectual self-governance? It is also debatable as to whether some concepts, such as self-regulation, are subordinate to, or inclusive of, metacognition (Veenman, Van Hout-Wolters, and Afflerbach 2006). These are questions generally bound up in cognitive science, and the hierarchy or relationship linking these ideas is far from clear.

Despite this confusion, metacognition makes regular and frequent appearances in the literature on critical thinking (see, for example, Chaffee, McMahon, and Stout 2004; Elder and Paul 2001; Mulnix 2010; Petress 2004; Scriven and Paul; van Gelder 2005 for a range of conceptualizations and instantiations). This may be because, while there is no suggestion that metacognition is synonymous with intelligence, it does seem to be the case that metacognition, or the skills of self-awareness and self-regulation one might associate with it, improves learning outcomes. Veenman (2006, 6), for example, claims that "on the average intellectual ability uniquely accounts for 10 percent of variance in learning, metacognitive skills uniquely account for 17 percent of variance in learning."

That metacognitive skills or strategies may contribute to academic success is not difficult to imagine if we accept that "a person who is metacognitive knows how to learn because he/she is aware of what he/she knows and what he/she must do in order to gain new knowledge" (Wilson and Bai 2010, 270). How this might work would presumably be a function of discipline or situational context, and involve a grasp of metacognitive strategies and skills, including knowing how and when to apply these.

What does seem consistent in discussions regarding metacognition is that the "meta" part of the word means that we create representations of our thinking, specifically our "first-order mental states" (Fletcher and Carruthers 2012, 12). These states are those desires and beliefs, including beliefs about the truth of factual knowledge, that we use to move through an inferential process to reach conclusions or direct actions. It is also the case, however, that we make higher-order representations, which may be about how these states or representations interact, including the dynamics of the learning and reasoning process. For example, when we perceive a chair we create a mental representation of it. We might also have a mental representation of the fact that our legs are sore, perhaps from a long day standing. Working with these two representations we could create a further representation of the chair being used to relieve the pressure on our legs, and so on until we take appropriate action. What's more, we may work with several representations of chairs and call into use an existing representation of an evaluation process to determine which chair would best suit our immediate needs. More formally, we might construct a mental representation of a valid deductive argument.

It is not the case, however, that higher-order representations are, simply by virtue of being higher order, consciously attended. We may drive a car using a remarkably complex suite of higher-order representations, including some very impressive future matching and evaluation processes, while chugging comfortably along in cognitive neutral, giving these representations little or no conscious attention. The use of the term "metacognition" for simply *having* both first-order and higher-order representations, while perhaps cognitively descriptive, does not seem able to account for the rich educational concept of consciously modifying or accommodating our thinking toward a specified end with a view to optimizing how we get there, let alone evaluating the process as one possible path among many. I suggest that we need to both have *and be consciously aware* of these representations to be metacognitive. We in education might find a definition of metacognition that goes along the following lines more informative and productive: metacognition is attending to mental representations such that the representations themselves, and their interactions, become objects of study.

I do not claim that this is ontologically the case, nor that a broader definition is not more descriptive of findings in cognitive science, but I will attempt to show that using this definition (or perhaps focusing on this aspect) has pedagogical implications that deliver insights into creating better critical thinkers.

This sounds, and is, a complex way of thinking about metacognition, but we can achieve much the same end using an idea grounded more in philosophy than in cognitive science. This is not to ignore the science of cognition, but rather to frame the concept of metacognition at a different functional level. While the cognitive science underpinning our understanding of metacognition continues to develop, how it is eventually understood may remain coherent with this higher function. In the same way as our idea of a car allows us to plan for transport needs even as the specific nature of cars changes significantly over time, progress may be made in the use of metacognition as an educational concept grounded in a functional understanding while our scientific understanding of its nature continues to develop.

Dennett's (1983) development and use of the intentional stance provides a useful conceptualization of the issue that focuses on the functional aspects of metacognition rather than the underlying cognitive processes studied by cognitive scientists. The intentional stance is one that is adopted to explain the behavior of complex systems (biological or otherwise) through the attribution to them of states of desires, needs, goals and ambitions—that is, to consider them as agents (Dennett 1988, 496).

A point to consider with regard to the intentional stance is that these states, such as we might ascribe them, are a consequence of an agent's place and purpose in the world. It also assumes the agent holds, and is subject to, a rationality that prescribes for it courses of action to take and courses of action to avoid. This is to say that the dynamic interactions of representations unfold in a way that is goal driven and geared toward a specified end. We automatically take the intentional stance toward others because we are naturally vigilant as to the manner in which other agents may benefit, inform, or deceive us, and the intentional stance is useful in predicting their behavior. It is very much to our advantage to understand what they may do in the future, based on their intentions.

Metacognition can be achieved by the deliberate and explicit adoption of the intentional stance toward oneself—an internal rather than external application. In considering our own drives, beliefs, desires, thoughts, and processes, we become the object rather than the subject, just as considering another agent would make them the object, for the purposes of anticipating or planning possible future events. Our own mental representations, or some of them, become explicit, thereby making them objects of study to better understand and direct the intentional systems that we individually are. There is an asymmetry

Visualizing thinking: A tool for developing metacognitive skils

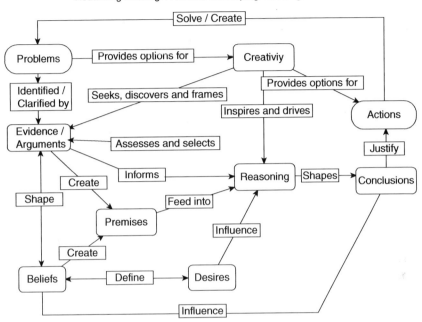

Figure 24.1 Mental representations and their interactions—a tool for examining thinking.

Note: There is no defined starting point.

between the external and internal intentional stance, as we ascribe states to others but experience our own; however, as we consider ourselves as agents, we can use our inwardly directed intentional stance to interrogate ourselves as to what these states might be. We adopt the intentional stance when we ask questions such as "what are the agent's beliefs in this matter?" or "how do these beliefs influence the agent's thinking?" and "what are the agent's goals in this situation and how can they best be achieved given the resources and opportunities available?" To replace "the agent" in these questions with "my," and to then engage with the question, is to become metacognitive.

In contrast to the asymmetry mentioned above, another result of this dual external and internal adoption of the intentional stance is the emergence of symmetry between how we think critically in processing and evaluating the beliefs and arguments of others and those of ourselves. It means the set of skills and abilities developed for external parsing can be applied with equal efficiency internally.

To gain a better understanding of this concept of representing mental states, figure 24.1 shows a visualization of thinking processes that I have trialed in critical thinking courses to promote metacognition and assist students to critically examining their thinking. I call this an MRI diagram (Mental Representations and their Interactions) in the hope, I suspect forlorn, that the analogy with the medical MRI (Magnetic Resonance Imaging) of making things clearer will stay with students.

This tool allows students to visualize their thinking. It captures in a diagram the nonlinear, temporally recursive processes that make up thinking about an issue. It is not suggested that this particular MRI diagram is definitive, or that it is comprehensive, it is simply presented as a *type* of diagram that assists critical analysis of thinking. Any number of similar diagrams could be constructed to accommodate different educational contexts.

The specifics of this diagram go beyond what might be considered necessary for critical thinking (the inclusion, for example, of "creativity" as a consideration), but it does show that such a representation is consistent with the intentional stance in that it incorporates states such as beliefs, goals, and desires, as well as how these may contribute to the reasoning process. It helps students to render explicit that which might otherwise have remained implicit in their thinking.

Visualizations of thinking processes are common, and they have significant educational benefits. Argument mapping, in which evidentiary links, premises, reasons, and conclusions are explicitly stated and diagramed to support or explain reasoning, have been shown to be efficacious in developing critical thinking skills (van Gelder 2005, 45). While argument maps focus on cognitive reasoning skills, figure 24.1 broadens the approach to include aspects that affect arguments and that are implied in the intentional stance.

Metacognition and critical thinking

Let me further explore the relationship between metacognition and critical thinking, and consider also the cognitive skills so often associated with the latter.

The American Philosophical Association developed, through a commissioned systematic inquiry, an expert consensus on what skills and affective dispositions are constitutive of critical thinking known as the Delphi Report. Cognitive skills and subskills agreed upon in this report include (see Facione 1990 for a summary):

1. Interpretation (categorizing, decoding, clarifying)
2. Analysis (examining, identifying)
3. Evaluation (assessing claims and arguments)
4. Inference (querying evidence, conjecturing alternatives, drawing conclusions)
5. Explanation (stating, justifying, presenting)
6. Self-regulation (self-examination and correction).

While the practice of all these skills may be considered as, or at the least to characteristically involve, the manipulation of mental representations, I contend that the last of the cognitive skills, self-regulation, recognizes the necessity of metacognition in critical thinking. Self-examination (a subskill of self-regulation) without metacognition seems untenable under the ME model so far described, considering the things to be examined are mental representations and their relationships.

I also propose that the remaining skills are best enacted in a metacognitive mode. In other words, metacognition is categorically distinct from these other skills, being more a type of thinking, or mode of cognition. Recall that, for our purposes, metacognition involves consciously attending to mental representations. The cognitive skills 1–5 above require working with these representations, even as they themselves are mental representations. Cognitive skills can be processed algorithmically, even the skill of evaluation, as feedback loops on flowcharts show, and while a thinker may be well versed in applying such skills, such application need not be beyond the capacity of a computer program. It seems critical thinking, inasmuch as we wish to distinguish it from algorithmic thinking, is metacognitive. Moreover, it is *metacognitively evaluative*, since good critical thinkers should presumably be adept at continually monitoring and evaluating the selection and application of their cognitive skills. It is the evaluative component that addresses the issue of "correction" in point 6 above, for one must both evaluate and then correct with a standard in mind.

As for the issue of consciously directed thinking, I make the point that consciously evaluating and consciously directing our representations are not exclusive, indeed they are necessary and sufficient for each other. To evaluate thinking is to compare it to a standard, the application of which requires directed representation. To consciously direct thinking is to work toward an end, the choice of which, and the path to it, being something that requires evaluation. Hence, to be metacognitively evaluative is also to direct, to some degree, one's thinking.

The metacognitive experience

Critical thinking is often said to be about *how* to think rather than *what* to think (see, e.g., Mulnix 2010). There is more to this than simple procedure, however. Critical thinking is, in the language of Ryle (1970), a knowing *how* rather than a knowing *that*. It should be noted immediately that use of the terms "how" and "that" reference the nature of the knowledge in question; however, English allows other terms to be substituted and keep this distinction, as the following examples will show. Knowing *how* to play the piano is different from knowing *that* downward pressure on the keys produces sound. Generally speaking, knowing *how* is experiential, knowing *that* is not. One of the ways of thinking about this is that "knowing *that*" type questions elicit propositional knowledge, that is, knowledge that can be true or false. For example, knowing the number of piano tuners in a city, knowing where piano tuners can be found, knowing why piano tuners may be necessary, or knowing how people usually find piano tuners is propositional knowledge. But knowing how to *play* a piano is nonpropositional knowledge.

This is a different epistemic categorization than is often used in education. More commonly a distinction between knowledge and skill is made, such that "knowledge" is material known to be true and "skill" the ability to apply the knowledge. This does not map well onto the knowing that/knowing how division, since knowing how to perform a chemical analysis, say, while arguably an application of knowledge (knowledge about mixtures, machinery, procedures, and the like), can also be propositional knowledge and can be carried out algorithmically—indeed computers routinely do so.

The significance of this for critical thinking is that knowing how to think is, in the model I am proposing, nonpropositional knowledge. Knowing how to manipulate and direct mental representations is nonpropositional knowledge in the same manner that knowing how to keep your balance while riding a bike is nonpropositional knowledge. This does not mean there is an absence of propositional knowledge. In both manipulating mental representations and in riding a bike, we must know *that* certain things are true; we must know *that* representations are present and *that* some rules governing how they interact, and we must know *that* pedals provide forward motion and *that* gravity is ever present, but ultimately this propositional knowledge is not sufficient for the job.

I suggest that the nonpropositional aspect of critical thinking, the consciously directed manipulation of mental representations toward an end that is itself represented, *is* the metacognitive aspect. It is the experience of operating metacognitively that we note, the experience of calling to attention and directing the interplay of our mental representations. Just as knowing *how* to ride a bike provides a conscious experience, the characteristics of which we can recall, so too the experience of being metacognitive is one with recognizable, recallable characteristics. In this sense, critical thinking becomes less a collection of cognitive skills and more of a craft, looking more like the Greek *techne* than *episteme*. As critical thinkers we analyze, infer, justify, and interpret as we do play a piano or ride a bike, as an experience not just as an algorithm. What ties the cognitive skills together and makes it an experience is the metacognitive evaluation of what we are doing. I am not suggesting that playing the piano or riding a bike must be done metacognitively, indeed they are often not done so. I am suggesting that the experiential, nonpropositional knowledge about these things is the sort of knowledge we have about thinking when we know how to be metacognitively evaluative.

Let me now give an example of metacognitively evaluative critical thinking. I have chosen piano playing and bike riding as instances of the use of propositional *and* nonpropositional knowledge. My claim is that being metacognitively evaluative is *also* an instance of the use of propositional and nonpropositional knowledge. I am relying on an analogy here, where the significant issue is the use of two kinds of knowledge. I know from my experience of analogies that an analogy is strengthened if the property being analogized (the use of two kinds of knowledge in this case) is present in cases whose other properties are quite different. While the piano player uses the propositional knowledge about written music, tempo, and the physical aspects of the keyboard, and the bike rider uses propositional knowledge about pedals, gears, and gravity, they both have nonpropositional knowledge that is more than the sum of their propositional knowledge. The propositional and nonpropositional knowledge in both cases is quite different, but they both use the two categories. This variation in particulars between the two cases allows me to be more confident in the application of this categorical distinction in the third case of metacognition. What I have done here is to use propositional knowledge—that an analogy is strengthened when the property being analogized is constant across instances where other properties vary—with nonpropositional knowledge about how to construct and analyze analogies to draw a conclusion. In this construction and analysis I have intentionally represented aspects of the concepts I wish to discuss, manipulated these concepts toward an end, chosen examples to instantiate them, attempted to communicate this concisely, and evaluated the possible effectiveness of my methodology.

"Higher-order thinking" and "problem solving," are often used interchangeably or in association with critical thinking and should be put in context using the ME model of critical thinking. I have already suggested that higher-order thinking is not the same as critical thinking, since higher-order representations need not be consciously attended to and hence are not metacognitive, as metacognition is understood here. Higher-order thinking is a necessary part of critical thinking, however, and to make explicit these higher-order representations and to understand and direct their interplay is to make them part of metacognitive evaluation, and hence critical thinking. Therefore discussion of higher-order thinking with students is a necessary part of creating critical thinkers, an imperative I will expand on in the next section. Having said that, it is important to understand that this definition of higher-order thinking is not the only one, and in some definitions higher-order thinking incorporates critical thinking (see e.g., Lewis and Smith 1993). Higher-order thinking as explained in the ME model of critical thinking, however, is more integrated and explicitly supports pedagogical imperatives, as I will show.

As for problem solving, the term is too vague. What constitutes a problem, after all, is open to interpretation. The range of definitions of "problem" covers

Dewey's organic interpretation of problems as disequilibria to the solving of deductive puzzles in mathematics and hence does not seem to be categorically clear in terms of common or academic understanding. Problems over this range of definitions could be solved through simple reflex physical action, intuitive or heuristic thinking, by the application of algorithms, or by thinking critically. I claim the phrase is unnecessary in the ME model of critical thinking beyond identifying a problem as an area of investigation to which cognitive skills may be applied. In the ME model, as evidenced in part by MRI diagrams, the issue becomes part of the set of mental representations to be recursively addressed. It is an important idea, but does not require categorization beyond common use.

I have considered metacognition as an application of the intentional stance to one's self. This is intended not only to produce an awareness of one's self in intentional terms and to link this to behavior, but also to make those intentional states, and the relationships between them, objects of inquiry. Such a view positions metacognition more centrally in critical thinking as a mode of operating in which other cognitive skills are best utilized. In the ME model of critical thinking, the application of cognitive skills *is* the manipulation of mental representations, and when this is done metacognitively and evaluatively critical thinking is occurring.

In summary, the key propositions of the ME model are:

P1: Cognition is the directed (not necessarily conscious) manipulation of mental representations.

P2: Metacognition is consciously attending to one's mental representations and their manipulations (to take the intentional stance toward one's self).

P3: Mental representations, and their manipulation, can be consciously directed and evaluated.

P4: To think critically is to be metacognitively evaluative.

P5: Knowing how to thinking critically is, in part, nonpropositional knowledge.

Pedagogical imperatives

An implication of the ME model of critical thinking is that just because you are thinking critically does not mean you are doing it well. Mental representations may be inaccurate or inconsistent, how they are directed or the rules followed to manipulate them may be sub-optimal or wrong, and standards of evaluation ill-conceived or inappropriate. This means, of course, that there is much to learn and much to teach about thinking critically. An advantage of the ME model is that two pedagogical imperatives emerge naturally from its definitions: the first imperative is to speak and plan in the language of cognition; the second is to shift the focus of learning and assessment from knowledge to inquiry. I present these below with justifications, showing how they relate to the ME model of critical thinking, and I include steps that teachers can take to facilitate their implementation. In a later section I will give examples of how these principles can be applied in a variety of learning experiences and assessment instruments:

1. *Speak and plan in the language of cognition.* Pre-service teachers often present me with plans that outline in great detail what they will be doing during the course of a lesson. But what the teacher is doing is only of interest insofar as it reveals what the students will be doing, in particular the details of student cognition. In the ME model, the creation of critical thinkers should begin by considering student cognitive activity and development, with the goal that students themselves become aware of this activity. One can consciously attend to one's own mental representations by having them pointed out, and one way point them out is to name them, and use these names when speaking about cognition. The names of mental representations, and the names for how they are manipulated, are generally contained in the language of the cognitive skills.

Establishing a common language around cognitive skills, with students and teachers both having a clear understanding of their meaning and use, allows the teacher to speak to what is happening cognitively in the classroom. Talking with students about their cognition help to makes this cognition an object of focus and study, and therefore provides a path to metacognition. Specific steps that assist to meet this end, and an explanation of how they are derived from the ME model of critical thinking, are presented in the following subsections. These steps are intended to work in concert and may mention different aspects of the same action, and for this reason there is some degree of cross-referencing in the text below. Claims made in these subsections are grounded in the assumptions of the ME model, which are outlined in propositions 1–5 above.

1.1. Develop a sound understanding of the nature and purpose of cognitive skills.

   Before cognitive skills can be taught to students, teachers must be aware of what these skills are and how they can be used. This may be complex, but it need not be hard. A sound knowledge of cognitive skills can be readily achieved through papers like the APA expert consensus mentioned earlier (Facione 1990). Some cognitive skills are listed and elaborated upon in this report, and examples are given as to how they may be used. While there is no definitive list of cognitive skills, and no universally agreed upon definition of each, this is not critical as 1.2 and 1.3 will show.

1.2. Articulate what cognitive skills the students will need as outcomes.

   The cognitive skills, as understood through 1.1, that are relevant to the course can now be determined. The broader the range of cognitive skills involved, the broader the cognitive experience for students. Some of these skills may reside better in learning experiences and others in assessment, some in laboratories and some in tutorials. To the extent that better critical thinkers can use a wider range of cognitive skills, it is desirable to explore as many options for including cognitive skills as possible.

1.3. Determine what the cognitive skills will look like in your course.

It is overly ambitious to imagine that the full scope of a cognitive skill, such as, say, justifying, could be attained in any one assessment piece, learning experience, or even course. It is therefore important to specify what justification will look like in your course; among all the possible ways to consider justification, what best sits in the subject context? Some circumstances may demand that justification means reduction to a sound deductive argument of some sort. Others might require only an appropriate sourcing of material or a legitimate appeal to an authority. Others may demand a demonstration of how an approved method has been applied in a given situation. In effect this limits the mental representations the students will be manipulating to a level that can be more easily articulated and assessed. These are all forms of justification and it is as important to determine the exact nature of the skill as it is to determine what skills will be used.

1.4. Place content in the context of cognitive skills.

There is nothing about the ME model of critical thinking that excludes or minimizes content. It rather asks of the teacher what will be done with the content. If the only skill is recall, then at least that can be made clear. But content can be analyzed, synthesized, categorized, inferred, evaluated, justified, explained, elucidated, represented, interpreted, and identified. Any course not claiming skills beyond recall would be a poor one. So questions can be asked about essential content in terms of the cognitive skills. For example, is the content of a particular lesson to be analyzed? If so, how? What aspects could be inspected and to what level might the analysis penetrate? The greater the degree of precision in the understanding of the analysis required, the clearer the mental representation of the skill in the mind of the student and the more precisely and effectively it can be manipulated.

1.5. Use criteria that address cognitive skills.

If there is a natural home for the language of what students are expected to do cognitively, it is in the course or task criteria. This claim follows from the purpose of criteria: to provide categories and standards of student performance. This is not to say that criteria must be written in the language of cognition. Criteria terminology may speak to physical skills, issues of clarity and accuracy, or of breadth and depth of treatment, just to name a few. While some of these may arguable reduce to cognitive skills, they may function perfectly well as they are. Feedback to students regarding their performance on a task is useful if it relates to the standards against which they are judged, and since we are interested in creating critical thinkers, it follows that a focus on cognitive skills in this area would be useful. One might, as an example, assign to an assessment task the three criteria of recall and identification, analysis, and evaluation, and then set standards as to how well these can be achieved specific to the task (see 1.3).

1.6. Give feedback in terms of cognitive skills.

Hattie and Timperley (2007, 81) have identified feedback as "one of the most powerful influences on learning and achievement." They have also defined feedback as "information about the content and/or understanding of the constructions that students have made from the learning experience" (2007, 82). The ME model suggests that, to create critical thinkers, providing students with feedback on their cognitive performance in the language of their cognition (including the "constructions students have made," which equates in the ME model to mental representations) is the best way to target the relevant cognitive skills.

1.7. Note how and where the methodology of your discipline embeds cognitive skills.

It is commonly asserted that studying a particular discipline develops critical thinking skills, and faculty web sites promoting critical thinking as a graduate attribute from a range of disciplines are legion. Insofar as critical thinking is commonly understood as, to some extent, the application of cognitive skills, the claims have some truth to them. Discipline methodologies have norms for cognition embedded in them, and these can easily be made explicit. For example, scientists generally seek to falsify hypotheses rather than confirm them. Not only does this falsification prevent putting effort into producing an infinite string of confirmations, each of which serves only to support the hypothesis (not prove it), it also works against the confirmation bias (that tendency to look only for confirming instances that support an existing bias). While the idea of falsification is a standard one, the cognitive aspects of it are not generally made explicit. According to the ME model, identifying and teaching cognitive skills in discipline methodology means that teaching critical thinking can be a fundamental part of discipline instruction.

2. *Shift the focus of learning experiences and assessment from knowledge to inquiry.* Bunnell and Bernstein (2012, 15) note the following for teachers in higher education institutions:

Traditional models of teaching in higher education position students as the receivers and reproducers of knowledge, and teachers select which information is received and later retrieved by their students. Quality of instruction is evaluated based on appropriate identification of the knowledge to be covered and skill in preparing organized and clear presentations of that knowledge.

This is consistent with the banking model of knowledge identified by Freire (1996), in which knowledge is held by the teacher and transferred to the student, who then adds this knowledge to an existing "bank."

In this model there is no necessary dynamic between knowledge and inquiry, and no imperative on the part of the student or teacher to evaluate the source or means of derivation of knowledge. This renders evaluation, or even inclusion, of inquiry as unnecessary at the best and educationally disruptive at the worst. Moreover, a logical consequence of accepting the banking model of education is to value didactic learning experiences above others. The rationalization of knowledge transfer is the driving and guiding pedagogical principle, and the mark of a good educator.

It is difficult to see how critical thinking is promoted, taught, or even valued within this model. What is also notable is how removed from students is the need to be metacognitively evaluative. As Bunnell and Berstein say (2012, 15), "Faculty members often struggle to view student work as more than merely a measure of the amount of information individual students can retain and reproduce". I call this "teaching by echo-location," in which teachers mark the progress of a lesson by receiving back from students what Ryle calls "strings of officially approved propositions" that were transmitted earlier (Peters 2010, 108).

As the ME model of critical thinking requires that students be metacognitively evaluative, it is necessary that they be provided with experiences to both use and evaluate cognitive processes, and to do so under conscious control. Recall that in the ME model, critical thinking also contains nonpropositional knowledge. Those wishing to learn to ride a bike or play a piano must be given experiences to do so. During these experiences feedback can be given not only by mentors and teachers, but also by perception; for example hearing in the case of playing the piano and overbalancing in the case of the bike. All of these combine to produce the nonpropositional knowledge of *how* these things are done. As the ME model draws an analogy with the nonpropositional knowledge acquired through playing the piano and riding a bike, it follows that thinkers need to be given opportunities to experience being metacognitively evaluative and to receive feedback during the process. This directly relates to the Hattie reference in 1.6, indicating feedback is effective when given explicitly and directed toward an end, in this case a cognitive end. It also reflects the findings of the Delphi Report, which states that "teaching cognitive skills also involves exposing learners to situations where there are good reasons to exercise the desired procedures, judging their performance, and providing the learners with constructive feedback regarding both their proficiency and ways to improve it" (Facione 1990, 15).

Shifting the focus from knowledge to inquiry in learning experiences and assessment provides opportunities to move into this metacognitively evaluative mode. I am not suggesting that inquiry displaces knowledge, simply that opportunities be sought where the focus can be shifted.

2.1. Target as wide a range of cognitive skills as possible.

As mentioned in 1.2, the broader the range of cognitive skills the greater the opportunity for students to exercise them, and to become familiar with their norms of operation. Consider a task in which two ways of thinking or operating are being examined, say two schools of sculpture in an art class. Rather than simply researching the particulars of the two schools, which requires a narrow range of cognitive skills, the students could be asked to observe two sculptures, one from each school. In doing so they could be *analyzing* aspects of the sculptures, *inferring* which tools and techniques were used, *comparing* and *contrasting* shape and style, *evaluating* the outcomes in terms of aesthetics or simplicity, *classifying* the two into broader categories, and so on. They could then *synthesize* all this to *present* their findings. The italicized words, which describe cognitive skills, could then be included in the rubric for the tasks, and they could also be mentioned explicitly in student instructions.

2.2. Integrate the cognitive skills to focus on higher-order thinking.

The range of skills utilized as outlined in 2.1 is important, but the ME model also suggests that making and working with higher-order representations and concepts, that is, dealing with complexity, also helps to create better critical thinkers. To continue the previous example of comparing sculptures, students could be asked to examine cultural conditions at the time of sculpting, and suggest how these might have influenced the artists' work. Greater complexity demands greater attention to cognitive detail, and this adds to the richness of the metacognitive experience.

2.3. Speak about inquiry in terms of the use of cognitive skills.

Following from increasing the range of skills taught and assessed as mentioned in 2.1, and to help achieve metacognition, the language of the task of feedback and of the criteria should be consistent and in terms of student cognition. Simple questions asked of students can do this. For example, asking what is being analyzed, why it is chosen for analysis, and how it is being analyzed breaks the skill down and makes the cognitive steps explicit. Feedback on the quality and nature of analysis can be done in the same way, and, as Hattie has shown in 1.6, is particularly effective when linked to performance criteria.

Critical thinkers

The vast majority of our thinking is done without consciously attending to it, and we operate at the level of heuristics and intuition most of the time (see, e.g., Kahneman 2011). There is recognition in the literature that the paradigmatic critical thinker can, to some degree, internalize the knowledge of critical thinking and perform effective, habitual thinking below conscious control. This is easily understood using the analogy of piano playing. While the beginner learns how to move her hands, how to translate symbols on a page into notes on the keyboard, and how to monitor and control timing, she is being metacognitive. Many actions, physical and mental, are subject to conscious control and evaluation. Over time these skills are removed slowly from conscious awareness and become more intuitive and habitual, though they may be recalled to conscious attention at any time. So too the ability to recognize a fallacy, to analyze text for logical inconsistency, to detect pseudoscience, or to evaluate an argument can become part of our heuristic response to stimuli.

This movement from the metacognitively evaluative mode of thinking into the intuitive, internalized, rule-of-thumb mode of thinking is a deliberate educational outcome and can be understood as the concept of mastery. In a trivial sense, one might master the mathematical times tables, but this is simply the skill of recall. To internalize and habitualize the use of a broad range of skills involving higher-order representations and to make from these heuristics or algorithmic approaches that rapidly decrease processing and response time would sit better with the common idea of mastery.

This progression from the metacognitive to the heuristic is not inconsistent with the assumptions of the ME model of critical thinking. To make a claim of conflict would be to confuse critical *thinking* with critical *thinkers*. Critical thinkers, by definition, can think critically, but they need not always do so. Over time critical thinkers may develop heuristics that are highly effective in problem solving, for example, without the need for the conscious attention necessary for a beginning critical thinker. Even though the use of heuristics may be successful on many occasions, while the process is not been reflected upon to ascertain its effectiveness it is not critical thinking in the ME model. But such heuristic thinking is not to be discouraged, as an experienced critical thinker might only spend a small amount of time thinking critically, and a large amount of time using effective heuristics developed through extensive metacognitive evaluation. Having a wide range of useful heuristics and the ability and willingness to metacognitively examine them as the need arises, and to know when this need has arisen, is critical thinking mastery. I am not excluding the assimilation and use of a wide range of propositional knowledge about critical thinking from the concept of mastery, just augmenting it with other skills.

In the movie *Shine*, which explores the life of the pianist David Helfgott, a memorable conversation takes place between the characters of Helfgott and his Professor. I believe this captures the concept of mastery very well (2013):

P: The page, for God's sake! The notes!
H: I'm sorry I was, uh, forgetting them, Professor.
P: Would it be asking too much to learn them first?
H: And-And then forget them?
P: Precisely.

Conclusion

Research in critical thinking forms a broad church. This is useful in that a range of thinking skills and their relationships is explored. It is less useful in the lack of any clear pedagogical direction that a more unified understanding may deliver. The metacognitively evaluative (ME) model of critical thinking provides unification, in that it can accommodate several understandings of critical thinking, including a focus on skills, metacognition and habitual critical thinking, and pedagogical imperatives drawn from the nature of metacognition and evaluation within the model.

Metacognition is proposed to be conscious attention to mental representations. This level of description allows space for the science of cognition to develop to some extent without invalidating the model. One path to achieving metacognition is the adoption of the intentional stance toward one's self, requiring an introspection of our goals, desires, and beliefs, as well as an understanding of our rationality and the norms of reasoning.

In terms of critical thinking outcomes, the pedagogical imperatives of the ME model are, first, the development and use of a language of cognition, to more effectively focus on the thinking skills of students, and to provide them with goals and feedback regarding reaching those goals, in terms designed to match their cognition. The second imperative is to focus on the (cognitive) process of inquiry rather than simply knowledge of content, with inquiry affording opportunities for students to use and develop their cognitive skills.

Experienced and effective critical thinkers have a wide range of cognitive tools, including well-developed heuristics and a comprehensive set of cognitive skills. They also have nonpropositional knowledge about critical thinking, including an ability to operate metacognitively and an ability to recognize when such metacognition is required.

Some aspects of the ME model of critical thinking help explain existing research in effective teaching (e.g., that of Hattie and Timperley mentioned earlier), but more work is needed in matching known educational success against models of critical thinking to test both their predictive and explanatory powers. In the meantime, work on critical thinking in all its manifestations and interpretations helps keeps a focus on student cognition, which is the core concern of educators.

References

Bunnell, S. L., and Bernstein, D. J. 2012. "Overcoming Some Threshold Concepts in Scholarly Teaching." *The Journal of Faculty Development* 26 (3): 14–18.

Chaffee, J., McMahon, C., and Stout, B. 2004. *Critical Thinking, Thoughtful Writing: A Rhetoric with Readings*. (third edition). Boston: Houghton Mifflin.

Dennett, D. C. 1983. "Taking the Intentional Stance Seriously." *Behavioral and Brain Sciences* 6 (03): 379–390.

Dennett, D. C. 1988. "Précis of the Intentional Stance." *Behavioral and Brain Sciences* 11 (03): 495–505.

Elder, L., and Paul, R. 2001. "Critical Thinking: Thinking with Concepts." *Journal of Developmental Education* 24 (3).

Ennis, R. H. 1964. "A Definition of Critical Thinking." *The Reading Teacher* 17 (8): 599–612.

Facione, P. A. 1990. "Critical Thinking: A Statement of Expert Consensus for Purposes of Educational Assessment and Instruction." *Research Findings and Recommendations*. 2012/12/09/22:50:47. http://www.eric.ed.gov/ERICWebPortal/detail?accno=ED315423.

Fletcher, L., and Carruthers, P. 2012. "Metacognition and Reasoning." *Philosophical Transactions of the Royal Society B: Biological Sciences* 367 (1594): 1366–1378.

Freire, P. 1996. *Pedagogy of the Oppressed*. (second edition). London: Penguin.

Hattie, J., and Timperley, H. 2007. "The Power of Feedback." *Review of Educational Research* 77 (1): 81–112.

Kahneman, D. 2011. *Thinking, Fast and Slow*. London: Allen Lane.

Lewis, A., and Smith, D. 1993. "Defining Higher Order Thinking." *Theory into Practice* 32 (3): 131–137.

Mulnix, J. W. 2010. "Thinking Critically about Critical Thinking." *Educational Philosophy and Theory*. http://www.tandfonline.com/doi/pdf/10.1111/j.1469-5812.2010.00673.x.

Paul, R. 1993. *Critical Thinking: How to Prepare Students for a Rapidly Changing World*. (first edition). Santa Rosa, CA: Foundation for Critical Thinking.

Peters, R. S. 2010. *The Concept of Education*. London and New York: Taylor & Francis Group.

Petress, K. 2004. "Critical Thinking· An Extended Definition." *Education* 124 (3): 461–466.

Ryle, G. 1970. *The Concept of Mind*. (New impression edition). Harmondsworth: Penguin Books Ltd.

Scriven, M., and Paul, R. *Defining Critical Thinking*. 2011/11/23/21:45:59. Available from http://www.criticalthinking.org/pages/defining-critical-thinking/766.

Shine Script—Dialogue Transcript 2013. October 30. Available from http://www.script-o-rama.com/movie_scripts/s/shine-script-transcript-david-helfgott.html.

van Gelder, T. 2005. "Teaching Critical Thinking: Some Lessons from Cognitive Science." *College Teaching* 53 (1): 41–46.

Veenman, M. V., Van Hout-Wolters, B. H., and Afflerbach, P. 2006. "Metacognition and Learning: Conceptual and Methodological Considerations." *Metacognition and Learning* 1 (1): 3–14.

Wilson, N. S., and Bai, H. 2010. "The Relationships and Impact of Teachers' Metacognitive Knowledge and Pedagogical Understandings of Metacognition." *Metacognition and Learning* 5 (3): 269–288.

Part VI

Critical Thinking and the Professions

As we noted in the Introduction to this volume, the importance of critical thinking is becoming increasingly clear outside the academy in the business world and the professions. The chapters in this section provide a useful overview of the different ways in which critical thinking extends beyond the academy and into the professional world.

Surveys of employers in the US business sector consistently show that a key skill demanded by employers is "critical thinking." A recent survey by *Forbes* using data from CareerBuilder and O*Net—the US clearinghouse of occupational information—identified critical thinking as the first-listed in a list of top ten skills that help people get hired. Among employers, and employer groups, there is a growing recognition that what is needed in graduates is not so much technical knowledge, but applied and transformative skills, especially skills in critical thinking.

As noted in the Introduction, recent survey data from 400 US employers has made this very clear. In data published in a major report by a consortium of US organizations in 2006 employers surveyed articulated the skill-set that they perceived was needed to succeed in the workplace in the new century. They applied this skill-set to recently hired graduates from three types of institutions: high school, two-year colleges, and four-year colleges. The highest-ranked skill as rated by employers was critical thinking, surpassing "innovation" and "application of information technology." Of the employers surveyed 92.1% regarded graduates from four-year colleges as being "deficient" in critical thinking. Critical thinking clearly matters in the professions.

In most universities, critical thinking is one of the graduate attributes most widely cited as essential for students emerging from universities. There is a growing recognition among employers that critical thinking skills are essential for corporate contexts. Unsurprisingly, perhaps, this is a matter of pragmatic necessity. It emanates not from a wish by companies to produce corporate philosophers, nor an altruistic desire to produce better graduates for the benefit of society, but simply from a wish to train people who are adaptable in the workplace, and who can "think on their feet." As noted by Halpern (2001): "Virtually every business or industry position that involves responsibility and action in the face of uncertainty would benefit if the people filling that position obtained a higher level of the ability to think critically...Critical thinking skills offer the greatest chance of success for creating and adjusting to change." This, it appears, is the driving force behind the attention paid by the business community on the development of critical thinking skills.

This section of the book offers five chapters, three of which cover critical thinking in the professions of accountancy, medicine, and science. The fourth chapter uses case studies on the use of social media in enhancing critical thinking in the fields of music, population health, and creative writing. The fifth looks at the role of critical thinking as a preparation for future practice in a variety of professional fields.

Sin, Jones and Wang investigate critical thinking as it applies to the profession of accountancy. They present a useful summary of recent qualitative studies focusing on how critical thinking is understood by accounting students, academics, graduates, and employers, triangulating this with data from interviews from seasoned accounting professionals. They identify five conceptions of critical thinking as it applies to the profession: identifying and promoting client outcomes; determining less-obvious implications of data; coping with complexity in social, organizational, and institutional scenarios; evaluating motivations of stakeholders; determining the relevance of abstract principles and concepts; and internalizing these concepts and principles into an overall disposition toward ethical practice. They view critical thinking in the context of the accounting profession "less [as] a cognitive ability than as a kind of purposeful, yet highly problematic doing."

Trede and McEwen take a very different approach to the topic. Rather than assume the value of critical thinking for the professions, they take to task the kind of criticality that genuine learning for future practice requires. Drawing on the critical pedagogy tradition, they argue that present practices of workplace learning and supervised work placements foster a culture of compliance, a readiness to accept and prepare oneself for professional "competence frameworks." Competence frameworks, of course, lead to fragmented assessment routines. This, in turn, leads to surface learning, memorization, recalling, and copying. This, they argue, does not encourage true critical thinking, and is anathema to the true purpose of higher education. They acknowledge that while students want to belong to a profession rather than critique it, it is also true that current university practices are driving students toward techno-instrumental, specialized knowledge dictated by professional bodies. Instead, they should be using the opportunities of workplace learning and placements to develop a broader sense of criticality that might be beneficial to the professions, and the society beyond. This "emancipatory" form of criticality equips students with skills to deal with complexity, diversity, and ambiguity, which is much needed in twenty-first-century learning. Learning for future practice, they argue, should retain the promise of the university as a place of liberal education.

Grace and Orrock look at critical thinking in the field of osteopathic medicine. They view critical thinking as being distinct from clinical reasoning in respect of the relative weight given to the meta-skills employed in clinical practice. The medical context is clearly distinguished by the relative neglect of the socially transformative character of criticality (given prominence by Trede and McEwen). Unsurprisingly perhaps, the medical context of critical thinking lies in largely technical, but also ethical, skills: forming reasons, gathering and analyzing information, proposing answers to questions, and enacting consequences, alongside important considerations of patient care.

Wilson, Howitt, Higgins, and Roberts investigate critical thinking in the context of scientific research. They compare extended scientific research projects in a variety of fields, assessed by means of traditional formal reports—the scientific journal article—with the more revealing format of a science blog. The scientific report genre typically omits from consideration any "thinking" behind the learning undertaken: misconceptions go unchallenged, mistakes are suppressed, and the "messiness" of scientific thinking is hidden from view. In their study, the use of a blog makes visible the critical thinking behind the science. Data from the blog entries provide evidence of how students' criticality matured during the process, and how this revealed a strong correlation between their critical thinking, their confidence as practicing scientists, and their associated sense of developing scientific expertise.

In keeping with the use of online tools, November offers an analysis of how critical thinking can be promoted by the use of social media in professional fields as diverse as music, population health, and creative writing. Like Wilson et al., she sees the online environment as a way of making critical thinking visible in respect of travelling with ideas, synthesizing multiple viewpoints, judging based on evidence, and fostering reflection.

Reference

Halpern, D. F. 2001. "Assessing the Effectiveness of Critical Thinking Instruction." *The Journal of General Education* 50 (4): 270–286.

25
Critical Thinking in Professional Accounting Practice: Conceptions of Employers and Practitioners
Samantha Sin, Alan Jones, and Zijian Wang

Introduction

Over the past three decades or so it has become commonplace to lament the failure of universities to equip accounting graduates with the attributes and skills or abilities required for professional accounting practice, particularly as the latter has had to adapt to the demands of a rapidly changing business environment. With the aid of academics, professional accounting bodies have developed lists of competencies, skills, and attributes considered necessary for successful accounting practice. Employers have on the whole endorsed these lists. In this way "critical thinking" has entered the lexicon of both accountants and their employers, and when employers are asked to rank a range of named competencies in order of importance, critical thinking or some roughly synonymous term is frequently ranked highly, often at the top or near the top of their list of desirables (see, for example, Albrecht and Sack 2000). Although practical obstacles, such as content-focused curricula, have emerged to the inclusion of critical thinking as a learning objective in tertiary-level accounting courses, the presence of these obstacles has not altered "the collective conclusion from the accounting profession that critical thinking skills are a prerequisite for a successful accounting career and that accounting educators should assist students in the development of these skills" (Young and Warren 2011, 859). However, a problem that has not yet been widely recognized may lie in the language used to survey accountants and their employers, who may not be comfortable or familiar with the abstract and technicalized pedagogical discourse, for instance, the use of the phrase "developing self-regulating critical reflective capacities for sustainable feedback," in which much skills talk is couched. Hence in this chapter we will focus on how practising accountants—often themselves employers or involved in recruitment—perceive and verbalize this apparent need when not influenced by, and perhaps confused by, the type of academic lexis typically presented to them in survey instruments. How is critical thinking conceptualized and how is it expressed in the language of the lifeworld?

It has been widely accepted since at least the 1980s that critical thinking is a key requirement for success in most practical and professional spheres, not just accounting. Indeed an ability to think critically has been recognized as one of the chief goals of tertiary education (Barnett 1997; Entwistle 1994). So it cannot be surprising that critical thinking skills have assumed a high profile in the literature on accounting education, where it is widely mooted that such skills are crucial for a successful transition from the classroom to the professional workplace. In the United States employers and professional bodies are nowadays at one on this (Accounting Education Change Commission 1990; Albrecht and Sack 2000; American Accounting Association 1986; American Institute of Certified Public Accountants 1999). In the late 1980s, the Big Eight accounting firms (as they then were) released an influential White Paper (Big Eight Accounting Firms 1989, 5), in which it was stated that "education for the accounting profession must produce graduates who have a broad array of skills and knowledge." The first of the essential skills to be mentioned in that document was an ability to use critical thinking and creative problem-solving techniques on unstructured problems in diverse and unfamiliar settings. In Australia and New Zealand the emphasis on soft skills, and on critical thinking skills in particular, is quite explicit in official documents of the professional accounting bodies. For instance (CPA Australia and the Institute of Chartered Accountants in Australia 2012, 5):

> The accounting profession needs graduates from diverse backgrounds with a range of competencies. It requires all graduates to have capacities for inquiry, abstract logical thinking and critical analysis, in addition to appropriate oral and written communication and interpersonal skills.

The development of critical thinking is the third of four main educational objectives identified in this document, which sets out the conditions of accreditation for accounting courses offered by tertiary institutions and indeed provides suggestions as to how this might be achieved (CPA Australia and the Institute of Chartered Accountants in Australia 2012, 5):

> 3. Encourage *innovation in teaching and learning* with particular focus on integrating the development of critical thinking, ethical judgment and communication skills of graduates.(Emphasis in original)

It is generally accepted that this emphasis on critical thinking skills—and critical thinking–related skills—by the professional bodies reflects the demands of employers, and much recent research supports this point of view. We cite a few studies here. Jackson et al. (2006) report that employers expect graduates to possess higher-order skills *such as critical thinking and analytical skills* (our emphasis). Freeman et al. (2008, 13) report that "developing theoretical understanding and critical thinking capabilities amongst other desirable graduate attributes…will allow graduates to be productive in their professional employment." And, while the focus is often on new graduates and their need to have acquired such skills, critical thinking and problem solving have also been identified as prerequisites to promotion in public sector accounting (see the US Partnership for Public Service 2007). The outstanding question is how best to teach this skill—if indeed it is a single skill—in postsecondary accounting courses, and numerous suggestions have been advanced (Camp and Schnader 2010; Jones and Sin 2003; 2004; Kimmel 1995; Lucas 2008; Sin, Jones, and Petocz 2007; Tempone and Martin 2003). We will address this question further below but will first try to develop a more coherent approach to the slippery concept of critical thinking itself, regarded as a cognitive ability.

Defining critical thinking—an academic exercise?

There is still considerable uncertainty as to the precise nature of critical thinking, both in the literature on accounting education and further afield, for example, in the various literatures of education and psychology. Indeed academics disagree at a very basic level: as to whether we are dealing with skills, dispositional traits, or values; and, if critical thinking consists essentially of skills, or a skill set, are these skills cognitive, behavioral, or cognitive-affective? Are they complex and content specific or simple and transferable? Kurfiss (1988, 2) suggested that critical thinking is an activity, that is, "an investigation whose purpose is to explore a situation, phenomenon, question, or problem to arrive at a hypothesis or conclusion about it that integrates all available information and that can therefore be convincingly justified." Kurfiss (1989, 42) also proposed that critical thinking involves two distinct phases: discovery and justification.

> In the discovery phase, we examine in search of patterns and formulate interpretations or hypotheses about what the evidence means. In the justification phase, we set forth our conclusions, reasoning, and evidence in an argument.

Paul (1989) made the important assertion that critical thinking differs according to the discipline and epistemic culture in which it is practised, and also emphasized its metacognitive (or reflective) nature. Peter Facione and colleagues have gathered evidence over a number of years to show that critical thinking is crucially a matter of the individual's personal psychological disposition (Facione

1990; Facione, Facione, and Giancario 1997; Facione, Sanchez, Facione, and Gainen 1995; Giancarlo and Facione 2001).

It is noteworthy that critical thinking (qua skill and/or qua attribute) is often represented as an aspect of another core capability such as creativity or problem-solving, or as a kind of hybrid ability, for example, "strategic/critical thinking" (American Institute of Certified Public Accountants 1999) and "analytical/critical thinking" (Albrecht and Sack 2000). This ties in with the fact that, in Bloom's *Taxonomy of Educational Objectives, Handbook 1: Cognitive Domain* (Bloom, Engelhart, Furst, Hill, and Krathwohl 1956), criticality or critical discrimination is implied by and subsumed under at least four main objectives: *applying knowledge in specific contexts, analysis, synthesis,* and *evaluation.* And when William Birkett (who drew on Bloom's schema) identified a core set of "generic skills" apposite to the accounting profession in Australia and New Zealand (Birkett 1993)—based on extensive surveys of practitioners at all levels and in several key subdisciplines—he too avoided mentioning critical thinking by name. However, critical thinking is implicated in a number of named skills and criticality does in fact receive one cryptic mention, in the injunction to "think and act critically." Birkett's systematic survey research was carried out at the specific behest of the professional accounting bodies (CPA Australia and the Institute of Chartered Accountants in Australia 2012) and has played a key role in promoting the generic skills agenda, particularly in providing the core curriculum generic skills list for professional accreditation of tertiary programs in Australia and New Zealand. (Birkett's set of core generic skills has more recently been incorporated in the accounting threshold learning outcomes in the latest accreditation guidelines, as per CPA Australia and ICAA [2012], to bring them into alignment with the new Australian Accounting Academic Standards.)

King and Kitchener (2004) have written for 30-odd years about *reflective judgment,* which is arguably another term for critical thinking—or, at least, a large component of that capacity. Meanwhile Barnett (1997) has expanded the concept of critical thinking in terms of *criticality* and *critical being,* that is, a set of values as much as a disposition and one which the traditional Western-style university, with its unique blend of research and teaching, was best positioned to inculcate. We note that, in this way, the ability to think critically is linked intrinsically to identity and, more specifically, professional identity. Our own data has convinced us that critical thinking involves more than a skill or set of skills and that, while it usually *manifests* as a psychological disposition and an inclination to adhere to certain values, it actually stems from one's sense of self-identity, a professional persona that has typically been cultivated over many years (Jones and Sin 2013; Sin 2011).

Before pursuing the definitional question any further on this abstract and context-free level, let us consider the discipline-specific nature of critical thinking, A broad survey of tertiary-level teaching staff in Australia (Anna Jones

2010, 8) shows clearly that attributes such as critical thinking and problem-solving are understood very differently in different academic domains:

> For example, in history, critical thinking is the examination of evidence (both primary and secondary sources), understanding the social context, acknowledging ambiguity and the multiple perspectives inherent in history and understanding the role of the historian in constructing a historical argument. In physics, by contrast, critical thinking requires an examination of the logic and accuracy of a model and for experimental physicists, the rigour of experimental technique. In economics, critical thinking is the use of the theoretical toolkit of the discipline to examine economic questions. In law, critical thinking involves the use of evidence and assumptions, questions of professional and social ethics, an awareness of the social context and a historical perspective. In medicine, critical thinking requires clinical reasoning that also utilizes an understanding of medical evidence and an awareness of the complex ethical issues associated with medical practice.

Critical thinking in business and organizational contexts — in which accounting, as a social practice, crucially takes place—means very different things for the different stakeholders. Clients, accountants, employers, and, outside the firm, the various regulatory bodies will have competing interests in any given situation and, consequently, will tend to conceptualize criticality in distinct ways. In this chapter we will draw on recent phenomenographical research (Sin 2010) to illustrate some of the differing conceptions of critical thinking that are held by accounting students, newly graduated accountants in their first year at work, and seasoned practitioners in a range of specialized sub-domains of accounting.

Critical thinking—by various other names

While there certainly remain intransigent terminological problems for academics, and while a range of cognitive abilities closely resembling critical thinking has been promulgated under a variety of superficially unrelated terms, the concrete meaning of the received term—"critical thinking"—becomes even harder to negotiate in discussions with accounting practitioners, whether relatively new or seasoned. We might say that the term "critical thinking" is not in their lexicon. The whole question of critical thinking and criticality more generally, and indeed the whole matter of necessary skills or abilities, is usually only adumbrated in the informal language of that lifeworld. Indeed such questions are often effectively obviated by a reliance on tacitly held beliefs about the ways in which workaday things are done—and get done. Skills, attributes, and abilities are discussed (if they are discussed at all) in nonacademic, idiosyncratic, extemporized, and often-inconsistent terminology. It is true that a variety of technical-seeming discourses, reflecting pseudoscientific fashions of thinking, have permeated the modern workplace, including the professional workplace, with apparently technical terms that are in fact quite fluid and ill-defined, in other words, by a variety of jargons (see Watson 2004, on managerial pseudo-jargon).

For the kinds of reasons discussed above employers may often fail to identify critical thinking as an explicit requirement when interviewed, though the skills they expect of graduates clearly implicate this ability. There is a clear disjunction between the findings of qualitative investigations, using semistructured interviews, and more quantitative research carried out using survey instruments designed by academics. In the latter, critical thinking will typically appear and will consequently be ranked (often highly). But, without the kind of involuntary suggestion involved in the presentation of a list of abstract terms, practitioners and partners find very different ways to talk about the attributes of successful practice. This is what we look at further below.

Unfortunately it is not always clear from the literature to what degree the language of the (academic) researchers might have influenced and been adopted by participants in particular studies and, moreover, to what extent their understanding of the terms used are in agreement with one another and with the conceptions of the researchers. For instance, the survey instrument developed by Albrecht and Sack (2000) and since adopted or adapted by others (see Kavanagh and Drennan 2008), is a typical example, containing numerous terms like "analytical thinking" and "critical thinking" that are the subject of live debate and competing definitions in the field of educational psychology and in the cognitive sciences more generally. It is not at all clear that any two people would assign them similar meanings. That said, employers typically identify "analytical skills" and "critical thinking skills" as essential abilities or traits in accounting graduates. However, they also frequently call for things like "business awareness" or "real life experience"—capacities that more clearly depend on biographical and contingent aspects of person's experience. Employers also typically demand high-level "communication skills," a keen "ethical awareness" and "teamwork" skills. Such labels, for all that they may appear reasonably straightforward and transparent, are just as context-dependent and as complex in operation as those mentioned earlier. In effect, the employers are asking for capabilities that are inextricably associated with an individual's unique knowledge base, life experience, and work experience (see Jones and Sin 2003).

We can approach the concept of critical thinking (and analytical-critical thinking, etc.) by examining self-regulating critical reflective capacities the kinds of issues, not to say problems, critical thinkers are said to be able to deal with. A number of scholars have suggested that critical thinking is called for when we are faced with the types of ill-structured problems, ambiguous problems, and ethical dilemmas that characterize professional work in today's world (beginning perhaps with Simon 1974). King and Kitchener (2004, 5) claim that ill-structured problems are characterized by two features: they cannot be defined with a high degree of completeness, and they cannot be solved with a high degree of certainty. Such problems represent the challenge that critical thinking and reflective judgement are generally thought to address. These are the contingencies that (according to Sarangi and Candlin 2010) surround professional practice and defy solution using the concepts and methods learnt as a part of professional education and training.

Facione et al. (1997, 3–4) state that in professional practice the problems to be solved vary along seven parameters:

> The parameters of professional judgment include the *setting* of the problem or decision which the professional is required to address. The problem or decision at hand can be described along several different dimensions. To oversimplify let us imagine that these continua are simple dichotomies. Even so, the problems or the decisions to be made can be (1) high stakes or low stakes, (2) time-constrained or unconstrained, (3) novel to the professional or familiar, (4) unexpected or planned, (5) requiring specialized knowledge or accessible by knowledge commonly shared, (6) requiring resolution by the professional working alone or working collaboratively, (7) well-structured/paradigmatic in the field or ill-structured and highly unusual. (Emphasis in original)

A familiar alternative in each of these seven parameters, if added together, constitutes a neat description of the types of challenge that accountancy professionals are likely to encounter almost on a daily basis in their offices today. However, although the nature of the challenge that necessitates the cultivation of critical thinking skills can be described in this way (for example), the nature of the skills required is considerably less certain.

Critical thinking—as conceptualized and verbalized by practicing accountants

Can researchers access the conceptions and expressions of practitioners? Some studies seem to have gone further in this direction than others. A good starting point is the detailed investigation carried out by Baril and colleagues (Baril, Cunningham, Fordham, Gardner, and Wolcott 1998), which focused on the public accounting profession. Their interviews teased out some of the language used by professional accountants to discuss critical thinking as they carefully avoided proposing terminology of their own devising. Thus their findings illustrate the diverse ways in which the concept was understood— and expressed—by members of that profession in the 1990s. For their investigation they obtained the formal employee performance appraisal documents from five of the then Big Six public accounting firms and identified sections or criteria addressing critical thinking. These contained evaluation items incorporating components of critical thinking, but the components were often combined with other aspects of performance such as an employee's ability to solve problems or to communicate selectively. Sometimes indeed assessment of the critical thinking was incorporated into the assessment of ill-defined personality traits such as leadership or professionalism. And sometimes they were described too broadly to be of use in the study. The researchers also conducted interviews with 20 professionals in seven public accounting firms, including all six of the Big Six accounting firms plus one regional public accounting firm. Their participants were selected from offices in six cities representing different parts of the United States. Participants had five to fifteen years of experience with their firms. Most of the interviewees were managers or senior managers, and the remainder were mostly partners. All interviewees had direct responsibility for formal staff evaluations, and all had conducted between five and fifty evaluations in the previous year. Because of the team approach used in public accounting, most interviewees had been involved in numerous additional evaluations. The researchers' findings revealed that the interviewees conceived of critical thinking as a broad set of competencies that included "cognitive attributes and characteristics" along with "non-cognitive attitudes and behaviours." They also identified a range of other, more-complex competencies. We reproduce their classification below (Baril et al. 1998, 392).

Cognitive attributes and characteristics:

- Recognizes problem areas
- Recognizes when additional information is needed
- Fits details into the overall environment; sees the "Big Picture"
- Transfers knowledge from one situation to another
- Anticipates, thinks ahead, plans

Noncognitive attitudes and behaviors:

- Exhibits initiative
- Exhibits curiosity
- Exhibits confidence
- Communicates clearly and articulately

Other competencies mentioned occasionally by Baril et al's interviewees were as follows:

- Displays creativity
- Accepts ambiguity
- Recognizes when there is more than one acceptable solution

- Makes qualitative judgments
- Displays rapid thought process
- Displays healthy skepticism; asks "why?" or "why not?"
- Challenges the status quo
- Determines the extent of what is reasonable; defines the limits of acceptability
- Recognizes personal limitations
- Exposed to diverse cultures, knowledge, and backgrounds
- Recognizes presence of biases

The "other competencies" listed above are clearly pivotal but require additional elaboration and have to be contextualized; we need to be able to understand how they are put into practice in concrete situations by professional accountants. For that reason we now turn to the voices of employers who are expert accounting practitioners, and newly graduated accountants—as expressed in the "native" idiom of the workplace. In the next section we summarize the results of a recent qualitative study by Bui and Porter (2010) that focused on interviews with accounting students, academics, graduates, and employers; we then present the voices of two experienced practitioners (based on Sin's interview data); finally we present voices of newly employed graduate accountants (based on interviews conducted by Wang).

Voices of employers and expert practitioners

Voices of employers

After examining the accounting program offered by one New Zealand university in some detail, Bui and Porter (2010) interviewed students, academics, graduates, and employers of graduate accountants with regard to the potential existence of an expectation-gap between the desired and perceived outcomes of that program. A total of 11 employers were interviewed. They were partners or human resource managers in firms that were classified as small to medium-sized or large. The researchers focused on having interviewees identify those competencies that they considered most important for successful practice. They were queried as to the competencies graduates from the case study program were thought to possess as well as those they were believed to lack. Crucially, there does not appear to have been any prompting regarding the terminology to be used by the participants. The results are thus couched in the language of the lifeworld. We reproduce the terms they used in our brief summary of the findings below.

One significant factor that quickly emerged in the study was the size of the firm. Small and medium-sized firms were concerned that entry-level accountants should possess what we might call routine skills: time management skills, teamwork skills, and so on. The larger firms not only wanted high-level presentations skills from the get-go, including advanced oral and written communication skills, but also valued highly "professional scepticism" in the graduates they employed. They saw "confidence" as a precondition for the adoption and exercise of a critical stance, though it was also seen to be integral to a type of professional demeanor that (it was believed) clients deemed trustworthy. Bui and Porter summarize this philosophy in their own words, writing that "confidence helps graduates to exercise professional scepticism and deal effectively with both prospective and current clients; it also enhances clients' trust in the firm's services" (2010, 35). Interestingly, representatives of medium and small-sized firms quite explicitly preferred to hire graduates with limited confidence "because, in their opinion, over-confidence hinders graduates from further learning and personal development" (Bui and Porter 2010). But this is a conundrum that we cannot explore further here.

Representatives of medium-to-large firms linked successful practice to high-level interpersonal and writing skills. The association of critical thinking with effective interpersonal skills has been made by Deal (2004). And there are cogent arguments to the effect that structured writing entails or indeed is a form of (critical) thinking (Dunn and Smith 2008). Moreover, recent research in linguistics has revealed that most structured types of writing entail complex interpersonal skills—in the form of predictions about their impact, attempts at persuasion, and attempted negotiations with the reader (Hyland 2005; White 2003). Not only is structured writing a challenging process of cognitive synthesis and creation (Bereiter and Scardamalia 1987; Flower and Hayes 1984; Hayes and Flower 1987), it is also a complex form of social engagement with imagined readers that fosters an ability to relativize the writer's own reality by imagining alternative contexts, meanings, and motives. In effect, writing is at least intimately associated with types of thinking that boost the mind's natural capacity for sustained, systematic, imaginative, and essentially critical thought (Galbraith 1998; Menary 2007).

Thus although terms like "critical thinking" were not mentioned by the participants in this research, representatives of the larger firms together drew a very clear picture of critical thinking in action—as a complex set of interrelated attributes and even an associated demeanor (on professional demeanor indexicals, see Jones and Sin 2013).

We next explore transcript of interviews with two expert practitioners (collected by Sin from 2009 to 2010, these have formed the basis of several publications). The interviewees were given all of the time they seemed to require, in extended and generally uninterrupted turns, to develop and verbalize their conceptions of the skills they deemed necessary for the practice of professional accounting in today's changing business environment. It will be seen, unsurprisingly perhaps, that the language they used to refer to critical thinking is not always that which an academic researcher might have used in designing prompts.

The voice of a senior accountant

Mark is a senior accountant with responsibilities for dealing directly with clients and overseeing the work of a number of other accountants in a public accounting and consultancy firm. The question to which he was responding in the passage quoted at length below focused on what he might look for in a graduate accountant. Mark identifies three capabilities that characterize the ideal accountant's attitude to financial data.

> A person who can actually sort of read the separate accounts and see what's in there and also not getting bogged into the numbers itself. And trying to analyse what it's telling you...what the figures are actually telling you and those sorts of things...you know...you should be able to read a set of financials and must be able to understand...get a story from it...is there a problem with their accounts? Or do these accounts look alright?

It has become something of a truism among professional accountants that one has to "go beyond the numbers" in order to fully comprehend the situation and prospects of a firm (and the popular American textbook *Accounting: What the Numbers Mean*, by Marshall, McManus, and Viele, is in its tenth edition). Although interpretive approaches have not gained widespread institutional or academic recognition (despite the idea being around since at least Morgan, 1988), the need to interpret the raw numerical data is a given in professional accounting practice. Accountants are nowadays expected to contribute to corporate strategy and decision-making, and hence to interpret financial statements in the light of these bigger-picture considerations. They have to identify the interrelationships between a company's core business, the market, current competition, the company's concrete operations, its past performance, and the nature and quality of management structure in the process of aiding and contributing to the decision-making process. In Mark's graphic idiom, the need is for people who don't get "bogged down" in the numbers and who can "get a story from it" (i.e., from the data). In his own way Mark is recognizing that numerical data—in their totality—implicate a scenario, a set of real-life conditions, which may or may not count as problematic from the viewpoint of a management team.

In the event, the interviewer asked Mark to elaborate on the meaning of the expression, "to not get bogged down by numbers" (we have italicized particularly telling phrases):

> The first thing is like looking at the number and, say, look at sales and gross profits and other things and *seeing something is wrong* with the gross profits.
>
> If you get a negative gross profit then some people give you a set of accounts with negative gross profits because they posted it into the wrong

set of accounts…sales must be in credits…income may be in debits sometimes…and sometimes expenses may be in credits…so it's not just numbers you got to read.

Mark suggests here that the accountant must envisage the possibility that the
accounts have been erroneously compiled. This is what might be referred to in
more psychological terms as having a critical imagination. Since accountants
qua financial officers are usually closer to the data source than other members
of a management team, they are well placed to assess its limitations, that is, to
tell which information is reliable and which is not.

A little later in the interview, Mark emphasizes that the accountant should
also consider the actual aims or purposes of the financial statement being
assessed:

> Are we actually looking at it to prepare a set of tax returns from it? Some
> people may look at it for management purposes. Some people look at it for
> financial reporting purposes. What sort of information you can summarize
> or expect…like tax returns…some of the accounts we need to expand them
> out or roll out for tax return purposes…whereas for publishing accounts you
> can summarize together…*knowing what the purpose of the accounts are for.*

Reflecting upon the purpose of the accounts can clearly assist one to evaluate
apparent anomalies. The facilitator prompts the interviewee further here, and
Mark obligingly elaborates by identifying a third, albeit very similar, ability of
the ideal accountant:

> And the difficulty most people find is basically *trying to understand the set of
> accounts…*

In response to a prompt, Mark elaborates:

> Understanding the set of accounts…you look at it, some people might know
> which companies are insolvent or going insolvent or is having credit cash-
> flow problems. You know you can look at the ratios, things like that and you
> can see whether they are under-funded or over-funded, whether they have
> cash flow to pay their liabilities before they fall due. Those sorts of things
> that we can look at. Maybe we find that they are spending too much more
> than their revenues. What cash reserves they have before the cash actually
> runs out at that rate. Different things really.

Again the emphasis is on seeing consequential meanings of and extrapolating
meaningful patterns from the figures so as to identify significant processes—
especially causal chains—in real-world situations. In other words, the real

financial and commercial implications of the numbers are only revealed to the persistently questioning, critical mind.

The voice of a senior auditor

Richard is a senior accountant who has specialized for many years in auditing. In addition to advanced technical accounting knowledge and skills, auditing is a specialization that requires sophisticated interpersonal communication skills (see Jones and Sin 2013), needed to cope with and overcome the multiple levels and types of ambiguity that are typically present in a given workplace (van Peursem 2005). A critical stance is a vital element in those skills. But along with the kind of skepticism often simply equated with criticality, auditors must also have the ability to cope constructively with different approaches to reporting, different interpretations of tasks and goals—sometimes producing apparently disparate or erroneous results—and with quite different perspectives on the nature of accounting work.

Richard began by emphasizing that professional accounting is not simply a matter of compiling and manipulating financial records (key phrases are italicized):

> You go from the start preparing the accounts and then all the way through to the *interpretation, consulting and trying to add value to the clients*, assisting the clients in adding value. So it's not purely just a bookkeeping role anymore because a bookkeeping role is actually diminished.

Then he focuses explicitly on "being critical":

> You have *to be critical*, look at things with *a questioning mind*. I think it's very important *to have a questioning mind.*...Not just working off a checklist, if you have one all the time you can get to the point where you just don't have to think about things. I think it's very important *to be questioning.*

It is only in retrospect—after engaging with the subsequent talk over three communicative moves—that the analyst understands that, when he speaks of being critical and having a questioning mind, Richard actually means something like *having an open and inquiring mind*. An open mind is not the same as a skeptical mind. What Richard actually means when he speaks of having a questioning mind emerges more fully when he describes his attitude not just to the accounts but also to the people behind the accounts, those responsible for drawing them up:

> There are so many different complications and a lot of things where *there's no right or wrong answer, just interpretation*. You've got to be able to lend your mind to twist and turn things and look at it from a couple of different angles.

Again:

> Everyone has their own interpretation particularly when people call up
> and they have questions and you understand the way someone else is
> going at it.

Richard describes his strategies for dealing with such situations, where he moves
rapidly beyond a critical evaluation of the facts to a quest for understanding:

> That's also what I found in audit, if you go and ask somebody how did they
> do it? Some people will go with the approach that the other person was
> wrong and they weren't doing it right, but the way to do it is how they did
> and question them and then they give you answers and then you bounce
> back and forth with them. That's the only way to get the answers and also
> to get the respect of the person that you're dealing with. If you deal with
> everyone and think that people don't know what they're doing then you're
> not going to get anywhere. It's very important to appreciate that everyone
> has different skills and you've got to come in with different people at their
> level, whether higher or lower level, you've got to grasp where people are at,
> where their understanding is at and then work from there.

Richard, a very experienced auditor, has learnt that respect for difference and
well-honed interpersonal communication skills are essential to the successful
enactment of his role. He has in this last excerpt very eloquently described
how he goes about his business and succeeds in it, in an unpretentious vernacular that captures his somewhat novel conception of criticality. What is novel
in it is his goal of being constructive in the face of inevitable diversity and
difference. Richard's emphasis on letting one's mind "twist and turn things"
and on looking at each situation from different angles is echoed below in our
discussion of open-mindedness (see our overview following the section on new
graduates' voices).

Voices of new graduates in practice

New graduates are often made keenly aware of their inadequacies, in terms of
both skills and personal attributes, on entering the workplace for the first time.
Their impressions provide an invaluable insight into the nature and range of
the skills and attributes that they feel they lack, and which they now consider
necessary for success in their new working environment. Their impressions
and conceptions, phrased in their own informal vernacular, amplify those of
the more senior accountants reviewed in the previous section. The term "critical thinking skills" was used as a prompt, to which the participants responded,

each in their own idiom. Our use of the term "conception" below reflects the phenomenographic methodology adopted in the research from which the extracts are based. Phenomenography is a form of interpretative analysis, well recognized in the field of educational research, in which the outcomes are "conceptions." A conception represents a subject's understanding of a concept at a given point in time.

In these interviews conducted with the new graduates in accounting (Wang 2007), five thematic foci emerged. Since they were all oriented to the topic (or concept) of "critical thinking skills," these foci of awareness add up to a single complex but largely consistent conception of critical thinking. Although these conceptions were formed just as the demands of actual practice, often in a new and challenging workplace, were being confronted by individuals who had only recently completed their technical cum professional "training," many aspects of their understanding resonated with the themes mentioned or elaborated upon by the much more experienced accountants discussed above.

Based on a close analysis of extended interview transcripts, we were able to identify five distinct "conceptions," each of which was oriented toward a specific theme: (1) client outcomes, (2) technical knowledge and experience, (3) defining and solving specific problems, (4) professional judgement, and (5) personal attributes and dispositions. Let us sample what the new graduates actually said.

Conception 1: Critical thinking skills are required to achieve client outcomes

A number of interviewees foregrounded the value of critical thinking skills for client outcomes. The type of outcome depends, naturally, on the role of the accountant and the task at hand, but several participants were in agreement that the desired outcome is always one in which financial value is added to the client's business and any benefits maximized. Dayna exemplifies this conception:

> You've really got to use those critical thinking skills in order *to advance your client's case*... You have your client's problem and I guess your aim is to, you know, help them minimize their tax obligations, maximize their tax benefits.

In a consulting role, Soolin also sees critical thinking skills in terms of optimizing financial outcomes for her clients:

> I think it is the ability to work beyond the square. For example, *you can work something out which can add value for your clients in addition to the normal accounting compliance work...show them how to make their business grow, increase sales, how to increase the net assets.*

Conception 2: Critical thinking skills build on specialized knowledge and experience

In this conception, experience and knowledge, including both technical disciplinary expertise and contextual knowledge, are viewed as prerequisites for—and the basis of —the successful application of critical thinking skills. Soolin makes this point in the following way:

> To be able to apply critical thinking skills *you need more knowledge and expertise in the area before you can do the job right.* Therefore, you need to read magazines, publications, books, papers, and to go to seminars on how to grow sales, how to manage money etc.

A lack of contextual understanding is seen as a serious impediment to applying critical thinking skills, but technical accounting knowledge is also indispensable, as Anne explained:

> If I don't understand *the correct accounting treatment,* I can't apply any critical thinking in these situations.

An ability to apply critical thinking skills is also seen to depend on prior experience. Shan touched on the need for professional skepticism with regard to client disclosure. Shan focused on the fact that clients are not always completely forthcoming and explained that experience is what allows an auditor to identify this:

> So you need your experience to apply critical thinking skills *to identify if there is anything* that is important for audit purposes *that the client didn't want to talk about. You have to dig a bit more in those situations.*

Villa makes a similar point about experience, using the popular idiom about thinking outside the square to indicate the role of imagination in critical thinking:

> So that's just experience, *after you've done a few jobs* you start to think outside the square, outside the normal routine. So that's part of my critical thinking skills.

Conception 3: Critical thinking skills are needed for identifying and solving problems

According to this conception the new graduates saw critical thinking skills as problem-solving skills. They talked about how critical thinking skills were related to the different stages of the problem-solving process. Anne's definition of critical thinking skills was clear-cut:

I define that as *problem solving skills.*

The statement that followed, by way of explanation, recalls what Mark said above about not getting bogged down in the numbers and specifies the type of ill-defined problem accountants typically need to solve:

> Because in accounting, you're not only dealing with pluses or minuses or only playing around with numbers. For example, if you have a problem in front of you, critical thinking is when you have to think further beyond the numbers. *What do the numbers mean or what can you see from the numbers?* Therefore, I think from an accounting perspective, critical thinking skills are problem solving skills.

For Kylie, critical thinking is essentially evaluative and is typically employed in the context of identifying, defining, and solving specific problems:

> In my point of view, critical thinking skill is like *an evaluation process* where you find the problem, evaluate the criteria and then make a reasonable judgement.

Finally, Terry describes how critical thinking skills not only involve but also go beyond the application of logic in solving problems—because the problems are typically *situated* problems:

> Critical thinking skills are how we come to—in a problem situation—how we come to a solution, how to solve it *in a way that is logical.* Sometimes *the circumstances make it very tough, that's where your critical thinking comes to apply.*

Clearly for Terry, as for other interviewees, critical thinking involves much more than a critical examination of financial data—though it may begin there. Terry also understands that the issues, in real-life situations, are rarely clear-cut. They are deeply enmeshed with the epistemological uncertainties and conflicting imperatives of workplace contexts.

Conception 4: Critical thinking skills are needed for the exercise of professional judgment

Several participants asserted that critical thinking skills are used in making judgments. Jarvic talks about the use of critical thinking skills in the exercise of professional judgment:

> Say, if I have to determine the level of testing for a particular client. *I have to use critical thinking skills to note* [= assess] *their materiality* so that I can

determine the level of testing. If I don't apply those skills then I may not know how far I have to go or test.

Sometimes, it is an ethical decision that has to be made. Kylie gives the following poignant and very true-to-life example:

> There is a sales person who is the boss's father-in-law. The boss wants to give him more commission. When I finished our sales report for example his sales figure was like $5,000 this month, but the boss wanted to give him more commission. He said "You can write down $7,000 on the sales report." See, *in this situation I need critical thinking skills to decide should I show the true value or just follow the boss's decision.*

Critical thinking skills are also applied when making judgments in the context of unequal power relations—a disparity that may be subtle or wholly tacit. Zen talks about her own experience:

> I *need to think about the position level of the person who gave me the work.* For example, I got work from a senior analyst and it was due very soon, so I had to really focus working on it. But I also got another piece of work from a partner. It was just a piece of research for a potential client...I should apply my critical thinking skills to set my priorities.

Finally, critical thinking skills are used to make judgments requiring empathy, as Anne relates:

> For example, the manager asked me to do A, B, and C for a job...I have to think about the quality required for A, B, and C, so *I think about the expectations of my manager.* I don't just do [the job]. I think critical thinking is *thinking a bit further into what or who you should prepare yourself for.*

Conception 5: Critical thinking skills are grounded in certain dispositions

In this conception, the participants' focus of awareness was on the dispositions that underlie critical thinking, such as dispositions to think in particular ways, or approach problems in specific ways. The following quotations illustrate this conception and the ways in which it was verbalized, and also provide examples of different dispositions identified as relevant. Being careful and a willingness to pay attention to details was emphasized by Jan:

> Yes, you have to *be very careful.* You have to pay attention to the numbers that you are entering. That's where the *attention to detail* comes in.

This disposition is, we would maintain, crucial. In all likelihood, it contains the seeds of higher level criticality. Next Travis, referring to an auditing context, identifies "professional scepticism" as an important disposition and, as it were, a prerequisite for critical thinking skills:

> Critical thinking skills, from an auditing perspective, are based around *professional scepticism*, you can't get away from it, *it's in everything we do—we don't trust.* It's not that we always think they're trying to cheat us—it is part of an unwritten law.

The above resonates with Richard's notions (above) about "being critical" and "having a questioning mind," though (as we will argue below) he had a considerably more democratic and understanding view of criticality, and ultimately a more effective one. Travis explains what professional skepticism means to him:

> You never take what you get for granted…It means *not to take everything at face value*…Once you've got some evidence then you might ask a question why this is not so? From that you might get an explanation and its true and it works out. But for the meantime, before you know it, *you have to be sceptical about what they're giving you.*

Anne, also an auditor, emphasized the need to connect or interrelate seemingly disparate information, though she (like Travis) has a black-and-white view of correct procedure:

> Sometimes it's very challenging for us to dig out what has been done wrongly by the client. Your clients provide you with information, for most of the time, what they have provided make sense…But the very important thing for us is *to relate all the sections together.*

Dayna focuses on lateral thinking, an aspect of creative thinking, employing an idiom common in jargon of management:

> Lateral thinking, *thinking outside the square*, critically thinking…it is important because you need to be able to think outside the square.

Critical thinking in accounting—a view from the inside

From the five conceptions outlined above, and the views of the employers and experienced practitioners discussed earlier, we see that critical thinking with its component skills is conceptualized by accountants less as a cognitive ability than as a kind of purposeful yet highly problematic doing. This involves a tolerance for complexity, conflict, and ambiguity, as practitioners grapple with messy and multifaceted problems that represent competing interests and imperatives. Critical thinking in these contexts typically involves at least the following: being able to identify and promote client outcomes; possessing an ability to determine the less-obvious implications of numerical data; an ability to cope with extremely complicated social, organizational, and institutional scenarios; evaluating the motives of key stakeholders as well as the relevance of abstractly formulated concepts and principles; and internalizing these abilities as attributes and ultimately a disposition that involves a commitment to honesty, accuracy, and ethical practice. However, the desirable traits also include *open-mindedness*—that is, the willingness to consider different viewpoints, different ways of doing things, and indeed different and often in some sense imperfect solutions.

What Richard meant when he spoke about the need for a questioning mind, that is, open-mindedness (see Richard's quotation on page 443), was recently expressed in somewhat more academic language by the knowledge management specialist, Olivier Serrat (2011, 1):

> Critical thinking, by its very nature, demands recognition that all questioning stems from a point of view and occurs within a frame of reference; proceeds from some purpose—presumably, to answer a question or solve a problem; relies on concepts and ideas that rest in turn on assumptions; has an informational base that must be interpreted; and draws on basic inferences to make conclusions that have implications and consequences.

As Serrat acknowledges, these are not easy principles to apply in practice. And as he further notes—echoing many comments made by employers and practitioners—they are ideally grounded in and spring from firmly embedded personal attributes, a disposition.

> These are tough intellectual standards. They spring from, and call for the development of, intellectual traits (or virtues) of humility, autonomy, integrity, courage, perseverance, confidence in reasons, empathy, and fair-mindedness. (Serrat 2011, 2)

Facione et al. (1997) have also emphasized the dispositional roots of critical thinking. Open-mindedness or fair-mindedness is not normally associated with critical thinking. However, Facione et al. (1997) have adopted a discourse of "open-minded inquiry" that resonates powerfully with our own data and analysis. On reflection it seems obvious that a generalized and open-ended type of criticality is precisely what is needed in typical contexts of professional accounting practice. Accounting professionals rarely have the luxury of taking sides, of pointing out that this is black and that is white. Typically, they deal with many-faceted "messy" problems, embedded in complicated social and organizational situations (Metlay and Sarewitz 2012; Thompson and Tuden 1959). People's reputations and ultimately perhaps their livelihoods are often what are at stake in problematic accounting situations. So the nature and degree of hardnosed criticism that they can display toward one or another actor or group must normally be mitigated by a variety of pragmatic concerns. And in many cases a desire to preserve relationships and maintain the organizational status quo—or at least the structural integrity of the organizations involved—must be balanced carefully against a practitioner's institutional and legal obligations.

There are contributions in the by now vast literature on critical thinking that support the definition that has been emerging here. Mezirow, for example, sees critical thinking as the process of becoming critically aware of people's tacit assumptions and expectations, including one's own premises and presuppositions, and being able to assess their relevance to a specific situation (2000, 4). Critical thinking is thus, as we might put it, a process of encompassing the range of different perspectives on a given problem situation, of weighing their differential relevance—and, of course, authority or "force"—and of finding solutions that go as far as possible toward satisfying the most relevant and pressing of a set of conflicting imperatives. Deal (2004, 9) sums this up as the ability to function "at the level of contextual relativism" and sees it as being based on the ability to empathize—cognitively as well as affectively—with one's interlocutors. And the mention of Deal allows us to emphasize once again the strong link between interpersonal skills—particularly communication skills—and what we can continue to call critical thinking.

Based on what employers and practitioners had to say, we can conclude that—for an effective auditor in particular, but also for effective practice in any type of professional accounting work—open-mindedness implies a (pre)disposition to believe that people and organizations are (1) honest, (2) know what they are doing, and (3) have the best interests of both their employers and clients at heart. A desire to build and maintain interpersonal relationships and to preserve organizational processes and structures where possible must be balanced against the need to identify lazy or dishonest individuals and/or ineffectual processes that cannot easily be reformed. But the accountant's—especially the auditor's—professional responsibilities should ideally be carried out in a climate of mutual respect and an expectation of, and a tolerance for, different ways of getting things done. The voices of the lifeworld as reported here have borne this out.

Easy or hard, critical thinking is what expert practitioners do most and do best. This ability, and the attributes or disposition in which it is grounded, are not acquired overnight; they are acquired through experience and through reflection on experience, and these skills presuppose the learning opportunities that are afforded by a prolonged intensive engagement with the dilemmas of a particular professional lifeworld.

How can critical thinking skills be fostered?

Globally, since as early as the 1960s, there has been a growing concern that accounting education overemphasizes the technical knowledge and abilities of graduates while neglecting other essential attributes and competencies. Since then professional accounting bodies and employer organizations have been calling for educational innovation and reform that will foster accounting graduates' "soft skills." The new requirements are clearly a function of the changing nature of the profession (Sin, Reid, and Dahlgren 2011). Accountants are nowadays expected to accept ever more responsibility for the business success of their organizations and to be increasingly proactive in safeguarding the potentially diverse financial interests of their clients. This was not the situation some decades earlier, when accountants had more restricted functions and responsibilities. As Hancock et al. (2009) put it, the difference between then and now is that "expectations regarding graduate 'soft skills' have evolved to include higher order skills like analytical and critical analysis, and to encompass an ability to engage with clients, to negotiate outcomes and to act strategically."

Responding to the new demands of accounting bodies and employer organizations, academics have been quick to come up with alternative instructional strategies. Some educators have turned to case-based teaching, role-playing, or the use of simulations to engage students more deeply in their own learning as well as to develop their creative and critical thinking skills (aims determined by the Accounting Education Change Commission 1990). Many innovative ideas have been systematically trialled, documented, and the results analyzed in print in relevant journals. Some innovations—though not many—have been explicitly designed to inculcate and/or assess critical thinking skills (Doney, Lephardt, and Trebby 1993 is an early example).

Theoretically, the individual practitioner—a professional accountant here—can engage in three distinct kinds of critique: informational-procedural critique, institutional critique, and social critique. In the first case, an individual critically analyzes financial information along with the procedures that have produced it. In the second case, a practitioner might evaluate critically the institutional mechanisms that constrain and facilitate professional practice. In the third case, a practitioner might engage in critically analyzing the social environment that frames institutions, practices, and procedures. Some academic accountants would encourage students to engage in this last type of critique (e.g., Craig and McKinney 2010), and see in such practices a way students can cope with the contradictions they detect between the rhetoric and the reality of capitalism, and the existence of socially bereft values in accounting curricula that simultaneously extol the virtues of principled or ethical practice. Institutional critique is healthy but is best left to senior practitioners with a good many years of experience in the profession. Where the new graduate needs to exercise her or his "professional scepticism" with most vigor and persistence is in the daily systematic scrutiny of financial data, the data manipulation processes, and the calculative procedures that they or their peers employ. This is the most basic form of the "questioning" referred to by Richard above.

This is moreover the type of activity that accounts for ninety percent of an accounting professional's working day—and especially the day of a newly graduate or an early career accountant. If we need to specify the cognitive processes that contribute to this process we can think of interpretation, evaluation, and reflective judgement. These are processes, or skills, that can be honed in the classroom (through a judicious mixture of the methods mentioned above) and, if conscientiously practised in the classroom, they can provide a solid foundation for the more mature capabilities and ultimately the unconscious habitus characteristic of the expert practitioner that only come with time and experience.

Acknowledgments

The authors gratefully acknowledge the financial support provided by the Department of Accounting and Corporate Governance at Macquarie University and constructive comments from colleagues, the anonymous reviewers, and the editors.

References

Accounting Education Change Commission. 1990. "Objectives of Education for Accountants: Position Statement Number One." *Issues in Accounting Education* 5 (2): 307–312.

Albrecht, W., and Sack, R. 2000. "Accounting Education: Charting the Course through a Perilous Future." Sarasota, FL: American Accounting Association.

American Accounting Association. 1986. "Future Accounting Education: Preparing for the Expanding Profession, Bedford Report." Sarasota, FL: American Accounting Association.

American Institute of Certified Public Accountants. 1999. " CPA Vision Project: 2011 and Beyond." New York: American Institute of Certified Public Accountants.

Baril, C., Cunningham, B., Fordham, D., Gardner, R., and Wolcott, S. 1998. "Critical Thinking in the Public Accounting Profession: Aptitudes and Attitudes." *Journal of Accounting Education* 16 (3–4): 381–406.

Barnett, R. 1997. *Higher Education: A Critical Business.* Buckingham: Society for Research into Higher Education and the Open University Press.

Bereiter, C., and Scardamalia, M. 1987. *The Psychology of Written Composition.* Hillsdale, NJ: Lawrence Erlbaum.

Big Eight Accounting Firms. 1989. "Perspectives on Education: Capabilities for Success in the Accounting Profession (White Paper)." New York: American Accounting Association.

Birkett, W. 1993. "Competency Based Standards for Professional Accountants in Australia and New Zealand." Melbourne: ASCPA and ICAA.

Bloom, B., Engelhart, M., Furst, E., Hill, W., and Krathwohl, D. 1956. *Taxonomy of Educational Objectives: Handbook I: Cognitive Domain.* Vol. 19: New York: David McKay.

Bui, B., and Porter, B. 2010. "The Expectation-Performance Gap in Accounting Education: An Exploratory Study." *Accounting Education: An International Journal* 19 (1–2): 23–50.

Camp, J., and Schnader, A. 2010. "Using Debate to Enhance Critical Thinking in the Accounting Classroom: The Sarbanes-Oxley Act and US Tax Policy." *Issues in Accounting Education* 25 (4): 655–675.

CPA Australia and the Institute of Chartered Accountants in Australia. 2012. "Professional Accreditation Guidelines for Australian Accounting Degrees." Melbourne and Sydney: CPA Australia and the Institute of Chartered Accountants in Australia.

Craig, R., and McKinney, C. 2010. "A Successful Competency-Based Writing Skills Development Programme: Results of an Experiment." *Accounting Education: An International Journal* 19 (3): 257–278.

Deal, K. 2004. "The Relationship between Critical Thinking and Interpersonal Skills: Guidelines for Clinical Supervision." *The Clinical Supervisor* 22 (2): 3–19.

Doney, L., Lephardt, N., and Trebby, J. 1993. "Developing Critical Thinking Skills in Accounting Students." *Journal of Education for Business* 68 (5): 297–300.

Dunn, D., and Smith, R. 2008. "Writing as Critical Thinking." In *Teaching Critical Thinking in Psychology: A Handbook of Best Practices*, edited by D. S. Dunn, J. S. Halonen, and R. A. Smith, 1st ed. Malden, MA: Wiley-Blackwell. 163–174.

Entwistle, N. 1994. *Teaching and Quality of Learning: What Can Research and Development Offer to Policy and Practice in Higher Education.* London: Society for Research into Higher Education.

Facione, P. 1990. *Critical Thinking: A Statement of Expert Consensus for Purposes of Educational Assessment and Instruction, "The Delphi Report."* Millbrae, CA: American Philosophical Association, The California Academic Press.

Facione, P., Facione, N., and Giancario, C. 1997. *Professional Judgement and the Disposition towards Critical Thinking.* Millbrae, CA: Californian Academic Press.

Facione, P., Sanchez, C. A., Facione, N. C., and Gainen, J. 1995. "The Disposition toward Critical Thinking." *The Journal of General Education* 44 (1): 1–25.

Flower, L., and Hayes, J. 1984. "Images, Plans, and Prose: The Representation of Meaning in Writing." *Written Communication* 1 (1): 120–160.

Freeman, M., Hancock, P., Simpson, L., Sykes, C., Petocz, P., Densten, I., and Gibson, K. 2008. "Business as Usual: A Collaborative and Inclusive Investigation of the Existing Resources, Strengths, Gaps and Challenges to Be Addressed for Sustainability in Teaching and Learning in Australian University Business Faculties." Scoping Report, Australian Business Deans Council. http://www.olt.gov.au/project-business-usual-collaborative-sydney-2006.

Galbraith, M. 1998. *Adult Learning Methods: A Guide for Effective Instruction.* Washington, DC: ERIC Publications.

Giancarlo, C., and Facione, P. 2001. "A Look across Four Years at the Disposition toward Critical Thinking among Undergraduate Students." *The Journal of General Education* 50 (1): 29–55.

Hancock, P., Howieson, B., Kavanagh, M., Kent, J., Tempone, I., and Segal, N. 2009. "Accounting for the Future: More Than Numbers, Final Report." Strawberry Hills: Australian Learning and Teaching Council Ltd. http://www.altc.edu.au.

Hayes, J., and Flower, L. 1987. "On the Structure of the Writing Process." *Topics in Language Disorders* 7 (4): 19–30.

Hyland, K. 2005. "Stance and Engagement: A Model of Interaction in Academic Discourse." *Discourse Studies* 7 (2): 173–192.

Jackson, M., Watty, K., Yu, L., and Lowe, L. 2006. "Assessing Students Unfamiliar with Assessment Practices in Australian Universities, Final Report." Strawberry Hills: Australian Learning and Teaching Council Ltd.

Jones, A. 2010. "Generic Attributes in Accounting: The Significance of the Disciplinary Context." *Accounting Education: An International Journal* 19 (1–2): 5–21.

Jones, A., and Sin, S. 2003. *Generic Skills in Accounting: Competencies for Students and Graduates.* Sydney: Pearson/Prentice Hall.

Jones, A., and Sin, S. 2004. "Integrating Language with Content in First Year Accounting: Student Profiles, Perceptions and Performance." In *Integrating Content and Language: Meeting the Challenge of a Multilingual Higher Education,* edited by R. Wilkinsoned. Maastricht: Maastricht University Press.

Jones, A., and Sin, S. 2013. "Achieving Professional Trustworthiness: Communicative Expertise and Identity Work in Professional Accounting Practice." In *Discourses of Ethics,* edited by C. Candlin and J. Crichtoned. Basingstoke, Hampshire, UK: Palgrave Macmillan.

Kavanagh, M., and Drennan, L. 2008. "What Skills and Attributes Does an Accounting Graduate Need? Evidence from Student Perceptions and Employer Expectations." *Accounting and Finance* 48 (2): 279–300.

Kimmel, P. 1995. "A Framework for Incorporating Critical Thinking into Accounting Education." *Journal of Accounting Education* 13 (3): 299–318.

King, P., and Kitchener, K. 2004. "Reflective Judgment: Theory and Research on the Development of Epistemic Assumptions through Adulthood." *Educational Psychologist* 39 (1): 5–18.

Kurfiss, J. 1988. *Critical Thinking: Theory, Research, Practice and Possibilities.* Washington: ASHE-Eric Higher Education Report No. 2, Associate for the Study of Higher Education.

Kurfiss, J. 1989. "Helping Faculty Foster Students' Critical Thinking in the Disciplines." *New Directions for Teaching and Learning* 37 (1): 41–50.

Lucas, U. 2008. "Being "Pulled up Short": Creating Moments of Surprise and Possibility in Accounting Education." *Critical Perspectives on Accounting* 19 (3): 383–403

Menary, R. 2007. "Writing as Thinking." *Language Sciences* 29 (5): 621–632.

Metlay, D., and Sarewitz, D. 2012. "Decision Strategies for Addressing Complex, "Messy" Problems." *The Bridge on Social Sciences and Engineering Practice* 42 (3): 6–16.

Mezirow, J. 2000. *Learning as Transformation: Critical Perspectives on a Theory in Progress.* The Jossey-Bass Higher and Adult Education Series, Washington, DC: ERIC Publications.

Morgan, G. 1988. "Accounting as Reality Construction: Towards a New Epistemology for Accounting Practice." *Accounting, Organizations and Society* 13 (5): 477–485.

Partnership for Public Service and Institute for Public Policy Implementation. 2007. *The Best Places to Work in the Federal Government.* Washington, DC: American University.

Paul, R. 1989. *Critical Thinking Handbook: High School. A Guide for Redesigning Instruction.* Washington, DC: ERIC Publications.

Sarangi, S., and Candlin, C. 2010. "Applied Linguistics and Professional Practice: Mapping a Future Agenda." *Journal of Applied Linguistics and Professional Practice* 7 (1): 1–9.

Serrat, O. 2011. *Critical Thinking.* Washington, DC: Asian Development Bank.

Simon, H. 1974. "The Structure of Ill Structured Problems." *Artificial Intelligence* 4 (3): 181–201.

Sin, S. 2010. "Considerations of Quality in Phenomenographic Research." *International Journal of Qualitative Methods* no. 9 (4): 305–319.

Sin, S. 2011. *An Investigation of Practitioners' and Students' Conceptions of Accounting Work.* Linköping: Linköping University Press.

Sin, S., Jones, A., and Petocz, P. 2007. "Evaluating a Method of Integrating Generic Skills with Accounting Content Based on a Functional Theory of Meaning." *Accounting and Finance* 47 (1): 143–163.

Sin, S., Reid, A., and Dahlgren, L. 2011. "The Conceptions of Work in the Accounting Profession in the Twenty-First Century from the Experiences of Practitioners." *Studies in Continuing Education* 33 (2): 139–156.

Tempone, I., and Martin, E. 2003. "Iteration between Theory and Practice as a Pathway to Developing Generic Skills in Accounting." *Accounting Education: An International Journal* 12 (3): 227–244.

Thompson, J., and Tuden, A. 1959. "Strategies, Structures, and Processes of Organizational Decision." In *Comparative Studies in Administration* edited by J. Thompson, P. Hammond, R. Hawkes, B. Junker, and A. Tudened. Pittsburgh: University of Pittsburgh Press. 195–216.

van Peursem, K. 2005. "Conversations with Internal Auditors: The Power of Ambiguity." *Managerial Auditing Journal* 20 (5): 489–512.

Wang, F. 2007. *Conceptions of Critical Thinking Skills in Accounting from the Perspective of Accountants in Their First Year of Experience in the Profession*, Macquarie University Honours Thesis.

Watson, T. 2004. "Managers, Managism, and the Tower of Babble: Making Sense of Managerial Pseudojargon." *International Journal of the Sociology of Language* (166): 67–82.

White, L. 2003. *Second Language Acquisition and Universal Grammar.* Cambridge: Cambridge University Press.

Young, M., and Warren, D. 2011. "Encouraging the Development of Critical Thinking Skills in the Introductory Accounting Courses Using the Challenge Problem Approach." *Issues in Accounting Education* 26 (4): 859–881.

26
Critical Thinking for Future Practice: Learning to Question

Franziska Trede and Celina McEwen

Introduction

Critical thinking can be used for many different purposes. For example, it can be used to develop technico-instrumental, specialized expert knowledge, and it can also be used to remind learners and teachers that specialized knowledge and instrumental skills have limits and need to be complemented by other, broader sets of skills. In this chapter, drawing on the critical theory of Habermas and other critical theorists, we discuss the latter purpose of critical thinking.

Although critical thinking is still listed as one of the key graduate outcomes, it remains contested and misunderstood especially due to current trends in university education to comply with competence frameworks and prepare students for the world of work and practice. This trend has reduced critical thinking to a standalone process, or even a measurable outcome in some instances, combining moral and quantitative reasoning with strategies to cope with and assess diversity (Bok 2006). Critical thinking is taught to students to help them evaluate arguments, write academically, and comply with the expectations of university learning. This rationalization of critical thinking has stripped it of the idea of cultivating receptive skepticism and challenging taken-for-granted assumptions (Brookfield 1987; 2005; 2012). The already tenuous identity of the university as a place for liberal education and critical thinking is being more insidiously challenged.

In this chapter, we explore critical thinking from a critical theory perspective and situate it in universities in the preparation of students for future professional practice. We are aware of a long tradition of fostering robust critical thinking in liberal studies and social sciences. With the more recent trend of educating for professional practice in universities there is, however, a risk of this tradition not being integrated into these new curricula (Parry 2010; Symes, Boud, McIntyre, Solomon, and Tennant 2000). Work-ready graduates and their employability have become a strong focus of the vocationalized universities (NCIHE 1997), with increasing demands by governments on these institutions to justify their existence and funding by contributing to the economic imperative through the development of a skilled workforce (DEEWR 2009). The new focus of universities on preparing students for future practice has ignited debates around definitions and pedagogies, and stimulated theorizing about professional practice (Billett 2010; Hager, Lee, and Reich 2012; Higgs, Barnett, Billett, Hutchings, and Trede 2012; Higgs, Fish, Goulter, Loftus, Reid, and Trede 2010).

"Practice" is a contested term that can be understood as a technical, professional, or social activity influenced by empirico-analytical, sociocultural, and/ or political knowledge, and interests (Green 2009). The way practice is framed and understood has implications for what is taught, as well as how and where it can be learned. It also has implications for what type of critical thinking is needed in university education and for future practice if we are to understand how to "do justice both to the challenges of practice in contemporary world but yet also do some justice to its liberal if not emancipatory promise" as Barnett (2010, 18) asks.

Just like "practice," "professional" is also a contested term that can refer to an approach as well as to a person. Green (2009, 6–7) defines four different meanings including *of* the profession, the practices of a given profession, an expert, and an ideal practice. As an ideal set of practices, "being professional" can be understood as adopting an objective, detached, and neutral approach or taking a discursively engaged position and reaching decisions based on reflexive and mindful deliberations (Macklin 2009).

The emphasis in university programs is on a curriculum design that complies with regulated national standards and competency frameworks. Such programs privilege technical and practical thinking. They also place a strong onus on students showing that they can "do" technical procedures and are "work-ready" as soon as they enter the workplace. This emphasis minimizes the need for students to show that they can "be" professional practitioners. When narrowly understood as preparing students for technical and procedural practices, a preoccupation with "work-readiness," can too easily neglect to nurture in students the types of reasoning, questioning, and becoming that shape future practice and, therefore, run the risk of turning students into unthinking technicians prepared for current, rather than future, professional practice.

Within the higher education context, Brookfield (2012, 28) explains that "the point of getting students to think critically is to get them to recognize and question the assumptions that determine how knowledge in that discipline is recognized as legitimate." In this chapter we argue for a critical thinking in universities that goes beyond academic reasoning to questioning the very roots of reasoning. First, we frame critical thinking within critical theory and its related philosophical stance, pedagogy, and tradition. We then explore current approaches to critical thinking by reporting on four research projects conducted by Trede between 2008 and 2012. Each study explored various aspects of preparation for future professional practice, including reflective practice from the students' perspective, professional development of academics, intercultural competence from academics' perspective, and assessment practices from workplace supervisors' (WPS) perspectives. We argue that a critical approach can be used within university curricula for professional practice to equip students with tools that will enable them to cope with the complexity, diversity, and ambiguity of learning to become a professional, as well as assist them with managing the increasingly frequent transition periods, uncertainty, and rapid changes they will face along their future career paths. We conclude with implications for critical thinking for future professional practice.

Framing critical thinking in critical theory

By locating critical thinking within the critical paradigm we define critical thinking as a practice that explores and leads to the implementation of other possibilities. Such critical thinking is underpinned by a social justice agenda and communicative questioning approaches that challenge self and others. It is a critical thinking that problematizes practice and theory and connects action with reflection. Critical thinking within critical theory means thinking autonomously as well as with others, without allowing others to think for us. It means questioning the traditions and motivations that shape practices in the first instance, and then participating in shaping other possibilities for future practices. Critical thinking is, therefore, for us, synonymous with radical thinking, because it means to go to the very ideological roots of why we think what we think and to expose why things are the way they are.

Underpinned by a philosophical stance

Critical theory is underpinned by the philosophical traditions of Aristotle, Marx, the Frankfurt School, and Habermas (Agger 1998). It emerged as a response to modernity and positivism with its claim that rational objective knowledge was superior to other ways of knowing. Such a detached, objectified, and one-dimensional view of the world would neglect cultural, ethical, and moral ways of knowing (Marcuse 1970). Following that, a critical perspective continued as a response to postmodernism with its claim that all interpretations are relative (Agger 1998). Later, critical perspectives were aligned with feminist, antiracist, and environmental movements critiquing dominant cultures, traditions, and practices (Kincheloe and McLaren 1996; Rasmussen 1996). Today, critical theory plays a role in commenting on how economic rationalism, globalization, social media, and the vocationalization of academia are affecting social conditions, and people's lives and identities (Newman 2009; Winter and Zima 2007). In the context of university education, we align our

Table 26.1   Overview of different types of critical thinking developed at university

TYPES / INFERENCES	Technical	Pragmatic	Political
Paradigms	Empirico-analytical	Interpretive	Critical
Knowledge acquisition	Transferred	Uncovered	Reframed
Reasoning	Objective, monologic	Intersubjective, dialogic	Disruptive, enabling
Pedagogy	Didactic, behaviorist	Collaborative, guided, experiential	Participatory, action-oriented, questioning
Context	Isolated	Connected	Networked
Professional practice	Traditional	Responsive	Changing
Professional culture	Assimilation	Tolerance	Inclusion
Professional identity	Compliant	Relativist	Deliberate

critical stance with Habermas (1972b, 4) who stated that the task of universities is not only to transmit technical, profession-specific knowledge and skills, but also to cultivate in students cultural awareness, political consciousness, and "action-oriented self-understanding."

Critical thinking is claimed by several philosophical paradigms and defined in relation to the interest and focus of diverse theorists (Trede and Higgs 2008). Table 26.1 offers a general overview of different types of critical thinking and their inferences. Critical thinking in action might manifest itself as a blend of types and this table should be interpreted as an overview only.

Within the empirico-analytical paradigm, critical thinking is used to pursue technical interests with the aim of prediction and control (Trede and Higgs 2010). Within the interpretive paradigm, critical thinking is based on communicative interests with the aim of finding common ground and shared understanding (Gadamer 1992). Within the critical paradigm, critical thinking has emancipatory interests as its focus with the somewhat utopian aim of reducing unnecessary constraints to enhance the common good based on robust debate and brutally transparent reasoning (Habermas 1972a).

The authors favor a critical thinking underpinned by the critical paradigm that does not negate technical and practical interests, but places political interest on an equally important footing, with an understanding that reason prevails over power and unreflected traditions (Habermas 1984). An awareness of interests enables critical thinkers to focus on reason rather than on unreflected tradition or authority. Habermas (1987, 25) theorized that "participants in argumentation have to presuppose in general that the structure of their communication...excludes all forces...except the forces of the better argument." Newman (2006, 14) described the meaning of critical thinking as:

> Analyzing human activity in terms of power, and refusing to take the words, ideas, injunctions and orders of others at face value. It meant not letting others make up our minds for us. It meant abandoning the search for some fixed set of principles, and adopting a stance of informed and continual critique. Critical thinking was not a neutral activity. Like the critical theory from which it sprang, critical thinking was associated with the pursuit of social justice.

Informed by a pedagogical approach

Critical pedagogies have framed critical thinking within the critical paradigm. Key thinkers in critical pedagogy include Freire (1972; 1973) and Brookfield (2005; 2012; 2013). Freire's seminal works on pedagogy can be summarized into three interwoven phases (dialogue, reflection, and action) that lead to praxis or what Freire calls conscientization. Dialogue here is underpinned by a theory of meaning making. Learners and teachers talk and listen to each other to gain a better understanding of the assumptions, backgrounds, and experiences that people bring to learning. Dialogue is closely linked to reflecting and exposing assumptions, concerns, and questions about self and others, especially when it is used to reflect "back to people their own discourse [and] enables them to become aware of their thinking and engage in discussion to better understand their situation" (Trede and Higgs 2010, 57). This process enables learners to become aware of their own assumptions and those of people around them, to highlight conflicts and paradoxes between their assumptions and actions, and to act and implement change.

Brookfield (2005) crystallized four methodological components of teaching critically. These are to foster (1) an understanding of the structures and regulations that underpin individual approaches to learning and teaching; (2) a good life; (3) an ethical and moral stance in questioning what appropriate practice actually means; and (4) full engagement in participatory and transparent discussions. Taking part in these processes is essential for enabling learners to identify key themes, issues, and contradictions; share information and ideas; and finally act on their newfound knowledge, as well as formulate further questions.

The key messages derived from these critical pedagogies are that critical thinking is a skill or a way of living brought alive within given social settings. Within this context, critical thinking is a learned way of questioning self and others, and considering future actions. Critical thinking requires the capacity to view oneself from a distance by questioning self and acknowledging other people's reality (Brookfield 2013; Habermas 1987). Critical thinkers have future interests in mind that open up new possibilities of interpreting and being in the world (Trede and Higgs 2010).

A didactic approach to teaching is often imposed on university students and academics, but there are plenty of ways to rethink or resist this. Teachers can start by drawing on learners' lived experiences, and learners can question what and how knowledge is being transmitted. This helps teachers and learners disrupt their routine way of reasoning by questioning and challenging their thinking. This approach has been coined uncomfortable learning (Cherry 2005) or reflection that interrupts routine thinking (Frank 2012). It suspends the immediate desire to follow routines, solve problems, and find quick solutions. That suspended desire also opens up possibilities to create new understanding and steer clear of narrowing critical thinking to quickly finding the best solution within an isolated context.

Cases of critical thinking at university

Dominant discourses emanating from universities globally present themselves as dialogical institutions where students will be nurtured to develop critical thinking qualities that will prepare them for work and the world of tomorrow. Indeed, as one university states online, "Good critical thinkers make good professionals. In the end, that's why you are at university" (University of Canberra 2012). Even if we subscribed to this statement, we might not be in agreement with what it means to be a "good" professional, and, hence, the type of critical thinking implied. These types of sweeping statements about critical thinking need careful deliberations to make them meaningful.

We find a number of limitations to this unreflected application of critical thinking, because it can imply critical thinking to be a problem-solving rather than a problem-posing approach. We illustrate this point with a discussion of several case studies conducted in a regional Australian university by one of the authors. These case studies used a Habermasian critical lens to examine the type of criticality and critical self-reflection adopted by students, academics, and WPS. They highlight the narrow focus of critical thinking around procedural, appreciative, divergent, and evidence-based practices rather than political, transformative, diverse, and judgment-based practices. These case studies are not representative of all university courses as they have a limited focus on educating for professional practice through workplace learning (WPL). Although there are other studies that have used a critical theory lens to explore criticality and critical self-reflection (Clouder 2000; Gonzalez 2003; Phelan, Sawa, Barlow, Hurlock, Irvine, Rogers, and Myrick 2006), we draw on these case studies for convenience, but also because we are closely connected with them and because they bring to bear for us the importance of locating critical thinking within a critical paradigm.

Procedural critical thinking

Trede and Smith (2012) conducted a study of students' perceptions of learning reflective practice while on mandatory work placements. They found that students were articulate about reflective practice and that they could easily identify its usefulness, but most often only from a technical and practical professional reasoning perspective. As mentioned earlier, reflection, or reflective practice, is closely associated with critical thinking. In our study, we found that reflection was predominantly applied to affirm current practices rather than question them. Though students reported that they valued academic preparation for reflection in the workplace, once there, they felt they depended heavily on their WPS's willingness to engage in reflective practice with them.

A majority of students engaged in and felt comfortable with reflections with their WPS when focusing on technical and procedural aspects of practice. Only a few students discussed reflective practice as a vehicle for developing critical and creative dimensions of their practice. Students avoided asking questions that might have exposed their gaps in knowledge. They also avoided asking critical questions that might make their WPS feel uncomfortable or even challenged. One physiotherapy student explained that she could only reflect as much as her WPS allowed her to: "It depends on the supervisor if I can use reflective practice with them or not. Some supervisors have a deeper level of thinking and can reflect. They make me ask questions and allow me to raise different points that even they were not aware of. They have more of a team approach, whereas others are more set in their own sort of ways." The nature of power relations between the WPS and the student, and the WPS's own critical dispositions, should not be ignored as it can enable or hinder critical thinking. To cope with or overcome hindering power relations, some students resorted to reflective self-talk. The danger of self-talk is that without critique from or disruption of their thought process by others, students may miss uncovering and challenging their assumptions, and struggle to develop a more reflexive type of critical thinking (Brookfield 2012). The nature of the WPS-student relationship, and the role that each partner plays, shapes the type of critical thinking that students learn and enact on placements. Therefore with a strong discourse on procedural and technical aspects of learning, as is currently the case in most universities and practice settings, critical thinking takes on the same narrow focus. WPS need to be aware of the consequences of their approach to reflective practice and critical thinking. To enable critical thinking on both sides, a student-WPS relationship must be built on curiosity, participation, dialogue, and a willingness to reconsider established professional practices.

Appreciative critical thinking

Trede (2010) undertook the evaluation of an online debate that engaged academics around issues pertaining to pedagogical practices that facilitate student learning in the workplace. Having sought participation from academic staff at a regional Australian university, Trede created a closed online forum for staff who responded to her invitation posted in the electronic university newsletter to debate, reflect, and engage in deeper meaning making on current practices of WPL for students. Participants recognized challenges and barriers to improving WPL programs, including staff's lack of time, resources, and formal policies. The evaluation highlighted that this recognition of barriers was, however, more "appreciative," or diplomatic and nonjudgmental, than a critical understanding of the situation, which allowed little attention to self-insights, leadership, and possibilities for improvement and change.

The online debate guidelines asked participants to be as open yet as skeptical as they could, focus on problem-posing and explore choices rather than immediately looking for solutions, and freely disagree and engage in vigorous debate, but treat all contributions respectfully. The guidelines also indicated that dialogue was for learning and transforming, not for preaching. Even so it proved difficult to shift the appreciative dialogues toward a more robust or critical debate.

A critical debate requires participants to be more explicit in expressing critique and more resilient in receiving it. Though there is a tradition of critically reviewing and problematizing each others' work, it did not spill into this debate to explore beyond individual teaching strategies and, for example, question existing power struggles between accrediting bodies and university course curricula designs. There seemed to be a greater tendency toward appreciating and affirming each others' work. To instill a critical approach in internal debates requires a cultural shift where academics engage with critical thinking themselves and consider wider power relations that shape university teaching and learning. Foundational requirements for such a cultural shift include devising and delivering courses that educate students for global citizenship rather than train them according to narrow competency frameworks dictated by professional bodies, working across courses and disciplines, and inviting students to take part in devising elements of the curriculum.

Divergent critical thinking

In a study that explored how academic staff prepared students for intercultural competence and global citizenship through international WPL experiences, Trede, Bowles, and Bridges (2013) found that though all the international programs they reviewed were procedurally well planned, they lacked an integrated intercultural learning approach. Students taking part in these international workplace experiences were informed about visa, health, and dress code requirements for their destination country. However, most participating academics omitted to include intercultural pedagogies into their programs. They did not purposefully prepare students for the new context within which they were going to be placed or equip them with strategies that might help them address potential professional, cultural, or ethical issues. They did not seek to develop intercultural competence and global citizenship in their students. They felt that the international experience was sufficient in itself to stimulate intercultural learning.

One of the very few instances when students were asked to critically think about cultural issues was in a course where they were required to write an assignment about the history of their destination country. The aim was to make students aware of the social, religious, and demographic differences between the host and the home countries, as well as highlight the existing international relations. However, the aim could have been for academics to raise students' awareness of cultural differences and then build upon that by getting them to critique and question current practices and think in terms of the relationship between the two countries, problematize their (or the university's) position as an international workplace student, and prepare them for dispositions of openness.

Clearly, this preparatory exercise was a missed opportunity to raise awareness beyond a simplistic fact-finding exercise about the host country's profile. Students could have been invited to assess their host country from diverse perspectives and consider issues associated with visiting and working in an international context. Academics could have made explicit use of critical pedagogical strategies to develop students' critical thinking capacity through intercultural experiences by asking them to discuss how the hosts might view the cultural, educational, and political impact of such placements on their workplace.

Evidence-based critical thinking

Another study by Trede and Smith (2014) explored assessment practices of physiotherapy WPS. Assessment practices are important to students as they drive what they learn, contribute to their overall university experience, and shape their professional identity and future practice. Not only do WPS provide feedback for learning, but they are also assessors who measure students' achievements. Playing a dual role of educator and assessor, which Boud (2000) coins the

"double duty," complicates the professional relationship WPS have with students. Although most academics experience this double duty as well, WPS take on the role as professional gatekeeper between students and their chosen professions. Assessment is the space in learning and teaching that imposes the strongest power differential on students (Tennant, McMullen, and Kaczynski 2010).

This study revealed that the WPS based their assessment practice closely on the national competence form of the Australian Physiotherapy Council (2006).

They welcomed this assessment tool because of its materiality, and because it described and prescribed what and how to assess. But they also voiced concern about the form, with some of the participating WPS saying they felt discouraged when trying to assess invisible or nonmeasurable aspects of practices not listed on the competence form, such as "soft skills" (e.g., developing supportive relationships, and emotional intelligence) and students' ability to question practice. Others felt that their judgment of students' capabilities for practice did not match the results of the rather evidence-based, fragmented assessment form. This often resulted in WPS disregarding their better judgment and awarding students marks according to the evidence-based formula that reduces assessment to visible outcome measures. The responsibility for assessment had been taken out of assessment practices. The well-intended competence form had reduced critical thinking about assessment practices to conforming with fragmented assessment elements. Competence assessment forms run the risk of falsely alluding to a more consistent and fair assessment practice when in fact they marginalize learning and critical thinking assessment.

Implications of critical thinking for future practice

Though our research into WPL experiences demonstrates that critical thinking is present within university curricula, it also highlights the fact that there are many missed opportunities by academics and WPS to help students develop an approach to critical thinking that would more adequately prepare them for future professional practice and reduce unnecessary constraints to existing practices.

The current dominant type of critical thinking leads students to adapt to criteria of employment (which do not necessarily lead to employment) (Lepage 1996), to cultivate a compliant technical practice, and to identify as self-administrating role models able to embody and enforce codes of conduct and the law (Bourdieu 1994; Wolfe 1993). The case studies discussed above show that a non-questioning approach will prepare students for factual knowledge and predictable times, while what we need to do is prepare them for uncertain times ahead, because, as Barnett (2010, 5) writes: "there is no stable world of practices to which higher education could 'correspond' even if it so wished."

We believe that WPL experiences are ideal opportunities to foster critical thinking, because they are the meeting place of academic and industry worlds. During WPL placements students are exposed to diversity (workplace cultures), complexity (the many interrelated factors and relationships that shape practice and, at times, competing interests), and ambiguity (the need to make judgments in the face of uncertainty), which cannot be simulated in the classroom. If well facilitated, WPL can enable students to reflect and critique professional expectations and practices. Such early exposure to critique and reflection sows the seed for lifelong critical thinking.

Critical thinking rooted in critical theory enables students to make sense of the diversity, complexity, and ambiguity they will encounter in the workplace, because it helps them understand professional practice as culturally and historically constructed. It equips students with the skills and capacity for ongoing learning and improvement. It also provides a framework for self-assessment and imagining other possibilities. Finally, it encourages students to make change happen and prepares them to solve elements of conflicts and contradictions within their professional practice. Such enabling kind of critical thinking would develop critical thinkers who have a greater awareness of options and the capacity to exercise choice, are able to participate more fully in their own practice, and in the making of history and transforming culture. A culture of questioning in critical thinking at university urgently needs to be addressed and reintroduced if universities are to truly deliver on their mission to educate future professionals ready for the future world of work.

Learning through questioning is immediately relevant and situated for the learner. It is the opposite of learning through being told. To question is crucial, because it creates the capacity to identify paradoxes, injustices, and contradictions leading to imagining other possibilities. Without questioning there would be no problematizing and deeper understanding, only memorizing, recalling, and copying.

The type of questions that enhance critical thinking from a critical theory perspective are not closed questions that require a "yes" or "no" answer. They invite others to explain and articulate themselves, and engage in critical dialogues. Critical questions are not immediately about gathering facts, but about problematizing events and situations. Critical thinkers ask questions to expose motivation and interests in order to get to the bottom of things. They ask questions about why things are the way they are. They open up rethinking and deliberate ownership of professional practice, culture, and identity.

In what follows, we examine critical approaches to questioning and the ways in which they allow students to acquire and understand technical tasks and roles (within and beyond occupation-specific skills and knowledge), to navigate and actively engage with workplace cultures and environments, as well as to develop a professional identity. We also highlight the ways in which the integration of this type of critical thinking in WPL provides an ideal opportunity to promote critical thinking outside of universities.

Questioning practice

Academics at university and practitioners supervising students in the workplace need to work with them to facilitate strategic questioning along the lines of: With whom, and why do I perform this task? What nontechnical skills and knowledge are essential for me to carry out this professional practice? How do I acquire those skills and knowledge? Are there other ways of doing this? What are the enabling and hindering environments (spaces and places) within the workplace to advance good practices in specific contexts and practice situations?

For the case study on assessment practices above, this questioning approach might have led the WPS to ask themselves what were the invisible qualities required to perform, as well as how, where, when, with whom, and why they were performed. This might have helped them adapt or use the assessment form differently. They could have also asked themselves what the form was helping them do better and what its limitations were. Such critique would have been the first step to enable these WPS to recognize the value of using a form to support the development of better-qualified future workers, leading to finding ways of working with or modifying the form to suit their assessment needs and support their professional judgment. It would also have been an opportunity to develop a better understanding of critical thinking strategies and how to apply them across professional practices, such as strategic planning, team building, and problem solving, to transform them from a mechanism of reproduction into vehicles of change (Brookfield 1987).

In the case of students' use of reflective practice in WPL, had they been guided toward using their reflective journal to move beyond the "what" and "how" they perform given tasks to the "why" and "what alternatives exist," they would have felt encouraged to enhance their ability to more purposefully practice and apply technical skills and knowledge.

Questioning culture

Becoming a professional member of a community of practice is another core aspect of preparing students for future professional practice (Loftus 2010). This requires building professional supportive relationships between students, academics, and practitioners. Academics and WPS can demonstrate inclusivity by allowing students to participate legitimately in practice and voice their questions and uncertainties about the context within which they are called to perform, rather than place them in a position of being passive observers. Examples of questions that will help students, academics, and WPS critically understand their community of practice include the following: What is the cultural and historical context of my profession? How connected is today's practice to past traditions? Who are the other members of this profession? Where are they located? Why has the profession developed in this way? Whose interests is this profession supporting?

In the context of our case studies, these questions could be translated into asking the debating academics, for instance, to examine the purposes WPL serves for universities and the workplaces. As for the assessment practices of WPS, the following questions could have been asked to encourage them to critically work with the form: How do other WPS assess? Do they also use forms? What is this form doing for me and others?

To enable students on international placement to make the most of their unique experience, strategic questions could have been used to help them prepare for, and then debrief from, their placement. Being exposed to different cultures often raises emotive and ethical dilemmas and presents questions about values and identity (Giroux 2005). International placements thus have the potential to strengthen learning and teaching in professional identity, and to better prepare students for multicultural environments where knowledge of cultural diversity enhances professionalism (Giroux 2005; Trede, Bowles, and Bridges 2013). By exploring professional practices to find common ground, students on international WPL placements would have been in a better position to become critically aware and engage with whatever cultural differences or fears they were confronted with. The following questions might have helped them do so: What is in this placement for me, the university, the host country, and the workplace? What are the tensions or dilemmas I am faced with? What are my own cultural biases and how might they affect my professional interactions in the host country?

This kind of critical thinking enables students to develop their ability to uncover power dynamics and political ideologies; actively shape their professional relationship with the WPS; engage more purposefully with workplace cultures; become aware of professional bodies, networks, and their current and potential position within their elected field of practice. It helps them cultivate an understanding of how, where, and when to seize opportunities.

Questioning identity

A questioning approach would also help students understand about "self" in relation to other(s). Developing a professional identity is complex because it goes beyond identifying with professional norms and practices. Developing undergraduates' ability to view their field of practice and professional identity critically means enabling students to grapple with the major issues they will be faced with in their future professions and as global citizens. We assert that "engaging with uncertainty and ambiguity is a complex skill that requires embarking on a critical learning journey, nurtured by a conscious approach to professional identity formation" (Trede and McEwen 2012, 29).

As a case in point, students could address the following questions: What are the different kinds of professional identities? What is their historical and geographical significance? How do I become a certain kind of professional? Are identities fixed or do we move through identities over time or depending on context? Do I need a professional identity to practice?

Applying this line of questioning to assessment could have led the WPS to highlight the tensions that exist between their different roles as practitioner, mentor, and assessor, and how to reconcile these roles under one professional identity. As for students on international WPL placements, academics could have used this opportunity to reflect on the ways in which their WPL experience might or did have an effect on the type of professional they were becoming and how it shaped their professional identity. For academics debating current practices of WPL, had they been asked to reflect on their appreciative stance and its relationship to their identity as lecturers and researchers, and the pedagogy they bring to their WPL program, they might have been more able to shift closer toward a critical debate. Similarly, students using reflective tools to enhance their WPL experience could have been asked by their lecturers to examine the relationship between different types and levels of reflection in practice, and the different kinds of professional identities required for different types of reflection. That might have made them more persistent in their use of reflective practice despite the lack of responsiveness from some of their WPS.

A commitment to a more "critical" type of critical thinking and to questioning professional practice, culture, and identity is essential if we are to educate for a future that is uncertain and fast changing. Of course, this means that we need academics and WPS who will implement a learning, teaching, and assessment curriculum that enables students to develop curiosity and skepticism by engaging in dialogue, by enhancing their questioning capacity, and inviting them to take responsibility for, and learn from, the consequences of their actions. It means that we need to have academics and WPS take part in the questioning process themselves in order to move from an appreciate stance toward full engagement in participatory and transparent discussions about their own professional practice, culture, and identity.

Having said that, we want to make it clear that we resist the temptation to locate critical thinking for future practice in a critical theory perspective as the only solution in university education. We acknowledge that there are challenges to implementing this type of critical thinking at university, because it assumes the existence of "workplace democracy," where workers have a certain degree of control over their practice (Brookfield 1987), or requires teaching defiance (Newman 2006), developing an understanding of when to comply and when to challenge. It also requires problematizing practice at a time when students and teachers are directed to predominantly learn and teach the "doing" and "knowing" of practice. Most students enroll in university courses to take on a profession or develop their career, and not for social justice reasons (James, Krause, and Jennings 2010). Undergraduate students commonly want to belong to a profession rather than critique it. It is easier to work with a definition of critical thinking that is unquestioning, and supports the existing culture, one that encourages reproduction and consumption, rather than disruption and uncomfortable learning. We want to stress that it is important to further explore this notion of whether there is a place for critical thinking at university, because the current trend is reducing universities' capacity to educate a more rounded critical global citizen (Faulkner 2011; Freedman 2011). We share the view of Solbrekke and Englund (2011) that it is important not to teach toward student satisfaction surveys, but toward cultivating critical future professionals who are capable of taking on social and moral professional responsibility.

Conclusion

Locating critical thinking within a critical philosophical stance and pedagogy invites students to explicitly understand the power relations and interests at play within their future professional practice. Future research should explore how well prepared academics and WPS are in facilitating critical thinking in their students. Our proposed questioning practice, culture, and identity approach to critical thinking needs to be tested with its stakeholders. Research with students could explore their experiences with this kind of critical thinking approach, and its potential and barriers, to better prepare them for future professional practices.

Universities need to produce competent professionals. But for students to find their place in the world of practice, they need to experience and find their own voice, which can most often only be developed through a process of problematizing, contextualizing, and situating professional practice, culture, and identity. To realize this form of critical thinking, we need to locate it within the critical paradigm so that it is not defined by default as a technico-rational educational practice, but rather as one entrenched in social justice and dialogical strategic questioning practice. It is important to note that integrating critical thinking at university is about increasing participation toward a more just society that is worth living in for all. To learn critical thinking and to critically understand is to learn about self in relation to something else whether it is, for example, a system, an authority figure, or a professional tradition.

A critical approach in university to preparing students for future practice, with critical thinking at its core, can equip them with tools to cope with the uncomfortable "messiness" of learning to become professionals, as well as to manage the increasingly frequent transition periods and rapid changes they will be faced with in their chosen field of practice.

References

Agger, B. 1998. *Critical Social Theories: An Introduction*. Boulder, CO: Westview Press.
Australian Physiotherapy Council. 2006. *Australian Standards for Physiotherapy: Safe and Effective*. Canberra: Australian Physiotherapy Council.

Barnett, R. 2010. "Framing Education for Practice." In *Education for Future Practice*, edited by J. Higgs, D. Fish, I. Goulter, S. Loftus, J.-A. Reid, and F. Trede. Rotterdam, The Netherlands: Sense. 15–25.
Billett, S. 2010. "Learning through Practice." In *Learning through Practice*, edited by S. Billett. Heidelberg: Springer, chap. 1. 1–20.
Bok, D. 2006. *Our Underachieving Colleges: A Candid Look at How Much Students Learn and Why They Should Be Learning More*. Princeton, NJ: Princeton University Press.
Boud, D. 2000. "Sustainable Assessment: Rethinking Assessment for the Learning Society." *Studies in Continuing Education* 22 (2): 151–167.
Bourdieu, P. 1994. *Raisons Pratiques: Sur La Théorie De L'action*. Paris: Editions du Seuil.
Brookfield, S. 1987. *Developing Critical Thinkers: Challenging Adults to Explore Alternative Ways of Thinking and Acting*. San Francisco: Jossey-Bass.
Brookfield, S. 2005. *The Power of Critical Theory for Adult Learning and Teaching*. Berkshire, UK: Open University Press.
Brookfield, S. 2012. *Teaching for Critical Thinking: Tools and Techniques to Help Students Question Their Assumptions*. San Francisco, CA: Jossey-Bass.
Brookfield, S. 2013. "What Does It Mean to Act Critically?" *Explorations in Adult Higher Education; An Occasional Papers Series, Our Values Our Goals* no. 2: 3–10.
Cherry, N. 2005. "Preparing for Practice in the Age of Complexity." *Higher Education Research & Development* 24 (4): 309–320.
Clouder, L. 2000. "Reflective Practice: Realising Its Potential." *Physiotherapy* 86: 517–522.
DEEWR. 2009. *Transforming Australia's Higher Education System*. Canberra: Department of Education Employment and Workplace Relations.
Faulkner, N. 2011. "What Is a University Education for?" In *The Assault on Universities: A Manifesto for Resistance*, edited by M. Bailey and D. Freedman. London: Pluto. 27–36.
Frank, A. 2012. "Reflective Health Care Practice: Claims, Phronesis and Dialogue." In *Phronesis as Professional Knowledge: Practical Wisdom in the Professions*, edited by A. E. Kinsella and A. Pittman. Rotterdam, The Netherlands: Sense. 53–60.
Freedman, D. 2011. "An Introduction to Education Reform and Resistance." In *The Assault on Universities: A Manifesto for Resistance*, edited by M. Bailey and D. Freedman. London: Pluto. 1–11.
Freire, P. 1972. *Pedagogy of the Oppressed*. Harmondsworth: Penguin.
Freire, P. 1973. *Education for Critical Consciousness*. New York: Seabury.
Gadamer, H. G. 1992. *Truth and Method*. Translated by J. Weinsheimer and D. G. Marshall. New York: Crossroad.
Giroux, H. A. 2005. *Border Crossings: Cultural Workers and the Politics of Education*. New York: Routledge.
Gonzalez, M. 2003. "Against Indifference." *Teaching in Higher Education* 8: 493–503.
Green, B. 2009. *Understanding and Researching Professional Practice, Professional Learning*. Rotterdam, The Netherlands: Sense.
Habermas, J. 1972a. *Knowledge and Human Interest*. Translated by J. J. Shapiro. London: Heinemann Educational.
Habermas, J. 1972b. *Towards a Rational Society*. London: Heinemann.
Habermas, J. 1984. *The Theory of Communicative Action (Volume 1): Reason and the Rationalization of Society*. Translated by T. McCarthy. Oxford: Polity.
Habermas, J. 1987. *The Theory of Communicative Action (Volume 2): The Critique of Functionalist Reason*. Translated by T. McCarthy. Oxford: Polity.
Hager, P., Lee, A., and Reich, A. 2012. "Problematising Practice, Reconceptualising Learning and Imagining Change." In *Practice, Learning and Change*, edited by Paul Hager, Alison Lee, and Ann Reich. Heidelberg: Springer, chap. 8. 1–14.
Higgs, J., Barnett, R., Billett, S., Hutchings, M., and Trede, F. 2012. *Practice-Based Education: Perspectives and Strategies*. Rotterdam, The Netherlands: Sense.

Higgs, J., Fish, D., Goulter, I., Loftus, S., Reid, J., and Trede, F. 2010. *Education for Future Practice*. Rotterdam, The Netherlands: Sense.

James, R., Krause, K.-L., and Jennings, C. 2010. *The First Year Experience in Australian Universities: Findings from 1994 to 2009*. Melbourne: Centre for the Study of Higher Education, The University of Melbourne.

Kincheloe, J. L., and McLaren, P. L. 1996. "Rethinking Critical Theory and Qualitative Research." In *Handbook of Qualitative Research*, edited by N. K. Denzin and Y. S. Lincoln. London: Sage. 138–157.

Lepage, F. 1996. "Fonction De L'art Dans Une Societe En Crise: Action Culturelle Et Education Populaire." Paper read at CRAJEP, Aix en Provence, France.

Loftus, S. 2010. "Exploring Communities of Practice: Landscapes, Boundaries and Identities." In *Education for Future Practice*, edited by J. Higgs, D. Fish, I. Goulter, S. Loftus, J. Reid, and F. Trede. Rotterdam, The Netherlands: Sense. 41–50.

Macklin, R. 2009. "Moral Judgement and Practical Reasoning in Professional Practice." In *Understanding and Researching Professional Practice*, edited by B. Green. Rotterdam, The Netherlands: Sense. 83–99.

Marcuse, H. 1970. *One Dimensional Man*. London: Sphere Books.

NCIHE. 1997. "Higher Education in the Learning Society. Report of the National Committee of Inquiry into Higher Education (the Dearing Report)." London: HMSO.

Newman, M. 2006. *Teaching Defiance: Stories and Strategies for Activist Educators*. San Francisco: Jossey-Bass.

Newman, M. 2009. "Educating for a Sustainable Democracy." In *Rethinking Work and Learning: Adult and Vocational Education for Social Sustainability*, edited by Peter Willis, S. McKenzie, and R. Harris. Heidelberg: Springer Science and Business Media, chap. 6. 83–92.

Parry, G. 2010. "The Dirty Work of Higher Education." *London Review of Education* 8 (3): 217–227.

Phelan, A. M., Sawa, R., Barlow, C., Hurlock, D., Irvine, K., Rogers, G., and Myrick, F. 2006. "Violence and Subjectivity in Teacher Education." *Asia-Pacific Journal of Teacher Education* 34 (2): 161–179.

Rasmussen, M. 1996. *The Handbook of Critical Theory*. Oxford: Blackwell.

Solbrekke, T. D., and Englund, T. 2011. "Bringing Professional Responsibility Back in." *Studies in Higher Education* 36 (7): 847–861.

Symes, C., Boud, D., McIntyre, J., Solomon, N., and Tennant, M. 2000. "Working Knowledge: Australian Universities and 'Real World' Education." *International Review of Education* 46 (6): 565–579.

Tennant, M., McMullen, C., and Kaczynski, D. 2010. *Teaching, Learning and Research in Higher Education: A Critical Approach*. London: Routledge.

Trede, F. 2010. "Enhancing Communicative Spaces for Fieldwork Education in an Inland Regional Australian University." *Higher Education Research and Development* 29 (4): 373–387.

Trede, F., Bowles, W., and Bridges, D. 2013. "Developing Intercultural Competence and Global Citizenship through International Experiences: Academics' Perceptions." *Intercultural Education* 24 (5): 442–455.

Trede, F., and Higgs, J. 2008. "Clinical Reasoning and Models of Practice." In *Clinical Reasoning in the Health Professions*, edited by J. Higgs, M. A. Jones, S. Loftus, and N. Christensen. Amsterdam: Elsevier. 31–42.

Trede, F., and Higgs, J. 2010. "Critical Practice and Transformative Dialogues." In *Education for Future Practice*, edited by J. Higgs, D. Fish, I. Goulter, S. Loftus, J. Reid, and F. Trede. Rotterdam, The Netherlands: Sense. 51–60.

Trede, F., and McEwen, C. 2012. "Developing a Critical Professional Identity: Engaging Self in Practice." In *Practice-Based Education Perspectives and Strategies*, edited by J. Higgs,

B. Barnett, S. Billett, M. Hutchings, and F. Trede. Rotterdam, The Netherlands: Sense. 27–40.

Trede, F., and Smith, M. 2012. "Teaching Reflective Practice in Practice Settings: Students' Perceptions of Their Clinical Educators." *Teaching in Higher Education* 17 (5): 615–627.

Trede, F., and Smith, M. 2014. "Workplace Educators' Perceptions of Their Assessment Practices: A View through a Critical Practice Lens." *Assessment and Evaluation in Higher Education* 39 (2): 154–167.

University of Canberra. 2012. *Critical Thinking.* University of Canberra 2012 [December 11, 2012]. Available from http://www.canberra.edu.au/studyskills/learning/critical.

Winter, R., and Zima, P. V. 2007. *Kritische Theorie Heute (Critical Theory Today).* Bielefeld, Germany: Transcript Verlag.

Wolfe, T. 1993. "Mau–Mauing and the Flak Catchers." In *The Purple Decades.* London: Picador. 201–232. Original edition, Farrar, Strauss and Giroux, New York, 1982.

27

Criticality in Osteopathic Medicine: Exploring the Relationship between Critical Thinking and Clinical Reasoning

Sandra Grace and Paul J. Orrock

Introduction

Critical thinking is an important graduate attribute that is universally cultivated in university courses, including professional entry courses in health. Osteopathic medicine is undergoing considerable change as it reevaluates its traditional foundations in the context of the evidence-based demands of contemporary health care (Fryer 2008; Licciardone, Brimhall, and King 2005) and thus offers a timely opportunity for practitioners and students to engage in critical thinking both within and about the field. This chapter discusses the relationship between critical thinking and clinical reasoning and what criticality means in osteopathic medicine, which may shed light on the relationship between critical thinking and clinical reasoning in health care more generally. We argue that critical thinking and clinical reasoning are primarily distinguished by context and by the metaskills that practitioners call on in the course of clinical practice.

Practitioners who undertake primary care roles assess and treat patients who may have an undifferentiated complaint at their first entry into the health care system (Australian Medical Association 2010). They have added responsibilities to liaise with practitioners from other disciplines and to refer patients for the most appropriate care. Practitioners have responsibilities to question their own notions of what counts as knowledge, their own social and personal epistemologies (Barnett 1997; Higgs and Bithell 2001), and recognize the patient as an expert in their own illness experience (sometimes called "narrative reasoning" [Jones and Rivett 2004]). If assumptions are uncritically accepted a self-confirming action cycle can develop (Brookfield 1987). Moreover, practitioners have responsibilities to engage with the wider social context in order to question dominant paradigms and taken-for-granted assumptions about the way our health system/society works.

Clinical practice involves challenge, uncertainty, unpredictability, and complexity. Practitioners draw on all the practice knowledge that they can muster, including what they have learned from past clinical experiences. It is their responsiblitiy to gather relevant information, and to interpret and engage with patients' world views, illness experiences, and expectations, often under time pressure. The clinical reasoning process must be sustained throughout the entire patient encounter, adding an extra challenge for practitioners who provide extended consultations or who see numerous patients. Moreover, the uniqueness of each patient encounter must be taken into account. Health practitioners learn to reconcile themselves to variable patient outcomes.

Osteopathy

Osteopathic medicine is a medical system of diagnosis and therapy based on a set of overarching principles: (1) structure and function are interrelated; (2) the body has its own repair, self-regulating, and self-healing processes; (3) the circulatory channels and nerves provide an integrating and supportive framework; (4) the musculoskeletal system's importance far exceeds the function of providing framework and support; and (5) there are somatic (body wall or frame) components to disease (Kuchera and Kuchera 1994). Osteopathy is practiced worldwide, predominantly in developed Western nations, and its practice varies from full medical scope in the United States to allied and adjunctive health in countries like the United Kingdom, Australia, and New Zealand. The osteopathic approach to health includes evaluating somatic tissues for signs of dysfunction that are treated with a range of manual therapies and adjunctive care, including exercise rehabilitation and nutrition advice.

Critical thinking

The extensive literature on critical thinking presents a range of definitions, predominantly those that describe critical thinking as skills and abilities for problem solving. Cottrell (2003; 2005), for example, described the central steps in critical thinking as (1) identifying assumptions, bias, and theoretical perspectives; (2) evaluating evidence, premises, and salience; (3) identifying arguments; and (4) developing conclusions. Such interpretations lend themselves to assessments using checklists, a range of which has been developed (LaFave 2014; Paul and Elder 2002). Other conceptions of critical thinking focus on attitudes or dispositions like willingness to engage in and persist at a task, flexibility and open-mindedness, and willingness to abandon nonproductive strategies in order to self-correct (Halpern 1999). Facione, Facione, and Giancario (1997) described a number of personal attributes that lead to a disposition toward critical thinking, including truth seeking, open-mindedness, analyticity, systematicity, confidence, inquisitiveness, and cognitive maturity.

Barnett (1997) drew our attention to the social responsibilities associated with critical thinking, arguing that the skills and attitudes that are required for critical thinking are inseparable from the wider social context in which they are situated. The contemporary social context in which professionals work and act is characterized by change—change that is driven by advanced technologies and our global connectedness (Barnett 1997; Kadir 2007). We are presented with vast volumes of information accumulating at an unprecedented rate. According to Barnett, critical thinking in the twenty-first century demands the ability to continually reflect on and appraise our thoughts and actions and to bring a commitment to social responsibility in this highly fluid world.

Historical development—parallels between critical thinking and clinical reasoning

Evolution of our understandings of clinical reasoning has paralleled that of critical thinking. Early conceptions of clinical reasoning focused on it as a cognitive process, including the hypothetico-deductive reasoning model (Elstein et al. 1978, Barrows et al, 1978), the pattern recognition model (Barrows and Feltovich 1987), the knowledge-reasoning model (Schmidt et al. 1990), and the intuitive reasoning model (Agan 1987). Increasing interest in medical sociology and the central role of the patient in all clinical encounters extended our understanding of clinical reasoning to take account of the social environment in which it occurred. Clinical reasoning models like the narrative reasoning model of Mattingly and Fleming (1994) and the collaborative reasoning model of Edwards et al. (2004) are examples of models in which patients' interpretations of their illness experiences are a prominent feature.

Criticality in clinical practice

Contemporary conceptions of clinical reasoning have also embraced the social ecology model of critical thinking (Barnett 1997). The wider social responsibilities associated with clinical practice are acknowledged and practitioners are encouraged to carry out wise actions not only within the confines of their clinics but also more generally in their professional and personal lives. The ethical reasoning model of Barnitt and Partridge (1997) and the multidisciplinary reasoning model of Croker et al. (2008) reflect our recognition of the social responsibilities of practitioners. Criticality in clinical practice involves entering into a "critical dialogue"with patients, practitioners, and others, that is, repeatedly "evaluating perspectives within context" in order to develop new insights (Forneris 2004). Health care systems in developed countries are facing a number of challenges that test their ability to deliver effective, efficient, and responsive services to the population. These challenges are well documented and include increasing demand for services (American College of Clinical Pharmacy 2000; Sibbald, Laurant, and Scott 2006), growing prevalence of chronic disease (Centre for Allied Health Evidence 2006; Queensland Health 2012), escalating service costs (Bosley and Dale 2008), diminishing workforce availability (Miers 2010; Richards, Carley, and Jenkins-Clarke 2000; Sadkowsky, Hagan, Kelman, and Liu 2001), and changing community expectations (Lewy 2010; Sadkowsky et al. 2001). The various reform agendas that have been established in response to these challenges require supportive legislative and industrial environments, a supportive professional environment, and supportive leadership (Australian Health Workforce Advisory Committee 2006). "Critical dialogues" with all stakeholders will contribute to a culture that fosters constructive change.

The relationship between critical thinking and clinical reasoning

Critical thinking and clinical reasoning have similar dimensions, contexts, metaskills, and purposes

Clinical reasoning has been described as extending critical thinking into practice (Mitchell and Batorski 2009). Indeed, many similarities can be drawn between the dimensions, contexts, metaskills, and purposes of critical thinking and clinical reasoning (table 27.1). Health practitioners use metaskills like knowledge generation and reflexivity (Christensen, Jones, Higgs, and Edwards 2008), applied in the contexts of clinical practice (e.g., practice knowledge and models, the workplace, the patient's context, and the practitioner's unique frames of reference). When health practitioners gather data about patients, analyze it, and develop working diagnoses, they draw on a range of strategies to deal with problems that they encounter in their daily practices. Some problems require a technical-rational decision while others, the "messy confusing problems which deny technical solutions" (Schön 1987, 3), require "a rich blend of biomedical, psychosocial, professional craft and personal knowledge together with diagnostic, teaching, negotiating, listening and counselling skills" (Jones and Rivett 2004, 3).

Critical thinking and clinical reasoning are distinguished by the context of clinical practice

Practitioners use metaskills like knowledge generation and reflexivity and may draw on emotional capability and practice model authenticity more than critical thinkers in nonclinical contexts. Smith et al. (2008), recognizing the decision-making capabilities of physiotherapists in acute care settings, included emotional capabilities like awareness of the impact emotions have on decision-making and the capacity to deal with problematic emotions when making the difficult decisions required for patient management, as a core component of clinical reasoning. The overarching remit of clinical practice is caring—doing what is within the scope of practice to care for patients; respecting their illness

Table 27.1 Comparison between critical thinking and clinical reasoning

	Critical thinking	Clinical reasoning
Dimensions	Acquire a deep understanding of and commitment to the tacit norms of a discipline, learn the challenges of contemporary work	Acquire a deep understanding of and commitment to the tacit norms of a health discipline, learn the challenges of contemporary work
	Gain an immediate sense of what citizenship might mean	Gain an immediate sense of what being a health practitioner might mean (e.g., having social and moral awareness of issues affecting the profession, the health care system, and broader social issues like chronic diseases, healthy aging, environment)
	Develop the powers of self-critique	Develop the powers of self-critique

Context	A set of problem spaces:
	 i) Knowledge
		A set of problem spaces:
		 i) Practice knowledge
		     Domain-specific conceptual knowledge
		     Domain-specific procedural knowledge
		     Dispositional knowledge

	ii) World	ii) Workplace context, practice models, patient's contexts
	iii) Self	iii) The practitioner's unique frames of reference
Metaskills	Including: Critical inquiry and reflection	Including: Critical inquiry and reflection
	Knowledge generation	Knowledge generation (including hypothesis generation/working diagnoses)
	Reflexivity Metacognition	Reflexivity Metacognition Emotional capability Practice model authenticity
Purpose	Wise action: Critically engage with the world and with oneself and with knowledge	Wise practice action: Critically engage with the patient's context, onself, and with practice knowledge to make decisions, individually and collaboratively constructed, that guide practice actions

experiences and treatment preferences; and dealing with their family and friends, other members of the health care team, and the organization. Health practitioners are indeed interactional—interactional "with their work environments, key players, situational elements pertinent to the patient" (Higgs and Hunt 1999), but more than that, they are interacting in a potentially emotionally charged environment. Practitioners not only manage their own emotional reactions to their patients' pain and suffering, but are also well aware that the outcomes of their critical thinking can directly affect their patients' safety and well-being.

Practice knowledge and practice models

Knowledge in critical thinking (table 27.1) becomes practice knowledge when applied in a clinical context. Billet (2009), summarizing two decades of research into expert performance, described three kinds of knowledge: domain-specific conceptual knowledge (concepts, facts, propositions); domain-specific procedural knowledge (knowing how to undertake strategic procedures); and dispositional knowledge (values and attitudes). Practice knowledge helps define professions and sets their boundaries. It is the basis of the professional curriculum—the formal (stated and endorsed) curriculum, the informal curriculum (unscripted interpersonal components, e.g., role modeling), and the hidden curriculum (influences of organizational structure and culture) (Hafferty 1998)—and it cultivates students' professional identities in preparation for practice life. Professional values and assumptions consciously or unconsciously guide the way health practitioners practise. Each profession is grounded in one or more practice models, such as the illness and wellness models; biopsychosocial model; metaphysical model; and practitioner-centered, patient-centered, and patient-empowered models (Trede and Higgs 2008). Clinical reasoning in a specific profession involves commitment to the practice models by which the profession identifies itself. By the time of graduation, many students will have assimilated the values of the profession and modified their own worldviews accordingly. For example, medical students' faith in complementary medicine was found to have declined during their medical education (Einarson, Lawrimore, Brand, Gallo, Rotatone, and Koren 2000; Furnham and McGill 2003). First year students typically displayed an openness to complementary medicine that had largely disappeared by the end of their training.

Clinical reasoning in osteopathic medicine

The practice model(s) adopted by a profession reflect its values. In mainstream health care, biomedical and/or biopsychosocial practice models predominate, reflecting the primacy granted to scientific evidence. By contrast, in homeopathy most practitioners identify spiritual and energetic dimensions in a metaphysical model of diagnosis and treatment. Osteopathy has evolved a combination of models to deal with the patients it serves. Osteopaths tend to see patients with complex problems who do not easily fit disease patterns (Licciardone, Brimhall, and King 2005; Orrock 2009). They have had to be inventive, and willing to embrace a wide range of etiological and treatment possibilities. Osteopaths, like all practitioners, must be able to think along multiple lines and often on different levels at the same time (Jones and Rivett 2004). Each patient consultation is unique and the clinical reasoning load it imposes is enormous. Practitioners need to be able to reflect *in* action (Schön 1987); to analyze and evaluate information, including evidence claims and the idiosyncratic complexities that both patients and practitioners bring to the encounter; to solve problems, and to provide a multiplicity of treatment and advice as part of the normal business of practice.

Nonmedical, primary contact health professionals in Australia, including osteopaths, initially prioritize a biomedical approach, including considering the possible presence of red flags that indicate serious underlying pathology requiring referral for medical or other assessment. Once a patient has been deemed suitable for osteopathic assessment, osteopaths attempt to make sense of the patient information by considering a range of practice models, that is, by constructing a working diagnosis. Subjective information collected from patients and objective data from tests and examinations may be considered in a context of connected functioning body systems (models), including biomechanical, biopsychosocial, energy-expenditure, neurological, nutritional, and respiraory-circulatory (see table 27.2).

Table 27.2 Practice models used in osteopathic medicine (Kuchera and Kuchera 1994)

Biomedical	The consideration of signs and symptoms in the context of defined diseases and a need for referral for further medical assessment and management (red flags). This is similar to any primary care practitioner.
Biomechanical	The assessment of the health of the musculoskeletal system, including how the structure (posture) and function are integrated. This is similar to other manual medicine practices, and is primarily a mechanical/orthopedic approach.
Biopsychosocial	A consideration of the psychosocial factors influencing the patient's health, including relational, occupational, and financial, and the need for multidisciplinary care.
Energy expenditure	An assessment of whether the patient has optimal energy utilization, and consideration of issues that may affect the healing process (e.g., relatively minor mechanical or immune dysfunctions).
Neurological	The assessment of function in the central, peripheral, and autonomic nervous systems, and the relationship of those systems to all tissues of the body.
Nutritional	A foundational dietary analysis for signs of deficiency or suboptimal nutritional status.
Respiratory/circulatory	The examination of the respiratory mechanism—ensuring that the function of breathing is optimal. An assessment of all tissues of the body for full blood supply and drainage. An assessment of the structural and functional relationship between the two systems.

Each practice model is associated with specific osteopathic treatment approaches that prioritize practitioners' choices of treatment techniques and the order in which they are applied. Conceptual shifts from one practice model to another are attempts to make sense of the information practitioners are presented with. Having a range of practice models enables practitioners to use an osteopathic lens to explore connections between seemingly unrelated things—a headache with a dysfunctional breathing pattern or with restricted movement in one knee, for example.

Table 27.3 illustrates various aspects of clinical reasoning in the context of osteopathic practice. The case of a 53-year-old female suffering from rheumatoid arthritis is used to highlight some of the contexts and metaskills involved in the clinical reasoning process. This very complex process has been simplified for the purpose of illustration.

This case example shows how critical thinking in osteopathic medicine has particular challenges. Osteopaths engage with a world that demands a rich body of scientific evidence to underpin its health practices. Osteopaths often have to contend with the inherent tensions between the lack of scientific evidence to underpin many of the practices of osteopathic medicine and strong clinical effectiveness. They may also contend with tensions between their own, other practitioners', and their patients' worldviews, particularly when patients coexist in several practice models simultaneously. For example, many patients of osteopathic medicine are also under the care of general medical practitioners and specialists and may also be consulting other practitioners like acupuncturists and naturopaths simultaneously. It is the responsibility of all practitioners to have an understanding of, or, at the very least, appreciate and respect, the key practice models of other health disciplines for the sake of their patients.

Implications for educating future osteopaths

Osteopathic curricula appear to have successfully embedded strategies that promote some of the dimensions, contexts, and metaskills of critical thinking, specifically taking the patient's context into account, critical appraisal of research evidence, and reflective practice. The World Health Organization's *Benchmarks for Training in Osteopathy* (2010) cites the "ability to appraise medical and scientific literature critically and incorporate relevant information into clinical practice" as a core competency, and includes clinical problem-solving and reasoning in its list of practical skills. Osteopathic students are also provided many opportunities to cultivate reflective practice (e.g., critical reflective journaling) designed to help them identify and challenge their own assumptions and beliefs (Brookfield and Preskill 1999; Francis and Cowan 2008). A

Table 27.3 Aspects of clinical reasoning: Case example from osteopathy

A 53-year-old female administrative assistant presents with long-term bilateral aching wrists, which have worsened in the last month as her workload has increased. She has felt increasingly tired during this time. She has a history of rheumatoid arthritis (RA), which was diagnosed in her early 30s. She is raising a teenage son as a single mother. She works 40 hours a week in a sedentary job, under deadline pressure. Her hobby is ceramics and pottery that she does in her limited spare time.

Practice models	Key considerations	Contexts	Metaskills
Biomedical diagnostic model	What is the evidence for RA? Consider previous medical diagnosis, expected course of condition, previous therapy, response to therapy including side effects	Practice knowledge	Reflexivity Critical inquiry and reflection
	Does the patient understand RA and its consequences?	The patient's context	Critical inquiry and reflection
	Are there any red flags? Referral necessary?	Practice knowledge Workplace context	Critical inquiry and reflection
	Need to develop rapport with patient	Patient's context Practitioner's frame of reference	Reflexivity Emotional capability
	Expected pattern with RA?	Practice knowledge	Critical inquiry and reflection
	Evidence for any other condition(s) (e.g., neurological)?	Practice knowledge	Critical inquiry and reflection
	Have I seen this context before?	Practitioner's frame of reference	Reflexivity
Biopsychosocial model	Consider stress in lifestyle	Patient's context Practice model authenticity	Critical inquiry and reflection Emotional capability
	Chronic pain often associated with depression. Any yellow flags?	Practice knowledge Practice model authenticity	Critical inquiry and reflection Emotional capability

continued

Table 27.3 Continued

The patient is deemed suitable for osteopathic care. On examination the patient is found to have swollen wrist joints and weak forearm musculature; restricted upper back, neck, and clavicle movement; restricted breathing.
Any recent diagnostic imaging of the wrists?

Possible osteopathic models	Key considerations	Contexts	Metaskills
Respiratory-circulatory model	Evidence for lack of ciriculatory drainage function? Plausible rationale for congestion?	Practice knowledge Practice models	Practice model authenticity Knowledge generation
Biomechanical model	Evidence for lack of mobility in related regions? Any recent diagnostic imaging of the wrists?	Practice knowledge Practice models	Practice model authenticity Knowledge generation
Biopsychosocial model	Evidence of stress and posture reducing capacity to heal?	Practice knowledge Practice models	Practice model authenticity Knowledge generation

The working diagnosis is RA with peripheral inflammatory nociception of radiocarpal joints. The likely cause of the patient's presenting complaint is autoimmune and inflammatory, precipitated and maintained by occupational stress and sedentary lifestyle.

Wise actions	Key considerations	Contexts	Metaskills
Patient education	Patient educated about RA and the osteopathic considerations. Introduced to stress reduction techniques and postural self-care. Treatment plan negotiated.	Practice knowledge Practice models Patient's context Practitioner's frame of reference	Practice model authenticity Emotional capability
Patient management	Commenced a low dose trial of manual therapy to drain fluid (respiratory-circulatory model), treat postural stiffness, structured graded strength and flexibility training for upper limbs, hydrotherapy in local pool, dietary analysis.	Practice knowledge Practice models Patient's context Practitioner's frame of reference	Practice model authenticity
	How many sessions before review? What is the prognosis? Any evidence from literature, experience, experts? Monitor patient's response to treatment.	Practice knowledge Patient's context	Critical inquiry and reflection Reflexivity Knowledge generation

Table 27.3   Continued

Wise actions	Key considerations	Contexts	Metaskills
	Communicate with patient's rheumatologist and general medical practitioner	Practice knowledge Workplace context	Critical inquiry and reflection
	Will the patient be compliant with exercise, given her stressful lifestyle? How to optimize compliance with exercise?	Patient's context Practice knowledge	Critical inquiry and reflection Reflexivity

The patient was reviewed after three treatments and reported a 40–50% improvement in pain and a similar gain in strength was observed. Stiffness was noted in her shoulders and neck. She was struggling with her workload and had no time for exercise. Her general medical practitioner supports continuation of osteopathic management.

Wise actions	Key considerations	Contexts	Metaskills
Patient management	Renegotiation of management plan Addition of biomechanical and biopsychosocial models of osteopathic care	Patient's context Practitioner's frame of reference Practice knowledge Practice models	Practice model authenticity Emotional capability
	Occupational stress— interact with employer? Ergonomic assessment required?	Workplace context Patient's context Practice knowledge Practitioner's frame of reference	Critical inquiry and reflection Reflexivity
	Long-term cost to patient of treatment— is there a lower cost option?	Patient's context Practitioner's frame of reference	Reflexivity
	Explore issues of stress with patient and commence coping-strategy education. Consider referral to psychologist. Review referral network for appropriate practitioner.	Practice knowledge Practice models Patient's context	Practice model authenticity Emotional capability Reflexivity Critical inquiry and reflection

A maintenance program of bimonthly osteopathic treatments and reviews was established. The employer has reduced her workload and she is able to go to the gym and pool twice a week after work. Her general medical practitioner and rheumatologist support her management plan. Her rheumatologist has reviewed and adjusted her medication.

high value is placed on work-based learning for its provision of authentic situations where students are called upon to engage in "adaptive and critical thinking" as they apply what they have learned in theory to practice settings (Billett 2009).

The social responsibilities associated with critical thinking, for example, being global citizens who have the potential to transform the societies in which they work, are not generally prioritized in osteopathic curricula. These responsibilities, however, are present in the generic graduate attributes of many universities. For example, at Southern Cross University, Graduate Attribute 2 requires "Creativity: an ability to develop creative and effective responses to intellectual, professional and social challenges" (Southern Cross University 2013). In contemporary society, many competing interests vie for students' attention, and therefore it cannot be assumed that students will keep up to date with current health news and research (including news and research about health issues outside their discipline). Devoting curriculum time for discussions about current health research and health issues raised in the media may orient graduates to the broad cultural, social, and political contexts in which they will function as health advisers and leaders in primary care practices.

Key recommendations for developing critical thinking in professional-entry health courses include active learning that encourages participation and cognitive engagement rather than passive reception of knowledge. In active learning teachers function as facilitators, guide discussions, model behaviors, and help students develop awareness of their own thinking (Brookfield and Preskill 1999; Kuhn 1999). In this model of pedagogy students can go beyond skills-based, discipline-specific ideas to embrace diverse perspectives and multiple discourses and be willing to risk critique (Barnett 1997). Moreover, embedding critical thinking in all approaches to teaching and learning may establish it as the normative approach to both health care practice and life beyond one's profession. Osteopathic curricula are designed to guide students through varying levels of complexity and challenges in practical and theoretical components of their courses. Curriculum review may be required to fully embrace critical thinking as a graduate attribute. For teachers, this calls for time to engage with curriculum and scholarly debate with colleagues, time to create safe and supportive environments for active learning, time to model wise action, time to address individual learning needs, and time for discussions with students. For students, this calls for time to develop independence in learning, to identify gaps in their own learning, to develop collaborative learning, to engage in scholarly debate (Brookfield and Preskill, 1999), and to acquire an orientation to the broader social responsibilities of practice.

Conclusion

Critical thinking and clinical reasoning have a complex relationship. Contemporary understandings of critical thinking and clinical reasoning have much in common: dimensions, contexts, metaskills, and purposes. In the process of both critical thinking and clinical reasoning questions are formed, information is gathered and analyzed, answers to questions are proposed, and the consequences enacted. However, critical thinking and clinical reasoning are distinguished by the nature of clinical practice. This includes the highly subjective nature of caring in the therapeutic relationship, the high stakes associated with decisions and subsequent actions that affect health care, and the dynamic interplay and complexity of collaboration with each patient. Such contexts call on metaskills like emotional capacity and practice model authenticity that may not be prevalent in nonclinical settings.

Health professions may be distinguished by the practice models to which they adhere. Criticality is particularly important in professions like osteopathy where the practice models of the profession have evolved to serve the complex conditions of many of their patients but may remain unsubstantiated by scientific evidence. It is the responsibility of osteopaths to evaluate perspectives in all contexts, including the tension between the demand for scientific evidence and strong clinical effectiveness or patient preference. Future research will endorse some traditional practices and reject others and emerging research must be appraised and integrated into practice where appropriate. This is the responsibility of all health practitioners of course, including those of well-established mainstream disciplines—to recognize and challenge unquestioned assumptions—if we are to achieve the best outcomes for patients and for our local and global communities.

References

Agan, R. 1987. "Intuitive Knowing as a Dimension of Nursing." *Advances in Nursing Science* 10 (1): 63–70.

American College of Clinical Pharmacy. 2000. "A Vision of Pharmacy's Future Roles, Responsibilities, and Manpower Needs in the United States." *Pharmacotherapy* 20 (8): 991–1020.

Australian Health Workforce Advisory Committee (AHWAC). 2006. "The Australian Allied Health Workforce: An Overview of Workforce Planning Issues." Sydney: AHWAC.

Australian Medical Association. 2010. *Primary Health Care—2010*. Available from https://ama.com.au/position-statement/primary-health-care-2010.

Barnett, R. 1997. *Higher Education: A Critical Business*. Buckingham: Society for Research into Higher Education and the Open University Press.

Barnitt, R., and Partridge, C. 1997. "Ethical Reasoning in Physical Therapy and Occupational Therapy." *Physiotherapy Research International* 2 (3): 178–194.

Barrows, H. S., Feightner, J. W., Neufield, V. R., and Norman, G. R. 1978. *An Analysis of the Clinical Methods of Medical Students and Physicians.* Hamilton, Ontario: McMaster University School of Medicine.

Barrows, H. S., and Feltovich, P. J. 1987. "The Clinical Reasoning Process." *Medical Education* 21 (1): 8–91.

Billett, S. 2009. "Realising the Educational Worth of Integrating Work Experiences in Higher Education." *Studies in Higher Education* 34 (7): 827–843.

Bosley, S., and Dale, J. 2008. "Healthcare Assistants in General Practice: Practical and Conceptual Issues in Skill-Mix Change." *British Journal of General Practice* 58: 118–124.

Brookfield, S. 1987. *Developing Critical Thinkers: Challenging Adults to Explore Alternative Ways of Thinking and Acting.* San Francisco: Jossey-Bass.

Brookfield, S., and Preskill, S. 1999. *Discussion as a Way of Teaching: Tools and Techniques for University Teachers.* Buckingham: Society for Research into Higher Education and Open University Press.

Centre for Allied Health Evidence (CAHE). 2006. *Systematic Review of the Literature on Support Workers in Community Based Rehabilitation.* Adelaide, Australia: CAHE.

Christensen, N., Jones, M., Higgs, J., and Edwards, I. 2008. "Dimensions of Clinical Reasoning Capability." In *Clinical Reasoning in the Health Professions*, edited by J. Higgs, M. A. Jones, S. Loftus, and N. Christensen. 3rd ed. Sydney: Butterworth-Heinemann. 101–110.

Cottrell, S. 2003. *Skills for Success.* New York: Palgrave Macmillan

Cottrell, S. 2005. *Critical Thinking Skills: Developing Effective Analysis and Argument.* New York: Palgrave Macmillan

Croker, A., Higgs, J., and Loftus, S. 2008. "Multidisciplinary Clinical Decision-Making." In *Clinical Reasoning in the Health Professions*, edited by J. Higgs, M. A. Jones, S. Loftus, and N. Christensen. 3rd ed. Sydney: Butterworth-Heinemann. 291–298.

Edwards, I., Jones, M., Higgs, J., Trede, F., and Jensen, G. 2004. "What Is Collaborative Reasoning?" *Advances in Physiotherapy* 6: 70–83.

Einarson, A., Lawrimore, T., Brand, P., Gallo, M., Rotatone, C., and Koren, G. 2000. "Attitudes and Practices of Physicians and Naturopaths toward Herbal Products, Including Use during Pregnancy and Lactation." *Canadian Journal of Clinical Pharmacology* 7 (1): 45–49.

Elstein, A. S., Shulman, L. S., and Sprafka, S. A. 1978. *Medical Problem Solving: An Analysis of Clinical Reasoning.* Cambridge, MA: Harvard University Press.

Facione, P., Facione, N., and Giancario, C. 1997. *Professional Judgement and the Disposition towards Critical Thinking.* Millbrae, CA: Californian Academic Press.

Forneris, S. G. 2004. Exploring the Attributes of Criting Thinking: A Conceptual Basis. *International Journal of Nursing Scholarship* 1 (Article 9).

Francis, H., and Cowan, J. 2008. "Fostering an Action-Reflection Dynamic amongst Student Practitioners." *Journal of European Industrial Training* 32 (5): 336–346.

Fryer, G. 2008. "Teaching Critical Thinking in Osteopathy: Integrating Craft Knowledge and Evidence-Informed Approaches." *International Journal of Osteopathic Medicine* 11 (2): 56–61.

Furnham, A., and McGill, C. 2003. "Medical Students' Attitudes about Complementary and Alternative Medicine." *Journal of Alternative and Complementary Medicine* 9 (2): 275–284.

Hafferty, F. W. 1998. "Beyond Curriculum Reform: Confronting Medicine's Hidden Curriculum." *Academic Medicine* 73 (4): 403–407.

Halpern, D. 1999. "Teaching for Critical Thinking: Helping College Students Develop the Skills and Dispositions of a Critical Thinker." *New Directions for Teaching and Learning* 80: 69–74.

Higgs, J., and Bithell, C. 2001. "Professional Expertise." In *Practice Knowledge and Expertise in the Health Professions*, edited by J. Higgs and A. Titchen. Oxford: Butterworth-Heinemann. 59–68.

Higgs, J., and Hunt, A. 1999. "Rethinking the Beginning Practitioner: Introducing the 'Interactional Professional.'" In *Educating Beginning Practitioners: Challenges for Health Professional Education*, edited by J. Higgs and H. Edwards. Oxford: Butterworth-Heinemann. 3–23.

Jones, M. A., and Rivett, D. A. 2004. *Clinical Reasoning for Manual Therapists*. Edinburgh: Butterworth Heinemann.

Kadir, A. 2007. Critical Thinking: A Family Resemblance in Conceptions. *Journal of Education and Human Development* 1 (2).

Kuchera, W. A., and Kuchera, M. L. 1994. *Osteopathic Principles in Practice*. Columbus: Greyden Press.

Kuhn, D. 1999. "A Developmental Model of Critical Thinking." *Educational Researcher* 28 (2): 16–46.

LaFave, S. 2014. *Critical Thinking Checklist: Steps in Argument Analysis* [January 3, 2014]. Available from http://instruct.westvalley.edu/lafave/CTchecklist.htm.

Lewy, L. 2010. "The Complexities of Interprofessional Learning/Working: Has the Agenda Lost Its Way?" *Health Education Journal* 69 (1): 4–12.

Licciardone, J. C., Brimhall, A. K., and King, L. N. 2005. "Osteopathic Manipulative Treatment for Low Back Pain: A Systematic Review and Meta-Analysis of Randomised Controlled Trials." *BMC Musculoskeletal Disorders* 6: 43.

Mattingly, C., and Fleming, M. H. 1994. *Clinical Reasoning: Forms of Inquiry in a Therapeutic Practice*. Philadelphia: F. A. Davis.

Miers, M. 2010. "Professional Boundaries and Interprofessional Working." In *Understanding Interprofessional Working in Health and Social Care*, edited by K. C. Pollard, J. Thomas, and M. Miers. Hampshire, UK: Palgrave Macmillian. 105–120.

Mitchell, A. W., and Batorski, R. 2009. "A Study of Critical Reasoning in Online Learning: Application of the Occupational Performance Process Model." *Occupational Therapy International* 16 (2): 134–153.

Orrock, P. 2009. "Profile of Members of the Osteopathic Association: Part 2—the Patients." *International Journal of Osteopathic Medicine* 12: 128–139.

Paul, R., and Elder, L. 2002. *Critical Thinking: Tools for Taking Charge of Your Professional and Personal Life*. Upper Saddle River, NJ: Pearson Education.

Queensland Health. 2012. Queensland Health Strategic Plan 2012–2016.

Richards, A., Carley, J., and Jenkins-Clarke, S. 2000. "Skill Mix between Nurses and Doctors Working in Primary-Care Delegation or Allocation: A Review of the Literature." *International Journal of Nursing Studies* 37: 185–197.

Sadkowsky, K., Hagan, P., Kelman, C., and Liu, C. 2001. *Health Services in the City and the Bush: Measures of Access and Use Derived from Linked Administrative Data*. Canberra: Commonwealth Department of Health and Aged Care.

Schmidt, H. G., Norman, G. R., and Boshuizen, H. P. 1990. "A Cognitive Perspective on Clinical Expertise: Theory and Implications." *Academic Medicine* 65: 611–621.

Schön, D. A. 1987. *Educating the Reflective Practitioner*. San Francisco: Jossey-Bass.

Sibbald, B., Laurant, M., and Scott, T. 2006. "Changing Task Profiles." In *Primary Care in the Driver's Seat? Organizational Reform in European Primary Care*, edited by A. Saltman, A. Rico and W. Boermaed. Berkshire, UK: Open University Press.

Smith, M., Higgs, J., and Ellis, E. 2008. "Factors Influencing Clinical Decision Making." In *Clinical Reasoning in the Health Professions*, edited by J. Higgs, M. A. Jones, S. Loftus, and N. Christensen. 3rd ed. Sydney: Butterworth-Heinemann. 89–100.

Southern Cross University. 2013. *Graduate Attributes*. [January 26, 2014] Available from http://policies.scu.edu.au/view.current.php?id=00091.

Trede, F., and Higgs, J. 2008. "Clinical Reasoning and Models of Practice." In *Clinical Reasoning in the Health Professions*, edited by J. Higgs, M. A. Jones, S. Loftus, and N. Christensen. Sydney: Butterworth-Heinemann. 31–42.

World Health Organization. 2010. *Benchmarks for Training in Traditional/Complementary and Alternative Medicine: Benchmarks for Training in Osteopathy.* Available from http://apps.who.int/medicinedocs/documents/s17555en/s17555en.pdf

28

Making Critical Thinking Visible in Undergraduates' Experiences of Scientific Research

Anna N. Wilson, Susan M. Howitt, Denise M. Higgins, and Pamela J. Roberts

Introduction

What does it look like when undergraduate students practice critical thinking in an authentic scientific research context? Do we provide such students with opportunities to show their critical thinking in action—or do we, for the most part, leave it hidden? And, if critical thinking can be made visible, how can we recognize and hence develop and assess it?

This chapter describes an attempt to reveal the dynamic processes of critical thinking as it happens, before endpoints or conclusions are reached, as students experience the unfamiliar context of scientific research. We start by considering what we might mean by critical thinking in this context. We then describe the processes used to attempt to make visible students' enacted, contextualized critical thinking. The examples we describe suggest students have opportunities to think critically in a variety of different ways; understanding this variation is crucial to recognizing and scaffolding the development of criticality. Finally, we consider what our data tell us about the nature of critical thinking as practiced by science students; its correlations with developing expertise and confidence; and implications for practice.

What do we mean by critical thinking?

The development of critical thinking may be seen as central to the broader higher education endeavor (Atkinson 1997; Barnett 1997; Barrie and Prosser 2004). This may be particularly true for students hoping to become professional/academic scientists, as the ability to make judgments about data, evidence, and hypotheses, and the quality and credibility of one's own work as well as one's peers, is crucial in the practice of science. Yet as will be evident from other chapters in this book, important questions remain as to what universities are doing (and what can be done effectively) to foster, recognize, and assess critical thinking in their undergraduates.

One problem is that precisely what is meant by critical thinking is often unclear, both to students and teaching academics. Many attempts have been made in the literature to pin down what seems to be an important but slippery concept—here we offer three examples that capture the general flavor.

Brookfield (1987) described critical thinking as involving four components:

- recognizing and challenging assumptions,
- challenging the importance of the context,
- being willing to explore alternatives, and
- being reflectively skeptical.

While Brookfield's work has concentrated mainly on the development of criticality with respect to the social and political environment, particularly in adult education outside the higher education context, these four components also apply to critical thinking in academic contexts.

More recently, Paul (2005) offered the following description of what a critical thinker *does*: "Critical thinking is the art of thinking about thinking in an intellectually disciplined manner. Critical thinkers explicitly focus on thinking in three inter-related phases. They *analyze* thinking, they *assess* thinking and they *improve* thinking (as a result)" (28). Hale (2010) summarized Paul's conception of critical thinking as comprising three elements:

- analysis and evaluation with a view to improvement,
- development of intellectual traits, and
- a process that is applied to one's own thinking, the thinking of others, and thinking within a discipline.

Rather than offering a definition of critical thinking, Barnett (1997) identified three different tiers of critical thought, with a widening focus on what one might be critical of:

- "critical thinking" as cognitive skills, usually involving problem-solving,
- " critical thought" as interchanges, debates, and standards within an intellectual field, and
- "critique" as metacriticism, involving the taking of a wider perspective, operating outside the discipline itself and sometimes directed at the rules of the discipline.

While differing in some respects, the emerging pattern is of a belief that critical thinking involves abilities such as analysis and evaluation, together with dispositions such as reflectivity, a willingness to challenge current or accepted thinking or practice, and a desire to seek improvement in one's own thinking or practice, that of the discipline or profession, or in society itself (Pithers and Soden 2000).

The importance of context

Despite our ability to identify common elements of critical thinking, it is important to recognize that it is also "irrevocably context bound" (Brookfield 1987, 18). Enacted forms of critical thinking will therefore be richly varied, informed by the epistemologies, aims, and character of the discipline providing the context for its exercise, as well as varying in scope as suggested by Barnett's three tiers (Barnett 1997). This variation, if left unexplored, poses challenges for universities claiming to develop (and possibly assess) critical thinking.

The descriptions of critical thinking given above are expressed in terms of generic skills and processes that are intended to be relevant across academic and workplace contexts. Some authors have, in addition, explored disciplinary differences in critical thinking—for example, Paul (2005) is concerned with disciplinary thinking; Moore (2011) describes academics' conceptions of critical thinking in philosophy, history, and literary studies; and Jones (2007) compares history and economics. McCune and Hounsell (2005) introduced the related idea of "ways of thinking and practicing" (WTPs) in a subject or discipline. Such WTPs include "the ways in which individual disciplines represent (or at least debate) the *nature* of knowledge in their domains, what counts as evidence and the process of creating, judging and validating knowledge…a more complex differentiated understanding of knowledge and its relationship to evidence " (Anderson and Hounsell 2007, 496). The idea of ways of thinking and practicing in a discipline thus incorporates discipline-dependent examples of the exercise of critical thinking given above, bringing the disciplinary context into sharp focus.

While these studies have explored what critical thinking means to expert academics, it is rare to find examples of what it looks like when students start to practice it in disciplinary contexts. In this chapter, we set out to do just this.

Grasping opportunities to develop critical thinking

These two ways of approaching critical thinking—generalist or specificist points of view—have given rise to significantly different ways of approaching the teaching and learning of critical thinking, together with sometimes heated debate (Atkinson 1997; Davies 2006; Ennis 1987, 1992; Facione 1990;

McPeck 1981, 1992; Moore 2011). In the present chapter, we don't want to take sides in this debate. Rather, we take the view that we would like to encourage critical thinking whenever the opportunity arises, and we hope that such opportunities arise at many points in an undergraduate's education. However, the explicit requirement to be critical, and scaffolded opportunities to develop appropriate forms of critical thinking, are likely to vary widely from discipline to discipline.

In science in particular, there is a need to make the nature and objects of appropriate critical thinking in undergraduate work more evident to students and staff alike. Critical thinking in the sciences is strongly associated with problem-solving, analytical thinking, the application of logic, and skepticism. It is also frequently described as a key learning goal within a science degree. For example, drawing on data from a study of WTPs in biology, Entwistle (2009, 60) quotes one bioscientist as describing his/her aims as "[to bring students to] challenge things, to question things, [to ask], 'Can both these people be right?' ... A good healthy dose of cynicism ... In the end of the day, it's you and your data, and you make up your own mind what you think." However, science students are typically expected to master a canon of received knowledge and skills, handed down by authority, over an approximately three-year period. Given the reductionist positivism that largely underpins scientific discourse, students are rarely asked to voice an opinion or encouraged to question the validity of the theories they set out to master.

One common context in which science students *are* expected to engage in critical thinking is authentic research projects, typically undertaken late in a degree. For example, in the UK and Australia, science students undertake a substantial, immersive project in the final year of an Honours degree. They also frequently have the opportunity to undertake smaller-scale projects, often in parallel with conventional coursework, in their third year of study. Such projects are likely to be exploratory or open-ended, with a looseness of structure that allows students to encounter surprises, obstacles, problems, ambiguities, uncertainty, and contradictions, and where resolution may be down to the student.

However, it is important to recognize that simply placing students in such contexts is not in itself a guarantee of opportunities to exercise and develop critical thinking; as has been recognized in previous work, scaffolding and opportunities for self-reflection and metacognition are critical, if not always delivered, elements (Pithers and Soden 2000). If critical thinking is to be actively fostered and developed, ways need to be found to make it visible, to characterize it in context, and to distinguish between different levels of sophistication, so that both students and academics can tell what has been achieved.

Designing a way to make the invisible visible

The data we draw on come from the TREASURE project—an ongoing, multi-disciplinary, multi-institution project based in Australia. The primary aim of the project is to better understand the learning that goes on when undergraduate students undertake extended research projects. Such projects, in principle, maximize the opportunities where critical thinking might be appropriately deployed, and indeed the development of analytical and critical thinking are among the most frequently cited motives for engaging undergraduate students in research. For this chapter, we draw on data from students in the sciences only.

We focus on projects that are assessed for grading, since under these circumstances explicit assessment of critical thinking is most likely to be sought, and so evidence of its taking place (or not) and its quality is highly desirable. Unfortunately, traditional approaches to assessing students' performance in science research projects work against the possibility of observing when, whether, and how students are thinking critically. Assessment tends to be predominantly based on a formal report written after the project work has been completed. Although this might be seen as an opportunity for students to display evidence of critical thinking, the opposite is more likely the case.

Science students' project reports follow the conventions of scientific journal article writing, which may be better seen as a form of writing involving the deliberate suppression of the active critical thinking processes involved in the discovery or creation of a result (Medawar 1963). The scientific report is constructed so as to present key findings as concisely and convincingly as possible. Student performance in research projects is thus usually assessed on the basis of polished, carefully constructed documents, showing few or none of the wrong turns taken, the misconceptions challenged, or the processes of diagnosis and resolution the student may have experienced before reaching a conclusion. The actual messiness of choices and judgments made in project design and data analysis—the interaction between observations, design, and hypothesis, and the evolving nature of underpinning research questions—is deliberately hidden in an attempt to create a comprehensible, logical narrative. While some critical analysis of the literature and final results may be evident in reports, the much larger scope for critical thinking during the process of research is typically hidden.

In describing requirements for successful strategies aimed at assessing critical thinking, Brookfield (1987, 19) suggests "that studying the dimension of action—what students do as well as what they say—is crucial." Thus we seek ways of gaining access to ongoing thinking processes *during* projects, rather than simply asking students to report on their project at the end. We wish to make the exercise and development of critical thinking visible while students engage in designing experiments or solutions to problems, executing those designs, troubleshooting, and reflecting.

Brookfield identifies another essential aspect of assessing critical thinking as using procedures that "allow learners to document, demonstrate and justify their own engagement in critical thinking" (1987, 20). Going further, Bean (2011) claims that it is the act of writing itself that often makes thinking explicit, both to students and academics involved in their assessment, providing opportunities for metacognition that might otherwise be missing.

To achieve these aims, we have introduced a blog-based system for students participating in our project to record their thoughts and experiences during their own research projects. We developed (in a process of codesign involving research project supervisors) a bank of prompt questions that students could choose to respond or add to. These were intended to direct students to focus on aspects of their experiences that offered opportunities for critical thinking and to encourage self-reflection:

- What did you do on your project [since your last post]?
- Did you make progress? If so, what allowed you to make progress? What kind of activities did you engage in that helped you make progress?
- Did you encounter any problems or obstacles? If so, what made them problems? How did you go about solving them?
- Has your research question changed? If so, how?
- How have the activities that you undertook [since your last post] helped you address your research question?
- Can you see any connections between your research activities and your other studies?
- Can you see ways in which you could apply what you have learned to other activities, in or out of university?

The subset of data described below comes from the first cycle of our study. Students from one of the participating institutions undertook substantial research projects over a full semester, typically working with an active researcher or research group. The students were in the later stages of their degrees, most in their third year of study. The project represented 25% of a full-time study load over a 13-week semester; normal coursework (the remaining 75%) continued in parallel.

The projects themselves were highly varied, including laboratory-based research in biochemistry, nuclear physics, and other fields; field-based studies in zoology and botany; hospital-based experiences in genetic counseling; and projects primarily involving literature review and research design. Despite this variation, all projects were treated as equivalent for the purposes of assessment by the institution. While exact assessment arrangements varied across the group, all projects were primarily to be assessed on a final report as described above. Blogs were either unassessed (and hence completely voluntary), or contributed to the final grade on the basis of upkeep rather than quality.

A total of 55 students had access to the blogging system. Of these, 25 made four or more posts, providing a substantial body of data recording their experiences and thinking during the course of their research.

Critical thinking, made visible

In many ways, the blogs exceeded our expectations. Students frequently treated them as effectively private, diary-like spaces where they recorded their experiences, thoughts, and feelings, providing rich and nuanced insights into their learning. Of course, the blogs are recorded after the fact of specific project experiences, and so present students' reports of and reflections on their activities rather than real-time records of thinking processes. However, they do have the advantage of showing thinking *within* the project, recording wrong turns and troubleshooting, and the associated emotional ups and downs, before a final result is achieved and history rewritten in the form of a final report. They also provide many explicit illustrations of metacognition, with students reflecting on their own learning and changes in their understanding.

Following the descriptions of critical thinking offered by Brookfield (1987), Paul (2005), and others (Pithers and Soden 2000), we analyzed the blogs for instances where students engaged in evaluation of their data or methods; recognized their own or others' assumptions; looked for constructive ways to solve problems; or explored alternative approaches or alternative hypotheses.

We focused on the range and variation in students' reports of their experiences, rather than the frequency with which particular types of experience occurred. Our approach is influenced by the intentions and methods of phenomenography (Marton 1981), looking for ways in which variation is manifested and for hierarchical relationships between different levels of sophistication. However, where phenomenography focuses on variation in conceptions of or ways of experiencing specific examples of learning—for example, learning to program (Bruce, Buckingham, Hynd, McMahon, Roggenkamp, and Stoodley 2006)—or personal development—for example, developing as an academic (Åkerlind 2005)—we focus on variations in enactment, or reported enactment, of a way of thinking. Our intention is to develop a framework for identifying and characterizing critical thinking that may help academics involved in the supervision and assessment of undergraduate research.

Developing a framework for recognizing and describing different forms of critical thinking

To better illustrate our approach, we provide two examples of the kind of fine-grained analysis that eventually allowed a more general structure to emerge. Consistent with our phenomenonographic influences, both examples show qualitatively different, and increasingly more sophisticated, ways of responding to aspects of the research experience.

Example 1: how students responded to problems in their research progress

Most students encountered and commented on some kind of problem during the course of their project. In some cases, their responses to such problems were entirely uncritical—they did not spontaneously notice anomalies in their data, for example, continuing to follow set procedures blindly (and unproductively). In other cases, students lacked the self-confidence to try to solve the problem themselves. Such responses form the first three levels presented in the first column of table 28.1. However, in other cases, students responded to problems with varying degrees of criticality, as described in the second three levels.

Example 2: what students noticed about the research environment

The academic or cultural research environment in which students found themselves offered another opportunity for critical thought. As described in the second column of table 28.1, three different types of uncritical response were evident in the data. However, some students did engage in a critical appraisal of aspects of the research environment, leading us to identify the three levels presented in the second half of the table.

Table 28.1 summarizes and compares the levels of thinking relating to both these examples and relates those where there is some evidence of criticality to Barnett's three tiers (Barnett 1997). We found very little evidence for thinking that could be characterized as critique (i.e., critical thinking that questioned the philosophy or way of thinking underpinning the processes of scientific research), except perhaps where students' understanding of the contingency of scientific knowledge develops. However, even in these cases, the criticality is directed more toward their own thinking, as realizations dawn about the nature of the discipline in which they are working, than at the discipline itself.

Building on these and other aspects of projects where students have opportunities to engage in critical thought (e.g., considering their own research design/methodology, scientific process and the evolution of scientific knowledge), we see a pattern emerge that suggests three broad levels of critical thinking in

Table 28.1   Examples of variation in students' responses to different aspects of their projects

	Response to problems	What is noticed about the research environment	Character of thinking
Absence of criticality	Student does not spontaneously notice problematic or anomalous data.	Not considered: not noticed as different from usual study environment.	Uncritical/procedural
	Student notices problem or anomaly; blames own practice, knowledge, preparation, or ability, or equipment failure; and gives up.	Research/researchers are seen as intimidating. The research environment is seen as complex, not something the student belongs to, requiring an expertise beyond their capabilities.	Aware of need to be critical, lacking self-confidence
	Student blames self and/or equipment and turns to supervisor for a solution.	Research/researchers are seen as awe-inspiring, something the student may hero-worship but not be part of.	Aware of need to be critical, lacking self-confidence, looking for a model
Exercise of criticality	Student independently identifies and explores factors contributing to failure with a view to finding an explanation or solution.	Research/researchers are seen as experts but not unattainable; participation is an exhilarating challenge to rise to.	Discerns important elements and attempts criticality.
	Student suggests coherent explanations, bringing together multiple factors in an integrated way and recognizing causal relations.	Research/researchers are recognized as fallible.	Engages in critical thinking (Barnett's first tier)
	Student suggests, and where possible enacts, solutions to problems.	Multiple possibilities are recognized, and the role of researcher style in determining research/practice directions is discerned.	Engages in critical thinking and critical thought (Barnett's second tier)

scientific research projects, corresponding to different stages in the process illustrated in figure 28.1.

In the following sections, we provide excerpts from the blogs illustrating each stage, while simultaneously highlighting how the object toward

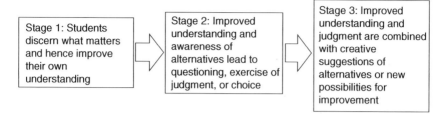

Figure 28.1 Different stages of critical thinking in undergraduate science research projects.

which the critical thinking is directed varies. Each student has been given a pseudonym.

Discerning what matters and improving the student's understanding

In the first stage, students think critically about existing disciplinary techniques, concepts, or approaches in a way that allows them to discern key features and hence achieve a deeper understanding. They may discern what matters by noticing variations between different circumstances, thus there is an element of comparison (often implicit) and evaluation in this stage. They may also recognize and challenge their own assumptions. However they do not challenge the assumptions or underlying intentions of the research team, or question whether there may be a better approach or way of understanding. This type of thinking may be seen as one that brings the student's thinking closer to that of the discipline.

For example, a variation in a standard procedure leads to this student's realization of the relevance of something she had previously not thought about at all:

> I have picked up little technique tricks in the process, and the logic behind each of them. For example, when using the solvent dichloromethane, it is important to pipette the liquid up and down before measuring the appropriate amount, due to the surface tension which could alter the quantity. Most of the time when working in the lab I am using simple liquids such as water, isopropanol and buffers, so I have never really considered the simple concept of surface tension to have an effect before. (Elizabeth)

Another post shows a student thinking critically about his data in order to discern a pattern and decipher the message they carry:

> I…find it really satisfying playing the detective role with my data, starting with a mish-mash of meaningless numbers and then figuring out what they are telling me. (Luke)

Another student relates how using a particular method in his project led to critical thinking about other learning experiences and a better understanding of the real practice of science:

> The biggest shock was the amount of troubleshooting required and tweaking of methods needed to get results. PCR, a reaction that we learn about in first year that seems so simple, proved to be a complex process requiring not only knowledge of how to adjust things when they do not work but also an element of luck. This differed significantly from my experiences in other courses where all the resources are laid out in front of you and the methods have been checked time and again to be successful. (Daniel)

As with the first quote above, this illustrates how research projects offer a context in which students can start to think critically about methods they might otherwise take for granted—but in the first case Elizabeth's focus remains on the technique itself, whereas in the third, Daniel relates his realization about a specific technique to the broader practice of science.

Perhaps the most sophisticated examples of critical thinking corresponding to stage 1 of figure 28.1 result in realizations about the way science progresses, as in the following:

> What is interesting when reading papers is that you can see the progression of thought within the scientific community on this question, which is something that is often hard to appreciate from lectures. For the cancer that I am looking into, Burkitt's Lymphoma, it has only recently been suggested that three genetic "hits" are required for a B cell to become malignant... it is hard to imagine that something that is being taught to you as widely accepted now, wasn't always known. This is something that comes through clearly from reading a range of papers, and trying to teach yourself through them. Early papers detail their new discoveries with obvious excitement, whilst the knowledge they are presenting is treated as assumed by more recent papers. In this context it is easy to appreciate how knowledge is accumulated. (Mary)

In these excerpts, we see an expanding focus from project-specific objects of thought (techniques and data) to what science is really like (troubleshooting and how scientific knowledge is created). Thus although they may not include elements of questioning and creativity, these instances of critical thinking are nevertheless important and valuable to the students' learning.

Critical thinking involving value judgments or choices between existing alternatives

In the second stage, students use their improved understanding to question and make active judgments about existing ideas or practice. These may be judgments of value or standard, or they may be choices involving explicit comparison between alternatives. If alternatives are considered, they are ones that the student has become aware of, for example, through reading or observation; that is, they are alternatives drawn from external (usually authoritative) sources. This type of thinking may be connected with Barnett's first tier, critical thinking.

The blogs reveal students exercising judgment about their own practice and that of others. In some cases, no alternatives are explicitly considered or choices between approaches clearly made, yet changes are made with the intention of improving outcomes. For example, one student describes his experience of background reading as follows:

> I ambitiously tried to start to read articles on recent clinical trials, it quickly became evident that my understanding on vaccination strategies was required first. This taught me to progress in a logical order and take it step by step. I have returned to some of the articles on clinical trials and it is abundantly evident how much I was missing in the first read through. (Mark)

It appears that Mark's initial judgment of his approach as inappropriate was followed by an immediate, obvious way forward, with no perceived need to consider alternatives.

Other posts reveal students thinking critically about the practice of experts they observe during their projects. In the following example, a student draws a comparison with her own previous observations in order to make a judgment:

> I found the change of environment altered the dynamics of the sessions. We were using a small pediatrics consultation room with a glass wall, which seemed odd, having the bed and such in the background and no real table and chair setup. I felt this made the consults seem less professional and less private. (Annika)

Critical thinking involving judgment and creativity

In the third stage, students use their improved understanding not only in the exercise of judgment, but also as a basis from which to put forward their own ideas and suggestions. These new ideas are proposed with the intention of improving practice or outcomes. This category adds an element of creativity to the questioning introduced in stage 2, and may be the way in which Barnett's second tier, critical thought, is manifested in the scientific context.

Most comments belonging to this category focus on the student's project. In the following example, Elizabeth builds on her observations to propose a new research question, and suggests a possible experiment intended to address it:

It is interesting that chiloglottone was found in high amounts in the sepals of *C. seminuda*; is it possible that if you remove the sepals, pollination will still occur? Or is it vital to the overall system? I think it would be interesting to remove certain parts of the floral tissues and see the "success" of the remaining parts that produce chiloglottone, possibly to view the differences, or roles each part plays in attracting a pollinator, or if it is simply a system to which enough pheromone is produced (and concentrated in the appropriate place) such that the pollinator is attracted and pollen is transferred. (Elizabeth)

In another case, we see a student respond to a surprising aspect of his data with a recognition of how the resulting new knowledge impacted on the method he had been using, and putting forward a revised approach in the light of his discovery:

Theoretically [the surprising factor] shouldn't have made any difference but from the empirical results it clearly did...knowing what I know now, if I was doing this again my approach would be quite different: instead of starting with just one particular metric and looking for broad correlations across a large sample set, I would instead start with just a handful of samples and all of the data points, looking for relationships and correlations and then slowly growing the sample set. (Ethan)

This type of comment shows not only critical thinking about observations and method, but also an awareness of how the two interact with each other. An even more sophisticated awareness of the interaction between observations, methods, hypotheses, and research design is evident in the following extract from a student's post describing her evolving project:

I initially set out to look at the costs and benefits to cockatoos of flocking with corellas...from the perspective of the cockatoos. As such I was only gathering data from mixed flock and cockatoo flocks. After a while, however, I began to suspect that the corellas may be benefitting more from the mixed flocking than the cockatoos: they seemed to be much more aggressive than the cockatoos...I hypothesized that when corellas associate with a cockatoo flock they may be benefitting from the vigilance of the cockatoos while experiencing a reduced level of aggression from that of their own flocks. Being surrounded by vigilant, non-mating cockatoos may also afford them more safety while they're courting and mating (activities where vigilance seemed especially low). The cockatoos, on the other hand, may be suffering from increased aggression when corellas are present, which may or may not be offset by the increased vigilance the extra birds afford. I decided that it would be interesting to examine the situation from the perspective of both species and see if one species was deriving a greater benefit than the other from the association. Of course, this meant that after a few weeks of gathering only cockatoo data I had to also start recording corella-only flocks. (Briony)

Instances of students imagining their own alternatives to the practice of others were much rarer. One of the few examples comes from a similar context to the excerpt about genetic counseling above, where another student not only discerns differences in practice but also relates them to her developing understanding and possible future professional practice:

I [have become] more observant of the way in which the counselors deliver information and how I think I might have done it. I am starting to be a bit more critical of the different counseling styles, which I think is good because it means I am starting to think more about the way in which information is communicated, which is a key aspect of genetic counseling. (Melissa)

Although she does not explicitly describe what she thinks she might have done, Melissa's comment implies that she has her own ideas.

Critical thinking and confidence

A key pattern that emerges from our data is a correlation between criticality and confidence. Students who engage in the exercise of judgment, choice, or creative thinking characterizing stages 2 and 3 in figure 28.1 appear to have also gained confidence in their own expertise. This confidence provides a basis from which to put forward ideas and opinions that are valid within the disciplinary context, facilitating deployment of a critical approach. This correlation can be seen in the following excerpts, which are taken from the blog of a single student, Magda. Early on in her project she recounts responding to a problem as follows:

I had an *E.coli* transformation fail, and purified some DNA samples, only to get rather low yields. My initial reaction to this was to blame myself for poor technique and look no further. (Magda)

Several weeks later, she responds to unexpected results in a radically different tone—note the confident use of sophisticated, technical language accompanying her own hypothesis about what might have happened:

The positive ligation mixture controls did not give the expected PCR bands for EITHER of my recombinant plasmids. I've hypothesized that my digestion of the initial plasmid pYM-N5 failed because the restriction sites were right next to each other. Perhaps my gel showing "successful lineariza-tion" was merely the result of one enzyme working giving a linear plasmid with one compatible and one non-compatible end. (Magda)

One of the key differences between these responses is a willingness to think for herself—indeed, her initial problem was only solved by turning to her supervi-sor for help. The apparent increase in Magda's critical thinking is coupled with increased fluency in the disciplinary discourse. It is evident that during the course of her project, Magda has acquired a substantial amount of disciplinary knowledge and technical expertise, possibly furnishing her with the confi-dence to make judgments and suggest her own hypotheses.

The shift from an uncritical approach to attempts at criticality, as exemplified in table 28.1, thus seems to be related to developing confidence. Confidence may also play a key role in determining which stage of figure 28.1 a student engages in in any given context. Where critical thinking is directed toward elements of the project, students are more likely to consider critically those elements they feel are under their control, or that they are capable of properly understanding/executing. Students who feel less sure of themselves are more likely to focus on highly specific, immediate aspects of the projects as separate tasks that they have to master, and to focus on achieving that mastery or improved under-standing: they seek improvements to their own practice, with improvement envisaged as better reproducing procedures and thinking defined by authority. Students who have developed a significant level of confidence with regards to their understanding of the project are more able to critically appraise it as an integrated whole, and seek improvements to its execution and scope for find-ings. Where critical thinking is directed toward the practice of individuals, the scope of criticality may be related to students' sense of relative equality, and therefore what they are eligible to judge and make suggestions about.

Precisely what triggers the development of enough confidence to think crit-ically in a previously unfamiliar research environment is not clear. It is an interesting but presently untestable question as to whether engaging in the blog process itself contributes toward students' sense that it is okay for them to reflect, make judgments, and speculate (although we may ourselves specu-late that the blogs provide a semi-private space where students feel more able to entertain such possibilities without fear of looking foolish or intellectually overreaching).

Implications for future research and practice

The blogs raise several interesting questions for future research. First, what does the critical thinking practiced by students look like in undergraduate research projects in other disciplines and fields? Can structures analogous to the stages illustrated in figure 28.1 be defined and compared? Second, what does such thinking look like in other learning contexts? For example, one might specu-late that more structured learning activities, perhaps explicitly focusing on the nature and practice of real science, might provide more and better oppor-tunities for metacritique to be exercised than research projects, since students might be encouraged to step back and view the discipline as a whole rather than find themselves deeply immersed within it. Finally, how easy is it for students to transfer their developing confidence and criticality to other areas or activities within the same discipline?

Our findings also have several implications for practice around the provi-sion and assessment of research projects. For one thing, our students may be thinking in surprisingly sophisticated ways about their research projects, but such thinking is hidden in most of our normal assessment processes. Where a final report would most likely start from the hypothesis that the student ended up with, the blogs allow us to see the processes by which hypotheses evolve and change in response to observations and active testing. Troubleshooting, liable to be unreported in formal scientific writing, is recorded as it happens, enabling us to see whether it has been undertaken in a purely procedural, uncritical way, or whether it has involved students critically evaluating their diagnostic and solution processes. Similarly, students would not typically explain the logic or even choice of standard techniques in a formal report, making it difficult to determine whether they followed them algorithmically or whether they reflected on and understood *why* they were doing what they were doing. Finally, our blogs reveal students reflecting on their own practice and that of others in a way that would be excluded in a report focused on the *results* of the project.

Our findings suggest that supervisors could look for and actively seek to encourage critical thinking directed toward a range of different objects, and at multiple levels. The possibility of students engaging in critical thinking, critical thought, and critique could be maximized by deliberate provision of opportunities to see science as more than procedural; by challenging students to go beyond instructions to think about why they are doing what they are doing (engaging in stage 1 of figure 28.1); by explicitly inviting them to choose between alternative techniques or approaches, or judge between different sug-gested interpretations (engaging in stage 2 of figure 28.1); by asking them to put forward their own ideas or suggestions (engaging in stage 3 of figure 28.1); and by deliberately finding ways to encourage a sense of relative equality, so that they are more likely to feel able to make critical judgments and put for-ward valid suggestions.

In these ways, perhaps we can provide opportunities for the capacity for crit-ical thinking to be exercised and strengthened, and encourage a propensity to use it as a constructive way of engaging with research. By helping students to see themselves as capable of learning enough to discern what is important, and make their own judgments, in contexts that initially appeared to be beyond them, we may make it more likely for them to believe themselves capable of similar development in future. That is, having gained enough confidence in their own knowledge and ability to analyze a situation or argument to think critically about a new idea/field once, we may hope that students will realize they are going to be able to do this again in other contexts.

References

Åkerlind, G. S. 2005. "Academic Growth and Development—How Do University Academics Experience It?" *Higher Education* 50 (1): 1–32.

Anderson, C., and Hounsell, D. 2007. "Knowledge Practices: 'Doing the Subject' in Undergraduate Courses." *Curriculum Journal* 18 (4): 463–478.

Atkinson, D. 1997. "A Critical Approach to Critical Thinking in TESOL." *TESOL Quarterly* 31 (1): 71–94.

Barnett, R. 1997. *Higher Education: A Critical Business*. Buckingham: Society for Research into Higher Education and the Open University Press.

Barrie, S. C., and Prosser, M. 2004. "Generic Graduate Attributes: Citizens for an Uncertain Future." *Higher Education Research & Development* 23 (3): 243–246.

Bean, J. C. 2011. *Engaging Ideas: The Professor's Guide to Integrating Writing, Critical Thinking, and Active Learning in the Classroom*. San Francisco: Jossey-Bass.

Brookfield, S. 1987. *Developing Critical Thinkers*. Milton Keynes: Open University Press

Bruce, C. S., Buckingham, L. I., Hynd, J. R., McMahon, C. A., Roggenkamp, M. G., and Stoodley, I. D. 2006. "Ways of Experiencing the Act of Learning to Program: A Phenomenographic Study of Introductory Programming Students at University." *Transforming IT Education: Promoting a Culture of Excellence*, edited by Christine Bruce, George Mohay, Glenn Smith, Ian Stoodley, and Robyn Tweedale. Santa Rosa, CA: Informing Science. 301–325.

Davies, M. 2006. "An 'Infusion' Approach to Critical Thinking: Moore on the Critical Thinking Debate." *Higher Education Research and Development* 25 (2): 179–193.

Ennis, R. H. 1987. "A Taxonomy of Critical Thinking Dispositions and Abilities." In *Teaching Thinking Skills: Theory and Practice*, edited by J. B. Baron and R. J. Sternberg. New York: W. H. Freeman. 9–26.

Ennis, R. H. 1992. "The Degree to Which Critical Thinking Is Subject Specific: Clarification and Needed Research." In *The Generalizability of Critical Thinking: Multiple Perspectives on an Educational Ideal*, edited by S. Norris. New York: Teachers College Press. 21–37.

Entwistle, N. 2009. "Teaching for Understanding at University. Deep Approaches and Distinctive Ways of Thinking." *Imprint* 11: 7.

Facione, P. 1990. *Critical Thinking: A Statement of Expert Concensus for Purposes of Educational Assessment and Instruction. Research Findings and Recommendations*. Newark, DE: American Philosophical Association.

Hale, E. S. 2010. *A Critical Analysis of Richard Paul's Substantive Trans-Disciplinary Conception of Critical Thinking*, Cincinnati, OH: Union Institute.

Jones, A. 2007. "Multiplicities or Manna from Heaven? Critical Thinking and the Disciplinary Context." *Australian Journal of Education* 5 (1): 84–103.

Marton, F. 1981. "Phenomenography—Describing Conceptions of the World around Us." *Instructional science* 10 (2): 177–200.

McCune, V., and Hounsell, D. 2005. "The Development of Students' Ways of Thinking and Practising in Three Final-Year Biology Courses." *Higher Education* 49 (3): 255–289.

McPeck, J. 1981. *Critical Thinking and Education*. New York: St Martin's Press.

McPeck, J. 1992. "Thoughts on Subject Specificity." In *The Generalizability of Critical Thinking: Multiple Perspectives on an Educational Ideal*, edited by S. Norris. New York: Teachers College Press.

Medawar, P. B. 1963. "Is the Scientific Paper a Fraud." *The Listener* 70 (1798): 377–378.

Moore, T. 2011. "Critical Thinking and Disciplinary Thinking: A Continuing Debate." *Higher Education Research and Development* 30 (3): 261–274.

Paul, R. 2005. "The State of Critical Thinking Today." *New Directions for Community Colleges* 2005 (130): 27–38.

Pithers, R. T., and Soden, R. 2000. "Critical Thinking in Education: A Review." *Educational Research* 42 (3): 237–249.

29

Using Social Media to Enhance Critical Thinking: Crossing Socio-Educational Divides

Nancy November

Introduction

Imagine the case of a first-year music student who is writing a review of an all-Beethoven concert that she attended the previous night (a typical assignment type for a music undergraduate). The student should first have carried out background research (reading/listening) on the performers and repertoire before attending the concert and taking notes. Ideally she will now not only summarize her experience, but will also take the critical "step back," to reflect on the event and offer evaluative comments that are supported by evidence. This will involve consideration of various parameters (musical, visual, social, etc.) and perspectives (audience reception, performers' expression and style, comparison with similar ensembles, etc.). Even if the repertoire is familiar, the task is complex, especially the reflective evaluation based on multiple perspectives. Typically students stumble over the latter, so that the shift out of descriptive mode, and out of the frame of their own unexamined reactions, never fully takes place.

For the finest critical thinkers on music, however, this stepping back is complete, the critical distance fully achieved. Examples include the authors of books such as *Boccherini's Body* (Le Guin 2006), *Bach's Feet* (Yearsley 2012), and *Schubert's Vienna* (Erickson 1999), who use, respectively, the ideas of corporeality, physicality, and perspectives on Vienna to illuminate musical subjects. These writers show how viewing a musical subject from outside the discipline can be an ideal way to achieve new insights and shift paradigms. As the first two titles suggest, music scholars also build on each other's approaches, often with a view to positioning themselves critically in relation to the discipline and its established knowledge.

So much for the experts: how can one encourage such lateral and reflective thought, and thoughtful collaboration, in undergraduates—who, after all, struggle with the basic mechanics of writing, and who are just starting to become literate within the discipline? It hardly seems reasonable to ask them to take a leap into yet another discipline.

There is a field, though, in which today's undergraduates have, in general, a high degree of competence already: that of social media. I will argue that, through their digital literacy, today's students often possess skills, attitudes, and conceptual understandings that are highly relevant for developing critical thinking dispositions. These include a readiness for travel (with ideas, subject, approaches), a strong sense of self in relation to an (online) audience, and openness to multiple perspectives. Educators can build on students' digital literacy to improve subject-specific literacy; but more significantly they can build on students' typical online behaviors and thinking dispositions to encourage modes of thought that are associated with critical thinking. Social media prove especially useful for helping students to synthesize knowledge, engage in knowledge-extending collaborations, self-reflect, and in general move outside the frame of their own unexamined opinions. I will discuss why this is so, and provide some case studies drawn from the undergraduate teaching of educators in three disciplines: music, creative writing, and population health.

Defining terms: critical thinking and social media

First, two key terms must be defined, critical thinking and social media, both of which have proven difficult to describe simply. Definitions of critical thinking are arguably the most useful when they are various and complex, that is, when one does not merely focus on the products of critical thinking, but acknowledges that it is a process. This chapter uses a new, threefold definition of critical thinking that encompasses (1) *attitudes*, such as openness to multiple ideas, flexibility/readiness to travel laterally with those ideas, and a desire to share knowledge; (2) *actions*, such as travelling with ideas, collaborating to obtain multiple viewpoints, making a judgment based on evidence, synthesizing complex data, stepping back to reflect; and (3) *outcomes*, such as obtaining distance, shifting paradigms, and having new insights.

This broad, process-oriented definition has several implications for educators. First, mindful of process, educators need to think about how to promote the *mindsets* in which broader and higher-level critical states can occur, and encourage a long-term propensity to these kinds of critical thought (Perkins, Jay, and Tishman 1993). This idea of establishing mindsets will be kept in the foreground when looking at how social media can help with critical thinking.

Then, one should also consider the objectives of critical thinking: what are the new insights or paradigm shifts about? From where is distance obtained? In promoting well-rounded critical thinking in students, educators need to think in terms of a suitably broad definition of these objectives of critical thought. They need not necessarily be confined to the knowledge of a particular subject area (e.g., critical thinking about the style of Beethoven's music). Students' critical thinking can encompass "their experiences in relation to knowledge, themselves and the world" (Barnett 1997, 109). Achieving such capacious criticality might seem daunting: it implies that teachers need to think of ways of promoting critical thought within, without, and about the discipline. However, one can start by ensuring that students are progressively developing an awareness of the discipline's discursive frameworks and constraints; the domains of application of its models; and its myths, tropes, and truisms (e.g., the popular myth of Beethoven as Western Classical music's "hero").

In achieving this well-rounded criticality, the educator's own approach is crucial: the messages that instructors send to students through traditional modes of teaching and research within the disciplines can inhibit students' development of desired critical perspectives. For instance, one can more easily convince students that multiple perspectives are vital for a well-rounded view of music history if well-structured opportunities for collaborative research are provided as an important component of courses, rather than primarily teaching in a (still typical) lecture-heavy fashion that promotes the view of lecturers as "oracles" and students as solitary "receptacles" of knowledge. Enabling social media to play a role encourages a more collaborative model of knowledge exchange and production, so that students are more aware of the contestable nature of knowledge.

We need a working definition of social media before considering in more detail how they can be used to promote critical thinking. Social media are means by which people create, share, exchange, synthesize, analyze, and critique information and ideas online, in virtual communities. These media are based on the developments in information and communication technologies known as Web 2.0, the so-called read-write web, meaning that they allow for—and indeed promote—multiple forms of interaction, rather than simply permitting passive viewing.

Six different types of social media have been identified (Kaplan and Haenlein 2010): collaborative projects (e.g., Wikipedia), blogs and microblogs (e.g.,Twitter, Blogger), content communities (e.g., YouTube, Scoop.it, Pinterest, Delicious), social networking sites (e.g., Facebook, Bebo, Google+), virtual game worlds (e.g., *World of Warcraft*), and virtual social worlds (e.g., Second Life). However, the boundaries between the different types have been increasingly blurred, and social media technologies (blogs, wall postings, music sharing, voice communications, image sharing, etc.) are increasingly being integrated through social network aggregation platforms.

In higher education, the most significant line that has been blurred, given the new affordances of Web. 2.0, is that between the social and educational spheres. Specific online tools, environments, and platforms have been designed with online learning (or e-learning) in mind, such as Moodle (an e-learning platform), Google Docs (for sharing and editing documents online), and Sloodle (a combination of Second Life and Moodle). Like the mainstream social networking tools and platforms, these virtual learning environments emphasize interaction, but they can be less accessible, or inaccessible, to those outside the class or institution, depending on settings and servers. In practice, and despite a good deal of debate and equivocation (see Roblyer, McDaniel, Webb, Herman, and Witty 2010, and below), increasingly many educators are integrating mainstream social networking sites like Facebook, and open-source software like Elgg (a social networking engine), into their courses.

Social media and critical thinking: an odd couple?

In this chapter it is not so much social media themselves but the way these media are used that is of interest. The argument here is that teachers can support students' journeys toward critical thinking, and achieve the desired outcomes or products of critical thinking, by taking advantage of *processes* that are involved in online social interaction. I am thus focusing on the attitudes and actions that students associate and carry out with social media.

At a basic but important level, social media can help educators to engage students in a sustained way, since students often have a desire to be online, coupled with well-developed digital media skills. Edmundson (2008) has characterized the online environment, from the student perspective, as a highly desirable space of unlimited potential, in which one takes pleasure in skimming through sites, subjects, and conversations. In the past the so-called Net generation (Tapscott 1998) has been most marked in the United States, but today students in many parts of the world are considered to be "digitally minded." Research into university students in the United Kingdom, Romania, Finland, and Hungary has revealed high levels of technology usage and competence, and a tendency to integrate the latest communication, information, and management technologies into their daily lives (Andone, Dron, Pemberton, and Boyne 2007).

College and university teachers have tended to assume that students' positive attitudes to and apparent aptitude for being online is a useful base for building their knowledge (Gonzales 2010; Hannon 2009). However, the translation from social to educational usage of digital media is not so straightforward. There has been active debate of the notion that today's students can be described as "digital natives" (see especially Bennett, Maton, and Kervin 2008; Helsper and Eynon 2009). Brookfield's (2006) injunction to "know thy students" applies here: the background of a given group of students will certainly vary, and they may have strengths and weaknesses in unexpected areas. Recent research helps with this issue, by developing tools for assessing students' digital skills (e.g.,Teo 2013). Research into undergraduate student cohorts at the University of Auckland (November and Day 2012) included course entry and exit "digital literacy" surveys and these produced results that fit with the general international trends cited above: the statistics show high levels of online usage and capability. However, some of the figures were unexpected. Ninety-five percent of students in a 2010 first-year population health paper (comprising 37 students) used a mobile music player—a figure unexpectedly high and comparable to that for music students. And an unexpectedly high number, 32%, of students in a 2010 first-year music paper (comprising 114 students) had taken a basic course in computer science (compare 10% in the health class). These data, as well as high Facebook and YouTube usage (91% and 97% in health and music respectively for both online platforms), provided the lecturers with background when designing courses that would engage students and build on existing digital skills.

Students, then, often possess digital literacy skills in relevant or surprising areas and show a high degree of use of social media. But what about their propensity to use that digital literacy in the service of critical thinking? In fact the use of social media to educational ends has been sharply criticized from various standpoints. Particularly important here is the idea that "being online" actually discourages critical thinking. The tantalizing yet daunting abundance of Internet hits on any given topic (e.g., "Beethoven" in Google gives 44,800,000 results in 0.33 seconds) is thought to inhibit "Net Generation" students from making critical judgments and indeed in thinking at all (Kek and Huijser 2011; Oblinger and Oblinger 2005). Edmundson (2008, B7) voices this concern when he wants students to move away from the online environment in order that they "stop and think." Educators like Andone et al. (2007) hope to use their students' enthusiasm to enhance and stimulate learning. Yet, like Edmundson (2008), they are concerned that "the trend away from predominately analytic knowledge towards primarily synthetic knowledge implies a loss as well as a gain" (52).

However, other educators have shown that being online does not necessarily mean dwelling in a limbo of possibilities, nor does it mean moving away from the analytical realm. A growing body of research has advocated social media to promote skills such as decisiveness and analytical judgment (see Ng, Goi, and Gribble 2008). One can also question the criticism of synthetic knowledge in this context, which is arguably more important than analytical knowledge for developing critical thinking. As traditional wisdom has it, there are two forms of intellectual tasks: making the simple complex and making the complex simple. Modern scholarship often focuses on the former, while the latter (synthesizing knowledge, reduction, abstraction) is arguably more difficult and more powerful. Thus, for example, an important first step in becoming critical of traditional views of Beethoven is to step back and see emerging themes, trends, and tropes in writings about this complex historical figure.

Most importantly social media enable a potentially crucial component in students' critical thinking processes, and one that has not been well emphasized in the past, especially in tertiary-level humanities education: collaboration. Social constructivism underlies theories of collaborative learning, and social constructivist thinkers have advocated: "Information must be shared, critically analyzed, and applied in order to become knowledge" (Garrison 1993, 201). This perspective has predominated in recent educational literature, particularly in the works of scholars who have advocated the development of what are known as "communities of practice" in higher education (a group that shares a craft or profession and the members of which learn from each other through sharing information and experiences; see Wenger, McDermott, and Snyder 2002). These educators have found that collaborative work not only yields a superior quantity and quality of knowledge, but also provides vital opportunities for students to engage in high-level thinking tasks such as judgment and reflection.

The online environment has been found to be something of an ideal place in which to foster this productive collaborative work (Gabriel 2004). The key disadvantages that have been documented for online asynchronous learning environments include the absence of visual cues, tone of voice, and immediacy of response, which can hinder good communication (Haynes 2002). However, the literature on cultures of online interaction suggests that for students who engage daily in online sociability, and for whom the boundaries between "face-to-face" and online experience are less sharply etched, these factors might not be perceived as problems (Anderson and Simpson 1998; Curtis and Lawson 2001; Flottemesch 2001; Gabriel 2004; Graham and Misanchuk 2004). As these writers observe, in asynchronous learning environments learners experience significant advantages in communication, such as time to think before responding, space to spell out one's ideas in words and opportunities to reread, reflect, and refine ideas. Even if asynchronous online environments are found problematic, the latest social media mitigate this by means of Internet telephony and audio/video file sharing.

Case studies: using social media to promote critical thinking

In *Higher Education: A Critical Business* (1997), one of Barnett's central messages to teachers is to stop trying to teach critical thinking, and provide opportunities for students to enact it. The most effective pedagogies, regardless of the discipline, take seriously the process nature of critical thinking, and place the students at the center of the process. Salmon's (2000; 2002) work demonstrates how to structure and channel online sociability to not only encourage students into discussion, but also lead them toward higher-order tasks. This structured approach, implied in the case studies below, is essential for helping students bridge what social constructivists term the "zone of proximal development" (Vygotsky 1978), the gap between that which they have already learned, unassisted, and that which they can achieve when provided with educational support.

This space, or zone, enables new levels of criticality (Wass, Harland, and Mercer 2011), not least because it lies outside a student's accustomed intellectual comfort zone. It can be reached through students' dialogue and conversation (with each other, with the lecturers, with the material), and through problem-based tasks that allow them to practise being a researcher. Researchers have shown that these student-driven, dialogic approaches can be far more effective than having students work alone through materials geared toward knowledge acquisition (Jonassen, Howland, Moore, and Marra 2002; Stahl, Koschmann, and Suthers 2006). Why? Because in order to attain high-level modes of thinking students often need to change their views of knowledge—from something static and non-contestable, to something to which they can contribute, on which they can reflect, and of which they are a part—and firsthand experience of knowledge creation and critique is required to make this shift in perspective effectively. In proposing a theory of online learning, Anderson suggests that "deep and meaningful formal learning is supported as long as one of the three forms of interaction (student-teacher; student-student; student-content) is at a high level. The other two may be offered at minimal levels, or even eliminated, without degrading the educational experience" (2008, 67). Social media are especially helpful in encouraging meaningful student-student and student-content interactions, as is demonstrated in the following case studies.

Synthesizing multiple perspectives in music in online discussions

Online discussion groups can provide an ideal place for the development of student-led questioning and evaluation; this can lead to high-level critical thought, including synthesis of multiple perspectives. Evidence of this is found in student data from surveys carried out before and after a first-year course in music history, "Turning Points in Western Music," which was run at the University of Auckland in 2008. The 123 students engaged in collaborative steps, carried out online, which led to the production of individual essays discussing the significance (in terms defined by the students) of selected landmark sound recordings in Western music history. Students were involved in various online interactions, including the group compilation of an annotated resources list, discussion in small groups (5 students or less), then critical reflections posted in large groups (around 20 students). After the course was over, the number of students who agreed or strongly agreed that interacting online helped them to learn had risen from 44% to 81% (123 and 103 student respondents to initial and final surveys, respectively). They were then asked which of the online interactions was most beneficial to their learning. The answer was group discussion, the activity involving the most action and interaction.

Why was action and interaction prized? The students valued the sharing of perspectives in motivating them to act critically. One student voiced a common sentiment, "In order to critically and effectively answer the question [posed by another student], a lot of research and thinking was required. This is valuable for the essay writing' (Student A, 2008). Students appreciated the division of labor, but also the capacity to create and share new ideas in the group discussion. Levels of interactivity were far lower in a fully offline version of the course, run in 2009, partly due to the fact that the only designated discussion times were in class; on the other hand, the 2008 students appreciated the ability to choose their time and place for online interactions. The higher level of interactivity in online discussion was not just a matter of convenience: there is an established culture of interacting online into which the teacher can tap, a desire to go online to read and respond, not only to make one's mark, but to connect with others, or "write on their walls," in the language of Facebook. The analogy is apt: 72% of students in the 2008 class used online communication via email, wikis, chatting, and blogs at least once daily, and 82% had registered with Facebook, Bebo, or another form of social communication.

The critical reflections phase, in which students posted to large groups their thoughts on what they had learned and experienced in the small-group discussions, engendered some of the highest-quality work. Many of the responses were lengthy and comprehensive, with analysis of various viewpoints and clearly worded conclusions. This decisive, action-oriented quality of the students' online discussion can be contrasted with characterizations of the online world as "limbo-like" and anti-analytical. Indeed, one can build on students' desires to respond online, feeding in new analytical affordances of the web that help students to synthesize multiple layers of feedback. The 2010 iteration of the course incorporated two online text analysis tools: Wordle (http://www.wordle.net/) and Helen Sword's Wasteline Test (http://www.writersdiet.ac.nz/wasteline.html). Students responded to each other's analyses; tried improving their own writing using the feedback; and finally commented on their peers' work. Another new step introduced in the 2010 course was peer reviewing of the students' final essays using the online peer review system Aropä (http://aropa.ec.auckland.ac.nz/src/aropa.php). The 2010 class average for the final essay was 81.63% (A-), a marked improvement from 2008, which was 74.08% (B). This was closely related to these students' improved abilities to synthesize multiple perspectives, an important part of the critical thinking process.

Using online presence to encourage critical reflection in creative writing

A course involving creative writing provides a good example of how students' heightened awareness of their presence in social media can be used to develop their critical reflections. "Writing and Audience" is a fully online second-year course run at the University of Waikato (approximately 45 students).

A fully online course, deploying asynchronous discussion groups, allows the instructor to promote the kind of critical reading that is central to the effective creative writer. At the most basic level, the fact that the entire course is conducted in writing and reading necessarily leads to improved skills in both areas, since students and teacher have to find ways to make themselves well understood, or at least understood well enough, in the absence of sonic and visual cues. They have to become more sensitive readers—an important aspect of critical thinking (Weinstein 2000).

Further, being online automatically gives students a sense of audience in their interactions, so that they need to think about how to situate themselves in relation to others, as they do in social media like Facebook. The course instructor for "Writing and Audience" observes: "The act of writing becomes a part of what they are doing... how they project themselves becomes very important to them." This ability to project oneself to an audience is crucial to any course that develops written communication, but often involves breaking a threshold in the students' understandings of how this works in practice. They struggle at the task: "It's a real jump for students to move from writing as a private activity," the instructor notes. "[If one want wants to become a professional creative writer] it's very important to break that barrier between private writing and actually saying 'well actually I'm a writer and I'd like to get it out there.'"

The course uses students' familiar (comfortable) modes of online interaction to break through this threshold. Students engage in two crucial activities that are quite unpalatable or unfamiliar to many undergraduates: considering writing as a slow, developmental process involving loops of self-reflection and feedback from others, and giving constructive negative feedback. But they carry out both processes by drawing on known spaces and modes of online interaction: the "blog" for a private exchange of ideas; the structured conversation for development of ideas and feedback; and the large group forum for presentation and praise. Further, the course draws on role-playing, familiar to many students from their online interactions, especially reader and commentator roles. Crucially, students are asked to frame peer feedback firmly as readers, rather than writers. Question prompts help frame critique, which helps them to gain perspective on their own writings: is the writer making the most of context to enhance the logic of the characters' actions? What's going on with time in this scene? More centrally: how does a shift of perspective from first person to third person allow this writer to move away from the confessional (subjective outpouring) toward more public ground in projecting the narrative? After this semiprivate online workshop, a creative writing webpage for the course provides a "public" forum for presentation and further feedback.

The course uses the "distanced" nature of the online asynchronous environment as a virtue to support the teaching of the "critical distance" that is so central in the development of the professional creative writer. The shift in roles from writer to reader, aided by the asynchronous interactions, helped students to step back from a subjective and simplistic expression of likes and dislikes to consider the process of communication in writing. Thus students moved through their "zones of proximal development": they obtained a critical perspective on themselves as writers that they would not have achieved alone.

Revising and rewriting: science knowledge building and critique on the Net

Wiki-based course glossaries provide a powerful means by which students can understand and reflect on the discourse of a new discipline, and start to see the constructed nature of a discipline's knowledge. Online class glossaries were used intensively by health students in a 50-student first-year course run at the University of Auckland in 2011. These glossaries were to be filled out by the students themselves, and helped them to come to terms with new disciplinary concepts as they arose in the lectures and course materials.

Course wikis such as these draw strongly on a pronounced behavior in students' social media interactions: the desire to travel with a concept or idea, which can involve extensive collaborative revising and rewriting (Bruce and Payton 1990; Docring 2007). In the health class, the glossary exercise led to a rich online resource with multiple perspectives on each glossary term. On average there were four postings per term or concept, and each term was viewed on average 43 times, with the number of views per term ranging from 8 to 149 (the 149 views related to the definition of depression, to which a lively discussion was attached). The glossary was a useful way for students to attempt definitions of words that would not normally feature in their everyday conversations. The postings were mostly informal: a mix of their vernacular use of medical terms. But gradually attempts at a more discipline-specific way of expressing their understandings emerged.

This wiki encouraged collaborative learning in multiple forms. Some students were testing their learning by asking questions, others were sharing their learning by exchanging comments and references, while yet others were simply observing and learning without making a contribution to the conversational

Figure 29.1 Social media and the critical thinking process.

thread. As in the music online discussion forums, students valued highly the sharing of multiple perspectives. One health student observed: "I liked the aspect of the [glossary] assignment that we could read all the other students' work and see the type of things that they were writing about and see the way they had written their assignment, and I think it was helpful for later work because I could then add to my explanations by kind of comparing other people's work" (Student A, 2011). In the very best cases, students extended themselves, showing higher-level self-reflection. For example, one health student observed: "Creating a glossary entry about a topic I was unfamiliar with challenged me to research it and associate terms more thoroughly, which broadened my knowledge. Writing a glossary comment reinforced this" (Student B, 2011).

Conclusion: exploiting social media for critical thinking

The foregoing case studies show that social media, far from inhibiting critical thinking, can be used to promote it. The flowchart in figure 29.1) presents a summary of what can be learned from this. Certain attitudes of students using social media (the left-hand box) can be turned to educational ends, to encourage socially based actions related to the critical thinking process: a desire to travel with ideas; a strong sense of online presence and self presentation; and an openness to multiple viewpoints. Especially important here is the desire for connectivity (meaning, in this context, online connection to each other and the larger Internet community), which leads easily to collaboration and is a crucial motivator for today's students. Connectivity lies at the heart of educational social media tasks that are designed as a carefully sequenced set of steps (Bruce and Payton 1990). While the box of this "attitude" can be considered a point of departure for designing these tasks, in practice the first step will always be to find out about the social media usage and attitudes of one's particular set of students.

This flowchart captures the process nature of critical thinking, with an emphasis on the actions and interactions (middle box) that can help to shift students through their zones of proximal development: travelling with ideas, synthesizing multiple viewpoints, judging based on evidence, and stepping back to reflect. What follows are comments and advice for teachers who are designing assignment sequences to encourage these actions, and the critical thinking to which they can lead.

Travelling In the glossary and online collaborative tasks, students can build on an enthusiasm to explore a wide range of content, travelling through formal readings, less-formal blogs and fan websites, and their peers' writings (Doering 2007). Students enjoy a high degree of flexibility in their online recreational worlds, but when making the shift to online work within a given discipline, a great degree of choice can engender anxiety. It is necessary to support them in the process of reflecting on and critiquing the resources that they use, for example, to model productive lines of questioning with prompts for discussion. In wiki glossary assignments, it is ideal to use a combination of student- and teacher-supplied terminology, and to make clear links between the teacher's use of discipline-specific vocabulary in the classroom and the glossary postings.

Role-playing When helping students to shift from the expression of subjective, unsubstantiated opinions into higher-order reflection and critique, and more formal writing, it proves ideal to give them teacher roles (as in the formal commenting and online peer reviewing tasks). These empower them as they develop a more academic voice and public presence (Barnard, Lan, To, Paton, and Lai 2009). As Garrison et al. (1999) observe, learning transactions comprise social presence, cognitive presence, and teaching presence, where the latter can be ably provided by the students, given the right framework. Helping students develop teaching roles within the social media framework alleviates the oft-cited problem that students resist the mixing of social and educational spaces. The fact is that students need to be able to see what to do with the social media, and why. When there are clear-cut, productive educational roles to try and the learning outcomes are explained and readily apparent, they understand the benefits they will receive from the socio-educational crossover, and appreciate the familiarity, flexibility, and new levels of insight (McDonald 2008).

Synthesis of multiple views As Warnock discusses, "Online, course 'talk' can become writing" (2009, 68–93). Mutatis mutandis, writing is a form of learning (Richardson 2003). Online writing and responses within a wiki glossary framework are ideal for airing and building on discipline-specific knowledge, and as a first step toward getting students to reflect critically on the discourse of the discipline. Glossaries allow for usefully multifaceted viewpoints (Dirckinck-Holmfeld and Lorentsen 2003), which encourage students to see knowledge as constructed, dynamic, and contestable, rather than static and reified.

Self-reflection Self-reflection and responsibility for learning can be achieved by encouraging students to ask their questions, develop their criteria for assessing data, and then reflect on their processes of enquiry and knowledge construction. This is arguably the most essential action in the critical thinking process: it perpetuates the process by encouraging students to see what is in it for them.

In summary, assignment design for improving and extending student critical thinking can be multilayered and designed to enable collaborative learning, such as that which is already taking place, albeit more casually, in our students' online worlds. This may seem a nice efficiency on the one hand, but on the other, the process of putting even a single unit from a single course into a social media framework can be extremely time consuming. The following points can help enhance teaching efficiency in this context:

Troubleshoot The technology usage should not pose a new barrier to student learning. Choose platforms that are fast, intuitive, robust, stable, and trustworthy for optimal usefulness.

Use peer review Integrate online tools and techniques that involve peer feedback and review, with appropriate attention to assessment rubrics (see Hamer, Kell, and Spence 2007, for example).

Know thy students Carry out a survey at the start of a course to find out which social media are the most used and best understood by a given cohort of students.

Know thy neighbor Many of the teaching practices described above translate across the disciplines, so that one can borrow the most effective and efficient practices of others, and collaborate on teaching-based research to create new pedagogies of critical thinking.

Acknowledgments

I would like to thank Adam Blake and Barbara Grant for useful advice about and resources for this chapter, my research assistant Aleisha Ward, and educators Karen Day, Jan Pilditch, and Tom McFadden for sharing their insights on teaching. Research for this chapter was funded through a Fulbright New Zealand Alumni Association alumni-initiated project grant.

References

Anderson, B., and Simpson, M. 1998. "Learning and Teaching at a Distance: A Social Affair." *Computers in NZ Schools* 10 (1): 17–23.

Anderson, T. 2008. "Towards a Theory of Online Learning." In *Theory and Practice of Online Learning*, edited by T. Anderson. Edmonton, AB: AU Press.

Andone, D. M., Dron, J., Pemberton, L., and Boyne, C. W. 2007. "E-Learning Environments for Digitally-Minded Students." *Journal of Interactive Learning Research* 18 (1): 41–53.

Barnard, L., Lan, W. Y., To, Y. M., Paton, V. O., and Lai, S.-L. 2009. "Measuring Self-Regulation in Online and Blended Learning Environments." *Internet and Higher Education* 12 (1): 1–6.

Barnett, R. 1997. *Higher Education: A Critical Business*. Buckingham: Society for Research into Higher Education and the Open University Press.

Bennett, S., Maton, K., and Kervin, L. 2008. "The 'Digital Natives' Debate: A Critical Review of the Evidence." *British Journal of Educational Technology* 39 (5): 775–786.

Brookfield, S. 2006. *The Skillful Teacher: On Technique, Trust, and Responsiveness in the Classroom*. San Francisco: Jossey-Bass.

Bruce, B., and Payton, J. K. 1990. "A New Writing Environment and an Old Culture: A Situated Evaluation of Computer Networking to Teach Writing." *Interactive Learning Environments* 1 (3): 171–191.

Curtis, D. D., and Lawson, M. J. 2001. "Exploring Collaborative Online Learning." *Journal of Asynchronous Learning Networks* 5 (1): 21–34.

Dirckinck-Holmfeld, L., and Lorentsen, A. 2003. "Transforming University Practice through ICT—Integrated Perspectives on Organizational, Technological, and Pedagogical Change." *Interactive Learning Environments* 11 (2): 91–110.

Doering, A. 2007. "Adventure Learning: Situating Learning in an Authentic Context." *Innovate: Journal of Online Education* 3 (6).

Edmundson, M. 2008. Dwelling in Possibilities. *The Chronicle of Higher Education*, March 14.

Erickson, R. 1999. *Schubert's Vienna*. Vienna: Lehner.

Flottemesch, K. 2001. "Building Effective Interactions in Distance Education: A Review of the Literature." In *The 2001/2002 ASTD Distance Learning Yearbook*, edited by K. Mantyla and J. A. Woods. London: McGraw-Hill. 46–61.

Gabriel, M. A. 2004. "Learning Together: Exploring Group Interactions Online." *Journal of Distance Education* 19 (1): 54–72.

Garrison, D. R. 1993. "A Cognitive Constructivist View of Distance Education: An Analysis of Teaching-Learning Assumptions." *Distance Education* 14 (2): 199–211.

Garrison, D. R., Anderson, T., and Archer, W. 1999. "Critical Inquiry in a Text-Based Environment: Computer Conferencing in Higher Education." *The Internet and Higher Education* 2 (2): 87–105.

Gonzales, C. 2010. "What Do University Teachers Think Elearning Is Good for in Their Teaching?" *Studies in Higher Education* 35 (1): 61–78.

Graham, C. R., and Misanchuk, M. 2004. "Computer-Mediated Learning Groups: Benefits and Challenges to Using Groupwork in Online Learning Environments " In *Online Collaborative Learning: Theory and Practice*, edited by T. S. Roberts. London: Information Science. 181–202.

Hamer, J., Kell, C., and Spence, F. 2007. "Peer Assessment Using Aropä." *Australian Computer Society*. http://www.cs.auckland.ac.nz/~j-hamer/peer-assessment-using-Aropa.pdf.

Hannon, J. 2009. "Breaking Down Online Teaching: Innovation and Resistance." *Australasian Journal of Educational Technology* 25 (1): 14–29.

Haynes, D. 2002. "The Social Dimensions of On-Line Learning: Perceptions, Theories, and Practical Responses." Paper presented at the Distance Education Association of New Zealand Conference, Wellington, New Zealand, April 10–12.

Helsper, E. J., and Eynon, R. 2009. "Digital Natives: Where Is the Evidence?" *British Educational Research Journal* 36 (3): 503–520.

Jonassen, D. H., Howland, J., Moore, J., and Marra, R. M. 2002. *Learning to Solve Problems with Technology: A Constructivist Perspective.* 2nd ed. Columbus, OH: Merrill/Prentice-Hall.

Kaplan, A. M., and Haenlein, M. 2010. "Users of the World, Unite! The Challenges and Opportunities of Social Media." *Business Horizons* 53 (1): 59–68.

Kek, M. Y. C. A., and Huijser, H. 2011. "The Power of Problem Based Learning in Developing Critical Thinking Skills: Preparing Students for Tomorrow's Digital Futures in Today's Classrooms." *Higher Education Research & Development* 30 (3): 329–341.

Le Guin, E. 2006. *Boccherini's Body.* Berkeley: University of California Press.

McDonald, A. 2008. "Facebook in the Classroom: Integration of Online and Classroom Debates into Courses." http://akoaotearoa.ac.nz/download/ng/file/group-3300/facebook-in-the-classroom-integration-of-online-and-classroom-debates-into-courses.pdf.

Ng, P. Y., Goi, C. L., and Gribble, S. J. 2008. "Adaptation of Google Group for Online Teaching and Learning." In *HERDSA 2008: Engaging Communities.* Milperra, NSW: HERDSA.

November, N. R., and Day, K. 2012. "Using Undergraduates' Digital Literacy Skills to Improve Their Discipline-Specific Writing: A Dialogue." *International Journal of the Scholarship of Teaching and Learning* 6 (2): Article 5.

Oblinger, D., and Oblinger, J. 2005. *Educating the Net Generation.* Edited by D. Oblinger and J. Oblinger. Louisville, CO: Educause.

Perkins, D., Jay, E., and Tishman, S. 1993. "Beyond Abilities: A Dispositional Theory of Thinking." *Merrill-Palmer Quarterly* 39 (1): 1–21.

Richardson, L. 2003. "Writing: A Method of Inquiry." In *Collecting and Interpreting Qualitative Materials*, edited by N. K. Denzin and Y. S Lincoln. Thousand Oaks: Sage. 499–541.

Roblyer, M. D., McDaniel, M., Webb, M., Herman, J., and Witty, J. V. 2010. "Findings on Facebook in Higher Education: A Comparison of College Faculty and Student Uses and Perceptions of Social Networking Sites." *Internet and Higher Education* 13: 134–140.

Salmon, G. 2000. *E-Moderating: The Key to Teaching and Learning Online.* London: Kogan Page.

Salmon, G. 2002. *E-Tivities: The Key to Active Online Learning.* London: Kogan Page.

Stahl, G., Koschmann, T., and Suthers, D. 2006. *Computer-Supported Collaborative Learning: An Historical Perspective.* Cambridge, UK: Cambridge University Press.

Tapscott, R. 1998. *Growing up Digital: The Rise of the Net Generation.* New York: McGraw-Hill.

Teo, T. 2013. "An Initial Development and Validation of a Digital Natives Assessment Scale (DNAS)." *Computers & Education* 67: 51–57.

Vygotsky, L. S. 1978. *Mind in Society: The Development of Higher Psychological Processes.* Cambridge, MA: Harvard University Press.

Warnock, S. 2009. *Teaching Writing Online: How and Why.* Urbana, IL: National Council of Teachers of English.

Wass, R., Harland, T., and Mercer, A. 2011. "Scaffolding Critical Thinking in the Zone of Proximal Development." *Higher Education Research & Development* 30 (3): 317–328.

Weinstein, M. 2000. "A Framework for Critical Thinking." *High School Magazine* 7: 40–43.

Wenger, E. C., McDermott, R., and Snyder, W. C. 2002. *Cultivating Communities of Practice: A Guide to Managing Knowledge.* Boston, MA: Harvard Business School Press.

Yearsley, D. 2012. *Bach's Feet: The Organ Pedals in European Culture.* Cambridge, UK: Cambridge University Press.

Part VII

Social Perspectives on Critical Thinking

In being critical, one is critical of something. Initially, in thinking about critical thinking, attention focused on propositions and arguments. To be critical was to attend carefully to the logical substance and rigor of propositions and arguments. But it was observed subsequently that, quite often, in exercising critical thought, one is critical of political or economic or social systems. This observation, therefore, profoundly opens the scope of critical thinking.

But there is a further twist, and a crucial twist at this, in this story. For there has been a steady stream of philosophical and social theoretical scholarship that has drawn attention to the social embeddedness of thought. Thought does not exist in the ether but emerges out of a particular constellation of interests, perspectives, and power-laden structures. And a yet further strain here, including those of Marx, the Frankfurt School of Critical Theory, and South American liberationalists (such as Friere), insists that not infrequently thought is contaminated, prejudiced, or corrupted by social interests.

The very notion of ideology, after all, refers to structured claims about the world that are systematically related to social interests. Ideology may offer us worthwhile and profound insights; it is not always pernicious. But the key point about ideology is that it is, at best, a partial view of the world, a partiality that not merely reflects but furthers certain interests, and characteristically interests that spring from positions of power.

The literature on critical thinking, accordingly, moved to embrace such social perspectives. The educational perspective ran something like this. If thought is often unduly distorted by its social context and social origins, a higher education in particular cannot but address the matter head-on. There is a link here with the earlier project of (Western) Enlightenment, in which it was considered that through the use of reason, humanity could obtain a measure of liberation from its self-enclosed (or more accurately its socially enclosed) fetters of the mind. Now, on this more critical perspective, it became imperative that a higher learning take on a responsibility for enabling students not just to see, but to see through, the ways in which thought was often far from pure and objective but tainted. Accordingly, higher education was enjoined here to take on a social-critical aspect, to enable students to gain some freedom (liberation indeed) from the myopia to which they might otherwise be in thrall. Perhaps the key term in this school of thought came to be that of "emancipation": critical perspectives could enable students to gain a measure of emancipation from the myopia of conventional frames of thinking.

This was a profound moment in the history of critical thinking as a pedaogogical matter. For now, instead of being concerned with texts as they presented themselves on the page, attention was to be paid to the social circumstances from which texts—in whatever form—had emerged. On this view, to be really critical was to be critical of the situations that had given rise to widespread and deeply held views and beliefs across society. This was a near-inversion of what had been the dominant view of critical thinking, from an internal orientation focusing on academic texts to an external orientation, focusing on social situations and social structures. Inevitably, this latter stance took on a political edge, being characteristically concerned to promote democracy, citizenship, human rights, and social justice. Critical thought, accordingly, came to be thought that sought to lay bare associations between power structures' systematic beliefs across society.

The chapters in this section have different stances in relation to this cluster of views on critical thinking. For Volman and Ten Dam, critical thinking offers a way of helping to promote citizenship among students. Setting off their proposals against "socio-constructivist approaches," Volman and Ten Dam argue for a "socio-cultural perspective." Key here is that students participate in "social practices," forming "communities of learning," such that their participation is experienced as "personally meaningful." "The social setting does not [in this approach], refer primarily to the group of students involved in…collaboration but to the activity itself." And this approach "is assumed to enhance critical thinking due to the fundamental intertwining of individual development and the cultural context." Here lies a focus "on critical agency," "connections [being] made between the learning process and the current and future situation(s) in which students can…apply the knowledge and skills they have acquired."

The chapter by Blackie, Leibowitz, Nell, Malgas, Rosochacki, and Young also aims at promoting critical citizenship but has a radically different starting point, namely that of post-apartheid South Africa. The authors offer their own stipulative definition of critical citizenship, "based on the promotion of a common set of shared values such as tolerance, diversity, human rights and democracy…and the imagining of a possible future shaped by social justice." They draw on this definition in conducting a study among staff and students in Stellenbosch University, once the academic center of apartheid, and where much effort has been put in to bring the university into a multicultural society (not least by its inspirational vice-chancellor, Professor Russel Botman, who died suddenly only very recently). As the authors observe, the concept of critical citizenship "remains contested," not least there being differences across the disciplines, evident both in the education of students and in the responsibilities taken on by universities. As the authors recognize, there remains a thorny matter as to "how one might usefully address issues of critical citizenship [especially] in a...scientific course, and indeed, whether one would want to."

The general term that has come to mark out forms of teaching and learning oriented toward social and critical action is that of "critical pedagogy," and Cowden and Singh pursue it in their chapter. Taking their cue from Freire, they explicitly look toward an idea of critical thinking "as a social practice," mindful that "processes or acts of learning...are themselves inherently political." Cowden and Singh trace the idea of critical thinking as emerging from Kant (and "the active self-construction of the subject") and then through Hegel (and the development of "critical self-consciousness," whereby "stable thoughts reveal their inherent instability") on to Dewey (where education was essentially connected with "democracy") and then Habermas (especially the idea of "communicative competence") and even Barnett (in his idea of "critical being"). Their view of critical pedagogy, accordingly, has at its heart "the idea that the production of knowledge and the identities of learners being themselves socially and ideologically mediated." It follows that there can be no didacticism here, but the development of "a critically conscious subject" through a "process where the teacher and the student are both engaged in teaching each other."

Stephen Brookfield's chapter is situated in the same terrain but it contains a particular sophistication. In opening up controversial issues, Brookfield observes that it is misguided of him to expect his students frankly to examine themselves unless he is prepared explicitly critically to examine himself in front of his students and so to model the kind of critical thinking being encouraged among the students. And here Brookfield recounts various quite personal matters and how he uses certain episodes critically to interrogate himself. This is far from unproblematic: on the contrary, "teachers who model criticality must expect to have, at least initially, less favourable student evaluations." We might conjecture, going beyond Brookfield, that such an outcome arises from the anxiety that some students might feel where they anticipate discomfort that might arise from forensic self-scrutiny (of the kind modeled by a teacher so willing to self-disclose as Brookfield).

The chapter by Szenes, Tilakaratna, and Maton stands somewhat apart from the others in this group. It takes a key concept from the "Legitimation Code Theory" that Maton has been developing over recent years, that of "semantic gravity." In a nutshell, semantic gravity refers to the extent to which propositions within a sequence are close to a particular context (and so have a strong semantic gravity) or are quite distant and, indeed, abstract (and so have a low semantic gravity). Szenes, Tilakaratna, and Maton show how it is possible diagrammatically to represent the flows of a set of propositions, according to its abstractness or context-boundedness; a sequence of ideas produced by a student is likely to exhibit highs and lows in this sense. The authors go on to observe that, empirically, therefore, it appears that processes of critical thinking among students are liable neither to be wholly context-dependent or to be entirely situated at a generic level. Further, they hypothesize that there may be different patterns in this respect across disciplines. The procedure that they deploy thereby prompts a move "beyond the false dichotomy of either genericism or subject-specificity by revealing both generic attributes and ways [they] may be realized differently in disciplinary contexts."

While the social dimension is evident in the examples on display—in social work and business studies—it could be instructive to apply this technique to curricula that were explicitly examples of critical pedagogy. In order for critical pedagogy to be effective, would there be likely to any characteristic patterns of semantic density? Does critical pedagogy—and the chapters here show it to be a family of differing teaching approaches—tend to sponsor generic forms of criticality or context-specific forms? For now, we have to leave the question hanging.

30
Speaking Truth to Power: Teaching Critical Thinking in the Critical Theory Tradition

Stephen Brookfield

Introduction

How critical thinking is conceptualized frames how it is taught. Prominent traditions in the critical thinking discourse are analytic philosophy and logic, natural science, pragmatism, psychoanalysis and psychotherapy, and critical theory. If your intellectual reference point is the hypothetico-deductive method, then the kinds of student behaviors you regard as examples of critical thinking will be very different from a colleague who views it as the analysis of language games. However, whatever discipline one teaches in—from statistics to theology, physics to romance languages—there is a common intellectual project regarding critical thinking across the disciplines. The point of getting students to think critically is to get them to recognize, and question, the assumptions that determine how knowledge in that discipline is recognized as legitimate. Sometimes the emphasis is on ferreting out the assumptions behind the arguments of experts in the field, sometimes on students themselves making clear the assumptions they operate under. But no matter what the discipline, all areas of academic study are constructed on assumptions regarding what scholars in those disciplines regard as legitimate knowledge.

In the English-speaking world the analytic philosophy tradition is the most influential intellectual tradition informing how critical thinking is understood and taught. When my children went through the North American education school system they were continually assessed (from the age of five) on their exercise of critical thinking. Boiled down to its simplest level this meant they were required to give reasons for any opinions, conclusions, or statements they made, whether these were in calculus, social studies, or science. Furthermore, these reasons were judged to be more or less valid according to the evidence adduced in support of them. They were taught to recognize logical fallacies; to distinguish between bias and fact, opinion and evidence, and judgment and valid inference; and to become skilled at using different forms of reasoning (inductive, deductive, formal, informal, analogical, and so on).

In this chapter I propose to shift the focus away from the analytic philosophy tradition and explore what critical thinking looks like, and how it is modeled, when teachers are informed by the tradition of the Frankfurt School of critical social theory (critical theory, for short). Although pragmatism holds that democracy is the political arrangement that best guarantees the intellectual openness necessary for the advancement of knowledge, critical theory's debt to Marxism, and its connection to democratic socialism, means it is a politically inclined critical tradition. Teachers informed by this tradition tie critical thinking to promoting a particular leftist conception of social justice and to uncovering and redressing power inequities. An example of critical thinking in this tradition is being able to detect and resist ideological manipulation, and to lay bare the abuses of power.

I begin the chapter by providing a synopsis of what comprises the critical theory tradition. Then I consider two questions: for teachers who work in this tradition, what curriculum does a critical thinking classroom explore? And, how do teachers model criticality for students?

The critical theory tradition of critical thinking

Critical theory is a term associated with thinkers from the Frankfurt School of critical social theory, such as Horkheimer and Adorno (1972), Marcuse (1964), and Habermas(1987). The theory describes the process by which people learn to recognize how unjust dominant ideologies are embedded in everyday situations and practices. These ideologies shape behavior and keep an unequal system intact by making it appear normal. As a body of work, critical theory is grounded in three core assumptions regarding the way the world is organized: (1) that apparently open, Western democracies are actually highly unequal societies in which economic inequity, racism, and class discrimination are empirical realities; (2) that the way this state of affairs is reproduced as seeming to be normal, natural, and inevitable (thereby heading off potential challenges to the system) is through the dissemination of dominant ideology; and (3) that critical theory attempts to understand this state of affairs as a prelude to changing it.

Dominant ideology comprises the set of broadly accepted beliefs and practices that frame how people make sense of their experiences and live their lives. When it works effectively it ensures that an economically unequal, racist, homophobic, and sexist society is able to reproduce itself with minimal opposition. Its chief function is to convince people that the world is organized the way it is for the best of all reasons and that society works in the best interests of all. Critical theory regards dominant ideology as inherently manipulative and duplicitous. From the perspective of critical theory, a critical person is one who can identify this manipulation and discern how the ethic of capitalism, and the logic of bureaucratic rationality, push people into ways of living that perpetuate economic, racial, and gender oppression. Teaching critical thinking, therefore, involves teaching people to see behind the apparently normal façade of daily life to realize how ideological manipulation works to keep people quiet and in line.

The critical theory tradition also defines a clear enemy to critical thought— the existence of dominant ideologies such as capitalism, White supremacy, patriarchy, heterosexism, and the concurrent process of ideological manipulation that ensures that the majority accept these ideologies unquestioningly. Those working within this paradigm are the most likely to be regarded as troublemakers who see power, race, class, sexism, ableism, and homophobia everywhere, even where they don't exist.

Teaching critical thinking informed by this tradition entails helping students to develop what C. W. Mills (1959) called a structural worldview, in which individual troubles are always analyzed as political phenomena. So, for example, layoffs are related to the workings of capitalism that seek to develop new markets, maximize profits, and cut labor costs. The progression of an individual illness and its treatment (or lack thereof) is explained by the way pharmaceutical companies and insurance companies structure and ration health care. A divorce is linked to the stress placed on a relationship by working three jobs to pay for education, health, care, etc. In this paradigm learning to think critically is a sort of ideological detoxification, in which the ideology of individualism—the belief that we all make our own destiny, are captains of our own souls—is revealed as a tool of false consciousness. False consciousness is the state of being duped by dominant ideology so that you think inequality is a normal consequence of Darwin's law of the survival of the fittest, and that prosperity comes to those who deserve it because of their extraordinary talents or because they work harder than everyone else.

One of the chief criticisms of the critical theory paradigm of critical thinking is that it too readily fixes on who are the good guys (critical theorists) and the bad guys (racist capitalists). It also is criticized for having a preconceived end point—democratic socialism—toward which it is working. This is in contrast to some of the other traditions where a condition of critical thinking is always to be open to new possibilities, and not to predetermine the end point of your critique. There is certainly some truth to this argument. Critical theorists are much more likely to call out what they see as an abuse of power, and they do regard certain ideologies (capitalism, White supremacy) as damaging, and certain enacters of those ideologies (bosses, managers, boards of directors) as agents of harm and destruction.

But the critical paradigm also has a self-critical strain within it; after all, the theory itself began as an attempt to reformulate Marxist thought in conditions Marx had not foreseen. The working class across the industrial world had not followed the Bolshevik revolution in Russia and overthrown capitalism. Indeed, they often seemed to be striving to become the bourgeoisie. Neither had Marx foreseen the rise of mass communications and the role that mass media played in carrying dominant ideology. As Bronner and Kellner(1989) observe, "Inspired by the dialectical tradition of Hegel and Marx, critical theory is intrinsically open to development and revision" (2). Books such as C. W. Mills's *The Marxists*(1962), Eagleton's *Ideology*(2007), and Bronner's*Critical Theory*(2011), all accept, provisionally, the basic accuracy and utility of explanatory frameworks drawn from the critical theory tradition, while at the same time doing their best to challenge these. So, in Marcuse's words, "critical theory is, last but not least, critical of itself and of the social forces that make up its own basis" (1964 72).

What does a classroom informed by critical theory explore?

Given that ideology and power are the conceptual foci of critical theory, these would similarly comprise the curricular elements of a critical thinking classroom informed by this tradition. Up to now I have cast a broad swathe as I've talked about dominant ideology. Specific ideologies that would be explored in a critical classroom are militarism, capitalism, White supremacy, patriarchy, heterosexism, and ableism. The nature of dominant ideology is that it is often unremarked-on, unnoticed, a part of the air we breathe. Being able to identify for students the particular ideology underlying a specific remark in discussion, or behind a common practice in an applied field of study, is made much easier if these ideologies are named and clarified. Consequently, a major part of a critical curriculum would be the identification of the different components of dominant ideology.

Militarism

As an ideology, militarism glorifies the use of force for force's sake; believes the only way to change behavior or maintain order is through the deployment of force; justifies the murder, torture, and rape of civilians as "collateral damage"; and emphasizes the constant stockpiling of weapons as the best way to secure peace. In a militarist ideology "the question surrounding war and violence is not whether they will occur, but rather who will survive" (O'Neill and O'Sullivan 2002, 174). In war zones militaristic values constantly inform everyday life. For example, in Mojab's(2010) anthology on *Women, War, Violence and Learning,*we see how misogyny and militarism intersect in post-Taliban Afghanistan (2010), how hard-fought gains in Iraqi women's rights have deteriorated since the American invasion (2010), and how Palestinian women in the Occupied Territories experience daily military obstacles to being educated (Shaloub-Kervorkian 2010).

In commenting on what can be learned from Mojab's work on women, war, and violence, Horsman(2013) documents the effects of living under militarism: the erosion of self-confidence, the development of feelings of blame and guilt people feel for their situations, depression, anxiety, post-traumatic stress syndrome, and insomnia. Crucially, for learning, there is a fear of being punished or humiliated for making mistakes. Educationally, Horsman argues, this calls for a considerable slowing down of what is thought to be a "natural" pace for the learning of war survivors, an attempt to create activities in which women can share their experiences in collaborative ways, and the use of visual and tactile modes of teaching.

Capitalism

Capitalist ideology as explored in this chapter is the economic version of the ideology of individualism, the belief that society works best and serves the interests of all when people are allowed to compete with one another, thereby ensuring that the best products rise to the top. By definition the "best" services are those that are most patronized. If companies go under or industries fail, this is explained by a Darwinian survival of the fittest argument. Because the free market is seen as entailing a natural checks and balances, whatever ends up succeeding in the free market must therefore be the best available product. This same view of life holds that individual competition creates a meritocracy in which the most talented rise to the top in any field of human endeavor.

This view of capitalism is not the classic Weberian view of capitalism that emphasizes the accumulation and investment of capital. Instead it is the free market, laissez-faire, libertarian ideology that regards government as something to be rendered obsolete as quickly as possible. This laissez-faire ideology successfully attaches itself to the coattails of two other high-octane ideological values—freedom and democracy. Hence, promoting unfettered capitalism is seen as the way to organize economic affairs to ensure political freedom and democracy. The same spirit of individual economic entrepreneurship that is trumpeted as being at the core of capitalism is paralleled in the values of free speech and liberty that democracy is held to guarantee.

Teaching to critique capitalism in the United States quickly earns you the label of Marxist, Socialist, or Communist. Because capitalism is successfully tied to American values, anyone who critiques it runs the risk, by association, of being tainted as un-American. This is a fantastically successful and impressive example of ideological manipulation. It means that discussions of socialism or Marxism are immediately off the table, no-go areas too dangerous to dabble in.

Critical theory's project was to seek to understand how capitalism could be replaced by democratic socialism, that is, by working class hegemony. Teaching that critiques and illuminates capitalism thus invariably raises the question of class analysis and here teachers, at least in the United States, face another ideological hurdle. Dominant ideology holds that under free market capitalism there is so much social mobility that the concept of class—with its notion of fixed social locations determined primarily by economics and one's material condition—is irrelevant. Shor's (1996) work in the 1980s and 1990s made explicit use of class analysis in describing his work with young adults at Staten Island Community College. Shor focused on unearthing shared attitudes among his students regarding their life chances, and their view of academic work, and showed how these were tied to broader working class culture. One of his most-storied exercises involved taking students to the college cafeteria and making the burgers and fries the objects of study as a way of teaching about globalization and the ways corporate power shaped our choices in food, as in life.

White supremacy

White supremacy is the ideology that places "Whiteness" as the preferred norm of what constitutes a fully realized life, "White people" as the "natural authorities" when it comes to making decisions on behalf of the collective, and White knowledge (and White forms of knowledge production) as the most valid produced by humankind (Colin 2010; European-American Collaborative Challenging Whiteness [ECCW] 2009). This ideology is usually implicit and is frequently denied by its perpetrators even as it's being disseminated. Very few Whites proclaim this ideology explicitly and many in fact condemn it. But its life is found in the small actions and spaces of everyday life, in racial micro-aggression; the small acts of racist exclusion committed in gestures, tone of voice, meeting behaviors, classroom interactions, media reports; and so on. Micro-aggressions are typically denied by their perpetrators, are labeled as unintentional, and often leave the recipient wondering "did that just happen?" When Whites try to recognize how White supremacy lives in them they have to peel back the multiple layers of the ideological onion. As the European-American Collaborative Challenging Whiteness (2009) observed, "We realized the potent irony: in trying to minimize our supremacist consciousness, we felt compelled to cast ourselves as superior (which is supremacist consciousness)" (266).

A major part of teaching to uncover White supremacy is exploring other racially based paradigms for creating knowledge and understanding the world, such as ones based on African, Native American, Indigenous, and Latino identity, or considering theoretical critiques such as those of critical race theory (Closson 2010) and postcolonialism(Alfred 2010). When these are made the subject of intentional inquiry then the epistemology of Whiteness (such as a dualistic right/wrong way of understanding complex questions, a privileging of rationality, a view of individuals as autonomous captains of their destinies) becomes more apparent (Paxton 2010; Shore 2010).

Patriarchy

As an ideology patriarchy holds that men have the rightful power to decide how women should behave and women should accede to that power. This is justified because men as decision-makers are held to be essentially rational, objective, and governed by reason, whilst women are believed to be moved primarily by emotion. This means women must therefore be considered to be illogical and unreliable decision-makers. This ideology hurts men as well as women. It denies men the expression of emotion whilst it justifies punishing women who challenge men's monopoly of control. Behind pornography and media depictions of women as sexual supplicants, to the rape of women because their dress and behavior means they "are asking for it," patriarchy is the dark angel that justifies symbolic and physical violence against women. Feminism challenges patriarchy by placing women's concerns and the centrality of gender at the forefront of analysis and undertakes a power analysis of gender-based inequality across personal and social relationships, work, politics and ideologies of sexuality.

Some proponents of feminism focus more specifically on what are often conceptualized as "women's issues" such as reproductive rights, rape, and sexual objectification via pornography. Others, notably materialist feminists, conduct a broader critique and dismantling of patriarchy and its ties to capitalism. In fact it is probably more accurate to talk of feminist *theories* and feminist *pedagogies* in the plural. Theorists such as Mojab (2005) and Hart (2005) insist that gender oppression be understood as intersecting with other forms of class and race-based oppression and argue that to separate them is empirically and theoretically untenable. This means that popular education—grass roots activism that works to empower the most marginalized—often has a feminist strain embedded within it (Manicom and Walters 2012).

Heterosexism

Our sexual practice and sexual identity, and the particular way we experience desire, exert a powerful influence on how we construct and navigate our lives. As an ideology heterosexism mandates sexual contact across gender as the norm, regarding same sex practices as deviant aberrations. When desire is expressed in relationships between males and females it is appropriate, but when expressed between people of the same sex, it is judged warped and perverted, a sign of antisocial behavior or sickness. The dominant, unquestioned belief that heterosexual relationships are the empirical norm leads to their being considered morally superior to same-sex relationships.

In recent years education has seen a foregrounding of Queer theory (Hill 2006; Hill and Grace 2010), a body of work that, in Hill's (2004) terms, is conducted in a spirit of wildness and mischief in its focus on bodily desire. Queer theory argues that pinning down one's sexuality in a fixed, static way is always complex, as in transgendered relationships, or in straight friendships between transvestites and cross-dressers. Classics in the field such as *Epistemology of the Closet*(Sedgewick 2008) "interrogate" (to use a favored term) dominant understandings and practices of sexuality. In other words, they question rigorously and continuously how certain ideas and behaviors become accepted as "normal" and others viewed as "deviant."

Queer theory rejects an essentialist epistemology that defines sexuality in a bifurcated, either/or, way as gay or straight, hetero or homo (Grace and Hill 2004). Instead, queer celebrates the idea of constantly shifting identities and broadens conceptions of sexual practices. Grace and Hill (2004) argue that Queer theory's radical inclusion connects it to theorizing in transformative learning whereby schemes and perspectives are gradually broadened to become ever more permeable and comprehensive. King and Biro (2006) extend this analysis to apply a transformative learning perspective to the development of sexual identity at the workplace.

Ableism

Rocco (2011) argues there is no issue of diversity, privilege, or human rights in the field of education that has been given less attention than disability. One reason for this is the acceptance of the ideology of ableism—"discrimination on the grounds that being able bodied is the normal and superior human condition" (McLean 2012, 13). Like heterosexism, ableism morphs a definition of normality (in this case with regard to human physical and mental functioning) into an assertion of superiority, often expressed in the guise of sympathy. Ableism infantilizes people with disabilities, viewing them as children who need to be spoken to loudly and slowly.

Those educators who teach against ableism work in three ways. First, they try to integrate the fields of disability studies and education, in the way that has happened with, say, critical race theory (Clark 2006; Ross-Gordon 2002). Second, they focus on exploring how education spaces and teaching practices can be changed to include proper attention to disability (Ostovary and Dapprich 2012). Third, they theorize disability as a dimension of structural exclusion and disenfranchisement to make it "become as visible as the race-class-gender triad" (Rocco and Delgado 2011, 9).

How do teachers model criticality for students?

Research into students' views of what teaching approaches most help them to learn to think critically emphasizes five themes (Brookfield 2012). First, students say they need to have their exposure to critical thinking sequenced so that early on they learn and apply mental protocols to situations and problems that do not represent their own mental models. Over time, the application is gradually brought to bear more and more on their own patterns of reasoning. Second, students say they experience critical thinking as a social learning process in which peers serve as critical mirrors reflecting back to them assumptions they did not realize they held. Third, students emphasize that understanding how different perspectives can be taken on an issue is helped when a course or module is team-taught. As faculty model an interdisciplinary conversation about a central idea, or as they apply different theoretical paradigms to understanding content, students say they see the importance of exploring different approaches and perspectives. Fourth, students say that the most dramatic leaps forward in their understanding of material are usually triggered by having to resolve some sort of disorienting dilemma, that is, having to reconcile two antithetical yet valid responses to a question, or having to incorporate into their frame of understanding a new piece of information that calls into question much of what they had previously held to be true.

The importance of seeing teachers model critical thinking, and having them name that process intentionally and publicly, is the fifth theme that emerges, and the one that students say is the most important in helping them understand both what critical thinking is and how it is done. When it comes to students learning how to practice critical thinking, it seems that they constantly look to their teachers to see what the process looks like. Furthermore, given the difficulties of this process, it's important that teachers earn the right to ask them to do it themselves by first modeling how they try to unearth and research their assumptions.

The modeling that students appreciate from teachers takes different forms. Most importantly, it seems that the more personal examples teachers give of how they try to think critically, the more students appreciate this. A teacher's early disclosure of a critical thinking experience can set a tone of openness that significantly influences students' readiness to delve into their assumptions. Before asking anyone to introduce themselves at the start of a class or workshop, teachers must introduce themselves. And it behooves them to use an autobiographical introduction that somehow incorporates allusions to one or two concepts drawn from critical theory.

One introduction I use is the one I included at the start of *Teaching for Critical Thinking* (Brookfield 2012) describing my history of depression. This is somewhat risky, but it's well suited to a course where examining power and ideology are the focus. I describe how I realized that one of the biggest obstacles to my seeking professional help was the way the ideology of patriarchy had lodged itself inside me. Because I had uncritically assimilated the notion that men were logical, rational, and could think their way out of problems, I refused to seek professional help. I thought all I needed to do was tell myself to "snap out of it." There was nothing to cause me to feel depressed so I should be able to reason my way out of it. Going to therapy and using pharmaceuticals were, to me, signs of weakness.

I explain how this refusal to seek help was itself an example of hegemony, of me acting in a way that seemed commonsensical and desirable but that actually was working against my best interests. These are all pretty personal disclosures and I only do it when it's appropriate. But in a session in which I'm trying to get students to see how dominant ideology is embedded in their daily thoughts and actions, I believe it's entirely appropriate to begin with an example of how one of those ideologies wrecked my life for several years. The trick is to make sure students see that I am using a personal story to teach about an important aspect of the material, not to portray myself as some sort of a heroic figure.

As a general rule I always try to introduce a challenging idea with a story of how it has somehow helped illuminate my life. If I can't think of such an example, I'd find it hard to justify trying to convince anyone else that the idea is important. When I teach about hegemony I use the example described at length in *The Power of Critical Theory* (Brookfield 2004) of how my acceptance of the notion of adult education as a vocation led to me to work all waking hours, visiting community groups or running educational clinics every night of the week. No surprise that I got divorced and that I've had three collapses at work that put me in the emergency room. When I teach about the unbridled consumerism that goes with capitalism, and the concept of commodity fetishism, I describe how I gladly take on extra consulting work to pay for the 24 guitars I own. There, I've said it! I own 24 guitars. Who can possible play more than one guitar at a time?

On capitalism I also talk about how the commodification of learning and teaching is lodged inside myself. For example, although I oppose the commodification of learning—the way people's creative engagement with new skills or ideas are turned into a thing that can be measured precisely—I commodify my own practice by believing I've done a good job only if I get a student rating of over 4.5 on a 5 point Likert scale evaluating my teaching. I talk about how the language of economic exchange has insidiously worked its way into my conversations, as when I say how important it is to "own" an idea, to get students or colleagues to "buy into" my suggestions for a course, or how I have "invested" in my relationships. I describe how I embody what Fromm (1956) called the social character of capitalism—someone who is punctual, organized, and orderly, who always comes in on time and who exceeds previously declared expectations. When I hand in a manuscript six months before the projected submission date, I am suitably pumped-up, reveling in my superhuman ability to beat deadlines!

After I have talked about how dominant ideology lives in me I then try to link this idea to the actual practice of the course or workshop people have signed up for. For example, when I teach graduate courses to educators, counselors, trainers, professional developers, and consultants, I often try to run the class as a professional development experience. But the university, and its wider accreditation body, requires written evidence that learning has happened. It doesn't matter how diligently people worked in classroom exercises, or how much they helped peers to learn; the way their learning is assessed is through a paper, an artifact, an object. So when I introduce the requirements for the paper I tell students this is an excellent example of what critical theory calls objectification and commodification. Learning—the creative flow of collaborative exploration—is required to be converted into a paper that follows APA guidelines and has a minimum word limit if it is to be taken seriously by the institution. I call the paper the "commodified artifact" for the course. This is such a good example, so close to home, that when students email me their papers the message header is often "Commodified Artifact"!

Finally, I am as explicit as possible in trying to name any positional power or authority I enjoy over students. I tell students I want to be as transparent as possible regarding how I exercise power and that the "elephant" in the adult educational classroom is teacher power. I say that I have an agenda I am working from and expectations about what should happen, and, if it's a course students are taking as part of a formal program, that I will be using certain criteria to assess their work.

This may all sound heavy-handed and authoritarian, as my attempt to show them who's the boss. But my intent in naming my power is not to intimidate, but to clarify that power is in the room. Part of teaching against power is to be aware of its presence. I try to be respectful and collegial; not to use sarcasm, ridicule, or bullying; to use participatory, democratic, and dialogic approaches; and to encourage students to take control over learning. But none of that alters the fact that I have the ability to exercise power, influence, and control, and that I can call on the full weight of institutional sanctions if I choose to. It's a major mistake in any organizational setting to be coy about your power. People know it's there and will talk about how you use it when you're out of earshot. Far better, in my view, to acknowledge that reality and to disclose the rationale behind your use of power.

Modeling criticality this way raises questions regarding assessment of both teacher performance and student progress. Teacher assessment protocols typically focus on clarity of exposition, use of multiple instructional approaches, organization of content, and so on. Assessing how well teachers model their own exercise of critical thinking is more subtle since, to beginning students, the modeling is often unclear. In evaluations of teacher performance students are not likely to welcome teachers who pose troubling and disorienting questions, or who take very different views on a question without saying which is the "correct" one. So teachers who model criticality must expect to have, at least initially, less-favorable student evaluations.

When it comes to assessing student learning a causal relation between a teacher's modeling and a student's understanding of criticality is hard to establish. Most of the standardized tests used by campuses to assess students' critical thinking, such as the Collegiate Assessmen of Acadaemic Proficiency (CAAP) test, document how well students understand analogy, make accurate inference, or identify logical fallacies. The student's developing capacity to understand how a teacher's actions model criticality, and how elements of that modeling can be incorporated into the student's repertoire of skills is missed entirely by such tests.

Conclusion: further research into integrating analytical and critical traditions

At the outset of this chapter I argued that the analytic philosophy tradition was the strongest one that informed the practice of critical thinking in North America. Although this chapter departed from that tradition by emphasizing critical theory, there are in fact points of connection between the two schools of thought. One promising direction for future research and study in critical thinking, therefore, is to see how these apparently disparate traditions might be integrated across a number of disciplines.

For example, although the analytic philosophy and logic tradition may seem to be primarily technical, concerned with the mechanics of putting arguments together and taking them apart, it is often linked to a moral purpose. Diestler(2009) argues that the reason for assessing the validity of arguments is so that one can spot manipulative, false reasoning and protect oneself against it. She, and others such as Bassham et al. (2007) maintain that a familiarity with language games helps one understand how language can be powerful and potentially misleading, derailing effective critical thinking. The analytic philosophy tradition argues that if one can understand how bias and prejudice masquerade as empirical fact or objective interpretation, one is better placed to know what to believe and what to do.

I believe that critical thinking in this tradition is potentially as political as the most critical elements of critical theory. George Orwell's essay "Politics and the English Language" (1946) makes this point supremely well. Propagandists and demagogues understand that language tricks are powerful tools in securing the consent of people to situations that are actually against their best interests. This is what hegemony is—thinking something, and acting enthusiastically on that thought as if it were the most obvious, commonsensical thing in the world, all the while being unaware that your actions benefit those who wish to keep you uninformed. Getting people to willingly agree to, even support, a situation that is hurting them is difficult and cannot be done with force, since outright and overt coercion is easily identified. But control how people think and how they perceive the world—particularly through the use of language and its juxtaposition with images—and you are well on the way to getting people to agree to things that will end up harming them.

Some of the most common language tricks used in this effort are as follows:

- Attaching an abstract argument to a highly personal, dramatic narrative—so people associate the argument with an easily remembered personal story. This is something I do myself in using depression to teach the concept of patriarchy!
- Repeating a distorted argument often enough so that it becomes fact and gains legitimacy through frequent repetition
- Taking one part of an opponent's argument out of context, changing its meaning, and highlighting it in such a way that it is presented as the main element in an opponent's platform
- Attaching powerful positive symbols and myths (democracy, patriotism, openness) to one's arguments and powerful negative symbols (communist, dishonest, unpatriotic) to one's opponent's arguments
- Representing one's argument as the will of the majority, as in prefacing a comment by saying "The American people will not stand for..." or "What the American people really want is..."
- Making inferences that are presented as indisputable truth rather than hypothetical predictions as in "This policy is bound to lead the country into bankruptcy"
- Choosing one example from a general category ("I know this Republican who...," "I have this English friend who...") and making an unwarranted generalization—portraying the specifics of their behavior as the behavior of the whole
- Using a revered authority as the justification for one's argument, as in "The Bible clearly tells us..." or "In the Constitution the founding fathers clearly believed that..."
- Stereotyping a whole class of phenomena such as people (rational Whites, dangerous Blacks, hardworking Asians, volatile Latins), or organizations (undemocratic unions, inflexible management, corrupt politicians, mendacious communists)
- Constructing an inaccurate analogy to discredit an opponent's argument, as in "Banning smoking is the first step to Nazism"

One could argue, in fact, that teaching people to recognize these language tricks is actually the most politically explosive thing a school can do. If students grasp how arguments are put together to persuade others of the inherent, objective rightness of a particular point of view, or how they are developed to achieve certain ends at the expense of others, then they are better placed to recognize when this is happening and to make an independent judgment whether or not these arguments should be accepted. The analytic philosophy and logic tradition is usually not attached to any particular political viewpoint, so it is both claimed by, and on occasion attacked by, different political orientations.

References

Alfred, M. V. 2010. "Challenging Racism through Post-Colonial Discourse: A Critical Approach to Adult Education Pedagogy." In *The Handbook of Race and Adult Education: A Resource for Dialogue on Racism*, edited by J. Johnson-Bailey, V. Sheared, S. A. J. Colin Jr. III, E. Peterson, and S. D. Brookfield. San Francisco: Jossey-Bass. 201–216.

Bassham, G., Irwin, W., Nardone, H., and Wallace, J. 2007. *Critical Thinking: A Student's Introduction*. New York: McGraw-Hill.

Bronner, S. E. 2011. *Critical Theory: A Very Short Introduction*. New York: Oxford University Press.

Bronner, S. E., and Kellner, D. M. 1989. *Critical Theory and Society: A Reader*. New York: Routledge

Brookfield, S. 2004. *The Power of Critical Theory: Liberating Adult Learning and Teaching*. San Francisco: Jossey-Bass.

Brookfield, S. 2012. *Teaching for Critical Thinking: Tools and Techniques to Help Students Question Their Assumptions*. San Francisco: Jossey-Bass.

Clark, C. M. 2006. "Adult Education and Disability Studies, an Interdisciplinary Relationship: Research Implications for Adult Education."*Adult Education Quarterly* 56 (4): 308–322.

Closson, R. B. 2010. "Critical Race Theory and Adult Education."*Adult Education Quarterly* 60 (3): 261–283.

Colin, S. A. J. 2010. "White Racist Ideology and the Myth of a Postracial Society." In *White Privilege and Racism: Perceptions and Actions*, edited by C. L. Lund and S. A. J. Colin III. San Francisco: Jossey-Bass.

Diestler, S. 2009. *Becoming a Critical Thinker: A User-Friendly Manual*. Upper Saddle River, NJ: Pearson/Prentice-Hall.

Eagleton, T. 2007. *Ideology: An Introduction*. London: Verso.

European-American Collaborative Challenging Whiteness (ECCW). 2009. "Challenging Racism in Self and Others: Transformative Learning as a Living Practice." In *Transformative Learning in Practice: Insights from Community, Workplace, and Higher Education*, edited by J. Mezirow and E. Taylor. San Francisco: Jossey-Bass. 262–274.

Fromm, E. 1956. *The Sane Society*. London: Routledge, Kegan & Paul.

Grace, A. P., and Hill, R. J. 2004. "Positioning Queer in Adult Education: Intervening in Politics and Praxis in North America." *Studies in the Education of Adults* 36 (2): 167–189.

Habermas, J. 1987. *The Theory of Communicative Action: Volume Two, Lifeworld and System—a Critique of Functionalist Reason*. Boston: Beacon.

Hart, M. 2005. "Class and Gender." In *Class Concerns: Adult Education and Social Class*, edited by T. Nesbit. San Francisco: Jossey-Bass.

Hill, R. J. 2004. "Activism as Practice: Some Queer Considerations." In *Promoting Critical Practice in Adult Education*, edited by R. St. Clair and J. Sandlin. San Francisco: Jossey-Bass. 85–94.

Hill, R. J. 2006. *Challenging Homophobia and Heterosexism: Lesbian, Gay, Bisexual, Transgender, and Queer Issues in Organizational Settings*. Vol. 112. San Francisco: Jossey-Bass

Hill, R. J., and Grace, A. P. 2010. *Adult and Higher Education in Queer Contexts: Power, Politics and Pedagogy*. San Francisco: Jossey-Bass

Horkheimer, M., and Adorno, T. 1972. *Dialectic of Enlightenment.* San Francisco: Jossey-Bass.

Horsman, J. 2013. Learning and Violence. http://www.learningandviolence.net/helpothr/drawing.pdf

King, K. P., and Biro, S. C. 2006. "A Transformative Learning Perspective of Continuing Sexual Identity Development at the Workplace." In *Challenging Homophobia and Heterosexism: Lesbian, Gay, Bisexual, Transgender, and Queer Issues in Organizational Settings,* edited by R. J. Hill. San Francisco: Jossey-Bass. 17–27.

Manicom, L., and Walters, S. 2012. *Feminist Popular Education in Transnational Debates.* New York: Palgrave

Marcuse, H. 1964. *One Dimensional Man.* Boston: Beacon.

McLean, M. A. 2012. "Getting to Know You: The Prospect of Challenging Ableism through Adult Learning." In *Challenging Ableism, Understanding Disability, Including Adults with Disabilities in Workplaces and Learning Spaces,* edited by T.S. Rocco. San Francisco: Jossey-Bass. 13–22.

Mills, C. W. 1959. *The Sociological Imagination.* New York: Oxford University Press.

Mills, C. W. 1962. *The Marxists.* New York: Dell.

Mojab, S. 2005. "Class and Race." In *Class Concerns: Adult Education and Social Class,* edited by T. Nesbit. San Francisco: Jossey-Bass.

Mojab, S. 2010. *Women, War, Violence and Learning.* New York, London: Routledge

O'Neill, E., and O'Sullivan, E. 2002. "Transforming the Ecology of Violence: Ecology, War, Patriarchy and the Institutionalization of Violence." In *Expanding the Boundaries of Transformative Learning: Essays on Theory and Praxis,* edited by E. O'Sullivan, A. Morrell and M. A. O'Connor. New York: Palgrave 173–183.

Orwell, G. 1946. "Politics and the English Language."*Horizon* 76 (April): 252–264.

Ostovary, F., and Dapprich, J. 2012. "Challenges and Opportunities of Operation Enduring Freedom/Operation Iraqi Freedom Veterans with Disabilities Transitioning into Learning and Workplace Environments." In *Challenging Ableism, Understanding Disability, Including Adults with Disabilities in Workplaces and Learning Spaces,* edited by T. S. Rocco. San Francisco: Jossey-Bass. 63–74.

Paxton, D. 2010. "Transforming White Consciousness." In *The Handbook of Race and Adult Education: A Resource for Dialogue on Racism,* edited by V. Sheared, J. Johnson-Bailey, S. A. J. Colin Jr. III., E. Peterson, and S. D. Brookfield. San Francisco: Jossey-Bass. 119–132.

Rocco, T. S. 2011. "Editor's Notes." In *Challenging Ableism, Understanding Disability, Including Adults with Disabilities in Workplaces and Learning Spaces,* edited by T. S. Rocco. San Francisco: Jossey-Bass.

Rocco, T. S., and Delgado, A. 2011. "Shifting Lenses: A Critical Examination of Disability in Adult Education." In *Challenging Ableism, Understanding Disability, Including Adults with Disabilities in Workplaces and Learning Spaces,* edited by T. S. Rocco. San Francisco Jossey-Bass. 132.

Ross-Gordon, J. M. 2002. "Sociocultural Learning amongst Adults with Disabilities." In *Learning and Sociocultural Contexts: Implications for Adult, Community, and Workplace Education,* edited by M.V. Alfred. San Francisco: Jossey-Bass. 96.

Sedgewick, E. K. 2008. *Epistemology of the Closet.* 2nd ed. Berkeley: University of California Press.

Shaloub-Kervorkian, N. 2010. "The Gendered Nature of Education under Siege: A Palestinian Feminist Perspective." In *Women, War, Violence and Learning,* edited by S. Mojab. New York: Routledge.

Shor, I. 1996. *When Students Have Power: Negotiating Authority in a Critical Pedagogy.* Chicago: University of Chicago Press.

Shore, S. 2010. "Whiteness at Work in Vocational Training in Australia." In *White Privilege and Racism: Perceptions and Actions,* edited by C. L. Lund and S. A. J. Colin. San Francisco: Jossey-Bass. 125.

31

Stumbling over the First Hurdle? Exploring Notions of Critical Citizenship

Elmarie Costandius, Margaret Blackie, Brenda Leibowitz, Ian Nell, Rhoda Malgas, Sophia Olivia Rosochacki, and Gert Young

Introduction

In South Africa the legacy of apartheid still lingers. It is most evident in the income disparity between white and black populations. The mean income of South African blacks (here the term "black" does not include those of colored [mixed race] or Indian descent) has increased at a greater rate in real terms since 1994 compared to other ethnic groups in the country. Nonetheless, the mean expenditure per black household is still approximately five times lower than that of a white household (Ozler 2007). Whilst a single measure cannot hold the complexity of the lived experience of a nation, it remains a fair indicator of the continuing economic disparity.

Evidence from the twentieth century suggests that it is a little naïve to presume that education is the magic bullet to establishing democracy let alone a more equitable society. Nonetheless, in the absence of good education for all citizens such ideals cannot be realized (Westbrook 1993). However, the lasting impact on the educational system is perhaps of greater concern. Whilst the first of the so-called free-borns (those born post April 1994) have reached higher education, the primary and secondary education systems remain profoundly troubled. The Bantu Education Act of 1953 established separate education systems for different ethnic groups, and by 1973 the average government spending on education per white citizen was 15 times that spent on black citizens (Thomas 1996). As a direct result the quality of education in the historically black universities was not equivalent to the quality of education in the white universities. The desegregation of the universities began in the 1980s, but even in the early 1990s black students remained in the minority at historically white universities.

The 1997 White Paper on the Transformation of Higher Education calls for institutions to contribute to "the socialization of enlightened, responsible and constructively critical citizens" (Department of Education 1997). Well over a decade later we must ask how far we have come in this endeavor. Stellenbosch University remains, in terms of demographic profile, the least transformed of the major universities. White people still comprise over 60% of the student body. The academic staff profile is even more skewed. The 2011 census data shows that South Africa's population is just 8.9% white (Ngyende 2012).

The predominant language at Stellenbosch University has been Afrikaans. This is partly because Stellenbosch was founded as an Afrikaans medium institution, and there has been substantial resistance to a shift to more teaching in English. For many black people there is a strong negative association with the use of Afrikaans, as it was seen as the language of the oppressor. The 1976 Soweto riots were a direct response to the forcing of black scholars to learn in Afrikaans (Mabokela 2000). In addition, H. F. Verwoerd, one of the major architects of apartheid, was a faculty member at the university when he and several others crafted the most comprehensive articulation of apartheid policy (Giliomee 2003, 375).

Furthermore, previous studies on the level of racial integration document some of the accepted wisdom that Stellenbosch is significantly behind the University of Cape Town in terms of racial awareness and integration. The following quote points to the lack of awareness and reflection in the student interviewed:

> There is not a race problem at Stellenbosch. At least not as far as we [white students] are concerned...Last year, SASCO [the black students' organization] had a toyi-toyi [protest] because they feel excluded. Nobody excludes them. They do it....We are not used to toyi-toyi and things like that here. They should try to be a part of this university. (Mabokela 2000, 73)

At the time that this research was conducted the Stellenbosch student body was still more than 77% white!

Against this backdrop a group of academics from various faculties and units across Stellenbosch University gathered to discuss ideas of critical citizenship. The group was formed following an internal teaching and learning conference at which one of the members discussed a community interaction project that had had significant impacts on the participating students. The project had been interpreted using the Johnson and Morris framework of critical citizenship (2010). The central desire of the critical citizenship group was to promote a "thick" version of the common good—described by Walker as "working for a genuinely inclusive and anti-racist higher education" (2005). After several conversations it became apparent that we needed to ascertain whether our presumptions of the attitudes held by students and academic staff were correct.

Critical thinking, defined by Ennis (2011) as a "reasonable reflexive thinking focused on deciding what to do or believe," must be at the heart of any educative process if it is to provide any kind of transformation in society. Glaser (1942) argues that having critical thinking skills in one context does not mean much if those methods of reasoning and enquiring are not applied to other areas of study and in everyday life. The transformative value of critical citizenship lies in its application in every aspect of life. Given the particular circumstances of being situated at the institution where the idea of apartheid was born, critical thinking should be indispensable to Stellenbosch University'sdegree programs. However, it is not clear that critical thinking on its own is sufficient for transformation in this context. As such, we have chosen to focus on the slightly broader concept of critical citizenship, which encompasses critical thinking but also addresses social issues more explicitly.

Critical citizenship has become a topic of interest in recent years, especially in higher education. The term, critical citizenship, remains contested in both the local and the global contexts. We have chosen to view the concept as the promotion of a common set of shared values such as tolerance, diversity, human rights, and democracy. This concept is clearly dependent on the development of discernment, reflection, and constructive critique. Research on education and teaching within a globalized and largely neoliberal capitalist context warns of the danger of omitting a reflective, ethical, and moral component from educational processes (Ennis 2011; Nussbaum 2010; Waghid 2004). Within our own context we are aware that a shift in the demographic profile is necessary and may be achieved through mechanisms such as quotas and access programs. Nonetheless such a shift is insufficient to guarantee real transformation.

Transformation cannot be brought about by social engineering; it requires the growth and development of individuals who pass through the institution. To us, a crucial part of this is the development of critical citizenship. It is perhaps more an ideal to be striven for, rather than a check-box to be ticked once certain minimum criteria have been reached. With the ideal of real transformation of the university culture in mind, we set out to explore the extent to which our presumptions of the depth to which the foundational attitudes necessary to foster critical citizenship were already present at Stellenbosch University. We set out to interview a small sample of academics and students from three different faculties to give a preliminary snapshot of the understanding of critical citizenship.

In this chapter we describe the actions we took, and present our findings of how students and lecturers understand and talk about critical citizenship. We begin by contextualizing critical citizenship education within the broader debates around global and national citizenship education, and consider its relevance within the context of South African higher education as it is currently shaped by discourses and policies of transformation.

Contextualizing critical citizenship education

Critical pedagogy, which has strong links to the concept of critical theory as developed by the Frankfurt School, is a philosophy of praxis espoused by theorists like Paolo Freire, Henry Giroux, and Jurgen Habermas. While there are certain overlaps with the concept of critical thinking, Johnson and Morris sketch out the fundamental differences between the two. The characteristics particular to critical pedagogy are (1) an emphasis on the political and ideological dimension of knowledge production, (2) a focus on the collective, (3) a focus on understanding subjectivity in critical thinking, and (4) an emphasis on praxis (Johnson and Morris 2010, 81–84). For Paulo Freire (1983), the project of critical pedagogy centered on the dismantling and deconstruction of structurally imbalanced power relations through critique and analysis. Johnson and Morris (2010, 80) refer to Burbules and Berk's concise formulation of the distinction between critical thinking and critical pedagogy: "Critical thinking's claim is, at heart, to teach how to think critically, not how to think politically; for critical pedagogy, this is a false distinction" (1999, 50). In order to explore the imperative of political engagement implicit in critical pedagogy, we follow Johnson and Morris's use of the distinctive features of critical pedagogy as a basis to elaborate a theory of critical citizenship. As a term that attempts to capture the relation between the individual, the state, and society, citizenship potentially provides illuminating links between critical thinking and political engagement.

Various arguments have been presented in favor of a renewed emphasis on teaching civic engagement and social change in higher education (Furco 2010; Hartley, Saltmarsh, and Clayton 2010). Citizenship has reemerged in educational literature as an influential yet contested term, which seeks to place the connection between the individual and society at the heart of educational debates (Nussbaum 2002; Osler and Starkey 2003; Schuitema, Ten Dam, and Veugelers 2008). The transition from strictly nationalist conceptions of political power to more complex transnational configurations is reflected in changing conceptions of citizenship. Current interest in both national citizenship education (Enslin 2003; Ramphele 2001; Waghid 2004) and global citizenship education (Andreotti 2006; Johnson and Morris 2010; Nussbaum 2002) stems from the ongoing project of global democratization and liberalization. Rather than working to promote national loyalty, contemporary citizenship education focuses on the curation of common supranational forms of citizenship based on shared values of democracy, human rights, and tolerance (Johnson and Morris 2010, 77)

It should be noted, though, that we need to be careful not to simply replace one hierarchical system with another. An uncritical incorporation of a drive to globalization under the guise of "citizenship" could well result in unintended domination. Various thinkers have espoused the idea that the apparent cultural and human interconnectedness of globalization disguises the deepening social and economic divisions on a global scale (Sen 2004; Wallerstein 2004). Calls for transnational conceptions of critical citizenship that are based on "universalizable" democratic principles should, consequently, be treated with caution, as they might function at an epistemological level as another form of domination and symbolic violence.

Education and notions of critical citizenship in South Africa

In the context of contemporary South Africa varying views of critical citizenship are present in debates on education (Enslin 2003; Ramphele 2001). Enslin reflects on the tension between the notion of citizenship articulated in the constitution, which was strongly informed by a participatory notion of citizenship that developed through the antiapartheid struggle, and a "more popular interpretation of citizenship as access to socio-economic rights" (Enslin 2003, 74). The relatively swift transition from an exclusive, racist, and discriminatory conception of citizenship to an inclusive, antiracist, and democratic notion of citizenship exists alongside what has now been recognized as a painfully slow social and economic transformation that still reflects entrenched racial hierarchies (Erasmus 2009; Reddy 2004; Seekings 2008; Vale and Jacklin 2009). The development of a robust democratic culture is not only undermined by persistent inequality, but also by "a popular preoccupation with entitlement to goods [which] erodes willingness to engage in active participation for the common good" (Enslin 2003, 79).

Critical citizenship of the type described above is more likely to be achieved through criticality, introspection, and compassion. Critical reflection and introspection are essential processes for realizing the progressive conception of critical citizenship that "provides a framework for a transformed citizen who will strive to overcome the past" (Enslin 2003, 76). Waghid (2004) argues that the form of liberal communitarian critical citizenship articulated in the new democratic educational reforms needs to be reinforced by compassion as well as sensitivity to the realities of others (Nussbaum 2002; Waghid 2004). While "liberal-communitarian values can result in pupils developing capacities for rational argumentation and deliberative engagement," this may not necessarily cultivate the virtue of compassion required for educational transformation (Waghid 2004, 535). Engagement and understanding, and indeed emotional maturity and compassion, are core components when practicing critical thinking and critical citizenship.

The link between compassion and critical citizenship may not be immediately obvious. Nussbaum explicitly makes the link in her criteria for education for critical democratic citizenship emphasize the importance of criticality and compassion. These criteria are (1) that develop a critical consciousness that allows students to question their own traditions while practicing a mutual respect for reason; (2) to think as a citizen of the world; and (3) to develop a "narrative imagination" that fosters a deeper understanding of difference and diversity and allows one to place oneself in the shoes of another (Nussbaum 2002). They form the basis of our understanding of critical citizenship.

Critical Citizenship Research Project

A group of lecturers from different faculties was formed with an activist aim of enhancing transformation at Stellenbosch University. The group spontaneously formed after a presentation on critical citizenship and transformation by one of the members at a local conference. Many conversations followed and this research project was the result of the conversations and collaborative desire to participate actively in actions to deepen transformation at Stellenbosch University. The working definition of critical citizenship that the group decided on was as follows:

> Critical citizenship is based on the promotion of a common set of shared values such as tolerance, diversity, human rights and democracy. As an educational pedagogy, it encourages critical reflection on the past and the imagining of a possible future shaped by social justice, in order to prepare people to live together in harmony in diverse societies. (Adapted from Johnson and Morris 2010, 77–78)

We realized that the educational outcome will only be achieved if a common pedagogical stance, not only of *what* we teach and learn but also of *how* we teach and learn, is held by a critical mass of students and lecturers. We are arguing that if our students are to graduate with some understanding of their responsibility as critical citizens, there must be academics in all faculties who operate from this vision of a critical citizenship pedagogical stance. The academics may articulate this vision in different ways, and the application will vary. The group argued that, before trying to "enhance" this vision among academics, it would be wise to begin by exploring the extent to which this stance was already present by investigating how students and lecturers understand and talk about critical citizenship.

Methodology

The research project was framed within the qualitative and interpretive paradigm, making use of the case study method. The research sample consisted of 15 lecturers and 24 students from three different academic faculties The faculties chosen were Theology, Arts and Social Sciences, and Natural Sciences. These faculties were chosen because members of the research group were present in each of these faculties. The diversity of subject choice was deemed sufficient to get a preliminary "snapshot" of the attitudes across the university. The students were all fourth- or fifth-year students (in the South African system that is students in an honors program or first year masters program). There was a mix of gender in all cases. We also tried to include people from different racial groups; however, this was not always possible. Nonetheless, both English and Afrikaans first language speakers were present in all groups. A total of nine individual interviews and four group interviews were conducted; these comprised four interviews with lecturers from Visual Arts, one group interview with third-year art students, five interviews with lecturers from Chemistry, one group interview with Chemistry students, one group interview with lecturers from Theology, and one group interview with Master's students from Theology. The interviews took place between April and August 2012.

During the interviews, the definition of critical citizenship that the research group used was given to the participants and they were asked to respond to it by stating whether they could relate to it and whether it was included in their curricula. The definition that we used was either read out at the beginning of the interview or sent to participants to read beforehand. Participants were also asked to share their own understanding of the term "critical citizenship" and the ways in which they understood it to be applicable, implicitly or explicitly, in their courses.

The data was captured by using electronic voice recorders, and transcribed. Inductive content analysis (Creswell 2005) was utilized for studying the data. The coding process started with an initial reading of the text, dividing it into segments and identifying codes, and then reducing or combining codes into themes.

The data gathered from the six groups can only be taken as an indication of possible perceptions and attitudes. It is highly probably that the people who agreed to participate in the research were self-selecting for those who were already more inclined to favor the notion of critical citizenship. The aim of the study inevitably was not to generalize but to provide an in-depth exploration of the phenomena that became apparent during the investigation. Given our own passion for fostering critical citizenship in our institution, we had also hoped that initiating the conversation around critical citizenship with those who participated in the research would stimulate further thought and discussion beyond the scope of the research.

Results and discussion

Many of the issues discussed in the above explanation of the notion of critical citizenship emerged during the interviews. The themes that emerged were clustered in two broad categories: (1) contesting notions of citizenship as a form of belonging or a set of responsibilities and (2) contesting opinions of the responsibility of the university to address social issues.

Contested notions of citizenship

While most students were both familiar with and positive about critical thinking as far as it applies to academic work, the notion of critical citizenship was often seen as something unfamiliar, external to the students and their own understanding of themselves. Most participants endorsed the idea of criticality, but there was a diversity of response to the word "citizenship." Some students took issue with the politically problematic associations of the term as too closely tied to the nation-state: "I am critical about the word citizenship. It is a nationalist concept that I do not necessarily agree with" (Theology Student; [Some quotations were translated into English where the original interview was done in Afrikaans]); or as a notion imposed by the West: "I think it is a very western term and to apply that to South Africa would be a challenge. Citizenship is more western, more our way of doing things, and one cannot apply that to other cultures" (Theology Student). While students saw these tensions as inherent in the term itself, their comments also suggest that the context of application is problematic. The equation of citizenship and nationalism could indicate a sensitivity about the segregationist nationalisms of South Africa's apartheid past, but it could also refer to a feeling of discomfort with contemporary South African nationalism. A student remarked: "I mean the fact that I am born here, you know I may not be interested in this country and the politics and its future…if I think of citizenship then I do not necessarily think in terms of South Africa" (Theology Student). Another student said that "there is a paradigm shift wherein I can say I am now a global citizen, or I am now a South African citizen" (Theology Student).

Both Enslin (2003) and Ramphele (2001) argue that South Africans are still struggling to articulate a shared understanding of citizenship as solidarity and belonging. The sense of social division emerged strongly from both academics and students. This is clear in the frequent use of "us and them" discourse:

> This is our culture and our people and it works and now other people come and bother us, and now we see after five years after being critically engaging with it, that wait it is not our country it is their culture. (Theology Student)

It is also clear in the treatment of citizenship as a status or set of responsibilities that one has the freedom to choose whether to embrace it. A student explained that one needs to decide individually upon the value of South African citizenship: "Then one have [*sic*] to decide if I am still interested to own my citizenship in this country, my brother is now a citizen of another country and my sister is on her way [to citizenship of another country]" (Theology Student). A theology lecturer commented: "Now, we are not citizens [whose] suitcases are packed for Perth or Toronto or wherever. With critical citizenship we mean we want to exercise our citizenship." Both comments suggest a position that assumes that "owning" one's citizenship and taking social responsibility by engaging in South African society are not the norm. It is perceived as somewhat exceptional or noteworthy to willingly accept the status of South African citizenship, as opposed to an obligation, responsibility, or even honor bestowed upon one by birth or political history.

Critical citizenship was often equated by the participants with a measure of social (and personal) responsibility—within a local, national, or global context: "[I]n a general sense I suppose that critical citizenship relates to an understanding of context and a sensitivity in terms of the way you negotiate your identity and your responsibility towards others and towards your environment" (Art Lecturer). However, relating to a broader, holistic notion of context was seen as problematic if it was at the expense of local context: "[t]here is a lot of talk about all these connections globally but I mean we don't have the connections locally" (Art Lecturer). The political and religious contexts in South Africa were sometimes seen as complicating factors in negotiating appropriate models of critical citizenship. For example, one lecturer argued that neither the consensus-based democracy popular in Africa nor the Christian religion is conducive to critical citizenship, which can be

> seen as disloyal: if you are not seen to be looking for consensus, you are seen as an agent with a hidden agenda. So how do you advocate for the advantages of critical citizenship within a milieu that is suspicious about it. And it can even be seen as un-Christian to be critical. (Theology Lecturer)

Others argued that such apparent contradiction conforms to current notions of citizenship:

> Critical citizenship would imply a certain critical loyalty. There seems to be a change in our thinking about citizenship, that it is about more than just obedience, it's about the ability to have a critical dialogue with your perceptions of society. (Theology Lecturer)

These various notions of critical citizenship could indicate the discomfort some students and lecturers experience when being identified as South African citizens. The upheaval one group experienced during the apartheid years and the loss of political power after 1994 could have resulted in some people struggling to associate with being a South African because of the loss of control and a feeling of being powerless, of now being controlled by the "previous enemy." The changes that the end of apartheid brought were liberating for many people, but difficult for others to accept. In the process of digesting this loss, they might find it hard to identify with the notion of being citizens of South Africa, and taking responsibility as such citizens. A global citizen has certain responsibilities, but local citizenship might require a more active participation and contribution locally.

The responsibility to address social problems in higher education institutions

A student observed that critical citizenship "does not just involve you or your career, I think it is more of how you even interact on a day to day basis.... I don't think it just involves what you do as a scientist" (Chemistry Student). The notion of personal responsibility as closely tied to conceptions of citizenship often emerged in the interviews. However, to what extent the university as a social institution could be held responsible to address social issues and foster critical citizenship was contested. Hence, the second question aimed to investigate the importance of critical citizenship or social responsibility within individual curricula and universities' institutional cultures. There was strong dispute about whether this imperative should be interpreted as an implicit ethos of the university, cultivated by individual attitudes and beliefs, or whether it should be a formalized educational venture.

A frequent remark was that critical citizenship is what is being done already in classrooms and on the campus without being forced:

> But I think that is part of what we do in class then, in a tertiary class... [to] bring various perspectives to the classroom for students to work with and [to] critically think about them to make more effective choices in how they are better citizens (Theology lecturer)

One of the art lecturers stated: "If you are teaching at a university, you are already a critical thinker."

A chemistry lecturer explained that "in a laboratory, I think is one of the most dynamic areas where critical citizenship happens spontaneously, without you having to force it."

While the "unforced" flourishing of critical citizenship qualities in students is ideal, it is more likely that the academic in this case is conflating "critical citizenship" with cooperation between individuals. This presumption of the spontaneity of development of critical citizens can quickly absolve the institution or the individual academics from taking responsibility for this process. As one Art lecturer put it: "By virtue of being academics and by virtue of working in an interdisciplinary way our research will be critical citizenship." Such comments indicate a rather naïve or "thin" (Walker 2005) understanding of critical citizenship among some of the lecturers.

Other academics emphasized the individual responsibility that lecturers need to take, and that leading by example is the only way to ensure social responsiveness: "It is more about the attitude that you cultivate, through the example that you set, the things that you say" (Chemistry Lecturer); "The only way I can make students more aware is to now and then mention something about society, and yes to give leadership and give examples" (Chemistry Lecturer); "And if it is sort of something that the lecturer takes on board, like as an approach and reinforces it sort of everyday, then it becomes, then it is not something extra, it is actually just my course really" (Chemistry Lecturer). Some academics argued strongly that the cultivation of critical and responsible attitudes should be reliant on the convictions or ethical behavior of the lecturers, and the formalization of such a concept is problematic:

> So the reason within the social sciences is that our students are sensitive to discourse of critical citizenship is precisely because it is not imposed on them. It is because it is generated out of the convictions of their lecturers and it is not packaged as critical citizenship. But had this been packaged as critical citizenship and had it been something even from within a deep sense of conviction that we addressed as such I think it would be somewhat artificial and I think to impose this on disciplines where to us it is not apparent is problematic and unnecessary. (Art Lecturer)

The notion that the university itself has an ethical and historical responsibility to foster critical citizenship emerged too: "The most important thing about a university is that it is a place of learning, that it is a knowledge facilitator, that it challenges the norm and the boundaries and universities come from philosophical thinking" (Art Lecturer). There were many lecturers who felt that such responsibility fell beyond their personal parameters as teachers. This constraint could relate to time: "To now go and address social problems in the class—I just do not have time for that. That is unfortunately how it is" (Chemistry Lecturer); but it could also relate to the scope of the discipline:

> If you mean politically, then definitely no, it's not going to find a place in a chemistry course, I…I don't feel that it will find recognition in a chemistry course, and I don't feel it's my task to cultivate it in a student in that way. (Chemistry Lecturer)

This is in direct opposition to Badat's argument that drawing strict distinctions between academic work and imperatives to address social transformation are problematic. Badat argues that the "powers conferred by academic freedom go hand in hand with substantive duties to deracialise and decolonize intellectual spaces" (Badat, cited in Hasan and Nussbaum 2012, 128). This raises a crucial question of how one might usefully address issues of critical citizenship in a purely scientific course, and indeed, whether one would want to.

It was noted that the university environment itself poses a challenge to address social issues adequately, as it is fraught with racial problems that themselves reproduce social tensions and inequalities. One lecturer argued that "these people in power don't realize that their decisions are racist" (Arts Lecturer), and that the university is

> a safe space for certain people *ja*. But not for everybody. I mean we had a first year fine arts student last year... [he] was doing brilliantly but he just opted out. It was just too difficult for him to be the only black student in the first year group. (Art Lecturer).

Another suggested that problems in the university are related to

> perhaps of a constructed space that greater Stellenbosch... I think that there is a historical effect on the placement and spaces, such as Kayamandi versus central Stellenbosch, that people are unable to sort of cross that divide. (Chemistry Lecturer)

Implications and conclusions

The research on which this chapter is based was aimed at exploring the perceptions and attitudes held by small groups of Stellenbosch University students and lecturers from the disciplines of Theology, Arts and Social Sciences, and Natural Sciences regarding the notion of critical citizenship pedagogy and its implications for the university. The research began by probing students' and lecturers' understanding of critical citizenship in order to get an initial snapshot of where the institution is in this regard.

The results of the research showed that the understanding of critical citizenship varied between the disciplines; students in the humanities seemed to find it easier to discuss this concept than those studying in the sciences. It is perhaps not surprising that students who are exposed to a critical discourse in the pursuit of their undergraduate degrees are more articulate and, by implication, more aware of the idea of critical citizenship.

It was also evident that critical citizenship tended for some to be regarded as an entity "out there," and was not seen as part of a dynamic process. This seems to imply that, while the skills necessary for individual critical thinking are developed to some extent through existing pedagogical structures, developing the practice of critical citizenship and the language required to engage with it, requires more specific attention. In the first instance this requires conscious engagement in conversation around these ideas across the university. The more the idea of critical citizenship becomes a part of the daily vocabulary, the more likely discipline-specific praxis is to emerge. As real transformation becomes part of the daily reality of academics, so the sensitive issues that emerged in the interviews, which relate to social division, race, and discrimination, could be addressed more directly and explicitly.

Any institution-wide effort to engage with critical citizenship will require buy-in from lecturers and other staff. Even though some lecturers felt that critical citizenship should not be explicitly taught, it appears that the first hurdle may be less of a stumbling block than we imagined it would be. Most of the lecturers interviewed were in agreement about the importance of critical citizenship, even though solutions are likely to be specific to faculties or disciplines. The real challenge will be the manner in which that is facilitated and communicated to the students.

Taking into consideration our context, as seen from the results of the research, a certain approach to teaching and learning or critical pedagogy should be explored and developed. Johnson and Morris (2010) developed a framework for critical citizenship education that includes critical thinking and critical pedagogy. We follow Johnson and Morris's line of thought, in which critical citizenship focuses on developing a certain stance or way of thinking, rather than focusing only on the content of the critical citizenship curriculum.

A critical pedagogy needs to evolve that enhances a critical consciousness (Freire 1983) in students and lecturers to go beyond received knowledge and to understand that knowledge is invested with personal, cultural, or political perceptions, and that knowledge could enable or disable power. Critical pedagogy encourages the development of context-specific educational methods where students and lecturers use dialogue to open up the critical consciousness with an emphasis on praxis. The role of contextual critical pedagogy as a guiding principle in the development of Stellenbosch University is of utmost importance to build the capacity of critical, engaged, and responsible graduates.

The aim of this research was to gain insight into the ways in which students and lecturers talk about critical citizenship at the institution, in order to inform the research group about how to take this issue forward, and what approach or strategy to follow to enhance critical citizenship education. As a team coming together from various faculties, we have been sharing our ideas about the importance of critical citizenship, but we still have a long way to go.

References

Andreotti, V. 2006. "Soft versus Critical Global Citizenship Education." *Policy & Practice: A Development Education Review* 3 (Autumn): 40–51.

Burbules, N. C., and Berk, R. 1999. "Critical Thinking and Critical Pedagogy: Relations, Differences, and Limits." In *Critical Theories in Education: Changing Terrains of Knowledge and Politics*, edited by T. S. Popkewitz and L. Fender. New York: Routledge. 45–65.

Creswell, J. W. 2005. *Educational Research: Planning, Conducting, and Evaluating Quantitative and Qualitative Research.* Upper Saddle River, NJ: Pearson Merrill Prentice Hall.

Department of Education (DoE). 1997. Programme for the Transformation on Higher Education: Education White Paper 3 (Government Gazette No. 18207). Edited by DoE, Pretoria.

Ennis, R. H. 2011. "Critical Thinking: Reflection and Perspective, Part I." *Inquiry: Critical thinking across the Disciplines* 26 (1): 4–18.

Enslin, P. 2003. "Citizenship Education in Post-Apartheid South Africa." *Cambridge Journal of Education* 33 (1): 73–83.

Erasmus, P. 2009. "The Unbearable Burden of Diversity." *Acta Academica* 41 (4): 40–55.

Freire, P. 1983. *Pedagogy of the Oppressed.* London: Continuum.

Furco, A. 2010. "The Engaged Campus: Toward a Comprehensive Approach to Public Engagement." *British Journal of Educational Studies* 58 (4): 375–390.

Giliomee, H. 2003. "The Making of the Apartheid Plan, 1929–1948." *Journal of Southern African Studies* 29 (1): 373–392.

Glaser, E. 1942. "An Experiment in the Development of Critical Thinking." *The Teachers College Record* 43 (5): 409–410.

Hartley, M., Saltmarsh, J., and Clayton, P. 2010. "Is the Civic Engagement Movement Changing Higher Education?" *British Journal of Education* 58 (4): 391–406.

Hasan, Z., and Nussbaum, M. 2012. *Equalizing Access: Affirmative Action in Higher Education: India, US, and South Africa.* New Delhi: Oxford University Press.

Johnson, L., and Morris, P. 2010. "Towards a Framework for Critical Citizenship Education." *Curriculum Journal* 21 (1): 77–96.

Mabokela, R. O. 2000. *Voices of Conflict: Desegregating South African Universities.* New York: RoutledgeFalmer.

Ngyende, A. 2012. *Census 2011.* Pretoria: Statistics South Africa.

Nussbaum, M. 2002. "Education for Citizenship in an Era of Global Connection." *Studies in Philosophy and Education* 21: 289–303.

Nussbaum, M. 2010. *Not for Profit: Why Democracy Needs the Humanities.* Princeton: Princeton University Press.

Osler, A., and Starkey, H. 2003. "Learning for Cosmopolitan Citizenship: Theoretical Debates and Young People's Experiences." *Educational Review* 55 (3): 243–254.

Ozler, B. 2007. "Not Separate, Not Equal: Poverty and Inequality in Post-Apartheid South Africa." *Economic Development and Cultural Change* 55 (3): 487–529.

Ramphele, M. 2001. "Citizenship Challenges for South Africa's Young Democracy." *Daedalus* 130 (1): 1–17.

Reddy, T. 2004. *Higher Education and Social Transformation: South Africa Case Study.* Cape Town: University of Cape Town.

Schuitema, J., Ten Dam, G., and Veugelers, W. 2008. "Teaching Strategies for Moral Education: A Review." *Journal of Curriculum Studies* 40 (1): 69–89.

Seekings, J. 2008. "The Continuing Salience of Race: Discrimination and Diversity in South Africa." *Journal of Contemporary African Studies* 26 (1): 1–25.

Sen, A. 2004. "How to Judge Globalism." In *The Globalization Reader,* edited by F. Lechner and J. Boli. Oxford: Blackwell. 16–21.

Thomas, D. 1996. "Education across Generations in South Africa." *The American Economic Review* 86 (2): 330–334.

Vale, P., and Jacklin, H. 2009. *Re-Imagining the Social in South Africa, Critique, Theory and Post-Apartheid Society.* Durban: University of KwaZulu-Natal Press.

Waghid, Y. 2004. "Compassion, Citizenship and Education in South Africa: An Opportunity for Transformation?" *International Review of Education* 50 (5–6): 525–542.

Walker, M. 2005. "Rainbow Nation or New Racism? Theorizing Race and Identitfy Formation in South African Higher Education." *Race Ethnicity and Education* 8 (2): 129–146.

Wallerstein, I. 2004. "The Rise and Future Demise of the World Capitalist System." In *The Globalization Reader,* edited by F. Lechner and J. Boli. Oxford: Blackwell. 63–69.

Westbrook, R. B. 1993. *John Dewey and American Democracy.* Ithaca: Cornell University Press.

32
Critical Pedagogy: Critical Thinking as a Social Practice

Stephen Cowden and Gurnam Singh

Introduction

Much of the literature on critical thinking focuses on the ways in which human beings develop the capacity, through complex cognitive processes and skills, to evaluate or make sense of information. Within the formal educational context, it is often associated with pedagogical strategies aimed toward nurturing and developing learners' capacity for logical enquiry and reasoning. Though such insights are clearly very important, a narrow focus on what might be termed the "science of learning" can result in a negation of an obvious but very important point, namely, to what end and for what purpose should we be seeking to nurture critical thinking. Put another way, what is the moral, ethical, and political dimension of learning to think critically? And it is this question that forms the main purpose of the present chapter. By invoking the idea of critical thinking as a social practice, we examine the educational approach known as critical pedagogy and consider its relevance to higher education today. Critical pedagogy in its broadest sense is an educational philosophy that seeks to connect forms of education to wider political questions by arguing that processes or acts of learning and knowing are themselves inherently political.

Perhaps the most important figure that is associated with developing the tradition of critical pedagogy is the Brazilian educationalist Paulo Freire (1921–1997). In the introduction to his famous book *Pedagogy of the Oppressed*, Richard Shaull has summed up his approach when he argues that the starting point for Freire is that education can never be neutral; it either acts to socialize the learner into the "logic of the present system" or it becomes the "practice of freedom." Freedom here is understood as the capacity of the learner to "deal critically and creatively with reality and discover how to participate in the transformation of their world" (Freire 1996, 16). In this sense Freire's approach contains three key elements: the availability of education opportunities to the broad mass of people; the social and psychological processes that reinforce acts of educational inclusion/exclusion, both within and outside formal educational institutions; and the pedagogical strategies deployed by teachers.

In line with a range of progressive thinkers from the Enlightenment onward, Freire believed that education needed to be made available to men and women from all strata of society, rather than just the social elite. But his most significant contribution concerns the "critical" element within "critical pedagogy" and the pedagogical practices he developed and then wrote about in his many books. He sought to embody a participatory egalitarianism on one hand, but at the same time to create a classroom in which students could think about their life and other people's lives in a new and deeply critical way. For Freire, genuine criticality could not coexist within educational processes that were purely instrumental; hence the question of understanding the underlying purpose of teaching and knowing is a crucial starting point. We would argue that these issues are still of crucial significance in talking about Freire's relevance today, even though it must at the same time be acknowledged that the form in which higher education is offered now has changed enormously since he was working in the field. We are currently living through a period in which higher education is being transformed from its older incarnation as an elite system serving the interests of a privileged few, to a massively expanded global system, which is drawing in hundreds of thousands of people across the world (Cowden and Singh 2013).

The terrain of education today would have probably been unrecognizable to Freire, and given that one of his major attacks on conventional systems was directed at the way they excluded all but the wealthy, it could be argued by defenders of the present arrangements that the availability of education has been substantially democratized. There is no doubt that purely in terms of access, certainly in most developed countries, we do now have something resembling a mass higher education system (see, e.g., Usher and Medow 2014). Yet, ironically, at the same time universities have become much less democratic, both in relation to their internal management structures and their accountability. The reason for this is that the rationale for their expansion has not been concerned with the idea of education as a social good, but rather as a lucrative globally salable commodity. This approach has fundamentally reshaped both the form and the content of higher education. In a detailed analysis of the current and future consequences of this approach on the UK higher education (HE) titled *Sold Out*, the Oxford academic Stefan Collini concludes that a system with "a very good record" in terms of "universally acknowledged creativity, streets ahead of most of their international peers" and in being a positive force "for human development and social cohesion" (2013, 12) is being transformed in the image of the financial institutions that so spectacularly demonstrated their incompetence in the banking collapse of 2007–2008. In a similar vein, Andrew McGettigan (2013) in his forensic examination of the funding of UK universities argues that the introduction of large fees coupled with the transfer of funding from the state and direct taxation to private finance and loans systems is comparable to the "subprime" mortgage market, creating new classes of students with high levels of debt and "subprime degrees." Moreover, this subordination of the university to the logic of finance capital poses serious challenges to the project of critial thinking. In this sense the need for an educational practice concerned with the liberation rather than the domestication of students is as great as it has ever been.

Against the backdrop of the wider context of HE this chapter is centrally concerned with setting out the distinctive contribution of critical pedagogy to the broader question of critical thinking. Much of our focus is on the work of Paulo Freire, but of course his approach does not emerge in a vacuum. For this reason we begin the chapter by revisiting the ideas of key figures within the European Enlightenment; postmodernist claims to his legacy notwithstanding, we feel we need to be absolutely clear that Freire's work stands on this legacy, though like Marx, one of his major influences, it was a legacy he both built on and challenged. We follow this with a discussion of a 1999 essay "Critical Thinking and Critical Pedagogy" by Nicholas Burbules and Rupert Berk that specifically compares critical pedagogy with other concepts of critical thinking. We conclude the chapter with a discussion of the importance of critical pedagogy in the context of the current reshaping of relationships between students and teachers in a neoliberal market model, arguing that Freire's work offers a framework for defending and expanding essential aspects of critical thinking that we regard as universal.

Theorizing criticality—a historical perspective

When we consider the history of the concept of "criticality," it is clear that it is crucial not just for theorizing the basis of education, but it is also, in a wider sense, deeply connected with a capacity for expression within a wider "public sphere," a space where ideas can be discussed and debated openly. The European Enlightenment, with its injunction that we "dare to know!" is crucial for initiating modern concepts of criticality. Immanuel Kant's 1784 essay *What Is Enlightenment?* famously defined "enlightenment" as the "exit of humans from their self-incurred immaturity" (Fleischacker 2013, 13), and this represents the elevation of a concept of criticality based above all on Reason. For Kant, Reason was a universal human capacity, and hence he regarded its denial as a denial of our humanity itself. He defines "thinking for oneself" as "seeking the highest touchstone of truth in oneself (i.e. in one's reason), and the maxim of always thinking for oneself is *enlightenment*" (Kant 1998, 146–147).

Reason in a Kantian framework is understood not just as a universal human capacity but as the capacity for critical engagement, which represents something much greater that the amount of information one possesses. "Becoming enlightened" involves liberating oneself at the level of thought and feeling, and Etienne Balibar has argued that "from Kant onwards...modern idealism is above all a theory of the active *self-construction* of the subject" (1994, xv). The corollary of this, Balibar argues, is the "autonomy of the political," which he characterizes as "reminiscent of a long tradition in the definition of *citizen-ship*...namely the emergence of 'we the people' as a political subject" (1994, x). The Enlightenment definition of criticality was thus inherently political and social, and connected with concepts of popular sovereignty, democratic citizenship, and, in their absence, revolution. These ideas were of course cru-cial aspects of the intellectual background of the French, American, and the Haitian revolutions, and which continue to be important to this day. This relationship between the capacity to use reason in a public and critical way remains as true of the *Declaration of the Rights of Man and of the Citizen* as it was adopted in August 1789 by the French National Constituent Assembly, as it is of contemporary struggles for genuine popular representation manifest in the Arab Spring.

But where does Reason come from? While the establishment of this prin-ciple was one of the most important legacies of the Enlightenment, Hegel's major contribution to this was the idea that it had to be accompanied by the development of "critical self-consciousness." As Pavlides (2010) argues, "Hegel attempted to demonstrate the active role which the human mind played in the evolution of civilization and, at the same time, he became aware of the contradictory essence of things as the moving force behind their transforma-tion" (83). It was by historicizing critical self-consciousness through his use of the "dialectical" method that Hegel established "the principle whereby stable thoughts reveal their inherent instability by turning into their opposites, and then into more complex thoughts" (Houlgate 2005, 38). In this way of think-ing Hegel demonstrated the importance of going beyond either/or forms of logic, thus overturning the perception that "things and concepts [were] either one thing or the other" (Houlgate 2005, 39). Hegel's approach radicalized criti-cality in the way it required a thinker to grasp "contradictions"—essential rela-tionships between things that only appeared to be opposed to each other, but were, at a deeper level, essentially related.

It has almost become a cliche to reiterate Marx's claim to have turned the con-cept of the dialetic "on its head," but as Cyril Smith has noted, it is more useful to think of Marx as taking the method Hegel used for understanding *philosophi-cal* contradictions as a means of understanding *real material* contradictions; in other words, Marx was "looking for the way to 'actualise philosophy'...Where Hegel's science sought to reconcile the conflicting forces of the modern world, Marx's science sets out from the necessity to actualise those conflicts and bring them to fruition" (1996, 147). This is demonstrated by the way Marx approached the question of religious belief. In common with most Enlightenment philoso-phy since Kant, Marx perceived uncritical religious faith as a major barrier to enlightened thought and existence. However rather than see this faith simplyas an "illusion" and thereby illogical, Marx argued that it needed to be understood as the inverted expression of real social contradictions:

> Religious suffering is at one time the expression of real suffering and a pro-test against real suffering. Religion is the sigh of an oppressed creature, the heart of a heartless world and the soul of soulless conditions. It is the opium of the people. (1975, 244)

Marx's description of religion as the "opium of the people" has often been misunderstood to mean that he was simply dismissive of religion. Rather, he saw it an analogous to an opiate, in that it dulled the pain of people's lives and allowed them to carry on, but without any fundamental change in the oppressive conditions in which they lived and worked. It was thus not the clarion calls for freethinking offered by Enlightenment philosophers that would undermine the appeal of religion, but "the abolition of religion as the illusory happiness of the people is the demand for their real happiness" (Marx 1975, 244). After working extensively on his critique of Hegel in his early writings, Marx shifted his focus toward understanding "political economy" where the material causes of the denial of people's humanity were to be found. This shift is captured in his eleventh thesis on Feuerbach: "the philosophers have interpreted the world, in various ways; the point is to change it" (Marx 1975, 423). This statement is remarkable for the way it encompasses criticality as a concept with inherently ethical, epistemological, and pedagogical dimensions, which themselves could only be realized through *praxis,* the unity of theory and practice.

In the twentieth century, as educational institutions expanded, debates around the significance of criticality moved more and more into the space of pedagogical practice. While the American pragmatist philosopher and educational reformer John Dewey did not see himself as a revolutionary in the way Marx did, he was equally concerned with the social implications of pedagogical practices. As Amsler notes, for Dewey, "[An] educator's decisions about what, how, why and where to teach can never be based on purely technical skill or theoretical knowledge. Instead they emerge from theorizing the particular form of democratic life, articulating the practical role that forms of education could play in this life" (Amsler 2013, 67). For Dewey education was not just about making a "good life," but also an essential component of a deepening practice of "democracy"that was predicated on the capacity of people at large being equipped with the skills to turn this into a reality. This is embodied in his oft-quoted statement that one should "cease conceiving of education as mere preparation for later life, and make it the full meaning of the present life" (Dewey 1916, 239).

The same questions about the social role of pedagogy are important in the early work of social theorist Jürgen Habermas, particularly his 1962 book *The Structural Transformation of the Public Sphere* (1989). This work drew heavily on the pessimistic analysis of mass popular culture in the work of his Frankfurt School colleagues Max Horkheimer and Theodor Adorno, but came to quite different conclusions. Returning to Kant's discussion in *What Is Enlightenment?* Habermas argued that the milieu of salons, coffee houses, and independent journals, which formed the context into which Kant's work was received, was very far from the context of the contemporary public sphere. He argued that this had developed primarily into a venue for entertainment where critical discussion was largely absent and social issues were framed in a language of "rational consensus" that was defined and dominated by powerful corporations and the simplistic slogans of political parties. As a result he argued that critical thinking had been "supplanted by manipulative publicity" (Habermas 1989, 178).

In order to prevent a resurgence of the sort of authoritarianism represented by both Nazi Germany and the USSR under Stalin, Habermas advanced the idea of "communicative competence." This concerned the capacity for a human subject to move beyond the dominant "rational consensus" and nurture a praxis whereby they could evaluate truth claims through a combination of reason, reflection, and critical thinking, the purpose of which was to unveil hidden forms of domination. Habermas used the term "ideal-speech situations" to characterize this ongoing struggle for reflective understanding. In ideal-speech situations people were not told what to think, but had the opportunity to participate in a genuine interaction in which it was possible for them to independently evaluate their understandings and views on a particular issue. These ideas have had a major influence in contemporary discussions of the social role of universities and the place of pedagogy within them. Ron Barnett's 1997 book *Higher Education: A Critical Business* is just one such discussion that develops a Habermasian defence of critical thinking in relation to HE in the UK. Barnett argues that it is not enough for university students to develop the capacity to reflect critically on knowledge; it is only through "critical reflection" and "critical action" that the learner can become a truly "critical being" capable of engaging "with the world and with themselves as well as with knowledge" (1997, 1).

Critical thinking and critical pedagogy

The far-from-exhaustive survey demonstrates just how central the relationship between ideas about criticality and a concept of "the public sphere" is, and even if such conceptualizations didn't explicitly articulate a pedagogical dimension, they certainly implied one. But what sort of pedagogy? This question is usefully explored by Nicholas Burbules and Rupert Berk's 1999 essay

"Critical Thinking and Critical Pedagogy: Relations, Differences and Limits." Their focus is a comparative analysis of the way the term "critical" functions within these two traditions of "critical thinking" and "critical pedagogy":

> Each invokes the term "critical" as a valued educational goal: urging teachers to help students become more sceptical toward commonly accepted truisms. Each says, in its own way, "Do not let yourself be deceived." And each has sought to reach and influence particular groups of educators, at all levels of schooling, through workshops, lectures, and pedagogical texts. They share a passion and sense of urgency about the need for more critically oriented classrooms. Yet with very few exceptions these literatures do not discuss one another. Is this because they propose conflicting visions of what "critical" thought entails? Are their approaches to pedagogy incompatible? (Burbules and Berk 1999)

They argue that both traditions deploy the term "critical" as characterized by the defence and expansion of spaces where students are able to reach independent judgments with regard to commonly accepted truth claims, and also argue for "a critical education [which] can increase freedom and enlarge the scope of human possibilities" (Burbules and Berk 1999, 46). But while critical thinking traditions focus on a concern with uncovering faulty arguments in logic, reasoning, and the use of evidence, critical pedagogy's primary concern is "with social injustice and how to transform inequitable, undemocratic or oppressive institutions and social relations" (Ibid.).

Burbules and Berk illustrate these differences using the example of research that purportedly demonstrates that African Americans are "less intelligent" than other ethnic groups, based on the fact that they score lower in IQ tests (1999, 54). Within the critical thinking tradition, concerns about whether such conclusions are justified would be addressed through methodological questions about the reliability of the instruments used to test intelligence; the validity of the findings; and the clarity of key terms, such as the concept of "intelligence." For critical pedagogy, while the latter questions would be important, the underlying problems are not just about methodology and evidence; they would be concerned with the wider context of IQ testing and the role of particular modes of inquiry with respect to power relations—in this instance the role of "intelligence testing" within a context of racist practice and ideology. Hence for critical pedagogy questions such as who is making these assertions about the relationship between "intelligence" and "race," why are they being made at this point in time, who funds this research, and who benefits from the promulgation of these findings are central.

While Burbules and Berk avoid presenting the two traditions as binary opposites, this example demonstrates the different ways in which pedagogy is conceived. Within the critical thinking tradition, this is based on positivist and "unbiased" modes of reasoning and inquiry that allow different truth claims to be evaluated. The distinctive feature of critical pedagogy, by contrast, lies not simply in the process of equipping learners with the skills that enable to them to think critically, but includes within this the idea that the production of knowledge and the identities of learners being themselves socially and ideologically mediated. In this sense the task is not one of seeking to be "unbiased"; instead we need to understand the way dominant frameworks define and constitute that which counts as "knowledge." Freire argues that the educator's knowledge is always inherently incomplete and therefore the "act of knowing" must be based on a critical dialogue between the teacher and student. What he is pointing to here is a way of understanding criticality as a process in which both the educator and the educated seek to "problematize" the basis of forms of existing knowledge, which could be personal and group experiences, "expert knowledges" based on existing research, policy, media perceptions, etc., with a view to looking at the way all these elements interact.

This points to the way the distinction between the two traditions outlined by Burbules and Berk can be read at two levels—that of epistemology and that of pedagogical practice. In terms of epistemology, the distinction between critical thinking and critical pedagogy can be read as a restatement of the differences between Kantian and Marxist approaches. Kant's work represents the beginning of classical liberal philosophy where the use of Reason acts as an expression of what Steutel and Spiecker have called "the autonomy of the individual" (2002, 63). For Kant, critical thinking is perceived as a necessary virtue of citizens and thus as a prerequisite for the sound operation of a society, which needs people who are able to participate in public debates about its overall direction and organization (Ibid.). By contrast Marx rejected the atomistic focus on "individual autonomy" as both philosophically confused and empirically false. He argued that as human beings are essentially social creatures, so social and economic theory must always engage with social totality: "Whenever we speak of production...what is meant is always...production by social individuals" (Marx 1975, 85). In other words Marx's conception of people working to create a material product is the same as people producing and reproducing particular sets of social relations. We can thus never be outside of social relations, whose shape and form have a profound influence on the forms of knowledge that are seen to be important or unimportant. Freire's conception of critical pedagogy draws on a similar understanding of the reproduction of social relations in schools and universities; hence their production of students whose "high level of intelligence" makes them fit to rule society; domesticated, unquestioning students whose knowledge never threatens the powerful; and poor "uneducable" students, excluded from participation in the system.

On speaking and remaining silent

This question of epistemology merges into the issue of the form of pedagogy. While there were egalitarian elements in Kant's thinking, he sees the pursuit of critical thinking as largely confined to formal, traditional intellectuals—those who, in Socrates's times, would have been deemed as "philosopher kings." A key theme in critical pedagogy by contrast is the need for an expanded and more egalitarian conception of intellectuality itself. The ideas of the Italian Marxist Antonio Gramsci are important here since he is one of the first people to theorize the role of "intellectuals" in the production and reproduction of power relations. Against the conventional understanding of intellectuals, whom Gramsci termed "traditional intellectuals," he counterposed what he called "organic intellectuals," who emerged from within and among the popular classes in society. The recognition of "organic intellectuals" linked together Gramsci's emancipatory vision of intellectuals and the idea of proletarian emancipation:

> For a mass of people to be led to think coherently and in the same coherent fashion about the real present world is a "philosophical" event, far more important and "original" than the discovery by some philosophical "genius" of a truth, which remains the property of small groups of intellectuals. (Gramsci 1984, 325)

While Paulo Freire also stands in a broadly Marxist tradition of social transformation, he develops this question differently from Gramsci through a focus on theorizing the mode of participation within the educational processes themselves. This expresses the way critical pedagogy seeks to foreground the impact of social relations of power, which could be at the levels of class, "race," and/or gender, and that act to silence those are who less powerful, in acts of what we might call (following Pierre Bourdieu) "symbolic violence" (Bourdieu and Wacquant 1994, 107–108). The point here is that the capacity of individuals to critically evaluate different truth claims takes place on a radically uneven terrain. Just as Marx argued that the religiosity of oppressed workers represented much more than their lack of enlightenment, so for Freire the passivity of the so-called uneducated cannot be seen as reflecting their lack of capacity for critical thought. Rather, this was an inevitable consequence of their construction within a political economy of *entitlement*; a question of who is and who is not allowed to speak. As he notes, oppressed people "suffer from a duality, which has established itself in their innermost being. They discover that without freedom they cannot exist authentically. Yet although they desire authentic existence, they fear it. They are one and the same time themselves and the oppressor whose consciousness they have internalized" (Freire 1996, 30).

For Freire, traditional didactic pedagogy produced silent, domesticated students for whom "learning" remained entirely separate to their consciousness and subjectivity, and he sought to challenge this by developing critical pedagogical methods that sought to give students the license to speak in their own voices and, in that process, develop critical insights into both themselves and the world they lived in. The distinction he develops is between what he calls "banking education" and "problem-posing education." Within banking education, students are conceived of as "receptacles' to be 'filled' by the teacher. Education thus becomes an act of depositing, in which the students are the depositories and the teacher is the depositor" (Freire 1996, 53). In contrast to this approach, Freire advocated a form of problem-based approach that sought to displace the traditional hierarchical model or teacher/pupil with a dialogical approach that enables both "the problems of human beings in their relations with the world [which] consists of acts of cognition, not transferals of information" (1996, 60–61).

As the quote from Freire above suggests, critical pedagogy involves a dialectic process where the teacher and the student are both engaged in teaching each other and learning from each other. This is not to deny the teacher's knowledge, but this needs to be understood not as a private accumulation, but as work whose inherent social collectivity is realized through engagement with students. Freire's concept of "dialogue" thus represents much more than the inherent value of people talking with each other; it involves a dialectical interchange between theory and experience. Equally it would be a mistake to think of critical pedagogy simply as encompassed by participatory teaching methods. For Freire, critical pedagogy was about the nurturance of intellectual capabilities not just as a tool for developing literacy and understanding, but also as a means of overcoming the "symbolic violence" that situates a person as not entitled to speak. In a book that offers one of the best accounts of Freire's philosophy, Jones Irwin notes that "problematization" is so crucial because "it avoids fatalism and determinism, aspects of behaviour which Freire sees as plaguing the oppressed and their conditions as well as their possibilities for overcoming oppression" (2012, 60). "Speaking" in this sense is linked with the discovery of a capacity for agency.

This idea has been developed in interesting ways by the black feminist writer bell hooks, who worked with Freire in the 1990s. In her book *Teaching to Transgress* hooks begins by reflecting on her own different experiences of pedagogical practices, the first in black-only classrooms that were based on an explicit basis of nurturing critical capacity, compared to being bussed into integrated classrooms:

All our teachers at Booker T. Washington were black women. They were committed to nurturing intellect so we could become scholars, teachers, cultural workers—black folk who used our "minds." We learned that our devotion to learning and the life of the mind was a counter-hegemonic act, a fundamental way to resist every strategy of white racist colonisation. (hooks 1994, 2)

However upon being bussed into integrated schools, she then had to learn that "obedience, and not a zealous will to learn was what was expected of us...We were always and only responding and reacting to white folks" (hooks 1994, 3–4). Hooks uses this starting point to develop an argument about the importance of critical pedagogy in creating a classroom in which the marginalization and silencing of women and black pupils was overturned. In this sense critical pedagogy explicitly seeks to enable a learner to move from self-objectification—at the level of class, "race," or gender—to being a "critically conscious"' subject. McLaren and Da Silva develop this point still further, noting that

> a major consideration for the development of contextual critical knowledge is affirming the experiences of students to the extent that their voices are acknowledged as an important part of the dialogue; but affirming these voices does not necessarily mean that the meaning students give to their experiences can be taken at face value, as if experience speaks romantically or even tragically for itself. The task of the critical practitioner is to provide the conditions for individuals to acquire a language that will enable them to reflect upon and shape their experiences and in certain instances transform those experiences. (1993, 49)

Concluding thoughts: criticality in the neoliberal world

The discussion throughout this chapter makes it clear that historically the ideal of education as a social good is inherently bound up with a concept of democratic citizenship. However, to come back to the present, the neoliberal model, which dominates the practice of universities across the globe, is based on a severance of this connection by promoting a narrowly instrumental notion of higher education. In that sense it represents a major breach with the classical liberal education tradition that has, until recent times, dominated the life of the modern university. Under this new political economy of higher education, students are increasingly treated not as people who are being invited to become members of an academic community, but rather as commodities acquiring a marketable value on the one hand and consumers of services on the other. Likewise, academic staff become less valued for their qualities as educationalists vested with a responsibility to nurture inquisitive critical thinkers, and more as "service providers."

While more traditional, socially elitist versions of academic education have been criticized for their lack of relevance, the new discourse of "relevance" now demanded of universities is one that, like the state itself, embraces a financially driven logic in which the demands of "the market" are paramount. Within this ideological context, the acquisition of knowledge and educational experience is presented largely as a commercial transaction, driven primarily for the benefit of individual students in terms of their employability in an increasingly ruthless labor market. As much as anything else, this undermines genuine criticality in universities, as open-ended educational processes are increasingly displaced by training the role of which is to produce new cadres of unquestioning domesticated students (Giroux 2007, 210). Alongside this, we are also seeing the managerialization of pedagogical practices whereby the sense of teaching as a craft and learning as a process of "drawing out" or self-realization is undermined and replaced, rather like a fast food, by a series of standardized prepacked curricula. Elsewhere we have described this is as analogous to a *Sat-Nav* educational experience (Cowden and Singh 2013) where students are increasingly being taught what to think, but not how to think (Canaan and Shumar 2008). Just as universities as institutions are becoming increasingly defined by the demands of the financial markets to which they are becoming ever more beholden, so also the student experience will come to be defined by cycles of debt to which students are bonded (Cowden and Singh 2013; McGettigan 2013). It is thus what we call the "social context of criticality" or this sense of criticality as a *practice* that we see as most threatened by the neoliberalization of education.

The full consequences of this are still to emerge, but we see already the emergence of a dangerous paradox. The lives of people across the globe are increasingly beset by deep underlying problems that urgently require new thinking, such as increasing ecological crisis resulting from global climate change; escalating social problems that are almost entirely a consequence of a growing chasm of social inequalities, both within and between nations; multiple forms of violence and conflict, particularly those associated with gender, class, and ethnic/communal divisions; and the rise of authoritarian religious fundamentalism and new forms of racialized nationalism. If the crises brought about by what David Harvey has called neoliberalism's "accumulation by dispossession" (2003, 158) are to be resolved in ways that do not destroy the social bonds that make societies viable and sustainable, we urgently need to nurture a socially engaged capacity for critical thinking. For all their problems, there is no escaping the fact that universities are unique in their capacity to contribute to this process. In the current climate it is not an exaggeration to assert that the defense of genuinely critical educational spaces is a defense of the idea of criticality itself. Moreover, in the face of the transforming of the mission of universities from democratic public institutions into businesses, it is equally important that new critical educational spaces, both physical and virtual, are developed outside the institutional structures. It is in this sense that we have sought to argue that critical pedagogy provides a means of nurturing critical-ity among students both as an intellectual pursuit and a social practice. And it is in the work of Paulo Freire that we see the most cogent articulation of a pedagogical project that is capable of enabling students to realize a deeper ethi-cal dimension to learning and education, without which we are impoverished both as individuals and as a society.

References

Amsler, S. 2013. "Criticality, Pedagogy and the Promises of Radical Democratic Education." In *Acts of Knowing: Critical Pedagogy in, against and beyond the University*, edited by G. Singh and S. Cowden. London: Bloomsbury.

Balibar, E. 1994. *Masses, Classes, Ideas: Studies on Politics and Philosophy before and after Marx*. London: Routledge.

Barnett, R. 1997. *Higher Education: A Critical Business*. Buckingham: Society for Research into Higher Education and the Open University Press.

Bourdieu, P., and Wacquant, L. 1994. *An Invitation to Reflexive Sociology*. Cambridge: Polity.

Burbules, N. C., and Berk, R. 1999. "Critical Thinking and Critical Pedagogy: Relations, Differences, and Limits." In *Critical Theories in Education: Changing Terrains of Knowledge and Politics*, edited by T. S. Popkewitz and L. Fender. New York: Routledge. 45–65.

Canaan, J. E., and Shumar, W. 2008. *Structure and Agency in the Neoliberal University*. New York: Bloomsbury.

Collini, S. 2013. Sold Out. *London Review of Books*, October 24,, 3–12.

Cowden, S., and Singh, G. 2013. *Acts of Knowing: Critical Pedagogy in, against and beyond the University*. London: Bloomsbury

Dewey, J. 1916. *Democracy and Education: An Introduction to the Philosophy of Education*. New York: MacMillan.

Fleischacker, S. 2013. *What Is Enlightenment?* Abingdon: Routledge.

Freire, P. 1996. *Pedagogy of the Oppressed*. 2nd ed. New York: Penguin.

Giroux, H. A. 2007. *The University in Chains: Confronting the Military-Industrial-Academic Complex*. Boulder, CO; London: Paradigm.

Gramsci, A. 1984. *Theory of Communicative Action*. Boston: Beacon.

Habermas, J. 1989. *The Structural Transformation of the Public Sphere: An Inquiry into a Category of Bourgeois Society*. Cambridge, MA: MIT Press.

Harvey, D. 2003. *The New Imperialism*. Oxford: Oxford University Press.

hooks, b. 1994. *Teaching to Transgress: Education as the Practice of Freedom*. London: Routledge.

Houlgate, S. 2005. *An Introduction to Hegel*. Oxford: Blackwell.

Irwin, J. 2012. *Paulo Freire's Philosophy of Education: Origins, Developments, Impacts and Legacies*. London, New York: Bloomsbury.

Kant, I. 1998. *Religion within the Boundaries of Pure Reason*. Translated by Di Giovanni. Cambridge: Cambridge University Press.

Marx, K. 1975. *Early Writings*. New Left Review ed. Harmondsworth: Penguin.

McGettigan, A. 2013. *The Great University Gamble: Money, Markets and the Future of Higher Education*. London: Pluto.

McLaren, P., and Da Silva, T. 1993. "Decentering Pedagogy: Critical Literacy, Resistance and the Resistance and Politics of Memory." In *Paulo Freire: A Critical Encounter*, edited by P. McLaren and P. Leonard. London, New York: Routledge.

Pavlides, P. 2010. Critical Thinking as Dialectics: A Hegelian-Marxist Approach. *Journal of Critical Education Policy Studies* 8 (2): 75–102, http://www.jceps.com/PDFs/08-2-03.pdf.

Smith, C. 1996. *Marx at the Millenium*. London: Pluto.

Steutel, J., and Spiecker, B. 2002. "Liberalism and Critical Thinking: On the Relation between a Political Ideal and an Aim of Education." In *The Aims of Education*, edited by R. Marples. London, New York: Routledge.

Usher, A., and Medow, J. 2014. *Global Higher Education Rankings 2010: Affordability and Accessibility in Comparative Perspective*. Higher Education Strategy Associates March 4, 2014. Available from http://higheredstrategy.com/publications/global-higher-education-rankings-2010-affordability-and-accessibility-in-comparative-perspective/.

33
The Knowledge Practices of Critical Thinking

Eszter Szenes, Namala Tilakaratna, and Karl Maton

Introduction

Which knowledge practices demonstrate "critical thinking" in higher education? A rapidly growing literature is addressing what kinds of "thinking" may be considered "critical." However, as yet, there is relatively little analysis of what could be called "actually existing 'critical thinking' in higher education," or the knowledge practices actors consider to be educational evidence of this capacity. The nature of the knowledge in, for example, what students write for tasks aimed at eliciting critical thinking, and what teachers reward in those assessments as evidence of critical thinking, remain underexplored. This chapter briefly illustrates how these knowledge practices can be analyzed in empirical research, drawing on the sociological framework of Legitimation Code Theory (henceforth "LCT").

We begin by arguing the need for the study of the knowledge practices in critical thinking to complement the existing focus of research on exploring cognitive processes of knowing. Second, we introduce LCT as offering conceptual tools capable of capturing the organizing principles of knowledge practices. For brevity, we focus on the concept of *semantic gravity*, which explores the context-dependence of meaning. Third, we enact this concept in illustrative analyses of two assessments ostensibly aimed at eliciting critical thinking: a high-achieving "critical reflection" essay from social work and a "reflective journal" from business. These texts are analyzed in terms of their principal stages, showing changes in the forms taken by the knowledge practices they express. We show that both examples of achievement in critical thinking are characterized by *waves* of semantic gravity, or recurrent movements between context-dependent meanings (such as concrete examples) and context-independent meanings (such as generalizations and abstractions), that *weave* together and transform these different forms of knowledge. We also highlight how this generic attribute is realized differently within the social work and business essays, revealing its subject-specific features. Last, we conclude by briefly discussing how studies using LCT are enabling the understanding of achievement and knowledge-building in ways that can foster students' skills in higher education.

Seeing knowledge practices

Critical thinking is becoming a key focus of research and policy in higher education. A voluminous literature is embracing such far-reaching issues as preparing tertiary students for lifelong learning, active citizenship, and employment. This significance is paralleled in policy by the inclusion of critical thinking in graduate attribute agendas by universities (Barrie 2004; Hammer and Green 2011; Moore 2013). Yet, it remains unclear what critical thinking refers to in terms of the knowledge expressed in pedagogic and assessment practices, that is, what is taught and assessed as evidence of critical thinking.

One reason is a "subjectivist doxa" endemic to much educational research: "the widespread belief that 'knowledge' entirely comprises a state of mind, consciousness or a disposition to act, is wholly sensory in source, and must be inextricably associated with a knowing subject" (Maton 2014b, 4). This doxa is reflected in the tendency to understand critical thinking as exclusively subjective states of consciousness and mental processes—a tendency possibly encouraged by the word "thinking." For example, well-known definitions include "reasonable reflective thinking focused on deciding what to believe or do" (Ennis 1993, 180), and "disciplined, self-directed thinking" (Paul 1990, 52). Similarly, the Delphi panel of 46 experts (Facione 1990) defined critical thinking as a set of cognitive skills (such as analysis, interpretation, inference, and self-regulation). Conversely, the notion of critical thinking as *also* involving the expression of forms of knowledge, such as in classroom discourse and student assessment, is largely obscured. Indeed, even when such educational practices are studied, they tend to be examined for outward signs of mental processes rather than as knowledge practices themselves (e.g., Hammer and Green 2011).

This emphasis on mental processes is echoed by the tendency of studies to focus on perceptions, such as academics' beliefs (Jones 2004) or participants' self-reporting of skills (see Taylor 2007). Moore (2013), for example, examined six academics' understandings of critical thinking at an Australian university, identifying judgment, skepticism, originality, sensitive reading, rationality, critical stance, and self-reflexivity. Similarly, studies of participants' perceptions of their critical thinking skills through interviews or questionnaires focus on such cognitive constructs as "abilities to identify issues and assumptions, recognize important relationships, make correct inferences, evaluate evidence or authority, and deduce conclusions" (see also Phillips and Bond 2004; Tsui 1998; 2000; 2002, 743). While offering insights into actors' *perceptions*, such studies rarely explore the nature of actors' *practices* in higher education.

Thus, what comprises critical thinking and how it is explored are both typically understood in terms of knowing processes located within the minds of knowers. In contrast, the knowledge practices held by actors in higher education to constitute demonstration of critical thinking in classroom discourse and assessments have been relatively neglected. Thus, what is required is a means for analyzing these knowledge practices. Moreover, such analysis needs to move beyond surface features of educational practices to explore their organizing principles, in order to show how these may differ across subject areas and stages of education. For example, a major focus of discussion on critical thinking concerns relations to disciplines. This debate has often polarized into arguments for critical thinking as *either* generic (Ennis 1985; 1997; Kuhn 1991) *or* subject-specific (Atkinson 1997; McPeck 1992; Moore 2011). As a growing number of scholars suggest (Davies 2006; 2013; Moore 2011), there is a need to move beyond this false dichotomy. Doing so in turn requires a means for systematically analyzing the organizing principles underlying knowledge practices, to show what features are generic or specific.

To explain this focus further is perhaps best achieved through illustration. As Moore and Maton (2001, 154) argue, "describing what is obscured by a blind spot is extremely difficult, for what you are trying to point to simply cannot be seen through the current lens." Accordingly, we shall introduce a framework for analyzing knowledge practices (LCT) and enact one of its concepts (semantic gravity) in analyses of student assignments judged by teachers in higher education to successfully exemplify critical thinking. Empirically, we analyze reflective assignments or "written documents that students create as they think about various concepts, events, or interactions over a period of time for the purposes of gaining insights into self-awareness and learning" (Thorpe 2004, 328). This form of assessment is becoming increasingly popular as a means of assessing critical thinking in applied disciplines, including business and management education (Carson and Fisher 2006; Fischer 2003; Swan and Bailey 2004), nursing (Epp 2008; Smith 2011), psychology (Sutton, Townend, and Wright 2007), social work and health sciences (Fook 2002; Fook and Askeland 2007), and teacher education (Hume 2009; Mills 2008; Otienoh 2009). Our examples are drawn from social work and business studies. We should emphasize: we are not concerned with determining whether these assignments demonstrate "thinking," "reflection," or other cognitive processes that are "critical" or otherwise. Rather, our aim is to briefly illustrate how a concept from LCT helps explore the nature of what has been judged by teaching professionals in higher education to demonstrate critical thinking in student writing in different disciplines. We thereby hope to illustrate how this approach can offer insights into how generic and subject-specific attributes of what is considered critical thinking can be analyzed, made explicit, and taught and learned.

Legitimation Code Theory and semantic gravity

Legitimation Code Theory (LCT) is a sociological framework for researching and informing practice (Maton 2013; 2014a; 2014b). It forms a core part of social realism, a broad "coalition" of approaches that reveal knowledge as both socially produced and real, in the sense of having effects (Maton and Moore 2010). LCT extends and integrates ideas from a number of approaches, most centrally those of Pierre Bourdieu and Basil Bernstein. This conceptual development has a close relation with empirical research. LCT is rapidly growing as a basis for studies of education at all institutional levels and across the disciplinary map—from primary schools to universities, from physics to jazz—in a widening range of national contexts, as well as beyond education. (For numerous examples of this body of work, see http://www.legitimationcodetheory.com.) The framework comprises a multi-dimensional conceptual toolkit, where each dimension offers concepts for analyzing a particular set of organizing principles underlying practices. Here, for illustrative brevity, we focus on only one concept: *semantic gravity* (Maton 2013; 2014a; 2014b).

Semantic gravity (SG) refers to the degree to which meaning relates to its context. Semantic gravity may be relatively stronger (+) or weaker (–) along a continuum of strengths. The stronger the semantic gravity (SG+), the more meaning is dependent on its context; the weaker the semantic gravity (SG–), the less meaning is dependent on its context. For example, the meaning of the name for a specific event in the academic subject of history (the 1917 Russian Revolution) embodies stronger semantic gravity than that for a kind of historical event (revolutions), which in turn embodies stronger semantic gravity than theories of historical causation. Semantic gravity thus traces a continuum of strengths with infinite capacity for gradation. It can also be used to analyze change over time by describing processes of *weakening* semantic gravity, such as moving from the concrete particulars of a specific case toward generalizations and abstractions, and *strengthening* semantic gravity, such as moving from abstract or generalized ideas toward concrete and delimited cases.

To analyze change over time one can trace profiles of the relative context-dependence of meanings (Maton 2013; 2014a). Figure 33.1 illustrates three simplified profiles: a "high flatline" (A1) of relatively context-independent meanings; a "low flatline" (A2) of relatively context-dependent meanings; and a "gravity wave" (B) of movement between stronger and weaker semantic gravity (and vice versa). These profiles also illustrate different ranges between their strongest and weakest strengths: A1 and A2 have much lower ranges than B.

This brief introduction is simplified and partial—semantic gravity is but one concept of this sophisticated framework. Nonetheless, it will suffice to illustrate how analyzing the organizing principles of knowledge practices may offer

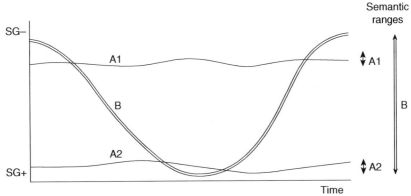

Figure 33.1 Illustrative profiles and semantic ranges.
Source: Adapted from Maton 2013, 13.

insights into what is judged as evidence of critical thinking. To do so we focus on written student assignments. The wider project from which we draw comprises analyses of model "critical reflection" essays from social work (Pockett and Giles 2008) and highly graded "reflective journals" from business studies (collected as part of an ongoing PhD study at a large metropolitan Australian university). To enable detailed illustrative analyses we explore here a single exemplary text from each subject area.

A "critical reflection" essay in social work

Our first text is a high-scoring "critical reflection" essay written by a final year undergraduate student in social work. The essay was published as a model answer in an edited collection titled *Critical Reflection: Generating Theory from Practice* (Pockett and Giles 2008). The purpose of the assignment was to prepare students to enact what is described as a process of critical reflection and thereby "create new professional knowledge" and develop "their emerging identity as 'new graduate social workers' about to enter the workplace" (Pockett and Giles 2008, xiv). To guide their writing, students were asked to

> select a critical incident from their field education experience and using Fook (2002, pp. 98–100), analyse the incident through the process of deconstruction and develop new practice theories as a form of reconstruction…identify, describe and critique key themes within a critical review of literature, and redevelop practical theory in relation to the critical incident. (Pockett and Giles 2008, xiv)

Students were required to "critically reflect on their learning" (Pockett and Giles 2008, xiv) based on Fook's (2002) model of critical deconstruction and reconstruction. This model comprises four stages:

1. *Critical deconstruction*: "searching for contradictions, different perspectives and interpretations" (92);
2. *Resistance*: "refusing to accept or participate in aspects of dominant discourses which work to disempower, or perhaps render a situation unworkable because of this" (95);
3. *Challenge*: "the identification or labeling of both the existence and operation of discourses and that which is hidden, glossed over or assumed" (96); and
4. *Reconstruction*: "formulating new discourses and structures" (96).

Uncovering one's own assumptions about social work practice through this kind of critical reflection is considered a highly valued skill for practitioners as part of fostering social justice (Brookfield 2001; Fook and Askeland 2007). In the assignment students were required to identify a difficult situation or "critical incident" that they encountered during their field placement and discuss that incident using Fook's model. Thus, to successfully demonstrate critical reflection, the incident must become an object of study to be analyzed by the student using ideas from social work.

To explore the model essay we shall begin with its basic structure. The essay comprises five stages that we shall term as follows:

- *Introduction*—in which the student discusses the importance of critical reflection for the subject area of social work;
- *Critical Incident*—where the student narrates an incident from her field placement when she was subjected to verbal sexual harassment;
- *Excavation*—in which the student deconstructs her own "dominant assumptions" by focusing on what she perceives as an inappropriate response to the incident, using "critical deconstruction," "resistance," and "challenge" from Fook's model;
- *Transformation*—where she draws on Fook's notion of "reconstruction" to discuss lessons learned from her experience and acknowledge the need to change her behavior in similar situations in future; and
- *Coda*—where she finishes the essay by emphasizing the role of critical reflection in enabling "self-transformation" in professional practice.

Figure 33.2 traces the profile of semantic gravity characterizing the knowledge claims expressed throughout the essay. One overarching feature to note

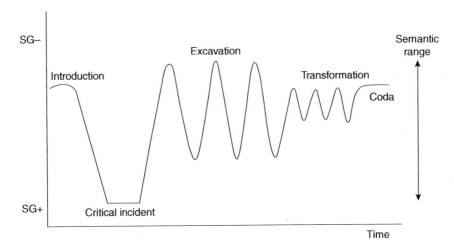

Figure 33.2 Semantic profile of a successful reflection essay in social work.

is the series of *gravity waves* characterizing the essay: recurrent movements are made between concrete particulars (such as an account of the "critical incident") and more generalized and abstracted concepts. The essay thereby weaves together meanings of greater and lesser context-dependence, empirical examples and theoretical constructs, and experiential and academic forms of knowledge. We now turn to explore the particular forms taken by this "semantic weaving" (Maton 2013; 2014a) by addressing in turn the key stages of *Critical Incident, Excavation,* and *Transformation.*

Critical incident

The essay begins by describing in general terms the technical concepts comprising a process of critical reflection in social work (thus the relatively high position of *Introduction* in figure 33.2). In the *Critical Incident* stage, the essay then comprises a short narrative of the student's difficult experience with a young male patient (Jared) who attended a drug and alcohol rehabilitation program. The student, a young female apprentice social worker, was subjected to verbal sexual harassment by the patient during her field placement. While the technical term "critical incident" frames the stage, this concept remains undefined by the student and embedded within the context of the particular case. The student provides an account of her concrete personal experiences that is highly contextualized; for example:

It was in this unit that my critical incident occurred ... I thought as I had established some rapport with the clients previously; I could get them involved [in preparing lunch]. I entered the lounge room where two of the boys were playing a video game. As I approached Jared, I asked "Jared, could you please give us a hand in the kitchen?" The answer was simple and encapsulated my critical incident: "I will if you give us a kiss." (Pockett and Giles 2008, 17)

The series of concrete contexts that give the narrative meaning represent relatively strong semantic gravity, which is maintained through this stage of the essay (*Critical Incident* in figure 33.2). This low gravity flatline works to ground the essay in the specific critical incident, and, as we now show, serves as the launchpad for a weakening of semantic gravity through introducing more detached, "objective," and theorized meanings.

Excavation

The knowledge claims comprising the incident are transformed by the student in the *Excavation* stage through their relation to the concepts of "boundaries," "gender," and "power." This creates a series of gravity waves (figure 33.2) as the essay moves between concepts and concrete examples generalized from the incident. The stage begins by introducing the concepts:

In my incident the emerging themes that I believe warrant further investigation relate to professional practice, namely the issue of boundaries, gender and power. (Pockett and Giles 2008, 17)

The student then strengthens semantic gravity slightly by relating these relatively abstract, context-independent terms to the concrete particularities of her critical incident. For example, the student negatively evaluates her assumptions about feeling obliged to maintain a professional persona while expecting clients to reveal their personal selves:

The irony of my distinction only becomes clear now. While I expect to be able to put on a professional "mask," consisting of the professional skills and knowledge of social work practice when working with clients, I expect clients like Jared to "bare all," to reveal to me their personal problems, issues and insecurities. (Pockett and Giles 2008, 20)

Though more context-dependent than technical concepts, this is not simply empirical description. While grounded in the specific events already recounted, the student is reflecting here on that incident, rising above the specific context to describe more generalized issues, such as feelings of expectations of which the encounter with Jared represents but one instance. Thus, as the profile of figure 33.2 shows, this represents weaker semantic gravity than the *Critical Incident* stage (the bottom of waves here are higher than those of that stage) but stronger semantic gravity than such highly abstract terms as "power" (represented by the peaks of waves).

As well as movements downward, the student also moves the knowledge being expressed back up the profile by transforming these generalized examples into the technical language associated with social work. For example, she redescribes her feelings in conceptual terms:

> Sommers-Flanagan and Sommers-Flanagan (2007) refer to this concept as "one-way intimacies" (p. 163), and as a necessary component of helping relationships. (Pockett and Giles 2008, 20)

A series of these shifts between generalized events and concepts throughout the *Excavation* stage create gravity waves with a high range. This creates the basis for the next stage of the essay in which these two forms of knowledge are transformed to become more closely woven together.

Transformation

In the final main stage of the essay the student draws lessons from the reflection process, such as identifying assumptions and her powerful position as a social worker, and proposes changes to her future practice. Despite the frequent use of personal pronouns (especially "I") that grounds the discussion in the experiences of the author, this stage exhibits weaker semantic gravity than either the earlier narrative of the *Critical Incident* stage or examples from that narrative woven into the *Excavation* stage. Meanings are no longer strongly grounded in the specificities of the case but rather refer to a greater range of potential future cases. Conversely, the focus on concrete practices restricts how high this stage reaches in comparison with the more theoretical parts of the *Excavation* stage. Thus, figure 33.2 locates this *Transformation* stage higher on the profile than *Critical Incident* but not lower than the peaks of *Excavation*. Simply put, as the profile of figure 33.2 shows, this stage is characterized by a closer weaving together of generalizable experiential meanings and conceptual terms; for example:

> I also acknowledge the intersection and overlap between "the personal" and "the professional" and that in any encounter I am not either one identity or the other. I am, for example, a "social worker," a "young person" and a "sexual being" (Stacey et al, 2002), just as the client has many identities, such as "offender,", a "young person," a "brother," and a "student." (Pockett and Giles 2008, 26)

As in the previous stage, *Transformation* involves movements up and down the profile. Exemplifying through contextualized meanings, such as the various identities of the author and the client, strengthen semantic gravity. In turn these various examples are then abstracted into the term "multiple and intersecting identities," weakening semantic gravity:

> There are multiple and intersecting identities which are interwoven and influence each other in any encounter. (Pockett and Giles 2008, 26)

By ending the stage with relatively weaker semantic gravity, the student moves beyond the immediate context of her field placement to demonstrate her ability to *re-examine* existing discourses and her own behavior. This weaker semantic gravity is continued in the final stage of the essay, the *Coda*, where the student concludes by re-examining the value of critical reflection for social work and creating "self-reflective" practitioners.

Overall, the essay begins by being grounded by the critical incident, after which the student shows her capacity to reconceptualize and recontextualize the meanings of this incident through successively weakening and strengthening semantic gravity, weaving together the case with concepts. These meanings are then generalized into future practice. Not only does the student bring together different forms of knowledge, but she also transforms them by theorizing concrete examples and exemplifying concepts—that is, semantic weaving achieved through waves. This offers insight into one potential characteristic of the basis of successful demonstration of critical thinking, as such "critical reflection" essays are held to involve (Pockett and Giles 2008). One issue it raises is whether this profile is reflected in other subject areas. To begin to address this question we now turn to business studies.

A "reflective journal" in business studies

Our second text is a high-achieving "reflective journal" from Business in the Global Environment, a core senior undergraduate Bachelor of Commerce unit. As we discussed earlier, this form of assessment is often claimed to provide a means for encouraging or enabling the demonstration of critical thinking skills. This specific assignment aims to develop students' reflective practice and their intercultural competence, defined in the Unit of Study Outline as "a dynamic ongoing interactive *self-reflective learning* process that *transforms* attitudes, skills and knowledge for effective communication and interaction across cultures and contexts" (Freeman 2009, 1; emphases added). To help students structure their journals, the following questions were provided:

1. Choose one behaviour that you thought was a strength or weakness and identify the "below the surface" value that underpins that behaviour.
2. Having identified the cultural value that you believe underpins your particular strength or weakness, now explain how and from where that cultural

value developed using the "core elements of culture" provided on page 50 of Solomon and Schell.

3. What does this teach you about the way you behave, and your expectations of others, when working in multi-national teams?
4. How might you integrate this awareness into future team work, either at university or in the workplace?

Understanding intercultural differences in business behaviors through this kind of reflective activity is considered an essential skill for working in multicultural organizations (Solomon and Schell 2009) and is one of the most important graduate attributes in business school curricula. In this task students were required to reflect on their experience of multinational teamwork by examining their visible and invisible values, beliefs, assumptions, and behaviors based on Solomon and Schell's (2009) model of intercultural competency.

The journal comprises three principal stages:

- *Excavation*—where the student identifies "individualism" as a "below the surface" value underpinning his experience of a group assignment;
- *Reflection*—in which the student concludes that valuing individualism over his Chinese peers' communitarianism led to his "discounting" of his collaborators' opinions; and
- *Transformation*—where the student pledges that in future teamwork situations his behavior will be guided by the intercultural competence skills he claims to have gained through this reflective process.

Figure 33.3 traces the profile of semantic gravity characterizing the knowledge claims expressed through the journal. Comparing this with figure 33.2

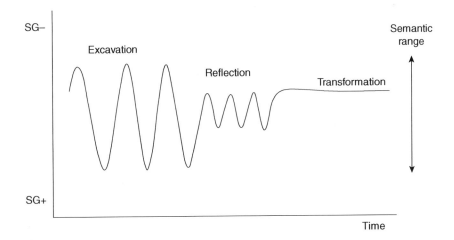

Figure 33.3 Semantic profile of a successful reflective journal in business studies.

highlights the similarities and differences between the two assignments. As in the social work essay, the business journal includes both flatlines (*Transformation*) and waves (*Excavation, Reflection*), and weaves together different forms of knowledge. There are also comparable stages: *Excavation* exhibits similar profiles of semantic gravity in both texts, while *Transformation* in the social work essay resembles the *Reflection* stage in the business journal (though here the *Reflection* stage replaces the *Critical Incident* and comes later). The overall profile also traces a different shape: the waves of semantic gravity in the business journal come earlier and are followed by a flatline. Moreover, this flatline is relatively high (a passage comprising consistently context-independent meanings) rather than the descriptive narrative with which the social work essay began. In short, as we shall show, though waving and weaving again feature in the business studies journal, the different functions they serve here create a different profile.

Excavation

In the first part of the journal the student uncovers a "below the surface value" he possesses—individualism—and outlines general features of Australian culture and history from which this value has evolved:

> Australia's history plays another role in Australia's core culture through its history of immigration (Encarta Encyclopaedia 2009a).... Some of Australia's national heroes are also responsible for developing individualism.

From such wide-ranging generalizations, exhibiting relatively weak semantic gravity, the journal shifts down to concrete examples such as:

> Sir Donald Bradman who is arguably the most famous sporting hero in Australia was made famous for his outstanding individual cricket batting record (*ESPN cricketinfo* 2009).

In turn, from the stronger semantic gravity characterizing these examples, the journal generalizes back up to the notion of "individualism":

> Individualism has consequently evolved from two main areas of core culture, its history and its heroes.

Thus, as figure 33.3 shows, the journal begins by weaving between concepts and cases, abstract ideas and concrete examples. This *Excavation* stage is thus similar in terms of its profile of semantic gravity to the same stage in the social work essay (see figure 33.2). However, in the other essay that stage worked to weave together the preceding empirical description of a critical incident with concepts; here the journal is attempting to ground an abstract idea (individualism) in the context giving the student's actions meaning (his Australian culture). By coming at the start, this stage also establishes from the outset the high semantic range the journal will traverse.

Reflection

In similar fashion to the social work essay's *Transformation* stage, the journal then works at weaving more closely together these extremes of strengths of semantic gravity, creating milder waves. This *Reflection* stage involves generalized cultural values, behavior, and communication styles that are mid-range: more context-dependent than individualism but less context-dependent than specific heroes and historical events. Moreover, these ideas are related to examples that are not simply narrated events but rather generalized through the student adopting a reflective voice. Nonetheless, while traversing less range, the stage again involves waving between stronger and weaker semantic gravity as it weaves together examples with concepts (see figure 33.3); for example:

> Analysing my behaviour and expectations of others with an open mind has led to some astonishing realizations. I was surprised that my long held belief that the vast majority of the world adopted individualism as a value was incorrect. Communitarianism which opposes individualism, emphasizes the need to focus on community interests over an individual's and is the value most widely adopted worldwide (Trompenaars & Hampden-Turner 2000, p.71).

From the mid-range point established by discussing his own (generalized rather than specific) "behaviour and expectations of others" and "realization," the student weakens semantic gravity by redescribing these meanings in terms of abstract concepts of communitarianism and individualism. In turn, the student then strengthens semantic gravity with a personal example of his experience of multinational teamwork, where he negatively judges his own directness toward his peers:

> My group had three members from China where communitarianism is generally valued and other cultural differences such as communication styles made their behaviour seem foreign to me (Trompenaars & Hampden-Turner 2000, p.71). My lack of cultural knowledge led to my use of a direct communication style which is in stark contrast to the commonly indirect Chinese style and may have offended the group members due to my unintentional effect of making them lose face to each other (Fox 2008, p.49–50).

This in turn is followed by a weakening of semantic gravity to describe "all situations":

> While we did receive a high distinction for the case study, clearly in almost all situations my current behaviour and expectation of others in a multi-national team will detract from team cohesion and the task at hand.

In creating milder waves of semantic gravity, the student demonstrates his capacity to weave together theoretical and practical cases; in being situated in the middle of the semantic profile, he also shows this weaving as not too abstracted from the kind of applied situations appropriate to conducting business.

Transformation

In the final *Transformation* stage the student pledges to apply his newly acquired knowledge or intercultural competence skills to future multinational teamwork situations. This stage is couched in terms of the concept of "intercultural competence":

> The development of intercultural competence is the key to overcoming my detrimental behaviour in a multi-national team situation.

Drawing on the process of reflection exhibited in the previous stage, the student provides a list of generalized skills he deems necessary for successful participation in teamwork situations. Despite the use of the personal pronoun "I," the discussion here has moved beyond contextualized meanings to a focus on generalizable practices. In contrast to the rest of the essay, repeated references to concepts from theoretical frameworks and a lack of references to the particulars of the case previous stages have discussed contribute to creating a high flatline (figure 33.3):

> I must acquire cultural knowledge regarding the preferred communication style, values, beliefs and even the core elements of [team members'] culture to ensure team cohesion (Matveev & Milter 2004, p.106). I need to develop behavioural modification skills and change my personality orientation so that I may use the cultural knowledge to facilitate better communication and display cultural empathy rather than embracing detrimental stereotyping (Matveev & Milter 2004, p.106). I must cease discounting behaviour and embrace the full potential a team can offer by facilitating all of the group ideas.

Though drawing on abstract concepts, this stage does not reach as high as the peaks of *Excavation* but approaches the peaks of waves of *Reflection* due to the generalized references to the student's own behavior and personality. The stage ends with references to highly generalized future teamwork situations woven together with the repeated use of the abstract concept of "intercultural competence" that frames the entire stage:

> Team members with even fundamentally different core cultures can work together in harmony and achieve far more than any individual if intercultural competence is embraced which is the view held by Associate Professor of Management Richard Milter (Matveev 2004).

Overall, the journal begins by establishing a wide range of semantic gravity, interweaving abstractions with concrete cases, before bringing these together in the discussion of a particular case, whose meanings are then generalized as rules for future practice. As in our previous example, the student thereby creates gravity waves that weave together and transform different kinds of knowledge, but he does so through a differently staged structure.

Conclusion

There is much emphasis in higher education research and policy on the importance of equipping students with critical thinking skills. Existing research focuses mainly on perceptions of staff and students of cognitively defined skills. Few studies explore student writing to examine the knowledge practices associated with what practitioners in higher education judge as successful demonstration of critical thinking. To do so, we drew here on the concept of semantic gravity from LCT to briefly trace the semantic profiles of high-achieving critical reflection assignments in social work and business studies. We conclude by considering what these illustrative analyses suggest about the knowledge practices of critical thinking and the usefulness of LCT for research into this area.

One aspect common to both texts is that they demonstrate mastery of semantic gravity, and specifically the capacity to create waves that weave together context-dependent and context-independent forms of knowledge, such as empirical cases and abstract concepts, transforming them into generalizable practices for future contexts. While, for reasons of space, we focused on two illustrative texts from social work and business, these characteristics are also being suggested by a rapidly growing number of studies using LCT to explore student work in a range of subject areas, including design (Shay and Steyn 2015), engineering (Wolff and Luckett 2013), English (Maton 2014b), environmental science (Tan 2013), jazz (Martin 2012), journalism (Kilpert and Shay 2013), physics (Georgiou 2015), sociology (Stavrou 2012), and teacher education (Shalem and Slominsky 2010). Such studies highlight that mastering semantic gravity to achieve a high range is crucial to achievement across the disciplinary map. In this chapter we are suggesting that waving, weaving, and a high range may also be generic attributes of knowledge practices associated with demonstrating critical thinking. Other possible generic attributes, at least of critical reflection assignments, include particular stages. Both assignments involved *Excavation*, in which students discuss their behaviors and beliefs to demonstrate what is termed as "critical reflection" by examining their assumptions, and *Transformation*, in which lessons learned about those assumptions are elaborated.

However, the analyses also revealed important differences between assignments. *Transformation* in the social work essay is characterized by mild waving, but in the business journal there is a high flatline of relatively weak semantic gravity. They also involve different additional stages (*Critical Incident* and *Reflection*) and in different orders such that the overall semantic profiles traced by the assignments differ, as shown by comparing figures 33.2 and 33.3. This highlights potential subject-specific differences for further study. For example, in less "applied" disciplines than social work and business, demonstration of critical thinking may be achieved by beginning and ending with more theoretical, abstract, and decontextualized meanings, which are applied to concrete examples, tracing a different profile than those explored in this paper (cf. Maton 2014a). In providing a framework for empirical studies of divergent subject areas, LCT thereby enacts calls to move beyond the false dichotomy of either genericism or subject-specificity by revealing both generic attributes and ways these may be realized differently in disciplinary contexts.

Of course, exploring semantic gravity does not by itself capture the knowledge practices associated with critical thinking. Neither is the concept of "semantic gravity" the whole of LCT: it represents but one isolated part of the framework. Studies are, for example, exploring in tandem the role of "semantic density" or the degree of condensation of meaning in knowledge (see Maton, Hood, and Shay 2015). What the necessarily brief analyses of this chapter demonstrate, however, is the potential value of using such concepts for research into critical thinking. Further, this is not confined to studies of student work. Research into pedagogic practices is revealing the significance of waves of semantic gravity (and semantic density) for cumulative knowledge-building in classrooms (Martin 2013; Maton 2013; Matruglio, Maton, and Martin 2013). This conceptual versatility offers great potential for not only research but also practical pedagogic outcomes. Not all students are able to demonstrate the mastery of semantic gravity that studies suggest is so highly valued across many academic disciplines, and the knowledge practices associated with critical thinking are rarely taught explicitly, leading to students feeling "lost" and "frustrated" (Moreno 2004). By making explicit the nature of knowledge practices that constitute a demonstration of critical thinking, such as waves of semantic gravity, LCT enables the possibility of designing pedagogic interventions for teaching the skills that achieve those practices (cf. Macnaught, Maton, Martin, and Matruglio 2013). Thus, LCT offers a framework that not only can be used to analyze the knowledge practices of critical thinking but also itself embodies those practices. Rather than either the high flatline of decontextualized and abstract discussions or the low flatline of empirical descriptions that remain locked into the specificities of their objects of study, LCT enables research to embrace a high range and to weave together theoretical concepts, empirical research, and practical outcomes. It thereby also enables the knowledge practices of critical thinking.

References

Atkinson, D. 1997. "A Critical Approach to Critical Thinking in TESOL." *TESOL Quarterly* 31 (1): 71–94.

Barrie, S. 2004. "A Research-Based Approach to Generic Graduate Attributes Policy." *Higher Education Research and Development* 23 (3): 261–275.

Brookfield, S. 2001. "Transformative Learning as Ideology Critique." In *Learning as Transformation*, edited by J. Mezirow & Assoc. San Fancisco: Jossey-Bass. 125–149.

Carson, L., and Fisher, K. 2006. "Raising the Bar on Criticality: Student's Critical Reflection in an Intership Program." *Journal of Management Education* 30 (5): 700–723.

Davies, W. M. 2006. "An 'Infusion' Approach to Critical Thinking: Moore on the Critical Thinking Debate." *Higher Education Research and Development* 25 (2): 179–193.

Davies, W. M. 2013. "Critical Thinking and Disciplines Reconsidered." *Higher Education Research and Development* 32 (4): 529–544.

Ennis, R. H. 1985. "Critical Thinking and the Curriculum." *National Forum: Phi Kappa Phi Journal* 65 (1): 28–31.

Ennis, R. H. 1993. "Critical Thinking Assessment." *Theory into Practice* 2 (32): 179–186.

Ennis, R. H. 1997. "Incorporating Critical Thinking in the Curriculum: An Introduction to Some Basic Issues." *Inquiry* 16 (3): 1–9.

Epp, S. 2008. "The Value of Reflective Journaling in Undergraduate Nursing Education: A Literature Review." *International Journal of Nursing Studies* 45 (9): 1379–1388.

Facione, P. A. 1990. *Critical Thinking: A Statement of Expert Consensus for Purposes of Educational Assessment and Instruction*. Millbrae, CA: California Academic Press.

Fischer, K. 2003. "Demystifying Critical Reflection: Defining Criteria for Assessment." *Higher Education Research and Development* 22 (3): 313–325.

Fook, J. 2002. *Critical Deconstruction and Reconstruction*. London: Sage.

Fook, J., and Askeland, G. 2007. "Challenges of Critical Reflection: Nothing Ventured, Nothing Gained." *Social Work Education* 26 (5): 520–533.

Freeman, M. 2009. "Embedding the Development of Intercultural Competence in Business Education." In *Final Report CG6–37*. Australian Learning and Teaching Council.

Georgiou, H. 2015. "Putting Physics Knowledge in the Hot Seat: The Semantics of Student Understandings of Thermodynamics." In *Knowledge-Building: Educational Studies in Legitimation Code Theory*, edited by K. Maton, S. Hood, and E. Shay. London: Routledge.

Hammer, S. J., and Green, W. 2011. "Critical Thinking in a First Year Management Unit: The Relationship between Disciplinary Learning, Academic Literacy and Learning Progression." *Higher Education Research and Development* 30 (3): 303–315.

Hume, A. 2009. "Promoting Higher Levels of Reflective Writing in Student Journals." *Higher Education Research and Development* 28 (3): 247–260.

Jones, A. 2004. "Teaching Critical Thinking: An Investigation of a Task in Introductory Macroeconomics." *Higher Education Research and Development* 23 (2): 167–181.

Kilpert, L., and Shay, S. 2013. "Kindling Fires: Examining the Potential for Cumulative Learning in a Journalism Curriculum." *Teaching in Higher Education* 18 (1): 40–52.

Kuhn, D. 1991. *The Skills of Argument*. Cambridge: Cambridge University Press.

Macnaught, L., Maton, K., Martin, J. R., and Matruglio, E. 2013. "Jointly Constructing Semantic Waves: Implications for Teacher Training." *Linguistics and Education* 24 (1): 50–63.

Martin, J. L. 2012. "Instantiation, Realisation and Multimodal Musical Semantic Waves." In *To Boldly Proceed: Papers from the 39th International Systemic Functional Congress*, edited by J. Knox. Sydney: International Systemic Functional Congress.

Martin, J. R. 2013. "Embedded Literacy: Knowledge as Meaning." *Linguistics and Education* 24 (1): 23–37.

Maton, K. 2013. "Making Semantic Waves: A Key to Cumulative Knowledge-Building." *Linguistics and Education* 24 (1): 8–22.

Maton, K. 2014a. "Building Powerful Knowledge: The Significance of Semantic Waves." In *The Future of Knowledge and the Curriculum*, edited by E. Rata and B. Barrett. London: Palgrave Macmillan.

Maton, K. 2014b. *Knowledge and Knowers: Towards a Realist Sociology of Education*. London: Routledge.

Maton, K., Hood, S., and Shay, E. 2015. *Knowledge Building: Educational Studies in Legitimation Code Theory*. London: Routledge

Maton, K., and Moore, R. 2010. *Social Realism, Knowledge and the Sociology of Education: Coalitions of the Mind*. London: Continuum

Matruglio, E., Maton, K., and Martin, J. R. 2013. "Time Travel: The Role of Temporality in Enabling Semantic Waves in Secondary School Teaching." *Linguistics and Education* no. 24 (1): 38–49.

McPeck, J. 1992. "Thoughts on Subject Specificity." In *The Generalizability of Critical Thinking: Multiple Perspectives on an Educational Idea*, edited by S. Norris. New York: Teachers College Press. 198–202.

Mills, R. June 2008. "'It's Just a Nuisance": Improving College Student Reflective Journals." *College Student Journal* 42 (2): 684–690.

Moore, R., and Maton, K. 2001. "Founding the Sociology of Knowledge: Basil Bernstein, Intellectual Fields and the Epistemic Device." In *Towards a Sociology of Pedagogy: The Contribution of Basil Bernstein to Research*, edited by A. Morais, I. Neves, B. Davies, and H. Daniels. New York Peter Lang. 153–182.

Moore, T. 2011. "Critical Thinking and Disciplinary Thinking: A Continuing Debate." *Higher Education Research and Development* 38 (4): 506–522.

Moore, T. 2013. "Critical Thinking: Seven Definitions in Search of a Concept." *Studies in Higher Education* 38 (4): 506–522.

Moreno, R. 2004. "Decreasing Cognitive Load in Novice Students: Effects of Explanatory versus Corrective Feedback in Discovery-Based Multimedia." *Instructional Science* 32: 99–113.

Otienoh, R. O. 2009. "Reflective Practice: The Challenge of Journal Writing." *Reflective Practice* 10 (4): 477–489.

Paul, R. W. 1990. "Critical Thinking: What, Why and How." In *Critical Thinking: What Every Person Needs to Survive in a Rapidly Changing World*, edited by A. J. A. Binker. Rohnert Park, CA: Center for Critical Thinking and Moral Critique, Sonoma State University.

Phillips, V., and Bond, C. 2004. "Undergraduates' Experiences of Critical Thinking." *Higher Education Research and Development* 23 (3): 277–294.

Pockett, R., and Giles, R. 2008. *Critical Reflection: Generating Theory from Practice;the Graduating Social Work Student Experience*. Sydney: Darlington.

Shalem, S., and Slominsky, L. 2010. "Seeing Epistemic Order: Construction and Transmission of Evaluative Criteria." *British Journal of Sociology of Education* 31 (6): 755–778.

Shay, S., and Steyn, D. 2015. "Enabling Knowledge Progression in Vocational Curricula: Design as a Case Study." In *Knowledge-Building: Educational Studies in Legitimation Code Theory*, edited by K. Maton, S. Hood, and E. Shay. London: Routledge.

Smith, E. 2011. "Teaching Critical Reflection." *Teaching Higher Education* 16 (2): 211–223.

Solomon, C. M., and Schell, M. S. 2009. *Managing Across Cultures—the Seven Keys to Doing Business with a Global Mindset*. New York: McGraw-Hill

Stavrou, S. 2012. "Réforme De L'université Et Transformations Curriculaires: Des Activités De Recontextualisation Aux Effets Sur Les Savoirs." Unpublished PhD thesis, University of Provence, France.

Sutton, L., Townend, M., and Wright, J. 2007. "The Experiences of Reflecive Learning Journals by Cognitive Behavioural Psychotherapy Students." *Reflective Practice* 8 (3): 387–404.

Swan, E., and Bailey, A. 2004. "Thinking with Feeling: The Emotions of Reflection." In *Organizing Reflection*, edited by M. Reynolds and R. Vince. Hampshire, UK: Ashgate. 105–125.

Tan, M. 2013. "Knowedge, Truth and Schooling for Social Change: Studying Environmental Education in Science Classrooms." Unpublished PhD thesis, Ontario Institute for Studies in Education, University of Toronto, Canada.

Taylor, E. W. 2007. "An Update of Transformative Learning Theory: A Critical Review of the Empirical Research (1999–2005)." *International Journal of Lifelong Education* 26 (2): 173–191.

Thorpe, K. 2004. "Reflective Learning Journals: From Concept to Practice." *Reflective Practice* 5 (3): 327–343.

Tsui, L. 1998. *Fostering Critical Thinking in College Students: A Mixed-Methods Study of Influences Inside and Outside of the Classroom*. Los Angeles: University of California.

Tsui, L. 2000. "Effects of Campus Culture on Student's Critical Thinking." *Review of Higher Education* 23: 421–441.

Tsui, L. 2002. "Fostering Critical Thinking through Effective Pedagogy." *Journal of Higher Education* 73 (6): 740–763.

Wolff, K., and Luckett, K. 2013. "Integrating Multidisciplinary Engineering Knowledge." *Teaching in Higher Education* 18 (1): 78–92.

34
Critical Thinking for Educated Citizenship
Monique Volman and Geert ten Dam

Introduction

Among the competences that are considered necessary for a democratic way of living are "consideration for others," "helping others," and "caring for each other" (Westheimer and Kahne 2004). In the past few decades, however, it has been emphasized that a democratic, pluriform society not only requires citizens to be prepared to make their own contributions to society, but also to do that in a critical way (see Ten Dam and Volman 2004; Wardekker 2001). Nowadays people are not expected to "know their place" but to "determine their own position." Of course, the extent to which a "critical" approach is valued and by whom differs, but "to be critical" has become an undeniable part of Western culture; a critical approach is frequently appreciated more than subservient accommodation. In this vein, definitions of "good citizenship" imply that citizens are willing and able to critically evaluate different perspectives, explore strategies for change and reflect upon issues of justice, (in)equality, and democratic engagement in addition to a capacity to function in a socially accepted and responsible manner within a community (Westheimer 2008). This also requires making choices and knowing why you are making that choice, respecting the choices and opinions of others, communicating about these, thereby forming your own opinion, and making it known. The interest since the 1980s in "critical thinking" as an educational goal reflects these new competences that citizenship in modern society demands.

In this chapter we focus on teaching critical thinking as a citizenship competence in higher education. First, we will give a brief overview of the various approaches to critical thinking and the strategies that have been proposed for teaching critical thinking in the last century. Building upon the premise that critical thinking can best be learned in meaningful contexts and in collaboration with others students (Ten Dam and Volman 2004), we then take a further step by focusing on the concept of communities of learners as a pedagogical concept. We differentiate between a socio-constructivist approach and a socio-cultural approach and explore their potential for learning critical thinking as a citizenship competence.

Teaching critical thinking

In the last decades of the twentieth century many people alleged that critical thinking contributes to the development of rational deliberation relevant to a democratic society (Lipman 1991; Weinstein 1991). From different perspectives, critical thinking was proposed as a valuable goal for education. From a philosophical point of view, critical thinking was primarily approached as the norm of good thinking, the rational aspect of human thought, and as the intellectual virtues needed to approach the world in a reasonable, fair-minded way (Ennis 1991; Gibson 1995; Paul 1992) Psychologists conceptualized critical thinking first and foremost as higher-order thinking skills and focused attention on the appropriate learning and instruction processes (Halpern 1998; Kuhn 1999; Pascarella and Terenzini 1991). Last, the concept of critical thinking functioned in "critical pedagogy." Here, critical thinking referred to the capacity to recognize and overcome social injustice (e.g., Burbules and Berk 1999; Giroux 1994; McLaren 1994). In particular the critical pedagogical point of view emphasized critical and democratic citizenship as an educational goal and focused on transforming society (Ten Dam and Volman 2004).

Three debates that characterized the literature on critical thinking of the 1980s and 1990s are still relevant if we consider critical thinking as a citizenship competence. First, it was discussed whether critical thinking must be understood as a set of general cognitive skills and dispositions (e.g., Ennis 1989; Paul 1992; Siegel 1992) or as skills and dispositions that vary in character across different domains (e.g., McPeck 1981; 1990). Second, the "rationalistic" foundations of the epistemology of critical thinking were called into question. It was argued that by focusing on logical thinking, critical thinking excluded other sources of evidence or forms of verification (experience, emotion, feeling) (Burbules and Berk 1999), and was thus gender, class, and culturally biased (Belenky, Clinchy, Goldberger, and Tarule 1986). Last, critical pedagogues argued that critical thinking took insufficient account of the social context (Giroux 1994). From the perspective of teaching critical thinking as a citizenship competence it was considered essential that a curriculum for critical thinking pays attention to the political effects of argumentation and reasoning.

In the literature several instructional strategies for enhancing critical thinking have been proposed and sometimes empirically studied, acknowledging the first issue mentioned above but tending to ignore the other two. These proposals vary from arguments on the starting points for critical thinking to complete instructional designs, and detailed descriptions of teaching strategies or characteristics of learning environments, empirically evaluated or not.

Guidelines for teaching concern ways in which teachers can motivate, activate, and instruct their students to argue logically and solve heuristic problems. Characteristics of instruction that are assumed to enhance critical thinking are promoting active learning; a problem-based curriculum; and stimulating interaction between students. In empirical research in which instructional variables are retrospectively correlated with students' critical thinking skills the importance of a large number of these characteristics is confirmed. This especially holds true for characteristics pertaining to stimulating the active involvement and contributions of students in the learning process, such as an elaborate interaction between students and between students and teacher, having students present their insights or formulate these in an essay exam (instead of testing through multiple choice exams) (see the studies reviewed in Tsui 1999). Another interesting finding is that students in higher education who follow a course of study that requires the integration of ideas and courses across disciplines, and students who follow courses with an interdisciplinary approach, tend to show greater gains in critical thinking than other students (Tsui 1999) (studies reviewed in Terenzini, Springer, Pascarella, and Nora 1995).

Whilst some scholars presuppose that critical thinking is the same across disciplines, and can therefore be learned in specially designed courses (e.g., Ennis 1989; Paul 1992), most argue that generalizable thinking skills do not exist, and thus critical thinking skills cannot be learned in isolation from a subject (Brown 1997; McPeck 1981; 1990). It proves impossible to demonstrate the effectiveness of courses or programs especially devised to improve critical thinking (see the studies reviewed in Tsui 1999). This may be interpreted as supporting the subject-specificity position, thus as arguing in favor of integrating critical thinking in the regular curriculum.

Moreover, the importance of using real-life problems is also often stressed. Many researchers agree that learning how to think critically should take place in the context of meaningful, rich, domain-specific subject matter (e.g., Angeli and Valanides 2009). In particular Brown (1997) has voiced the opinion that critical thinking must be taught in the context of specific subject matter in such a way that transfer to other domains is possible. She argued that we cannot expect children to progress in the development of thinking unless we give them something meaningful to think about. On the one hand this is supposed to be motivating and stimulates students' active involvement. On the other hand real-life problems are precisely the kind of ill-defined, messy, complex problems for which critical thinking is needed anyway (see also Halpern 1998; Kennedy, Fisher, and Ennis 1991). Little guidance was given, however, on how to enhance students' critical thinking competences in a meaningful, subject-oriented way.

Since the beginning of the century socio-constructivist and sociocultural perspectives on education have gained influence and pedagogical approaches have been elaborated, among others by Brown herself, that provided such guidance. It has been argued that, if education is to further the critical competence of students, learning contexts must be chosen that appeal to students and that invite them to engage in critical agency and to reflect upon it. Recently, the concept of communities of learners has been proposed as an instructional format that fits these criteria.

Community of learners as a way of enhancing critical thinking

Many educationalists have built upon the "community of learners" work of Ann Brown and Joe Campione (1990; 1994) (see, e.g., *Journal of Curriculum Studies*, 36 [2], a special issue devoted to the topic). Their model "Fostering a community of learners" (FCL), is aimed at developing deep understanding and critical thinking skills. Students are engaged in a recurring research cycle that involves the following steps: (1) the conduct of research in small groups on central topics for a subject area with each student specializing in a particular subtopic; (2) the sharing of what has been learned in the small group and with the other groups; this can be supported with instructional methods based upon the principles of cooperative learning and include reciprocal learning and the jigsaw method; and finally (3) work on a new "consequential task" that requires the students to combine their individual learning in such a manner that all class members come to a deeper understanding of both the main topic and subtopics. The work of Bereiter and Scardamalia on knowledge-building communities (Bereiter and Scardamalia 1987; Scardamalia and Bereiter 1991) has also been a source of inspiration for enhancing critical thinking. The basic principle is that learners should be engaged in meaningful learning and problem solving while working on authentic problems.

Both models of fostering critical thinking in communities of learners can be considered socio-constructivist approaches. They share a number of characteristics. First and foremost, critical thinking is learned in a meaningful setting (i.e., a context that requires the solution of "real" problems). Students are asked, for example, to study the problem of water pollution in different places within the city of Amsterdam (Beishuizen 2008). The problems to be solved should be (1) related to the subject matter, (2) engaging and thus concern "genuine issues" and have a clear purpose for students, and (3) usually require research-like activities.

Second, critical thinking is enhanced by a social setting in the sense of collaboration between students. Collaboration presumably activates students due to the need to interact with each other, increases the availability of resources since knowledge is distributed among the participants, and ideally results in better arguments and solutions due to the interaction between participants.

Third, critical thinking should be embedded in the basic concepts underlying particular disciplines. In communities of learners based upon the work of Brown, for example, the aim is for students to work in such a manner that they develop a critical understanding of so-called big ideas (e.g., evolution for the discipline of biology) (Campione, Shapiro, and Brown 1995).

Authors working from a socio-constructivist perspective basically argue that students working in a community of learners tend to develop a deep level of subject understanding, a critical attitude, and critical thinking skills as a result of solving real problems in a collaborative manner (see, e.g., Beishuizen 2008). More specifically, in socio-constructivist theories of learning, knowledge is assumed to be something that emerges during a process of active construction and is promoted by interaction.

As might be expected, the realization of an effective community of learners from such a perspective requires careful organization and design of the learning environment and also careful guidance of the learning process. That is, the learning environment associated with an effective community of learners is not a roughly organized environment in which learners simply do some things together but, rather, a setting with often rather detailed instructional formats with a focus on critical thinking and knowledge building (see Edwards 2005).

With regard to the concept of "community" itself, the multi-voicedness of the group is emphasized and valued as this stimulates exchange of viewpoints and critical reflection. The different voices are predominantly elaborated in terms of cognitive abilities. It is argued that students can profit most from each other's knowledge and skills when the knowledge and skills of the class are diverse (Brown and Campione 1994).

Community of learners as a way of enhancing critical citizenship

In our view, critical thinking requires not only higher-order thinking skills, but also a caring attitude, empathy, and commitment (see Noddings 1992). To prepare students for this, instructional designs are needed that do not capitalize on arguing (as in a psychological point of view), nor on the cognitive activity of analyzing social problems (as in critical pedagogy), but contribute to the ability as well as the readiness of students to participate independently in a meaningful and critical way in concrete social practices and activities.

While considerable similarities can be detected with the line of thought outlined above, there are authors who accentuate the importance of learning to participate in their elaboration of community of learners. Vygotskian approaches such as those of Lave and Wenger (1991), Wells (1999), and Rogoff (Gutiérrez and Rogoff 2003; Rogoff, Paradise, Mejia Arauz, Correa-Chávez, and Angelillo 2003) can be mentioned in this connection. In such sociocultural approaches to learning to think critically the notion of "community" is inherent in the definitions of both knowledge and knowing. Exemplary is the view of Wells who construes "knowing" as "the intentional activity of individuals who, as members of a community, make use of and produce representations in the collaborative attempts to better understand and transform their shared world" (1999, 76). Compared to the socio-constructivist approaches discussed above, the conceptualization of learning communities from a sociocultural perspective entails a different set of emphases.

A meaningful setting is—just as for socio-constructivist communities of learners—a key characteristic of sociocultural communities of learners as well. However, when viewed from a sociocultural perspective, a meaningful setting does not entail a specific problem-solving context but, rather, participation in "social practices" in the sense of historically and culturally evolved constellations of human activities that have a particular value and meaning within society (e.g., business, art, care). "Meaningfulness" is, also a two-sided concept, the other side being that participation in such practices is experienced as personally meaningful by students themselves (see Leont'ev according to Van Oers 2009; Van Oers and Wardekker 1999). In a university context, activities organized in such a manner that students can learn something from them are of primary concern (e.g., by taking part in the organization of a symposium, by being a member of the editorial board of a journal, or by joining a research team). By participating in such practices students explore different meaningful roles, such as the role of an organizer, an editor, a researcher, or a lawyer (see Wells 1999). That is, as "legitimate peripheral participants" (Lave and Wenger 1991), students can assume a variety of roles in social practices. Essential in any role is that students' own questions are used as a starting point for learning.

The social setting does not, in a sociocultural approach to communities of learners, refer primarily to the group of students involved in the collaboration but to the activity itself. The social setting encompasses knowledge, concepts, instruments (tools), and so forth. The resources that the students call upon are themselves social products and are meaningful within the activities of the community (Rogoff, Goodman Turkanis, and Bartlett 2001). Students can master these tools by putting them to use within the relevant setting and with an image of the goal to be achieved ("prolepsis"), for example, using guidelines for interviews as tools in a qualitative research project. In the instructional format for such a social setting, the teacher or other more capable adult or peer plays a critical role in the support of the participation of students. That is, the support for learners can be explicitly provided in the form of "scaffolding," which entails helping students to perform tasks that they are not yet capable of performing on their own or—in Vygotskian terms—perform tasks within their "zone of proximal development" (Van de Pol, Volman, and Beishuizen 2010). While the social situations that students participate in are "pre-arranged" to make them suitable for learning objectives, the instructional format is generally less fixed than in socio-constructivist approaches.

When advocates of a sociocultural approach to communities of learners claim that such communities enhance learning, they mean the quality of the participation of students in social practices. This includes, but is not restricted to, disciplinary practices. Learning involves becoming a member of particular communities (Lave and Wenger 1991; Wells 1999). Student learning is, therefore, not so much aimed at the building of a shared knowledge base that encompasses "big ideas" but, rather, at a different type of outcome. The outcome might be, for example, being able to run a store with one's own products, to publish a journal, or to organize a symposium. The acquisition of knowledge and skills is perceived as a "by-product" of these activities. Not only do knowledge and skills undergo development but also the manner in which the student participates in an activity and—in this connection—the identity of the student, which is the motor for subsequent learning processes. Learning is identity building (Wells 1999). "School concepts," or what Vygotsky calls "scientific concepts," play a role in this in a manner similar to how "big ideas" function as "tools for thinking." Such concepts can be distinguished from the "everyday concepts" that children spontaneously acquire. Some authors further emphasize the fact that the participants in an activity can and should learn to be critical participants. The focus of learning should be on transformation (Edwards 2005; Engeström 1999). Thus not only are the rules for participation in an activity important but also, in particular, the degree of freedom associated with students' participation in the activity (Van Oers 2010).

Viewed from a sociocultural perspective, the reasoning about the manner in which a community of learners fosters critical thinking (i.e., the theory of learning) differs from that associated with a socio-constructivist perspective. The ideas about what exactly the outcomes of learning should be also clearly differ. A community of learners from a sociocultural perspective is assumed to enhance critical thinking due to the fundamental intertwining of individual development and the cultural context:

- Students participate in a social setting, that in itself has historically and culturally evolved, and that requires particular knowledge and skills (i.e., the requirements of the activity drives students' development).
- Students actively negotiate the meaning of cultural tools, which are thus acquired ("appropriated") in such a manner. Cultural tools are not only acquired by students, however, but can also be transformed; this constitutes the dynamic character of social practices.

Just like in a socio-constructivist community of learners, the collaboration must be carefully structured in order to promote critical thinking in the sense of critical *participation*, and this requires well-thought-out learning environments and guidance by competent adults. Wells has described some examples of such communities of inquiry. Van Oers (2010) further shows how communities of learners can be designed in such a manner that student activities retain an element of play. The rules for communication inside classrooms and some recommendations for the role of teachers in "dialogic teaching" have been outlined by Mercer (2005).

Also within sociocultural approaches of communities of learners the multivoicedness of the community is recognized. On the one hand, the social practices that are represented in schools are neither homogeneous nor neutral. All learning content refers to social positions and has particular cultural meanings. On the other hand, learners themselves belong to different social groups. As a consequence, they relate differently to learning content and to learning itself. Social identities are thus developed in "learning through participation in social practices" (Volman and Ten Dam 2007). From a sociocultural approach of communities of learners, however, not much attention has been given until now to the question of how differences between students can function as a potential for learning to participate critically.

Conclusion

In this contribution we departed from the premise that critical thinking is an essential competence required by citizens to participate in a modern, democratic society; critical thinking enables citizens to make their own contribution to society in a critical and aware manner. We discussed the concept of "community of learners" as a promising pedagogical approach for promoting critical thinking in education, as it has the potential to overcome the limitations of the "instrumental" and "higher-order skill" perspectives of critical thinking. In particular the sociocultural interpretation of "community of learners" focuses on critical agency. The concept of "participation" is a key concept here. In the participation approach the educational objective must not be formulated exclusively in terms of critical thinking but rather in terms of acquiring the competence to participate critically in the communities and social practices to which a person belongs. This competence includes knowledge and skills and the willingness to use these (agency).

The learning process for this "critical competence" occurs by being actively involved in meaningful social practices. From this perspective, the objective of critical thinking can never be realized by means of special "programs for critical thinking" in which the relevant skills are taught as technical skills. If learning must be meaningful to the individual in order to contribute to identity development (Wardekker 1998), it is essential that connections are made between the learning process and the current and future situation(s) in which students can and want to apply the knowledge and skills they have acquired (see, e.g., Lave and Wenger 1991).

This does not mean that the instructional designs and procedures discussed in part II, "Teaching Critical Thinking," are of no value at all. Even though they are based on different theoretical frameworks, they provide useful guidelines for promoting "logical thinking" (philosophical approach) or "higher-order thinking skills" (psychological approach). These guidelines can be included in elaborations of instructional strategies to be used in the context of communities of learners.

In our view, the idea of a community of learners provides ways to shape education in such a manner that it not only contributes to the pedagogical goals of achieving a deep level of subject understanding and critical thinking skills but also promotes a willingness and capacity to act in diverse social practices on the basis of these competences. This is particularly relevant at a time when economic profit is given priority over teaching students how to think critically and introducing them to complex global questions, as Martha Nussbaum argues in her pamphlet *Not for Profit* (2010). This also transcends a mere theoretical discussion. The issue at stake is how to strengthen the role of higher education in contributing to the development of "educated citizenship." How can education serve as a solid basis for a democratic society and involve students in meaningful educational practices aimed at enhancing the quality of their participation in society?

References

Angeli, C., and Valanides, N. 2009. "Instructional Effects on Critical Thinking: Performance on Ill-Defined Issues." *Learning and Instruction* 19 (4): 322–334.

Beishuizen, J. J. 2008. "Does a Community of Learmers Foster Self-Regulated Learning?" *Technology, Pedagogy and Education* 17 (3): 183–193.

Belenky, M. F., Clinchy, B. M., Goldberger, N. R., and Tarule, J. M. 1986. *Women's Ways of Knowing: The Development of Self, Voice, and Mind*. New York: Basic Books.

Bereiter, C., and Scardamalia, M. 1987. *The Psychology of Written Composition*. Hillsdale, NJ: Lawrence Erlbaum Associates.

Brown, A. 1997. "Transforming Schools into Communities of Thinking and Learning about Serious Matters." *American Psychologist* 52 (4): 399–413.

Brown, A. L., and Campione, J. C. 1990. "Communities of Learning and Thinking, or a Context by Any Other Name." In *Developmental Perspectives on Teaching and Learning Thinking Skills (Contributions in Human Development, 21)*, edited by D. Kuhn. Basel: Karger. 108–126.

Brown, A. L., and Campione, J. C. 1994. "Guided Discovery in a Community of Learners." In *Integrating Cognitive Theory and Classroom Practice: Classroom Lessons*, edited by K. McGilly. Cambridge, MA: MIT Press/Bradford Books. 229–270.

Burbules, N. C., and Berk, R. 1999. "Critical Thinking and Critical Pedagogy: Relations, Differences, and Limits." In *Critical Theories in Education: Changing Terrains of Knowledge and Politics*, edited by T. S. Popkewitz and L. Fender. New York: Routledge. 45–65.

Campione, J. C., Shapiro, A. M., and Brown, A. L. 1995. "Forms of Transfer in a Community of Learners: Flexible Learning and Understanding." In *Teaching for Transfer. Fostering*

Generalization in Learning, edited by A. McKeough, J. Lupart, and A. Marini. Hillsdale, NJ: Lawrence Erlbaum Associates. 35–68.

Edwards, A. 2005. "Let's Get Beyond Community and Practice: The Many Meanings of Learning by Participating." *The Curriculum Journal* 16 (1): 49–65.

Engeström, Y. 1999. "Activity Theory and Individual and Social Transformation." In *Perspectives on Activity Theory*, edited by Y. Engeström, R. Miettinen, and R.-L. Punamaki. Cambridge: Cambridge University Press.

Ennis, C. 1991. "Discrete Thinking Skills in Two Teachers' Physical Education Classes." *The Elementary School Journal* 91: 473–486.

Ennis, R. H. 1989. "Critical Thinking and Subject Specificity: Clarification and Needed Research." *Educational Researcher* 18 (3): 4–10.

Gibson, G. 1995. "Critical Thinking: Implications for Instruction." *Reference & User Service Quarterly (RQ)* 35: 27–35.

Giroux, H. A. 1994. "Toward a Pedagogy of Critical Thinking." In *Re-Thinking Reason: New Perspectives in Critical Thinking*, edited by K. S. Walters. Albany: SUNY Press. 200–201.

Gutiérrez, K., and Rogoff, B. 2003. "Cultural Ways of Learning: Individual Traits or Repertoires of Practice." *Educational Researcher* 32: 19–25.

Halpern, D. 1998. "Teaching Critical Thinking for Transfer across Domains: Dispositions, Skills, Structure Training, and Metacognitive Monitoring." *American Psychologist* 53 (4): 449–455.

Kennedy, M., Fisher, M. B., and Ennis, R. H. 1991. "Critical Thinking: Literature Review and Needed Research." In *Educational Values and Cognitive Instruction: Implications for Reform*, edited by L. Idol and B. F. Jones. Hillsdale, NJ: Lawrence Erlbaum and Associates. 11–40.

Kuhn, D. 1999. "A Developmental Model of Critical Thinking." *Educational Researcher* 28 (2): 16–46.

Lave, J., and Wenger, E. 1991. *Situated Learning. Legitimate Peripheral Participation.* Cambridge: Cambridge University Press.

Lipman, M. 1991. *Thinking in Education.* Cambridge: Cambridge University Press.

McLaren, P. 1994. "Foreword: Critical Thinking as a Political Project." In *Re-Thinking Reason: New Perspectives in Critical Thinking*, edited by K. S. Walters. Albany: SUNY Press. 9–15.

McPeck, J. 1981. *Critical Thinking and Education.* New York: St Martin's Press.

McPeck, J. 1990. "Critical Thinking and Subject Specificity: A Reply to Ennis." *Educational Researcher* 19 (4): 10–12.

Mercer, N. 2005. "Sociocultural Discourse Analysis: Analysing Classroom Talk as a Social Mode of Thinking." *Journal of Applied Linguistics and Professional Practice* 1 (2): 137–168.

Noddings, N. 1992. *The Challenge to Care in Schools: An Alternative to Education.* New York: Teachers College Press.

Nussbaum, M. 2010. *Not for Profit: Why Democracy Needs the Humanities.* Princeton: Princeton University Press.

Pascarella, E., and Terenzini, P. 1991. *How College Affects Students: Findings and Insights from Twenty Years of Research.* San Francisco: Jossey Bass.

Paul, R. 1992. *Critical Thinking: What Every Person Needs to Survive in a Rapidly Changing World.* Santa Rosa, CA: Foundation for Critical Thinking.

Rogoff, B., Goodman Turkanis, C., and Bartlett, L. 2001. *Learning Together: Children and Adults in a School Community.* New York: Oxford University Press.

Rogoff, B., Paradise, R., Mejia Arauz, R., Correa-Chávez, M., and Angelillo, C. 2003. "Firsthand Learning through Intent Participation." *Annual Review of Psychology* 54: 175–203.

Scardamalia, M., and Bereiter, C. 1991. "Higher Levels of Agency for Children in Knowledge Building: A Challenge for the Design of New Knowledge Media." *Journal of the Learning Sciences* 1 (1): 37–68.

Siegel, H. 1992. "The Generalizability of Critical Thinking Skills, Dispositions and Epistemology." In *The Generalizability of Critical Thinking*, edited by S. P. Norris. New York: Teachers College Press. 97–108.

Ten Dam, G., and Volman, M. 2004. "Critical Thinking as a Citizenship Competence: Teaching Strategies." *Learning and Instruction* 14 (4): 359–379.

Terenzini, P. T., Springer, L., Pascarella, E. T., and Nora, A. 1995. "Influences Affecting the Development of Students' Critical Thinking Skills." *Research in Higher Education* 36: 23–39.

Tsui, L. 1999. "Courses and Instruction Affecting Critical Thinking." *Research in Higher Education* 40: 185–200.

Van de Pol, J., Volman, M., and Beishuizen, J. 2010. "Scaffolding in Teacher-Student Interaction: A Decade of Research." *Educational Psychology Review* 22: 71–297.

Van Oers, B. 2009. "Developmental Education: Improving Participation in Cultural Practices." In *Childhood Studies and the Impact of Globalization: Policies and Practices at Global and Local Levels—World Yearbook of Education 2009*, edited by M. Fleer, M. Hedegaard, and J. Tudge. New York: Routledge. 293 – 317.

Van Oers, B. 2010. "Children's Enculturation through Play." In *Challenging Play: Post Developmental Perspectives on Play and Pedagogy*, edited by L. Brooker and S. Edwards. Maidenhead: McGraw-Hill.

Van Oers, B., and Wardekker, W. 1999. "On Becoming an Authentic Learner: Semiotic Activity in the Early Grades." *Journal of Curriculum Studies* 31 (2): 229–249.

Volman, M., and Ten Dam, G. 2007. "Learning and the Development of Social Identities in the Subjects Care and Technology." *British Educational Research Journal*, 33 (6): 845–866.

Wardekker, W. 1998. "Scientific Concepts and Reflection." *Mind, Culture and Activity* 5: 143–153.

Wardekker, W. 2001. "Schools and Moral Education: Conformism or Autonomy?" *Journal of Philosophy of Education* 35: 101–114.

Weinstein, M. 1991. "Critical Thinking and Education for Democracy." *Educational Philosophy and Theory* 23: 9–29.

Wells, G. 1999. *Dialogic Inquiry: Towards a Sociocultural Practice and Theory of Education.* Cambridge: Cambridge University Press.

Westheimer, J. 2008. "On the Relationship between Political and Moral Engagement." In *Getting Involved: Global Citizenship Development and Sources of Moral Values*, edited by F. Oser and W. Veugelers. Rotterdam/Taipei: Sense. 17–29.

Westheimer, J., and Kahne, J. 2004. "What Kind of Citizen? The Politics of Educating for Democracy." *American Educational Research Journal* 41: 237–269.

Contributors

Richard Andrews is deputy vice-chancellor for research and innovation at Anglia Ruskin University, Cambridge, and previously was professor in English at University College London's Institute of Education. His research focuses on argumentation, writing development, e-learning, and English studies. As well as having taught on the MA in English Education, he has supervised research students in the above fields and in second-language learning, supplementary schooling, and student voice. He is the author of *Argumentation in Higher Education* (2009), *Re-framing Literacy* (2011), and *A Theory of Contemporary Rhetoric* (2014); he is currently working on *A Prosody of Free Verse: Explorations in Rhythm* and the second edition of the *SAGE Handbook of E-learning Research*. He is co-series editor of the Cambridge School Shakespeare and chair of its international advisory board for Cambridge University Press.

Sharon Bailin is professor emeritus in the Faculty of Education, Simon Fraser University. She has written extensively on critical thinking and on creativity, and is the author of *Achieving Extraordinary Ends: An Essay on Creativity*, and, along with Mark Battersby, *Reason in the Balance: An Inquiry Approach to Critical Thinking* (2010). Bailin is one of the originators of a conception of critical thinking that has formed the foundation of a major curriculum project for K-12 schools, and she currently directs a masters program on critical inquiry for educators at Simon Fraser University. Bailin is a past president of the Philosophy of Education Society and the Association for Informal Logic and Critical Thinking.

Maha Bali holds a PhD in Education from the University of Sheffield in the United Kingdom, and her thesis is titled *Critical Thinking in Context: Practice at an American Liberal Arts University in Egypt*. Her research interests include critical thinking, higher education, faculty development, intercultural learning, e-learning (including MOOCs) and citizenship education; she has recently published articles in the journals *Teaching in Higher Education* and *Hybrid Pedagogy* tackling issues of digital pedagogy for the global South. Two of her articles "Critical Citizenship for Critical Times" and "Unveiling Prejudice" written for *Al-Fanar Media* were among the top four best opinion articles published in the magazine for the year 2013. She is currently associate professor of Practice at the Center for Learning and Teaching at the American University in Cairo.

Ronald Barnett is emeritus professor of Higher Education, Institute of Education, University College London, UK. He is an honorary research fellow in the Department of Education, University of Oxford, UK, and a fellow of the Academy of Social Sciences, the Society for Research into Higher Education, and the Higher Education Academy. His recent books include *Being a University* (2010) and *Imagining the University* (2013).

Mark Battersby holds a doctorate in Philosophy from the University of British Columbia and is a retired professor of Philosophy from Capilano University. He has also taught critical thinking at the University of British Columbia, Simon Fraser University, and Stanford University. Battersby is the founder of the British Columbia Association for Critical Thinking Research and Instruction. He also led a provincial curriculum initiative focusing on learning outcomes in higher education. He has written numerous articles on critical thinking and is the author of *Is That a Fact?* (2009), and, along with Sharon Bailin, *Reason in the Balance: An Inquiry Approach to Critical Thinking* (2010).

Margaret Blackie holds a doctorate in Chemistry from the University of Cape Town, South Africa. She is a senior lecturer in the Department of Chemistry and Polymer Science at Stellenbosch University and a research associate in the Faculty of Theology. She has research interests in medicinal chemistry, science education, and Ignatian spirituality.

Tracy Bowell is senior lecturer in Philosophy at the University of Waikato in Hamilton, New Zealand, and holds degrees in Philosophy from the universities of Sussex, Cambridge, and Auckland. Her research interests include critical thinking and argumentation, Wittgenstein, and feminist philosophy. She is the author of a number of papers in these areas, and also of (with Gary Kemp) *Critical Thinking: A Concise Guide*, the fourth edition of which was published in 2014.

Eva M. Brodin is an associate professor in Educational Sciences and a senior lecturer at the Department of Psychology at Lund University in Sweden. She also works as an educational developer at Lund University within the field of doctoral education across all faculties and holds a research position at the Centre for Higher and Adult Education at Stellenbosch University in South Africa. Brodin has conducted research on critical thinking in higher education since her doctoral studies in 2002–2007, and on doctoral students' learning since 2008. In 2012 Brodin and co-author Liezel Frick gained the Emerald Literati Network Award for Excellence for their article on conceptualizing and encouraging critical creativity in doctoral education. Brodin has continued to investigate and broaden this field of research.

Stephen Brookfield is the John Ireland Endowed Chair at the University of St. Thomas in Minneapolis-St. Paul, Minnesota. He has written, cowritten, or edited seventeen books on adult learning, critical thinking, teaching, discussion and critical theory, six of which have won the Cyril O. Houle World Award for Literature in Adult Education. In his spare time he plays in *The 99ers*, a pop punk, rock-and-roll band.

Lorelle Burton is professor of Psychology at the University of Southern Queensland. Lorelle has been an invited assessor for national teaching excellence awards and grants and has led numerous national collaborative research projects on student transition. She is an internationally recognized psychology educator and her current research focus involves leading cross-community collaborations to promote community capacity building and well-being.

Elmarie Costandius holds a doctorate in Curriculum Studies from the University of Stellenbosch, South Africa, and is a lecturer in Visual Communication Design at the Visual Arts Department at Stellenbosch University. In 2012 she received a teaching fellowship from the Department of Education, and in 2013 a teaching award from HELTASA/Council of Higher Education (CHE) in South Africa. Her research focuses on social responsibility and critical citizenship in art education.

Sharon K. Chirgwin has a PhD in Science and postgraduate qualifications in Indigenous Knowledges, which has enabled her to publish in a wide range of areas in the last six years including Ethno-ornithology, Indigenous Research and Nursing Training. She currently works with postgraduates at Menzies School of Health, Darwin, providing research training and public health perspectives to health professionals working with Indigenous Australians. She was formerly coordinator in the Research, Teaching and Learning Division of Batchelor Institute of Indigenous Tertiary Education, where she provided training and supervision for Higher Degree by Research students.

Stephen Cowden is a senior lecturer in Social Work at Coventry University, where he has worked since 2001. Originally from Melbourne, Australia, Stephen has lived in the United Kingdom for over twenty years. He completed his PhD at the University of Kent in 1999, which looked at Australian nationalism and the construction of discourses of "white" identity in Australian literature. His research in social work is concerned with issues of social justice, social work ethics, and the impact of neoliberalism on social welfare and education, and he has published in all these areas. He recently published a jointly edited book with Gurnam Singh in 2013 titled : *Acts of Knowing: Critical Pedagogy in, against and beyond the University.*

Martin Davies is an honorary principal fellow in Higher Education in the Graduate School of Education at the University of Melbourne, and a senior learning advisor at Federation University Australia. He holds doctorates from the University of Adelaide (2002) and Flinders University (1997). His recent books are *Study Skills for International Postgraduate Students* (2011) and (with M. Devlin and M. Tight) *Interdisciplinary Higher Education: Perspectives and Practicalities* (2010).

Yu Dong holds a doctorate in Logic and Philosophy of Science from McMaster University (1993). He has taught critical thinking in Canada and China. He is the author of *Principles and Methods of Critical Thinking* (2010) and was past editor-in-chief of *China Newsletter on Critical and Creative Thinking Education*. He is currently leading and developing a nationwide training curriculum for teachers of critical thinking and subject-matter courses in higher education in China

Peter Ellerton is director of the University of Queensland Critical Thinking Project and is currently completing a PhD in Educational Philosophy. He has worked for many years as a curriculum head of science, mathematics, and philosophy in high schools and is a consultant to the International Baccalaureate Organization in the design and implementation of science curricula. He won the 2008 Australian Skeptics prize for Critical Thinking for his work in developing critical thinking educational resources and has appeared on ABC Radio National's *Philosopher's Zone* and *Ockham's Razor*. He was left unimproved by being faith-healed on national television.

Robert H. Ennis is a professor emeritus of Philosophy of Education at the University of Illinois. He has also been a professor of Philosophy of Education at Cornell University, president of the Association for Informal Logic and Critical Thinking (AILACT), president of the Philosophy of Education Society, director of the Critical Thinking Projects at Cornell and Illinois, and a fellow at the Center for Advanced Study in the Behavioral Sciences, Stanford. He received an award "recognizing his profound leadership and influence on analysis and research in critical thinking" at the Sixth International Conference on Thinking, and has authored, coauthored, or coedited over eighty publications concerning critical thinking. Ennis has done empirical and conceptual research in deduction and other aspects of critical thinking, and he is currently working on critical thinking assessment, critical thinking across the curriculum, and causality.

Sandra Grace is an associate professor in Osteopathic Medicine at Southern Cross University and a research fellow at the Education for Practice Institute, Charles Sturt University. Her roles include enhancing the scholarship and practice of osteopathy through teaching, student supervision, research, and publications. She has extensive experience as a practitioner in private practice and as a curriculum developer, teacher, and clinical supervisor in higher education. Her research interests include models of primary care, interprofessional learning and practice, and practice-based education.

Paul Green earned his doctorate in philosophy from the University of California, Irvine, in 1997. He is associate professor of Philosophy at Mount Saint Mary's University in Los Angeles, California. His current scholarship focuses on pedagogy and the philosophical foundations of pedagogy.

Phil Griffiths is a lecturer in political economy at the University of Southern Queensland and holds a doctorate in political science from the Australian National University. His key areas of expertise in research include Australian political history, along with class analysis, racism, and the history of the labor movement, as well as his research on student development of critical thinking in the humanities.

Benjamin Hamby holds a PhD in Philosophy from McMaster University, in Hamilton, Canada. His thesis, *The Virtues of Critical Thinkers*, is an inquiry into the character of the ideal critical thinker. He has published articles on argumentation theory and critical thinking theory and pedagogy, and he has instructed at San Francisco State University, McMaster University, and served as an adjunct professor with Central Texas College on deployed United States Navy ships. He was awarded the Association for Informal Logic and Critical Thinking (AILACT) Essay Prize in 2011 and 2012, the McMaster University Teaching Assistant Excellence Award in 2012, and is the author of a supplementary textbook on critical thinking titled *The Philosophy of Anything: Critical Thinking in Context* (2007). Dr. Hamby is currently a full-time lecturer in the Department of Philosophy and Religious Studies, at Coastal Carolina University, in Conway, South Carolina.

Sara Hammer is a lecturer in learning and teaching at the University of Southern Queensland and holds a doctorate in Applied Ethics from the Queensland University of Technology. Key areas of expertise in her research and practice include the development and assessment of graduate attributes, including both lifelong learning and critical thinking.

Maralee Harrell holds a doctorate in Philosophy and Science Studies from the University of California, San Diego (2000), and is currently associate teaching professor of Philosophy at Carnegie Mellon University. Presently she is the director of Undergraduate Studies in the Philosophy Department, on the editorial board of *Teaching Philosophy*, and the author of "Argument Diagramming and Critical Thinking in Introductory Philosophy" (*Higher Education Research and Development*, 2011).

Denise M. Higgins is an educational developer, currently working in the College of Asia and the Pacific at the Australian National University on curriculum development for online and blended learning. She has managed a number of large-scale, externally funded educational research and development projects, including "Teaching Research—Evaluation and Assessment Strategies for

Undergraduate Research Experiences" (TREASURE) led by Susan Howitt and Anna Wilson.

David Hitchcock is professor emeritus of Philosophy at McMaster University in Hamilton, Canada. He is the author of *Critical Thinking: A Guide to Evaluating Information* (1983) and coauthor with Milos Jenicek, MD, of *Evidence-Based Practice: Logic and Critical Thinking in Medicine* (2005). He coedited with Bart Verheij *Arguing on the Toulmin Model: New Essays in Argument Analysis and Evaluation* (2007). He has published articles on inference evaluation, relevance, the concept of argument, practical reasoning, instrumental rationality, argumentation schemes, *ad hominem* arguments, and the effectiveness of computer-assisted instruction in critical thinking. He was the founding president (1983–85) of the Association for Informal Logic and Critical Thinking.

Susan M. Howitt holds a doctorate in Biology and a Master of Higher Education from the Australian National University. She is the recipient of several teaching awards and is currently deputy head of the Biology Learning and Teaching Centre at the Australian National University. Her research interests are in the area of student learning about research, conceptual understanding, and developing higher-order thinking skills.

Henk Huijser has a background and PhD in screen, media, and cultural studies, and he has published widely in both the areas of media and cultural studies and learning and teaching in higher education. He joined Batchelor Institute of Indigenous Tertiary Education in June 2012 as an online learning specialist after two years at Bahrain Polytechnic in the Arabian Gulf. Henk is currently senior lecturer in the subject of Flexible Learning and Innovation, and Higher Degrees by Research Coordinator.

Alan Jones is a senior research fellow in the Department of Linguistics at Macquarie University, Sydney. His research interests include the specialized discourses of disciplines and professions, text-oriented discourse analysis, critical discourse analysis and discourse theory. He is particularly interested in strategic uses of polysemy and ambiguity. Alan has carried out research and copublished with professionals in the fields of physics, contract law, and accounting.

Anna Jones holds a doctorate in Higher Education from the University of Melbourne and is currently professor of Education at the Centre of Research in Lifelong Learning at Glasgow Caledonian University. Her research interests include critical thinking, graduate attributes, academic policy and practice, and medical education.

Justine Kingsbury is senior lecturer in Philosophy at the University of Waikato in Hamilton, New Zealand. She works in a number of different areas: aesthetics, philosophy of mind, informal logic, and metaphilosophy. She has written a number of articles in these areas, and recently coedited *Millikan and Her Critics* (2013).

Masuo Koyasu holds a PhD in Psychology from Kyoto University, Japan (1997), and is currently professor of Psychology at the Graduate School of Education, Kyoto University. He also holds a position as the president of Japan Society of Developmental Psychology (2008–2014). His research interests have focused on how young children develop an understanding of others' minds, as well as children's abilities in hypothesis-testing, understanding nonliteral expressions, perspective-taking, and human figure drawing. He has published many psychological papers in such journals as *British Journal of Developmental Psychology*, *Journal of Social Psychology*, *Psychologia*, and so on.

Takashi Kusumi holds a doctorate in Psychology from Gakushuin University, Japan. He is currently professor in Cognitive Psychology at the Graduate School of Education at Kyoto University in Japan. He has been working on psychological research related to thinking, language, and education. In particular, he has been researching the cultivation of critical thinking and scientific literacy. He edited the following Japanese books, *Practical Intelligence of Professional Experts* (2012), *Development of Critical Thinking in Higher Education* (2011), *Thought and Language: Theories and Applications of Cognitive Psychology* (2010), and *New Directions in Metaphor Research* (2007).

Joe Y. F. Lau obtained his PhD from MIT, and is currently associate professor at the Department of Philosophy, the University of Hong Kong. His textbook on critical thinking, *An Introduction to Critical Thinking and Creativity: Think More, Think Better*, was published by Wiley in 2011. He was a university teaching fellow in 2006. In 2011, he received the Arts Faculty Knowledge Exchange Award for his online Web site *Critical Thinking Web*, at http://philosophy.hku.hk/think.

Brenda Leibowitz holds a doctorate from the University of Sheffield in Education. She was director of the Centre for Teaching and Learning at the University of Stellenbosch for ten years. In January 2014 she moved to the University of Johannesburg to take up a chair in Teaching and Learning in the Education Faculty. Her key task is to encourage the scholarship of teaching and learning among academics. Her research interests focus on enhancing teaching and learning and social justice in education. She is a coeditor of the *International Journal of Academic Development* and convener of the national CHE-Heltasa annual teaching awards.

Stephen M. Llano, PhD, is director of Debate and and assistant professor at St. John's University in New York City. He has taught or lectured on the art of debating worldwide, most notably in Japan, Ukraine, and Morocco. He studies the relationship of debate to the field of rhetoric, and the pedagogy of argumentation. He frequently publishes in the *Monash Debate Review*, and blogs about pedagogy and argumentation at sophist.nyc.

Jason M. Lodge, PhD is a psychological scientist and research fellow at the Science of Learning Research Centre, University of Melbourne. Jason's research concentrates on the psychology of learning. Specifically, he explores the technological, cognitive, and emotional factors that influence learning, thinking, and the development of expertise. His research has implications for the development of learning technologies and for informing educational practice and policy.

Beatrice Lok holds a doctorate from the University of Cambridge (2008) and is a postdoctoral fellow at the Centre for Learning Enhancement and Research in the Chinese University of Hong Kong, in Hong Kong. She is interested in the areas of second-language learning motivation, second-language development of non-native English speaking students, and quality enhancement in higher education.

Rhoda Malgas is a lecturer at the Department of Conservation Ecology and Entomology at Stellenbosch University. She is interested in advancing student learning that goes beyond the content of sustainability science and that includes students appreciating their own position as the citizens they wish to serve. She comanages an internationally funded program that aims to advance graduate learning for sustainable agriculture in South Africa.

Emmanuel Manalo holds a PhD in Psychology from Massey University, New Zealand. He is a professor at Kyoto University's Graduate School of Education, in Japan. His research interests include student learning strategies (including the use of diagrams, critical thinking, and memory strategies), teaching and learning in higher education, and language learning. He is author of several books, including *The Business of Writing* (2009) and *Thinking to Thesis* (2004), and numerous research papers in edited book series like *Diagrammatic Representation and Inference*, and journals like *Learning and Instruction* and *Journal of Applied Research in Memory and Cognition* (*JARMAC*).

Karl Maton is associate professor of Sociology at the University of Sydney and honorary professor at Rhodes University (South Africa). Karl is the principal author of Legitimation Code Theory (LCT), which is being widely used by researchers in education, sociology, and linguistics, including studies of "critical thinking". Karl recently coedited *Social Realism, Knowledge and the Sociology of Education* (with Rob Moore, 2010) and *Disciplinarity: Functional Linguistic and Sociological Perspectives* (with Fran Christie, 2011). Karl's book *Knowledge and Knowers: Towards a Realist Sociology of Education*, which sets out key ideas from LCT, was published by Routledge in 2013; a primer of research studies using

LCT, *Knowledge-Building: Educational Studies in Legitimation Code Theory,* is being published by Routledge this year.

Celina McEwen holds a PhD in Performance Studies/Anthropology from the University of Sydney and she is an adjunct research fellow at the Education For Practice Institute at Charles Sturt University, Australia. She has been awarded a research fellowship with the Research Institute for Professional Practice, Learning and Education at Charles Sturt University. Her research interests are in the areas of critical university studies, social and professional practice, intercultural competence, and cultural development.

Yasushi Michita holds a Master of Arts degree in Experimental Psychology from Hiroshima University (1988), and he is currently professor at the Faculty of Education in University of the Ryukyus in Okinawa, Japan. He has been working on psychological research related to critical thinking and school education. He has won the Professor of the Year Award at the University of the Ryukyus three times. He has written articles and books about critical thinking, including a book for researchers, *Development of Critical Thinking in Higher Education* (2011), and a book for nonacademic people, *The Strongest Critical Thinking Maps* (2012).

Ian Nell holds a doctorate from the University of Pretoria, South Africa, and is an associate professor in Practical Theology at the Faculty of Theology at Stellenbosch University, South Africa. He coordinates the Masters of Divinity program at the Faculty and has written numerous research papers and book chapters on religious leadership, practical theology, and homiletics. His research interests include the integration of graduate attributes in the programs of the Faculty of Theology and he has received a research grant for the project from the Centre for Teaching and Learning at Stellenbosch University.

Nancy November lectures in musicology at the University of Auckland. Her research and teaching interests center on the music of the late eighteenth and early nineteenth centuries: aesthetics, analysis, reception history, and social history. She recently completed a book titled *Beethoven's Theatrical Quartets Opp. 59, 74, and 95* (2013) and is currently producing new editions of Beethoven's middle-period string quartets, to be published by Henle (Munich), as part of the Beethoven Werke series. Her latest book project, titled *Cultivating Chamber Music in Beethoven's Vienna*, engages her twin interests in social history and chamber music making around 1800.

Erin O'Connor, PhD, is a lecturer in Psychology at the Queensland University of Technology in Australia. Erin was awarded the Australian Psychological Society's Early Career Teaching Award in 2012 for innovative curriculum design and her research focuses on adult learning and decision making within higher education and health contexts. Erin is also a practicing psychologist and has completed postdoctoral training in multidisciplinary learning at the University of Minnesota, USA.

Paul J. Orrock is an osteopathic clinician and academic. He is senior lecturer and inaugural head of the osteopathic program at Southern Cross University. Paul has a Masters degree by research and is currently studying for his PhD looking at pragmatic clinical trial methodology. Paul is active in clinical and educational research, the results of which have been published and presented internationally.

Pamela J. Roberts has worked as an academic developer at the Australian National University, where she convened the Graduate Certificate in Higher Education, and she is now at the University of Canberra. She has a background in engineering and is currently undertaking a PhD in higher education.

Chris Robinson holds a PhD in American Studies from the University of Kansas. His primary research focuses on jazz criticism, black music, and critical race and gender theory. He is also a freelance jazz writer, and his work has been published in *Downbeat, Earshot Jazz, Jazz Perspectives,* and the *Grove Dictionary of American Music.* He is a copywriter at Duke University Press.

Sophia Olivia Rosochacki holds an undergraduate degree in Graphic Design, with majors in English Literature, Art History, and Philosophy, an honours degree in Political Philosophy and Social Theory and a masters degree in Sociology. For the last five years she has been involved in numerous programs and initiatives based in the Western Cape, both NGOs and through universities, using the arts as a means to engage with social issues. She has been involved in teaching, program development, project coordination, as well as research and writing on this topic. She has taught art subjects at Stellenbosch University, both practical and theoretical, and developed community-based projects. She now works as the research manager at the African Arts Institute in Cape Town.

Rhonda Shaw is course director for the School of Psychology, Charles Sturt University (CSU). Her PhD, which examined age-related differences in verbal, visual, and spatial working memory, was conferred in 2007. Before taking up the position of course director she worked as a lecturer in psychology at CSU teaching foundations of psychology, cognition, biopsychology, health psychology, and geropsychology. Rhonda has a special interest in teaching first-year students and transition to university. She is currently developing an online learning environment aimed at teaching students the skills required to study at university. Rhonda received the Vice Chancellor's award for leadership excellence in 2011 for her work in course design and for community engagement.

Samantha Sin holds PhDs in Education from Macquarie University, Australia, and Linköping University, Sweden. She is a senior lecturer in the Department of Accounting and Corporate Governance at Macquarie University. In 2004, she was awarded the Macquarie University Outstanding Teacher Award and the Teaching Innovation Award. Her research interests are in professional ethics, curriculum development, assessment, and the assurance of learning. She actively researches in both Accounting and Higher Education and has published her research in monographs, book chapters, and international research journals.

Gurnam Singh is a principal lecturer in Social Work at Coventry University in the United Kingdom. He completed his PhD from the University of Warwick on *Anti-Racist Social Work* in 2004. His teaching and research interests center on critical pedagogy and critical practice, specifically in relation to questions of social justice, human rights, and anti-oppression. He has published widely on all these and related issues. In 2009 in recognition of his contribution to Higher Education he was awarded a prestigious National Teaching Fellowship from the UK Higher Education Academy. He recently published a jointly edited book with Stephen Cowden in 2013 titled *Acts of Knowing: Critical Pedagogy in, against and beyond the University*.

Eszter Szenes is an associate lecturer at the Learning Centre and a PhD candidate at the Department of Linguistics at the University of Sydney. Her PhD research explores the language of academic success and the basis for legitimation of knowledge claims in undergraduate business reports using Systemic Functional Linguistics and Legitimation Code Theory as informing theories. She was involved in the planning and development of the unit Critical Thinking in Business at the Sydney University Business School, which prompted the ongoing interdisciplinary research project "Knowledge Practices of Critical Thinking in Applied Disciplines" with A/Prof Karl Maton and Namala Tilakaratna.

Yuko Tanaka holds a doctorate in Educational Studies from Kyoto University (2009), and is currently project researcher at National Institute of Informatics of Research Organization of Information and Systems in Tokyo, Japan. She specializes in educational psychology and cognitive science. Her research interests have been on the individual cognitive process of critical thinking, and critical thinking of crowds. One of her most recent papers, which examined the possibility of using social-technology in helping us to think critically, received the Best Paper Award at the 46th Hawaii International Conference on System Sciences in 2013.

Geert ten Dam is professor of Education of the University of Amsterdam. She was also the president of the Education Council of the Netherlands. Her research focuses on social differences and social inequality in education in relation to learning and instruction processes. In particular she supervises projects aimed at enhancing citizenship in both vocational and general education. Ten

Dam is the project leader of the Dutch participation in the International Civic and Citizenship Education Study 2016.

Keith Thomas holds a doctorate from Deakin University (2004) and is director of Learning and Teaching at the College of Business, Victoria University, in Melbourne. His earlier career was in the Australian military, and he has continuing links via the Defence Leadership and Ethics Centre, in Canberra. His research interests include critical studies in higher education, in areas such as staff development and teaching strategies to embed graduate capabilities into the curriculum, and in leadership and decision making. He is on the editorial board of one international journal and is a regular reviewer for a number of other international journals.

Namala Tilakaratna is a PhD candidate in Linguistics at the University of Sydney. Her research explores the construction and management of national identity in the English language curriculum in Sri Lanka. She has published on language in the context of media repression during the Sri Lankan ethnic civil war in *Language and the Human Sciences* and published a white paper on "A Principles-Based Approach for English Language Teaching Policies and Practices" for the TESOL International Association with Ahmar Mahboob. She is currently involved in an ongoing interdisciplinary project on 'Knowledge Practices of Critical Thinking in Applied Disciplines' with A/Prof Karl Maton and Eszter Szenes.

Franziska Trede holds a PhD in critical practice in the health sciences from the University of Sydney (2006), and is currently deputy director of the Education For Practice Institute at Charles Sturt University, Australia. She has been awarded two research fellowships with the Research Institute for Professional Practice, Learning and Education at Charles Sturt University. Her research interests include professional practice, professional identity development, capacity building through deliberate pedagogies, and critical voices in higher education. She has published over 70 book chapters and scholarly peer-reviewed journal papers and she has been an invited speaker at higher education institutions in Austria, Italy, Norway, Germany, United Kingdom, Canada, and New Zealand.

Tim van Gelder is principal fellow in the School of Historical and Philosophical Studies at the University of Melbourne, and a principal in van Gelder and Monk, a Melbourne-based consulting firm specializing in critical thinking and applied epistemology. He played a key role in the development of argument mapping and its use in critical thinking instruction, and was responsible for a series of argument-mapping software programs. With van Gelder and Monk he provides services including training in argument mapping and related techniques such as decision mapping and hypothesis mapping for many large organizations including law firms, intelligence agencies, and corporations. He is on the editorial board of *Informal Logic.*

Iris Vardi has a PhD in Education from the University of Western Australia (2003) and is currently principal consultant and director of Vardi Consulting. She is a recipient of the 2012 Higher Education Research and Development Society of Australasia (HERDSA) Fellowship. Her research and scholarship in higher education is focused on improving student outcomes and capabilities, particularly in the areas of assessment, feedback, tertiary literacy, and critical thinking. She is the author of the guide, *Effective Feedback for Student Learning in Higher Education* (2012), and guide, *Developing Students' Critical Thinking in Higher Education* (2013).

Monique Volman is professor of Education and director of the Educational Sciences Program at the Research Institute Child Development and Education of the University of Amsterdam. Main areas in her research are learning environments for meaningful learning, diversity, and the use of ICT in education, issues that she approaches from a sociocultural theoretical perspective. In her work she aims to build bridges between educational theory and practice by applying and further developing methodologies in which teachers and researchers collaboratively develop and evaluate theoretically informed innovations. Her research has been published in (among others) the *Journal of the Learning Sciences, Review of Educational Research, Learning and Instruction,* and *Teaching and Teacher Education.*

Zijian Wang was an associate lecturer in the Department of Accounting and Corporate Governance at Macquarie University. The research area for his honours thesis was critical thinking for the accounting profession. Zijian has now moved onto a career in commerce and business.

Milton W. Wendland holds a PhD in American Studies as well as a juris doctorate in law. He is currently on the faculty at the University of South Florida, Tampa, where he teaches and directs the undergraduate internship program in the Department of Women's and Gender Studies. Both his legal and academic work focus on LGBT cultural issues.

Danielle Wetzel holds a doctorate in Rhetoric from Carnegie Mellon University (2005) and is currently teaching professor of English and Rhetoric at Carnegie Mellon University. Presently she is also the director of the First-Year English Program, Chair of the Task Force on Non-Tenured/Untenured Writing Program Administrators for the Council of Writing Program Administrators, and the author of *Adaptation across Space and Time: Revealing Pedagogical Assumptions* (forthcoming).

Anna N. Wilson is a former nuclear physicist and award-winning university teacher who has worked in universities in the United Kingdom, United States and Australia. Her interest in student learning led her to education research, initially exploring undergraduates' experiences of doing research as part of their degrees. Combined with her experiences of working with issues of policy and implementation in various educational development roles, this led to a broader interest in the teaching and learning of the higher-order cognitive skills and dispositions encapsulated by many of the commonly claimed graduate attributes and generic skills. Anna is currently undertaking a second PhD in education at the University of Stirling in the United Kingdom.

Peter A. Williams holds a PhD in American Studies from the University of Kansas. His research explores questions of identity and power in US popular culture and history. His dissertation, "Weird Bodily Noises: Improvising Race, Gender, and Jazz History," explores the significance of identity and embodiment in jazz and other improvised cultural practices. His work has been published in *The Encyclopedia of Jazz Musicians* at jazz.com and the edited volume *Bodies & Culture: Discourses, Representations, Communities, Performances*. He is currently a lecturer in American Studies at the University of Kansas.

Gert Young did his undergraduate studies in Theology and postgraduate studies in Biblical Languages as well as Political Sciences, and is currently a senior advisor for the Extended Degree Programmes at the Centre for Teaching and Learning at Stellenbosch University.

Index

3P model, 250

Aberdein, A., 234, 244
ableism, 531, 532, 536
Aboulghar, M., 320, 332
Abrami, P. C., 183, 189, 191, 208,
    209, 210, 291, 292, 293
academic aptitude, 398
academic literacy, 340
academic writing, teaching of,
    225, 398
Academically Adrift, 3, 23, 104, 108, 109,
    120, 183, 191, 230, 348, 387
Accounting Education Change
    Commission, 432, 452, 453
ad hominem arguments, 240, 610
Agan, R., 477, 487
Agger, B., 459, 471
Aiken, W. M., 31, 45
Åkerlind, G. S., 497, 507
Albrecht, W., 431, 432, 434, 436, 453
Alexander, P. A., 169, 181
Alfred, M. V., 534, 542, 543
algorithmic capitalism, concept of, 23
alternatives, seeking of, 32
Alvarez, C., 189, 191
ambiguity, tolerance of, 429, 435, 467,
    469
ambiguity and relevance, fallacies of.
    See separate entries
Ambrose, S. A., 110, 116, 118, 119, 120
American Accounting Association, 432,
    453, 454
American College of Clinical Pharmacy,
    487
American Council on Education, 107
American Institute of Certified Public
    Accountants, 432, 434, 453
American Philosophical Association, 10,
    181, 211, 291, 293, 414, 454, 507
American universities, 107, 319, 325
American University in Cairo, 318, 319,
    320, 331, 333, 334, 605
Amsler, S., 563, 571
analogy, arguments from, 287
analytic philosophy, 529, 530, 540, 542
anchoring and adjustment heuristic, 395

Anderson, B., 522
Anderson, C., 171, 182, 493, 507, 514, 515
Anderson, T., 522
Andone, D. M, 512, 513, 522
Andreotti, V., 548, 557
Andrews, R., 28, 49, 52, 53, 54, 56, 57, 59,
    61, 62, 248, 262, 605
Angeli, C., 595, 601
Annis, D., 216, 230
apartheid, system of, 526, 545, 546, 547,
    552, 553
Arab cultures. *See* Islam, Islamic
    scholarship and culture
Arendt, H., 194, 266, 268, 269, 271, 273,
    281
argument analysis, 19, 137, 185, 190, 193,
    194, 213, 214, 216, 217, 221, 222, 229,
    255, 256, 373. *See* argument mapping
argument diagramming. *See* argument
    mapping
argument mapping, 184, 185, 193, 213,
    214, 217, 218, 219, 223, 226, 227, 228,
    229, 240, 241, 290, 292, 293, 616
argumentation. *See* critical thinking,
    argumentation
Aristotle, 49, 55, 142, 151, 376, 459
Arts, the, 231, 332, 550, 551, 556, 605,
    607, 611, 613, 614
Arum, R., 4, 23, 101, 104, 107, 108, 109,
    120, 183, 189, 191, 217, 230, 338, 344,
    348, 385, 387
Asgharzadeh, A., 320, 332
Asian University for Women, 300, 315
Association for Critical and Creative
    Thinking Education (ACCTE),
    353, 366
Association for Informal Logic and
    Critical Thinking (AILACT), 605,
    608, 609, 610
Association of American Colleges and
    Universities, 315
assumptions, 12, 20, 33, 42, 58, 99, 103,
    159, 164, 165, 170, 173, 175, 176, 178,
    179, 180, 181, 194, 248, 262, 320,
    323, 355, 356, 391, 435, 457, 458, 461,
    463, 475, 476, 480, 482, 487, 492, 497,
    500, 529, 537, 574, 580, 581, 583, 588

65879594R00360

Made in the USA
Middletown, DE
05 March 2018